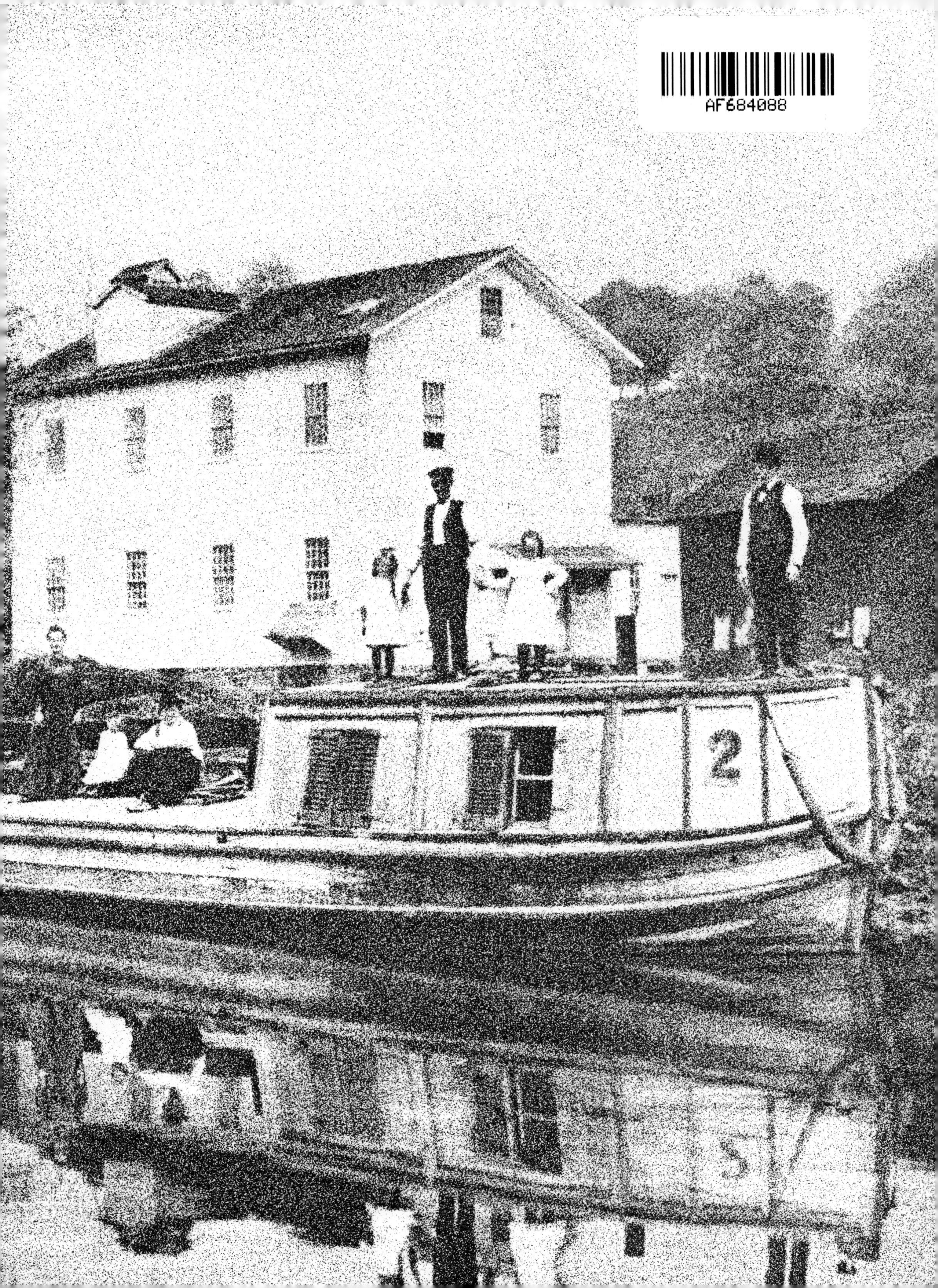
2

A Photo Album of Ohio's Canal Era, 1825–1913

A PHOTO ALBUM OF OHIO'S CANAL ERA, *1825–1913*

by JACK GIECK

with an introduction by
GEORGE W. KNEPPER

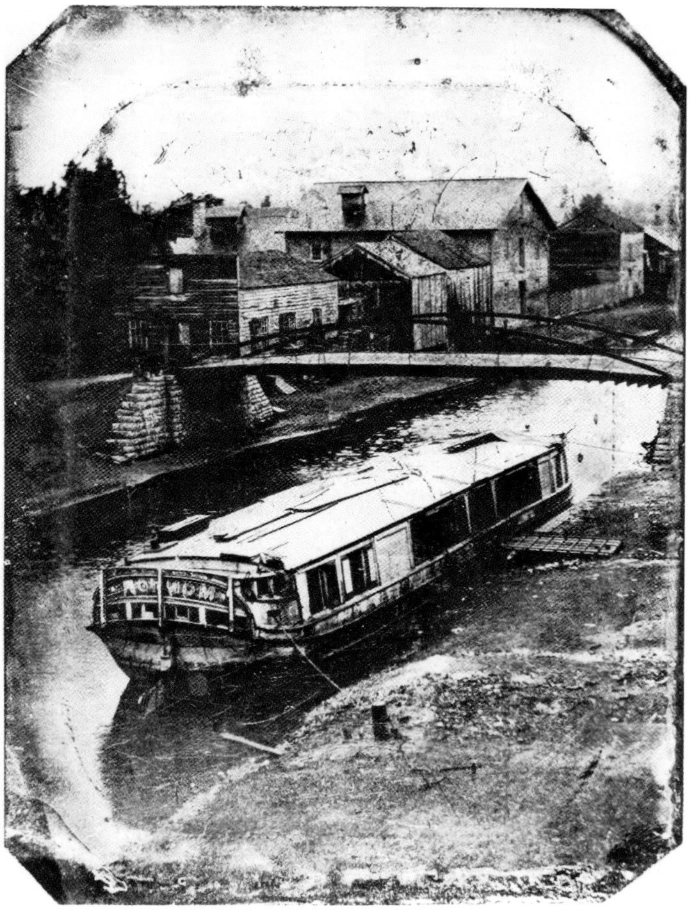

The Kent State University Press
Kent, Ohio, and London, England

Library of Congress Catalog Card Number 87-29776
ISBN 0-87338-353-2
Manufactured in the United States of America

Second printing 1988

This publication is made possible in part under grants from the Ohio Humanities Council and the GAR Foundation.

The paper in this book meets the guidelines for permanence and durability of the Committee on Production Guidelines for Book Longevity of the Council on Library Resources.

Library of Congress Cataloging-in-Publication Data

Gieck, Jack.
A photo album of Ohio's canal era, 1825-1913.

Includes index.
1. Canals—Ohio—History. 2. Canals—Ohio—Pictorial works. I. Title.
HE395.034G54 1988 386'.48'09771 87-29776
ISBN 0-87338-353-2 (alk. paper)

British Library Cataloging in Publication data are available.

Dust jacket photo is by *Akron Beacon Journal* photographer Paul Tople. In a 1986 Sunday magazine feature, for which the newspaper asked its photographers to choose their own favorite pictures, Mr. Tople selected this one as his. It shows two Canal Fulton mules and their driver towing the authentically built *St. Helena II* down the section of canal restored by civic groups in the town. (Courtesy *Akron Beacon Journal*.)

In commemoration of the twenty-fifth anniversary of the founding of the Canal Society of Ohio

Contents

Preface

Although it is entitled "A Photo Album . . . ," this book necessarily includes a number of woodcuts, metal engravings, and other examples of earlier graphic arts, since it was not until 1837 that Louis Daguerre produced the first photographic image. Indeed, some of the pictures in this volume are as old as photography itself. It follows that any book of canal-era photographs owes its greatest debt to those intrepid nineteenth-century photographers who created many of these pictures. "Took" just won't do as a description of their craftsmanship, since the popular Kodak camera with George Eastman's "you push the button—we do the rest" advertising was not to appear until 1888.

The canal photographer's camera was a large wooden box—only a little smaller than a breadbox—resting on a sturdy wooden tripod. It had a large if rather primitive lens on the front, but no shutter or iris. Behind his tripod, the photographer pitched his darkroom tent, into which he unloaded his bottles of chemicals, a small worktable, and assorted paraphernalia.

To make a picture, the early photographer first had to make his *film*, or its equivalent. In the tent he unwrapped a polished glass plate, over which he poured a mixture of viscous egg albumen and iodine. In the 1850s the albumen was replaced by collodion (which is a solution of gun cotton, or cellulose nitrate, dissolved in alcohol and ether). After the surface of his plate had dried to a tacky layer, he dipped it into a solution of silver nitrate to "excite" the emulsion; and he could actually watch the formation of a creamy precipitate of light-sensitive silver iodide in the sticky layer. This operation, taking about five minutes, was carried out under orange light, to which his emulsion was insensitive. By today's standards, the resulting "film" had an ASA rating of less than *one*!

Not quite dripping wet, the plate was inserted into a light-tight plate holder, after which the photographer could emerge from his tent and install both the holder and its encapsulated wet plate in the back of his camera. To save a few steps, some entrepreneurs built their rigs so that the back of the camera was accessible from inside the tent.

After pulling the dark slide from the front of his plate holder, the artist simply removed the lens cap for anywhere from six to twenty seconds, depending on the lighting—first warning any people in the scene to freeze in position lest they register as a ghostly blur. It is this forced, teeth-gritting immobility that accounts for the strange expressions that appear on the faces of some of this book's human subjects—as, for example, the fishermen next to the lock spillway pictured on page 118.

Replacing his lens cap and dark slide, the photographer returned to his tent and removed his still-wet glass plate from the wooden holder. Over its surface he poured a solution of pyrogallic acid (an organic acid made from "nutgalls"—a resinous excrescence sometimes found on oak trees) or ferrous sulphate, either of which served as a reducing agent, or "developer."

Under his candle-lit safelight the artist could watch his image begin to appear. When he judged it to be fully developed, he rinsed the plate in clear water (which he had to carry into the field with him), and then fixed the image in "hyposulphite of soda" (sodium thiosulphate), or sometimes in potassium cyanide! Once again, he washed the chemicals from his emulsion in clear water. Holding the fixed plate between thumb and forefinger, he rotated it over the candle flame until dry. While it was still warm, he varnished it.

The next day, weather permitting, the photographer printed his pictures in sunlight, on a paper he also sensitized himself. He first soaked his paper in a weak solution of common salt (sodium chloride); when dry the paper was bathed in a strong solution of silver nitrate, precipitating light-sensitive, insoluble silver chloride in the fibers of the paper. Sunlight darkened the paper wherever it was not impeded by relatively opaque portions of the negative glass plate against which it was pressed. Development was usually carried out in salt water, which reduced the sensitivity of unaltered silver salts to daylight.

One of these "salt prints," circa 1859, from the vault of The Western Reserve Historical Society, is reproduced on page 59; it exhibits remarkable resolution of detail in the buildings bordering the canal in Cleveland.

Instead of making a print, the photographer could place his glass plate against a dark surface (e.g., black velvet) and view his picture directly—the highlights of the perceived positive reflecting off the bright, developed silver surfaces. This technique was similar to how earlier "Daguerreotypes" (and later tintypes) were viewed. Daguerreotypes utilized a metal plate coated with silver instead of a glass plate, employed a different developing process, and required an exposure time of ten to twenty *minutes*.

Only two of the photographs in this collection are copies of Daguerreotypes: the portrait of Edward H. Gill (page 220), which could have been made no later than 1842, and the portrait of Sebried Dodge (page 217), made in 1845. To put these remarkable photographs in perspective, it should be noted that Louis Daguerre and Joseph Niepce had divulged their revolutionary process to a joint meeting of the Paris Academy of Science and the Academy of Fine Arts ("a chemical and physical process which gives Nature the ability to reproduce herself") on August 19, 1839. The detail apparent in the eyes and skin texture of these portraits identifies them as Daguerreotypes rather than paper-negative "Calotypes," which contemporary process (invented by English scientist William Talbot) produced a much coarser image. Two possible Calotypes are identified in the photo captions.

In researching antique photographs of a given era, one finds runs of pictures that suggest they were probably the work of the same photographer. One such artisan appeared in the Akron area in 1858, when albumen emulsions were beginning to give way to collodion. Someone else (with a penchant for vertical formats) was setting up his tripod on the commercial waterfront of Lake Erie in the Port of Cleveland about the same time. We don't know who shot the impressive Pennsylvania Railroad Bridge picture near Bedford in 1864—in which the train must have stopped on the stone trestle so the lensman could avoid blur during his ten-second-or-so exposure (see page 270). Another photographer showed up in the 1890s in Navarre, where he couldn't get enough of Lock 6—photographing it summer and winter. Dayton had one who worked in large plate formats, producing especially sharp pictures documenting the decline of the canal in that area during the period 1904–11. We are indebted to all of them.

This album presents dozens of photographs which were originally made utilizing the pioneer imaging processes described. Indeed, some of these predate Mathew Brady's famous Civil War photos. But although there are many third-generation copies of early canal pictures in existence, locating and reproducing publication-quality prints for this book was only possible through the unremitting efforts and encyclopedic minds of such Canal Society stalwarts as former president Terry Woods, for many years chairman of the Historical Acquisitions Committee of the Canal Society of Ohio; fellow photographer and Society trustee Frank Trevorrow, who edited the Society's *Towpaths* quarterly for a number of years and whose personally executed drawings and photographs appear throughout the book (and who entrusted us with dozens of his irreplaceable fine-grain negatives to avoid generation loss in picture quality); trustee John Droege, whose recent retirement as a research chemist made time available for another kind of research; and especially the indomitable Edith McNally, current editor of *Towpaths*, whose investigative efforts produced dozens of pages of indexed camera-ready source material, plus some negatives from her own collection—her contribution also improved both the accuracy and historical depth of many of the picture captions. All are also authors in their own right. I am grateful to each of them.

As an appendix to the book for the convenience and guidance of the reader, Frank Trevorrow has graciously permitted the reprinting of his Physical Data and Tables of Distances, originally compiled for his 1973 book, *Ohio's Canals*, which data were the result of his researching state records going back to 1832. For this work, Trevorrow has updated his data in collaboration with Canal Society trustee Barnett Golding of Columbus, who expanded the scope of the tables to include most of Ohio's feeder canals.

Our primary picture sources were the archives of the Canal Society of Ohio and the Summit County Historical Society, including the remarkable collection of the late Carl Pockrandt. These collections are housed in The University of Akron's Archival Center under the professional care of its dedicated director, John Miller—who not only devoted personal weekends to this project, but who even provided us with the luxury of student assistants on several occasions. Other archival sources include The Western Reserve Historical Society, the Ohio Historical Society, Roscoe Village Foundation, the Canal Visitor Center of the Cuyahoga Valley National Recreation Area (which also provided a copy of the National Park Service's "Historic Structures Report"), the Cincinnati Historical Society, the Middletown Historical Society, Flesh Public Library in Piqua, the Troy Historical Society, the Toledo/Lucas County Public Library, The Cleveland State University Library, a number of nineteenth-century Ohio county atlases, archival engravings assembled by Dover Publications, and individual collections of Canal Society members.

The look and flavor of the book are the work of the talented staff of The Kent State University Press, whose contributions in editing, design, and production are apparent on this and ensuing pages. Actually, the KSUP staff edited the book's second generation—after a first pass through my wife's perceptive eyes and critical gray cells. Vicki has not only put up with Ohio's canals ever since we moved to Akron, but for almost four decades has driven me and my cameras to hundreds of remote locations around the state in pursuit of the fading images of Ohio's remarkable canal era.

Finally, I want to publicly express my sincere thanks to George Knepper, professor of history, Fulbright Scholar, award-winning author, and good friend, who somehow found time to write the introduction to this book. Before dipping into the pictures, the reader is strongly urged, indeed *commanded*, to read Knepper's concise, factual, lyrical overview of Ohio's canal era, putting these images into a historical perspective beyond the objectivity of a canal fan.

Jack Gieck
Akron, Ohio

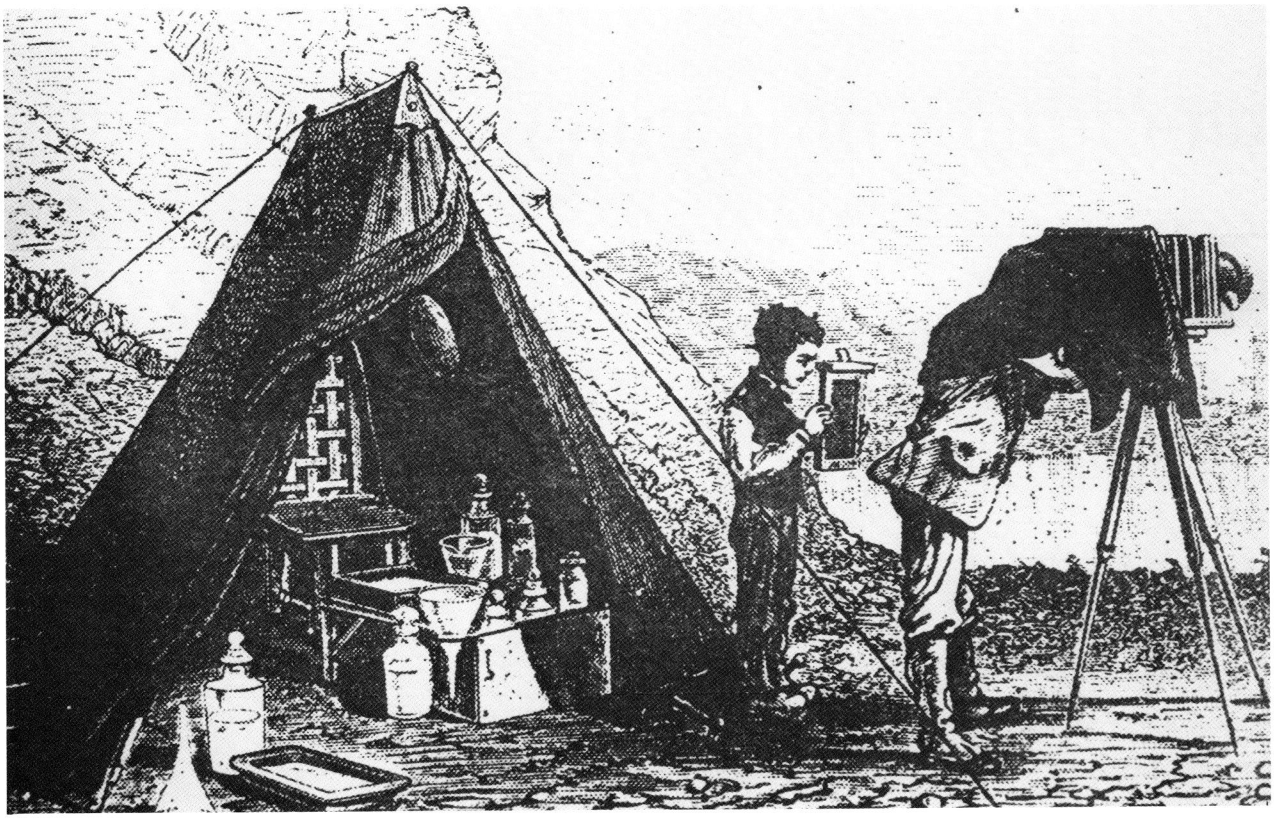

Introduction: Ohio's Canals

Long before the French and Indian War gave Great Britain control of the Ohio Country, that distant land had worked its magic upon the imagination of eastern colonials. Not until they had won their independence from Great Britain, however, were Americans to find these lands open to settlement. Just fifteen years after the first legal settlement was made at Marietta in 1788, Ohio became the seventeenth state. A rapid influx of settlers brought the new state's population to 231,000 in 1810. With removal of the Indian threat in the War of 1812, Ohio's population shot up to 581,000 in 1820.

Easterners had heard stories about this forested wilderness before migrating to it, but few could appreciate how isolated Ohio was from the settled East. Newcomers felt neglected behind the Appalachian barrier, ignored by the federal government which should build roads and canals connecting the West to eastern markets. But the federal government was involved in only one such improvement, the National Road. By 1817 it stretched across the mountains from Cumberland, Maryland to Wheeling on the Ohio River, but it would be another sixteen long years before construction would proceed to Columbus. The National Road was one way for an Ohio farmer to get goods to an eastern market provided he could first get them to Wheeling. Livestock was driven on the hoof over this and other routes, and pack animals with small whiskey kegs slung over their backs could use the forest roads and trails, but such trails were not suitable for handling the bulk of Ohio's commerce.

Ohio's only outlet to national and foreign markets was the Ohio-Mississippi route to New Orleans, over a thousand dangerous miles away. Few farmers and craftsmen could afford the time and risk of taking their own cargoes to New Orleans. Most downriver business was carried on by commission merchants who assembled cargoes in Ohio and transported them to the New Orleans market. Since all the West was directing its agricultural produce to New Orleans at the same time, that market was frequently glutted. Prices were low and spoilage was high. Even after the advent of steamboats on the western waters following the War of 1812, the New Orleans market was a clumsy one to depend on.

Clearly, if Ohio were to grow and prosper (and that was the only acceptable alternative to optimistic frontier types who had gambled everything on it), better and cheaper routes must be developed to eastern markets. The interior of the state would never get its share of population unless it were made accessible. New settlers would have to be convinced that they would have access to the outside world. This central fact of early Ohio life encouraged the state's political and business leaders to ponder ways of ameliorating the problem of isolation.

Hope came from forces outside Ohio. In 1817 the first earth was turned on the Erie Canal, New York State's ambitious project for connecting Lake Erie with the Hudson River and New York City. It would marry the commerce of the West to the New York market. The Erie Canal took eight years to complete, but long before it was finished in 1825, Ohio had taken positive steps to emulate the Empire State, to build a canal system of its own which would tie the state's commerce directly to the New York market.

In the forefront of Ohioans pushing the virtues of a canal system was Governor Ethan Allen Brown of Cincinnati. Early in his terms (1818–22) he lobbied for action with such vigor and success that he was later to be called the "Father of Ohio's canals." With Brown's support, the state legislature created a state canal commission which was empowered to hire surveyors and engineers to study possible routes for connecting Lake Erie with the Ohio River. The seven commissioners were remarkable public servants. Two of their number, the so-called acting commissioners, were men of exceptional quality. They were Alfred Kelley, a young lawyer, banker, and land speculator from Cleveland, and Micajah Williams, a Quaker businessman from Cincinnati. Seldom has any state been so well served.

Canals can go only where there are natural watercourses nearby to keep the canal channel full. This fact limited the principal north-south routes to three prospects. In the east the line of the Muskingum, Tuscarawas, Cuyahoga (or, alternatively, the Black); in center state the Scioto, Sandusky; and in the west the Great Miami, Auglaize, Maumee. Each of these was an ancient Indian pathway which involved crossing the height of land separating the northward from the southward flowing rivers. For the canal builders, a critical concern was whether adequate water could be supplied where the canal crossed the divide. Many assumed that the Scioto-Sandusky route would be picked since it was centrally located, served the state capital, and accommodated the politically potent Scioto Valley.

In 1822 the canal commissioners hired James Geddes, an experienced Erie Canal engineer, to survey possible routes. To the surprise of nearly everyone, Geddes's measurements indicated that there was insufficient water on the divide separating the Scioto and Sandusky. However, there was adequate water on the Portage Summit (Akron) between the Tuscarawas and the Cuyahoga where a natural spring-fed lake lay right atop the route's highest elevation.

As Geddes's preliminary findings became known and potential routes were identified, political lobbying was stimulated as each section of the state jockeyed for advantage. In light of the bitterness of the political contest it is rather remarkable that the legislature was able to pass a bill establishing a route; but they did. In February 1825 the legislature authorized construction of the Ohio & Erie Canal. Commencing at the Ohio River, it was to run northward up the Scioto Valley to a point eleven miles south of Columbus. There the route bore east across the Licking Summit, passed through Newark on its way to the Muskingum, the Tuscarawas, and thence to Lake Erie. The commissioners knew in February that the northern portion of the route was to cross the Portage Summit and follow the Cuyahoga to Cleveland, but they kept people guessing about this portion of the route for three months hoping to lure landowners to offer free right-of-way as an inducement for building the canal through the Cuyahoga Valley.

Political realities dictated several compromises. Southwestern Ohio, the state's most populous region, demanded and got a canal from Cincinnati to Dayton (the Miami Canal) with the expectation that as development proceeded in northwestern Ohio it would be extended to Lake Erie. Marietta and Zanesville, angry at being bypassed, were mollified by assurances that the Muskingum River would be canalized and thus made navigable year round for commercial vessels. Columbus was accommodated by an 11.7 mile connection or "feeder" canal to the main line of the Ohio & Erie. There was nothing to offer people in the vicinity of Sandusky and Steubenville that could placate them. Both regions harbored longstanding animosities to the canal enterprise.

Essential though they were, not even these compromises were enough to guarantee passage of the canal legislation. Victory was finally assured by attracting critical votes from the Marietta region. To secure them, canal backers had to vote for a new tax in support of public education, a cause close to the hearts of Marietta representatives. The deal was struck—canal votes in return for school votes.

While routes were being resolved, attention was given to means for financing construction of the canals. There was little prospect for federal aid. The project was so massive, and the state still so poor, that there was no real prospect of private interests undertaking the task. So Ohio borrowed once more from New York's lead. A Canal Fund Commission was established to market state-backed canal bonds in eastern money markets. The fund commissioners decided to secure money as it was needed by floating bond issues sequentially. As work progressed and the completed portions of the canals started to carry traffic and generate income from tolls, later bond issues would become more attractive and sell on more favorable terms. And so it proved.

In 1825 the commissioners—Ethan Allen Brown, Ebenezer Buckingham, and Allen Trimble (soon to be replaced by Simon Perkins)—sold the first bond issue of $400,000, in its entirety, to the New York financial house of Rathbone and Lord. Interest on the bonds was slightly above 5 percent, payable semiannually. The Ohio bonds proved attractive because New York investors were doing well with Erie Canal bonds for which they found a ready market, because New York would capture much of the western trade as soon as transportation was improved, and because the state of Ohio pledged its good faith and credit (i.e., its taxing power) behind the bonds.

While bond sales provided by far the major portion of support, help came from two other sources as well. As the canal commissioners anticipated, landowners donated property to the state in an effort to ensure that the canal would service their interests. An excellent example occurred along the Portage Summit. To guarantee that the canal would follow a route to his benefit, Simon Perkins and an associate, Paul Williams, had a plat surveyed for a new town which came to be called Akron. The canal would pass through this town. Perkins offered to the state the canal right-of-way, land for two turning basins, and one third of the town lots. Perkins (not yet a fund commissioner) was a friend of Alfred Kelley and probably had an excellent idea that his offer would help the commissioners choose the Portage Summit route. He was right! Perkins's Akron was a child of the canal, and both the town and the donor profited by it.

The other source of help was the federal government. It had been unenthusiastic about supporting internal improvements partly because eastern and southern congressmen did not see such support working to the advantage of their regions. The East had already paid for much of its own improvements, and the South had placed its dependence on its extensive system of navigable rivers. However, the rapid population increase in the West after the War of 1812 gave that region much increased political power in Congress, and the West spoke as with one voice on the central importance of internal improvements. In 1828 Congress granted 500,000 acres of federal land in northwestern Ohio to the state. Money from its sale was to be used for canal building, specifically for extension of the Miami Canal to the Maumee and thence to Toledo. The work was completed in 1845 and the Miami & Erie Canal now provided the state with a second through connection between river and lake.

Construction on the first sections of the Ohio & Erie and the Miami canals started as soon as the first bond issue was marketed. The commissioners wisely decided to build the canals in sections as money became available. The first sections built were those which connected directly to market outlets. If construction funds dried up, the completed sections would be useful. On July 4, 1825, at Licking Summit south of Newark, the first shovelful of earth was turned by Governor DeWitt Clinton of New York, the man who was so influential in making the Erie Canal a reality. Work commenced on that spot because it involved a deep cut which would require a long time to complete. Two weeks later, on July 21, Governor Clinton participated in a similar groundbreaking ceremony for the Miami Canal at Middletown. At the same time work commenced on the northern section of the Ohio & Erie from Portage Summit to Cleveland and on the southern section of the Miami Canal between Cincinnati and Middletown. When construction funds ran short temporarily all available labor was concentrated on the Portage Summit-Cleveland stretch, but work was soon resumed along other stretches of the system.

If a great construction project were being undertaken today, it would probably be bid out to a single contractor who would hire such subcontractors as were required. There were no contractors in those times capable of undertaking such a massive project. Instead, the canal commission acted as the prime contractor. Under its direction specifications and cost estimates were drawn up for each section of the work. A section usually consisted of a half-mile portion of the channel or a single lock. These were small enough units to be handled by small contractors. Even an unskilled farmer might bid successfully on a section or two of channel work. The commissioners were lucky in their timing. Just as the bids went out, contractors were available from the Erie Canal where work was just completed. There were enough bidders on most jobs that the commissioners could choose their contractors with some care. Most bids came in under cost estimates.

The main channel was specified as twenty-six feet wide on the bottom, forty feet wide at the waterline, and a minimum of four feet deep. A channel this size permitted two boats to pass easily, but it was small enough to conserve water. On each side of the channel, woods and underbrush were to be cut back twenty feet to minimize the possibility of deadfalls blocking the channel. On the side nearest the river, a ten-foot towpath was to be built. Locks, gates, sluices, diversion dams, aqueducts, reservoirs, and other more sophisticated features were carefully specified. This work went to experienced contractors.

With scores of contractors, many of them inexperienced, working on these projects, careful supervision was essential. In the early years that supervision came from Alfred Kelley, who was responsible for the northern and central portions of the Ohio & Erie, and Micajah Williams, whose responsibility was the southern portion of the Ohio & Erie and the entire Miami Canal, at that time limited to the stretch between Cincinnati and Middletown. Both of these remarkable men had spent months working with Erie Canal engineers until they knew every detail of canal construction and operation. These men literally got on their horses and rode the construction routes, inspecting the work and making certain it was up to specifications. In the later stages of canal building, this task was performed by regional engineers responsible to the Board of Public Works which succeeded the canal commission. Some engineers were competent and careful while others lacked the dedication brought to the task by Kelley and Williams who neglected their own business and personal interests to see the task through.

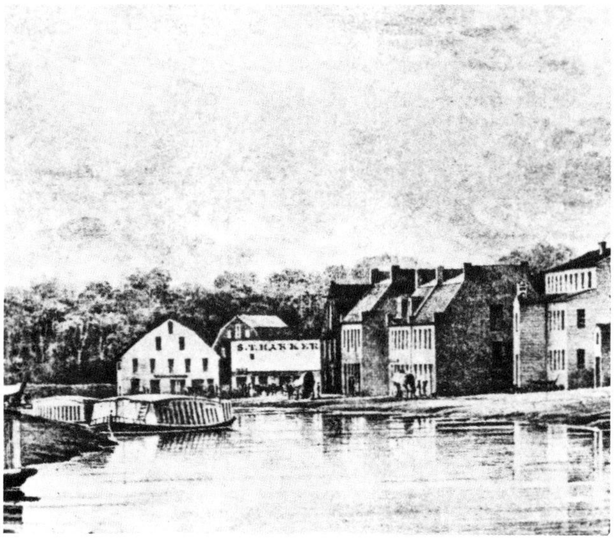

Initially the canal commissioners had not planned to build as well as they did. Encouraged by the availability of funds and caught up in the enthusiasm of the time, they upgraded the quality of materials and workmanship in the canals. The result was an engineering work of high order. Inexperienced contractors occasionally failed to complete a job, or a contractor might take his pay from the state and disappear without paying his workers. In such crises the commissioners stepped in to rectify matters. State workers were employed to finish the work. On one occasion, convicts from the state penitentiary were used in the Columbus vicinity.

Labor on the canals was provided by local farmers and their sons working in the off season and by thousands of immigrants, primarily Irish but including some Germans. Many of the itinerant workers came to Ohio from New York. They lived along the routes, concentrating in small "shanty towns" in places like Cleveland, Akron, and Cincinnati. Akron's shanty town was nicknamed "Dublin." Poor, uneducated, sensitive to slights, and easily aroused to belligerency, the Irish were social pariahs among people of "quality." Were it not for their labor—twelve hours a day, often in stinking ooze up to the waist—the canals could not have been built so expeditiously. Thirty cents a day and a gill of whiskey was their usual wage. Many died from malaria, typhoid, and smallpox, and from occasional epidemics such as the cholera scourge of 1832. Too much alcohol, poor nutrition, accidents, and murderous fights took many lives. Many a canal worker, we are told, was placed in an unmarked grave alongside, or occasionally in, the canal bed.

The cost of construction rose as major canal projects in Pennsylvania and elsewhere competed for contractors and laborers. Philadelphia had watched with alarm as the Erie Canal threatened to syphon off the western trade to its competitor, New York. To respond to this threat, Pennsylvania embarked on an expensive canal scheme of its own at the same time Ohio was starting work on its system. Baltimore also feared loss of the western trade. Some came its way via the National Road, but local merchants pressed for extension of the Chesapeake & Ohio Canal and ultimately the Baltimore & Ohio Railroad.

In constructing the canals top priority was given to the thirty-eight-mile section connecting Portage Summit and Cleveland. Summit Lake, athwart the continental divide at Akron, was 966 feet above sea level; Lake Erie at Cleveland was 573 feet above sea level. Thus the canal builders had to raise the boats some 395 feet in the first thirty-eight miles south of Lake Erie. The most precipitous difference in elevation was at the Portage Summit. Here a staircase of locks was required, seventeen of them in a mile and a half. The average lock had a lift of nine feet, and forty-two locks were required between Portage Summit and the canal basin in Cleveland. Two more locks were completed later at Cleveland to drop the canal to the Cuyahoga River level.

On July 3, 1827, a gala party including Governor Trimble and the canal commissioners boarded the *State of Ohio*, built at Akron, and arrived in Cleveland on July 4 to a glorious reception. This was the first canal in the West. Its impact on the surrounding area was so immediate and encouraging that it gave a great boost to the funding and completion of the remainder of the route. In 1832 the entire 308-mile route of the Ohio & Erie was open to traffic. There were trouble spots. The extreme southern connection with the Ohio River was never free of trouble, and from time to time waters impounded in the Portage Reservoirs south of Akron and in Buckeye Lake (the old Buffalo Swamp) south of Newark were required to keep the channel full at its high points. Similar problems affected the Miami & Erie Canal. Water levels were maintained along the high point by waters impounded in Grand Lake St. Marys, Loramie Lake, and Indian Lake.

Canal building and maintenance was an uncommonly discouraging business. Each spring, freshets, or major floods, wiped out portions of the work. A sudden local cloudburst could put stretches of the canal out of business for weeks. Locks and aqueducts needed constant surveillance. During periods of excessive dryness sections had to be closed due to low water. In winter, parts of the system would be frozen for four or five months. Sections would be drained for repair by the "state boats" during this down time.

Travelling on the canal could be quite pleasant and restful were one to take passage on a well-managed packet boat with congenial companions, watching the countryside pass by at a stately four miles an hour. Or it could be a nightmare: overcrowded vessel aswarm with fretful infants and obnoxious children (or adults for that matter); hatches closed in bad weather creating a stifling, claustrophobic atmosphere; minimal facilities for personal hygiene and sleeping; straw tick mattresses reeking of perspiration and effluvia.

It was freight, however, not passenger traffic which made the canals most valuable. The usual freight boat was sixty to seventy feet long and fourteen feet wide. It could carry up to sixty tons of cargo and was pulled by two or three horses or mules. Loading facilities were located at basins along the route. These basins were lined with warehouses for goods of every description.

Despite disasters and discomforts, the state had wrought well. State enterprise was not confined to the canal plan as laid out in 1825. The system was being constantly enlarged with some of the new additions proving much more costly to build than were the original routes.

To assist in these costly ventures the state, in 1837, passed a Loan Law (or "Plunder Law" as its enemies called it) permitting the state to invest public money in private improvement projects. Counties and cities could also invest in private schemes. The state grab bag was now open to anyone clever enough to reach in. For example, private promoters of the Pennsylvania & Ohio (Mahoning) Canal from Akron to the Ohio River in Pennsylvania got $400,000 in state support. Other private companies also tapped the public till. Nascent railroad companies were especially vigorous in seeking public support. The effect was to balloon the state debt at a time when such indebtedness was the subject of heated political controversy. The law was repealed in 1840, but not until the state debt had risen dramatically.

Waiting in the wings was an even greater threat to the long-range health of the canals. As noted, the state and various localities were using public money to subsidize private railroad companies. Local Sandusky businessmen, embittered by being left out of the canal bonanza, urged the state to build a railroad to serve that city. Ultimately the road was built by private investment supplemented with public funds. The first route completed from Sandusky ran southward to Springfield and was later extended to Cincinnati with financial help from that city. It was clear that this route would compete with the southern end of the Miami & Erie. It appeared that the state, in its eagerness to improve transportation, was in danger of competing with itself.

Did Ohio's canal system work? Did it accomplish for the state and its people what its backers claimed it would?

Certainly the Ohio & Erie worked well from 1827 until about 1840 when it lost its monopoly to competing routes. In that part of northern Ohio through which the canal ran there was a steep rise in real estate prices even before construction started. Construction brought money to the area in the form of wages and the purchase of supplies and services for those building the ditch. The opening of the canal from Portage Summit to Cleveland resulted in the rapid growth of commercial wheat farming. Before the canal opened, wheat in the Akron area sold in a glutted local market for about twenty cents a bushel. After the canal opened, Akron area farmers were getting at least seventy-five cents or more for a bushel of wheat. Most farmers sold to local commission merchants who assembled cargoes for shipment to the New York market where, after adding in transportation costs, the brokers sold the wheat at an even higher price. Soon after this trade commenced, much local wheat was milled into flour locally at the large new mills erected along the canal route to take advantage of excess water power provided by the canal. As the line of the canal moved south with completion of new stretches, these experiences were repeated. In central Ohio there was still cheap government land available and it was now feasible to clear and plant this land since the produce therefrom could get to a profitable market. Regions around Massillon, Canal Dover, Coshocton, and Newark all replicated the experience of the Akron area.

New settlers poured into the canal's service area. Population swelled in the 1820s and 1830s, far outdistancing growth rates in other parts of the state. Cleveland emerged from its early doldrums and began to fulfill its promise as a commercial center. It dominated the western wheat trade until the railroad era realigned trade routes. Vigorous towns—Akron, Massillon, Canal Dover, Coshocton, and Newark, for example—developed all along the Ohio & Erie. However, they were more than canal towns. Each developed a manufacturing base which endured after the canals fell into disrepair. They stand in contrast to towns like Boston, Peninsula, Clinton, Canal Fulton, Roscoe, and many others which lacked the manufacturing base. As the canal lost importance, so did these towns. Today they remain pleasant backwaters. The same phenomenon of growth and/or stagnation can be seen along the Miami & Erie.

The terminal port cities—Cleveland, Portsmouth, Cincinnati, Toledo—prospered from the canal trade. Much of the produce of the Scioto Valley flowed to Portsmouth and of the Miami Valley to Cincinnati from which places there were steamboat connections with markets along the Ohio-Mississippi waterways. Much Portsmouth trade went by steamboat to Pittsburgh and thence to Philadelphia. The direction and volume of trade was manipulated by the rate-setting policies of the Board of Public Works which supervised the canal system from the 1830s. Hence the flow of goods was not a free flow responding to natural market forces. The board acted to favor Ohio interests but, as in any such effort, their policies sometimes backfired or had unanticipated consequences which ultimately worked to Ohio's disadvantage.

The canals carried an astonishing variety of goods. The main products shipped from Cleveland were wheat, corn, coal, flour, and pork. Scores of other commodities such as potash and lumber were shipped in lesser amounts. Shipments from Portsmouth were principally corn, pork, and bacon bound for southern markets. Products from the East came into Ohio via these same ports. "Merchandise" —clothing, farm implements, books, hardware, glass, furniture, processed foods, and much more—dominated this inflow. Salt was the next largest item followed by processed lumber. Some transportation costs on eastern goods were reduced as much as 90 percent by the canals. Goods coming into Ohio were more abundant and less costly than before. Over time, prices leveled out but, early in the canal era, the Ohioans living along the canal routes enjoyed the best of all economic worlds; they sold dear and bought cheap!

Ohio's canals reached the height of their influence between 1827 and 1850. In the decade of the fifties, competition from the railroads became severe. Ohio developed more miles of track than any other state during that decade. Railroads were independent of watercourses; they could go anywhere. The basic north-south orientation of canal trade routes gave way to east-west rail lines. Railroads were easier to build than canals. Trains were faster than canal boats and they ran the year around. The canal's only distinct advantage was cost. Certain bulk goods could be moved more cheaply by boat than by rail. In the fading years of the canal system, coal, lumber, sand, gravel, and cargoes of this sort were carried on the short stretches still open to traffic.

When all the romance of the great canals is stripped away, were they worth the trouble and expense? The answer is a resounding yes. At the time of their inception no one could have known how soon they would be outclassed by the railroads. The canals opened up Ohio to new markets and new settlers. They poured large sums of money into the internal economy of the state. Deposits of bond money in state banks strengthened those institutions. The canals fostered agriculture and industry. They tied Ohio's natural water routes—Lake Erie and the Ohio River system—into an effective transportation network. Above all, they gave Ohioans a surge of confidence and optimism when they most needed it.

Building on that confidence, Ohio rushed to maturity. By 1850 it was the third most populous state. By 1860 it led the nation in value of its agricultural products, and it was approaching third place in industrial production. It had more miles of railroad track than any state. Without the stimulus provided by the canals, these accomplishments would have been delayed. Indeed, Ohio might have developed in quite a different way.

Ultimately the canal system fell into disarray. It never fully paid for itself. The undisciplined and expensive extensions and "improvements" of the late 1830s and the 1840s enfeebled the state's position by greatly increasing debt and saddling the state with costly maintenance and repairs for the benefit of a marginal traffic. Ultimately tolls failed to cover expenses, and Ohio had to use public money to pay its canal debt.

Moreover, the solid direction, proprietary oversight, and planning that the canal commission brought to the early stages of development were dissipated by the bureaucratic behavior of its successor body, the Board of Public Works, which was not strong enough to resist ill-advised demands for preference.

In 1861 the state leased the canals to a private operator who allowed the system to deteriorate further. Large sections were inoperative; some sections were abandoned, and adjoining landowners moved onto state property without challenge. After the state reclaimed the canal system in the 1870s, feeble efforts were made to arouse support for an overhauling of the system. Nearly every governor in the late nineteenth and early twentieth centuries asked the legislature to proceed with that task. But all efforts were fruitless.

The final blow was struck in 1913 when a massive statewide flood wiped out the functioning remnants of the system. It was their death knell, and the canals passed into the history, the folklore, and the legends of Ohio.

George W. Knepper
Stow, Ohio

Abbreviations

BECHS	Buffalo and Erie County Historical Society
CSO	Canal Society of Ohio
CHS	Cincinnati Historical Society
CVNRA	Cuyahoga Valley National Recreation Area
OHS	Ohio Historical Society
RVF	Roscoe Village Foundation
SCHS	Summit County Historical Society
TRS	Troy Historical Society
VHS	Virginia Historical Society
WRHS	The Western Reserve Historical Society

A Desperate Need

1

When Ohio became a state in 1803, the land was still a forested wilderness. In all of Ohio there were only about seventy thousand people. Moreover, the state's few inhabitants were scattered through the countryside in a literally random manner—since many of them (in the Firelands and the Western Reserve) had actually drawn their tracts of land in a lottery. The plight of these early Ohio settlers has been described by the noted Ohio historian Harlan Hatcher: "Each family, or each little party, fought its way through the wilderness until it came to its assigned spot; and there, in the vast loneliness, perhaps fifteen or twenty miles from the nearest neighbor, . . . it chopped out its own hole in the forest, erected a cabin, planted a few acres of corn or wheat, and struggled to keep alive."[1]

But getting the products of their labors to market was next to impossible for these pioneers. Ohio's few roads were really only muddy trails; and when the spring rains came most of these became impassable. It would be nearly three decades before the young nation's first primitive steam locomotives would pull onto a few miles of track laid by the Baltimore & Ohio Railroad—and half a century before railroads would become a viable transportation system.

Although flour sold for $8.00 a barrel in New York City, the same barrel brought only $3.50 in Cincinnati—itself a river town, with access to markets from St. Louis to New Orleans. Farmers upstate salted down their pork and beef, and some managed to sell it for two or three cents a pound. Eggs were sold at three for a penny; chickens went for a nickel apiece. Some farmers converted their grain into fermented mash, to distill a product having a much higher cash value per pound.

Salt, coffee, and imported staples were expensive. Most families had real coffee only on Sundays—getting along on a concoction made from

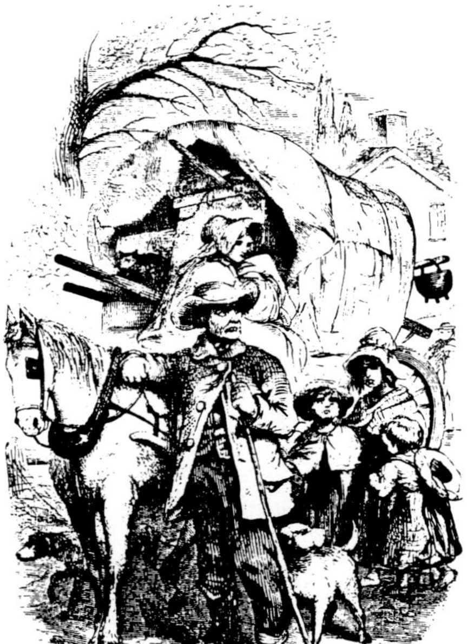

In the early years of the nineteenth century many Ohio settlers migrated from the east to claim their lottery-drawn plats of land in Connecticut's Western Reserve, or in the Firelands—property assigned to families whose homes had been burned by the British during the Revolutionary War.

1. Harlan Hatcher, *The Western Reserve: The Story of New Connecticut in Ohio* (Indianapolis: Bobbs-Merrill Co., 1949).

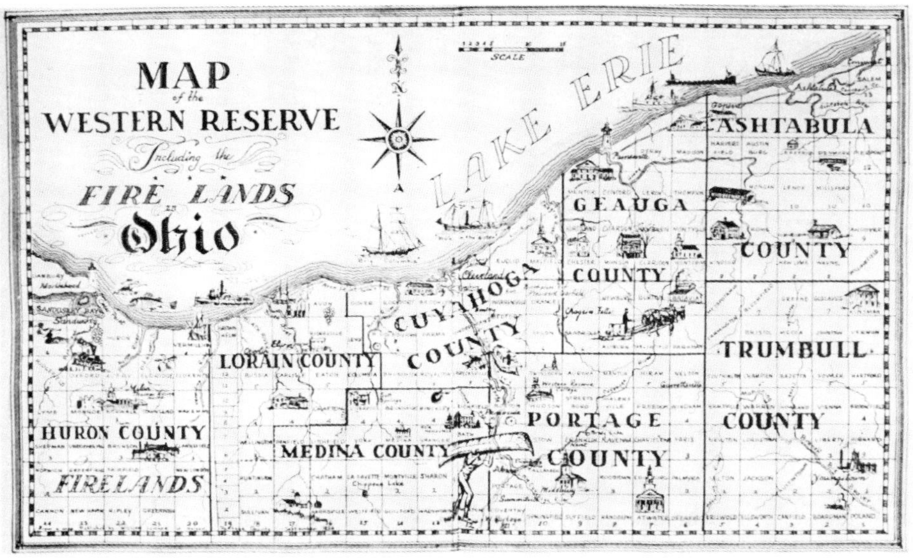

The original Western Reserve, including the Firelands.

burnt rye or wheat during the week. Window glass was a luxury few could afford.

Some Ohio farmers actually drove their livestock overland to Philadelphia and other eastern markets. Others who lived near tributaries of the Ohio River built rafts or flatboats and floated their produce down the Ohio and Mississippi, all the way to the international port of New Orleans. There, they broke up their boats and sold them for timber or firewood. But those who pursued this adventure found they were not alone in discovering New Orleans as a marketplace—having been preceded by entrepreneurs from many other river valleys in the Mississippi Basin. As a result, the New Orleans market was frequently swamped, and many Ohio farmers spent half the winter in the city, waiting for prices to go up enough to pay for their trip home. Usually they walked back, often along the Natchez Trace, where professional bandits hid out to relieve them of any profits they might have earned.

It was George Washington who first envisioned an inland waterway linking Lake Erie with the Ohio River, and he actually suggested such a plan in a letter to the Virginia House of Burgesses before the Revolutionary War—not as a shortcut between these bodies of water, but as a desperately needed inland transportation system. But it was not until 1816—the year Ohio's capital moved from Chillicothe to Columbus—that the cause of such a canal was taken up in earnest by Judge Ethan Allen Brown of Cincinnati who stumped the state and was elected governor in 1818 on a canal-building platform. The net product of his efforts was a bill passed by the Ohio state legislature on January 31, 1822, to establish a commission of seven members to direct surveys, hire a competent civil engineer, and survey five possible routes. Brown became a United States senator later that year, but continued to serve the cause of Ohio canals as one of the seven appointed canal commissioners.

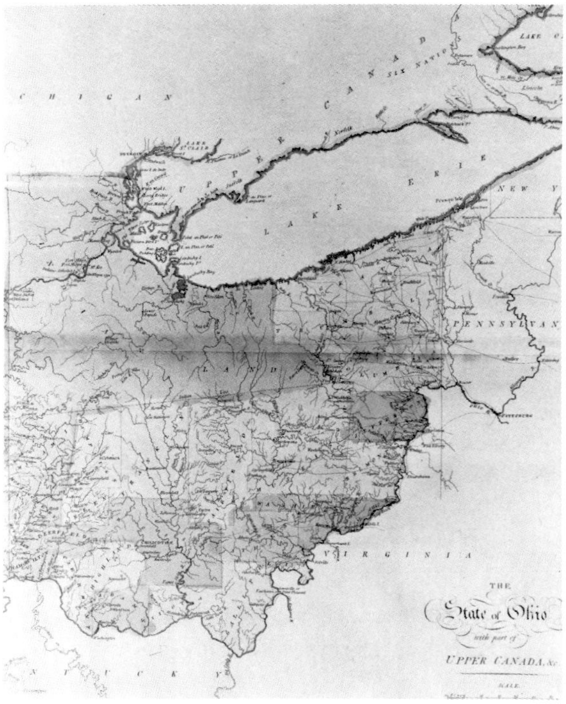

An 1814 map of the recently admitted state of Ohio, made by engraver Samuel Lewis. Nearly half the area of the state is labeled "Indian Land."

Three years later, upon hearing the news that on February 4, 1825, the state legislature had actually passed its famous Canal Act, authorizing a bond issue "to provide for the internal improvement of the State of Ohio by navigable canals," the citizens of Muskingum County held an "illumination." Lighted candles were set in all the store windows in Zanesville, and a contemporary account describes the festive parade that marched over the snowy streets most of the night. A band of musicians with violins and drums led two sleds of celebrants drawn by sixteen yoke of oxen; and the shouting, singing citizenry, warmed by brandy and corn liquor, trudged noisily along behind.

". . . Each family, or each little party, fought its way through the wilderness until it came to its assigned spot; and there, in the vast loneliness, perhaps fifteen or twenty miles from the nearest neighbor . . . it chopped out its own hole in the forest, erected a cabin, planted a few acres of corn or wheat, and struggled to keep alive" (Harlan Hatcher, *The Western Reserve*, Bobbs-Merrill Co., 1949).

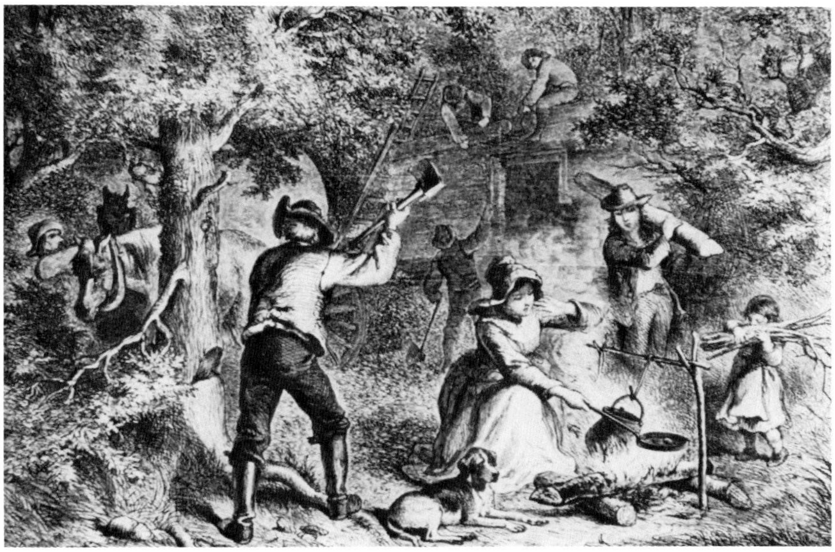

Candlemaking the hard way—by dipping wick strings again and again into a vat of tallow; candles were a luxury used only on special occasions. Most pioneer families went to bed at sunset and got up with the dawn.

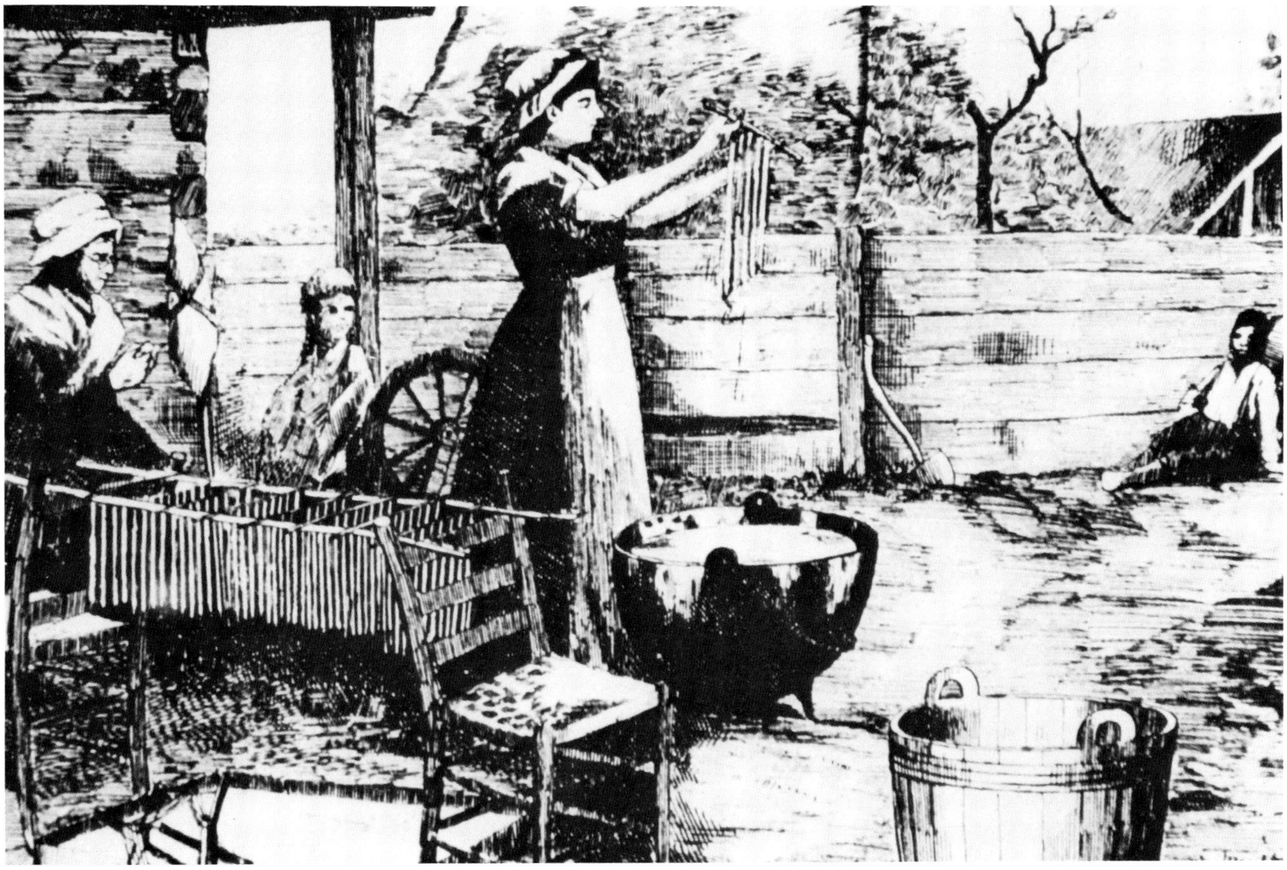

Construction: The Men and the Means

2

The Fourth of July crowd at Licking Summit milled about in the hot afternoon sunshine. Most of those present had driven over from nearby Newark, Ohio, but some had booked stagecoach passage from as far away as Cleveland to hear famed Erie Canal pioneer, New York's Governor DeWitt Clinton, on the same platform with Ohio's Governor Jeremiah Morrow, and that brilliant young orator, Thomas Ewing. Resplendent in their blue uniforms decked with yellow braid and sporting red plumes on their dress hats, the Zanesville Artillery Company had marched over thirty miles for the occasion, sleeping in a barn at Nashport along the way. The artillery horses, tethered on both sides of the colorful speakers' platform, chewed frothily at their bits and lashed their tails at the persistent flies.

Straining to make himself heard above the confusion, Governor Clinton concluded his speech at last and, accepting a ceremonial shovel passed up to the platform by a sergeant of the Chillicothe Blues, descended the steps and plunged it into the earth. As the first spadeful of dirt landed in the wooden wheelbarrow a roar of approval rose from the crowd—but was almost immediately drowned out by a rolling thunder of field cannon as a loading and firing contest commenced between the various artillery companies present.

While the cacophony continued, Governor Morrow took the spade from Governor Clinton and lifted the second shovelful of earth into the wheelbarrow. Suddenly, a scuffle broke out among several prominent citizens over who would have the honor of digging the next spadeful of dirt. Then, fortunately, a second spade appeared—and a third—and in a moment the wheelbarrow was full. And as the prize cargo was wheeled away by a beaming militia captain, many in the cheering throng "wept manly tears of joy."

Thus, July 4, 1825, began an era that would transform Ohio from an agrarian frontier state into one of the world's great industrial centers; construction of the Ohio & Erie Canal[1] was under way. In similar ground-breaking ceremonies two weeks later at Middletown, the Miami Canal—which would ultimately become the Miami & Erie—was launched on July 21.

•

Serious planning of the ambitious projects had begun three years earlier when James L. Geddes, one of the nation's most prominent civil engineers, was engaged by the Board of Canal Commissioners. Geddes, who had made his reputation on New York's Erie Canal, began surveys of five possible canal routes in April 1822, and in the following eight months surveyed some nine hundred miles of lines. Although the commission hoped to prove the practi-

1. In both historical literature and in public documents, the names of Ohio's canals have been represented by several styles, viz. *Ohio-Erie, Ohio & Erie*, and *Ohio and Erie*. A case can be made for the historical correctness of using the conjunction "and," rather than a hyphen or an ampersand; however, this style has often led to confusion with New York's Erie Canal, and the assumption that the "Miami and Erie" or, especially, the "Ohio and Erie" were somehow western branches of the New York waterway. In fact, the "Erie" reference intended in both names is *Lake* Erie. In an effort to minimize future confusion, this work has opted for the ampersand; Ohio's two main trunks are referred to as the Ohio & Erie Canal and the Miami & Erie Canal, with similar styles for Ohio's feeder canals.

cality of a "diagonal canal," running from Cincinnati eastward to the Scioto River valley and then north to Lake Erie via the Muskingum River (passing through as many settled regions of the state as possible, to garner political support), Geddes showed that it would probably be impossible to engineer the water supply necessary to maintain such a canal and operate the locks. Instead, he concentrated his on-site instrument studies on two other possibilities: a central canal running directly northward from the Scioto Valley across the portage to the Sandusky River; and a meandering route that stretched from the Scioto eastward to the upper Muskingum River, and then north to Lake Erie by way of either the Black or the Cuyahoga River—a route that eventually became that of the Ohio & Erie Canal.

Because of his New York commitments, Geddes was unable to finish the planning job, which was taken up by two of the canal commissioners, Cleveland lawyer Alfred Kelley, and businessman Micajah T. Williams of Cincinnati. Conscious that political and economic interests were as important as engineering considerations, Kelley and Williams eventually selected two compromise routes.

The Miami Canal, laid out by Williams, would connect influential Cincinnati and Dayton, with the promise that the route would later be extended up the Maumee Valley to Lake Erie at the present site of Toledo, ultimately becoming the Miami & Erie Canal. Kelley's proposal, the Ohio & Erie, would begin on Lake Erie at Cleveland and travel generally south through the Cuyahoga and Tuscarawas valleys to Coshocton, where the Tuscarawas River joins the Walhonding to become the Muskingum; from Coshocton it would travel southwest, paralleling in turn the Muskingum and Licking rivers, Little Walnut and Big Walnut creeks, reaching the Scioto River twelve miles south of Columbus; from here it would follow the Scioto Valley south to connect with the Ohio River at Portsmouth. A short Columbus Feeder would link the state's capital with the canal's main trunk. Many other feeder canals—some of them privately built—would eventually be added to the system, including the Wabash & Erie, the Pennsylvania & Ohio, the Walhonding, the Hocking, the Sandy & Beaver, and the Muskingum River Improvement, or "canalization."

In 1823, Alfred Kelley gave up his law practice for the next eight years to manage planning and construction of the Ohio & Erie Canal. Micajah Williams assumed a similar role to supervise work on the Miami. Since the Ohio & Erie assignment was a vastly bigger job, Williams also agreed to take on its southern end, from Chillicothe to Portsmouth, leaving Kelley responsible for the northern and central sections.

Much of the detailed survey and layout work required was delegated to such well-known engineers as David S. Bates, Samuel Forrer, Sebried Dodge, and Richard Howe, each of whom brought substantial experience to the project. In an age when an engineering education in America was rare (except in the military services), and the analytical art of stress analysis was still being developed, Ohio managed to attract a number of talented engineers from Europe to design its canal system. One of these was John Roebling, a German hired by another European, Edward H. Gill, chief engineer of the Sandy & Beaver Canal. Roebling not only laid out canal routes; he introduced twisted iron-filament cable as a safer substitute for hawsers braided from hemp—and later created a design to suspend the center span of an aqueduct from his wire cable attached to a pair of towers. Two decades later Roebling designed the Brooklyn Bridge.

•

To provide the most practical routes from the standpoint of hydraulics, the canals would follow the relatively flat, though tortuous, river valleys, using a system of locks to compensate for changes in elevation, especially when crossing over the divides between the valleys. With a few exceptions, the rivers themselves could not be used because they were shallow, with wide seasonal fluctuations in water volume, and often with treacherous currents.

Detailed specifications were laid down. The canal was to be no less than 26 feet wide at the bottom, with sloping sides, to provide 40 feet of width at the water level; it was to be a minimum of 4 feet in depth. A 10-foot towpath was to be provided on one of the banks. In practice, the ditch was dug 5 to 12 feet deep, with the width varying from 50 to 150 feet. To make them as waterproof as possible, the banks and bottom were lined with clay—but not before grubbing and clearing the banks of all porous stone, brush, and other vegetable matter that might contribute to leakage.

Locks were built of cut stone: usually sandstone on the Ohio & Erie; limestone on the Miami & Erie—quarried as near as possible to the canal, tuckpointed with lime mortar (although a few locks in heavily timbered areas along the Miami & Erie were constructed entirely of wood). Lock gates were made of a hardwood—often white oak. Hinges and other hardware were of wrought iron. Aqueducts of timber and culverts of stone spanned rivers and streams which intersected the routes.

As the contractors hired on their hands, Irish and German immigrants who had worked on the Erie Canal migrated westward from New York. These laborers worked from sunrise until sunset for an average wage of thirty cents a day—supplemented during the first year by a daily ration of whiskey. Villages of rough-built worker shacks sprang up along the route; some were to become major cities.

Labor was scarce. To get started on the Columbus Feeder, the confinement sentences of forty-five convicts from the nearby penetentiary were commuted to "hard labor on the Columbus Lateral Canal." Hundreds of small contracts were let for the project, frequently for "sections" of half a mile each. Some farmers contracted to dig the canal through their own acreage. At Zoar, the communal society of German Separatists was paid $21,000 for digging the ditch through the Zoarites' farmland. The actual digging was done by the men; the women hauled away the dirt in tubs—thus paying off their heavy mortgage and saving the struggling community.

Since the gigantic project would take many years to finish, a major effort was launched, largely for political reasons, to complete the first thirty-eight miles of the Ohio & Erie Canal south of Cleveland, linking it with the shanty town of Akron, where fifteen locks would be constructed in a single mile. Harlan Hatcher describes the scene: "By Thanksgiving Day, 1825, about two thousand men were swarming in the Cuyahoga Valley. They looked like a colonial army assembled along a 38-mile front to throw up breastworks for an impending engagement. It was indeed a struggle for survival. All along the valley they cleared out the trees, brush, roots, grass, and herbage so that the canal banks, as the contracts specified, might 'unite securely with the solid earth beneath.' "[2]

Between 1825 and 1847, the state of Ohio constructed 813 miles of canals, 32,903 acres of reservoirs, 29 dams, 294 lift locks, 44 aqueducts, and hundreds of smaller structures at a total cost of $15,967,653, of which $2,257,487 was realized from the sale of federal lands donated by Congress; $710,000 was received from donations and bonuses on sales of bonds. Net cost to the state was thus just over $13 million.[3]

2. Hatcher, *The Western Reserve*.
3. Ohio Board of Public Works, Ohio Canal System Condensed Statistics, 1935.

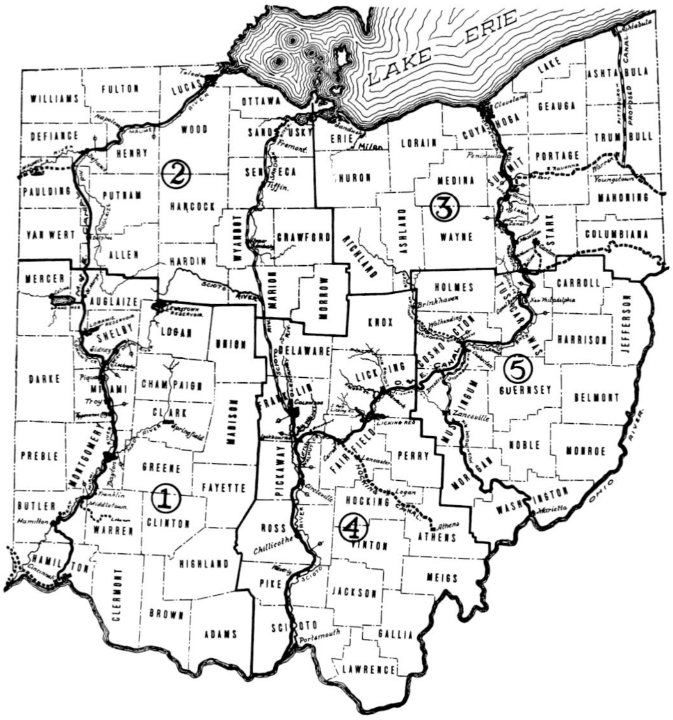

Ohio's canal system, including feeders and some proposed canals never built, as summarized in a 1935 map published by Ohio's Board of Public Works. In the western part of the state, the Miami & Erie Canal at first connected Cincinnati and Dayton and was later extended up the Maumee Valley to Toledo; it was completed in 1845. Farther east, the Ohio & Erie went north from Portsmouth, through Chillicothe and Circleville to a point twelve miles south of Columbus; a feeder canal connected the main trunk with the state's capital. From there it followed river valleys northeast to New Philadelphia, where it turned north through Massillon, Canal Fulton, and Akron, reaching Lake Erie near the mouth of the Cuyahoga River at Cleveland. The Ohio & Erie was completed in 1832. A simplified map of the system appears at the beginning of chapter 6. (Ohio Board of Public Works)

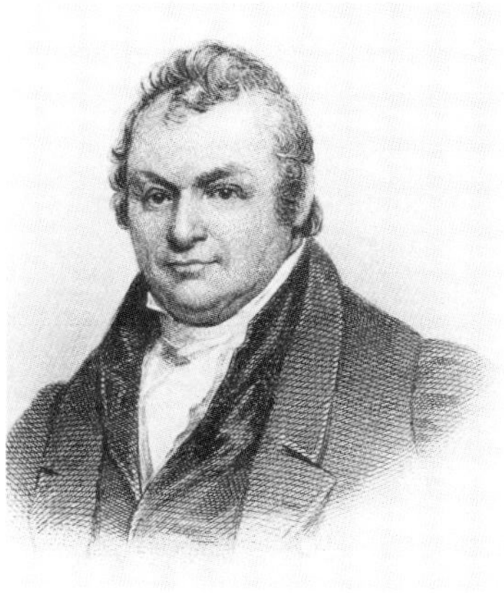

Civil engineer James L. Geddes, who had established his professional reputation on the Erie Canal, surveyed some nine hundred miles of potential lines for Ohio's canal system in eight months during 1822—at times packing his surveying gear through primeval wilderness. His studies included the identification of watersheds adequate for canal operation. (Engraving by H. B. Hall for Chas. B. Stuart's *Civil and Military Engineers in America*, D. Van Nostrand & Co., 1871, courtesy Cleveland Public Library.)

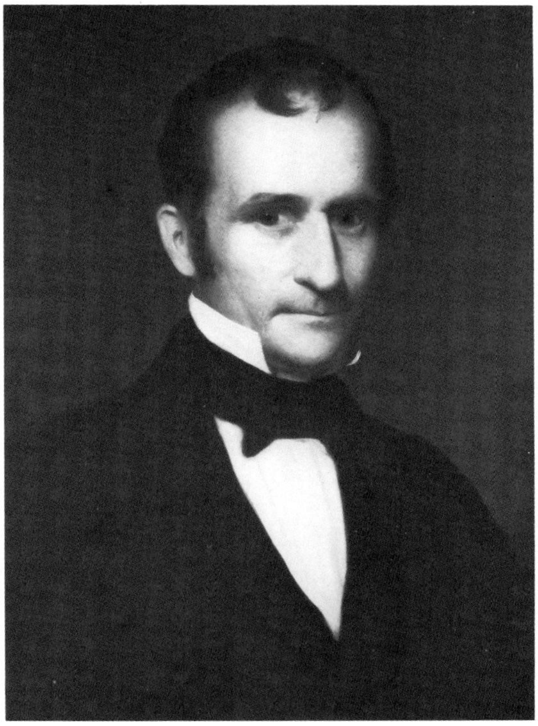

Canal commissioner Alfred Kelley, who gave up his Cleveland law practice for eight years to survey and manage construction of the northern and central sections of the Ohio & Erie Canal after completing Geddes's plan. (OHS)

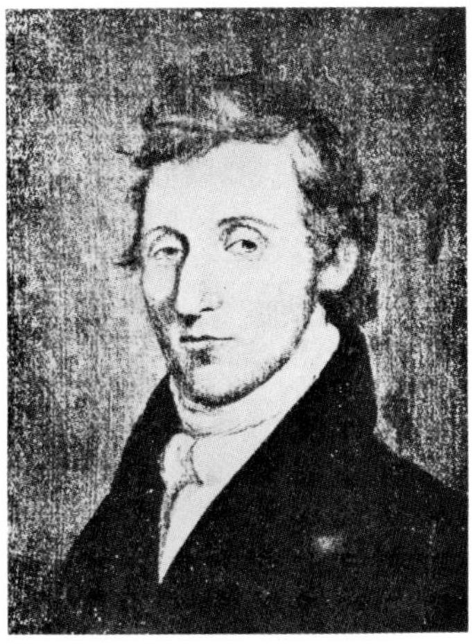

Canal commissioner Micajah T. Williams, a Quaker businessman from Cincinnati, participated in an 1823 survey of the Sandusky Valley route; later that year he made a personal inspection of the Erie Canal. Williams conceived the plan for the Miami & Erie Canal and supervised construction of its first segment, the Miami Canal between Cincinnati and Dayton. He was also in charge of the building of the southern section of the Ohio & Erie Canal between Chillicothe and Portsmouth. (CHS)

Irish and German immigrants who had worked on the Erie Canal migrated westward to make up the main work force hired by Ohio canal contractors. Here, they dug almost a thousand miles of ditch with picks and shovels; their only auxiliary power came from yokes of oxen and the sometimes unwilling efforts of the ubiquitous mule—the unsung hero of Ohio's canal era. After helping to build the canal system, mules provided the motive power to operate its boats. (George Eastman House, Rochester)

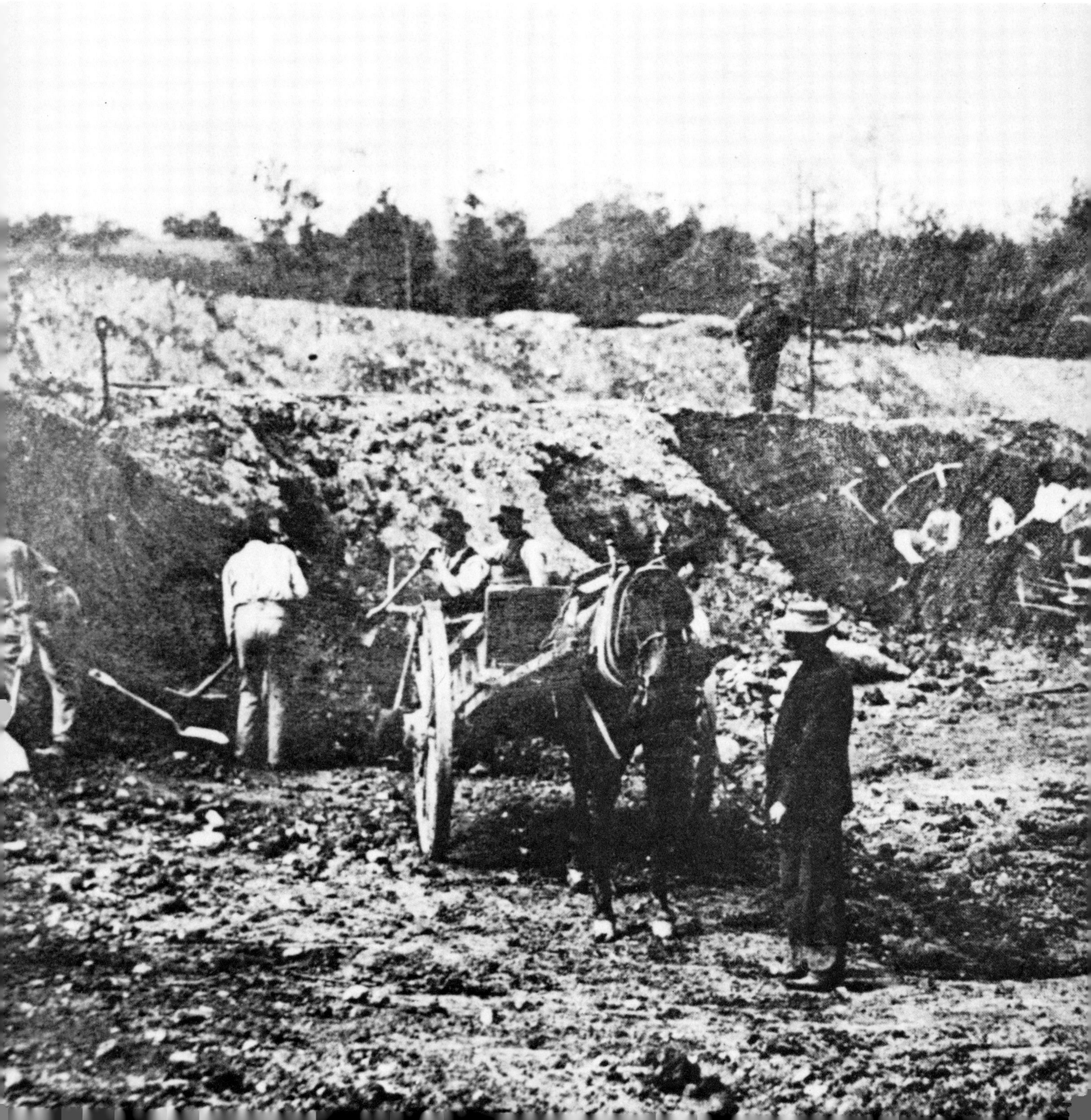

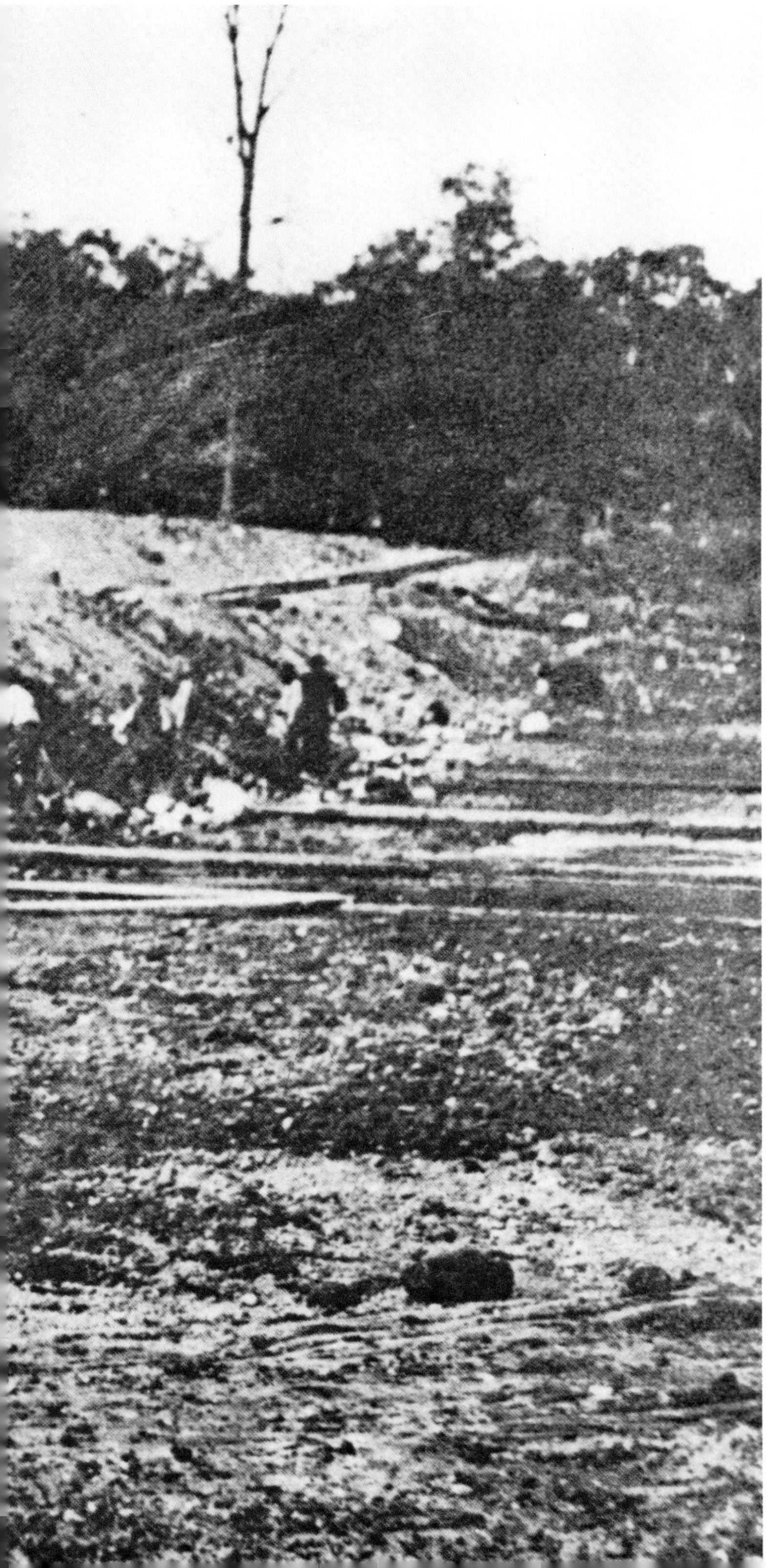

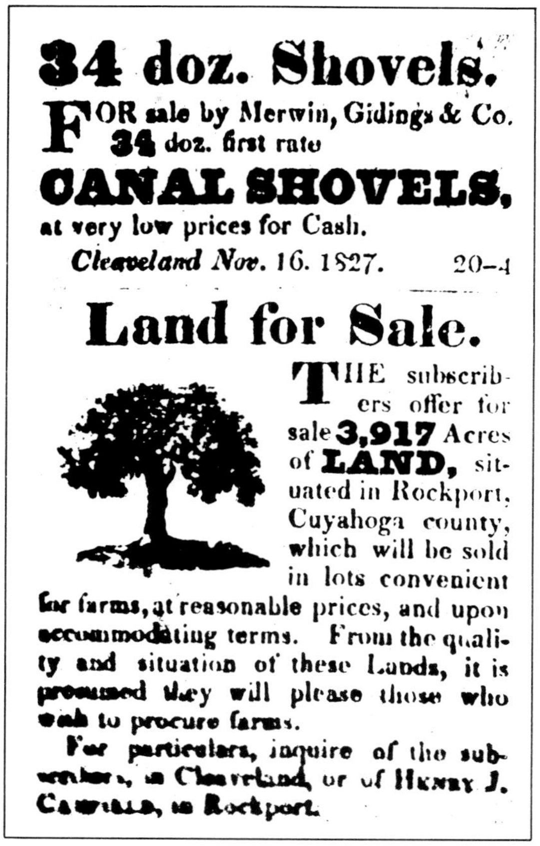

34 doz. Shovels.

FOR sale by Merwin, Gidings & Co.
34 doz. first rate

CANAL SHOVELS,

at very low prices for Cash.

Cleaveland Nov. 16. 1827. 20–4

Land for Sale.

THE subscribers offer for sale **3,917** Acres of **LAND,** situated in Rockport, Cuyahoga county, which will be sold in lots convenient for farms, at reasonable prices, and upon accommodating terms. From the quality and situation of these Lands, it is presumed they will please those who wish to procure farms.

For particulars, inquire of the subscribers, in Cleaveland, or of HENRY J. CANFIELD, in Rockport.

Advertisement in the *Herald*, Cleveland, Nov. 16, 1827. (WRHS)

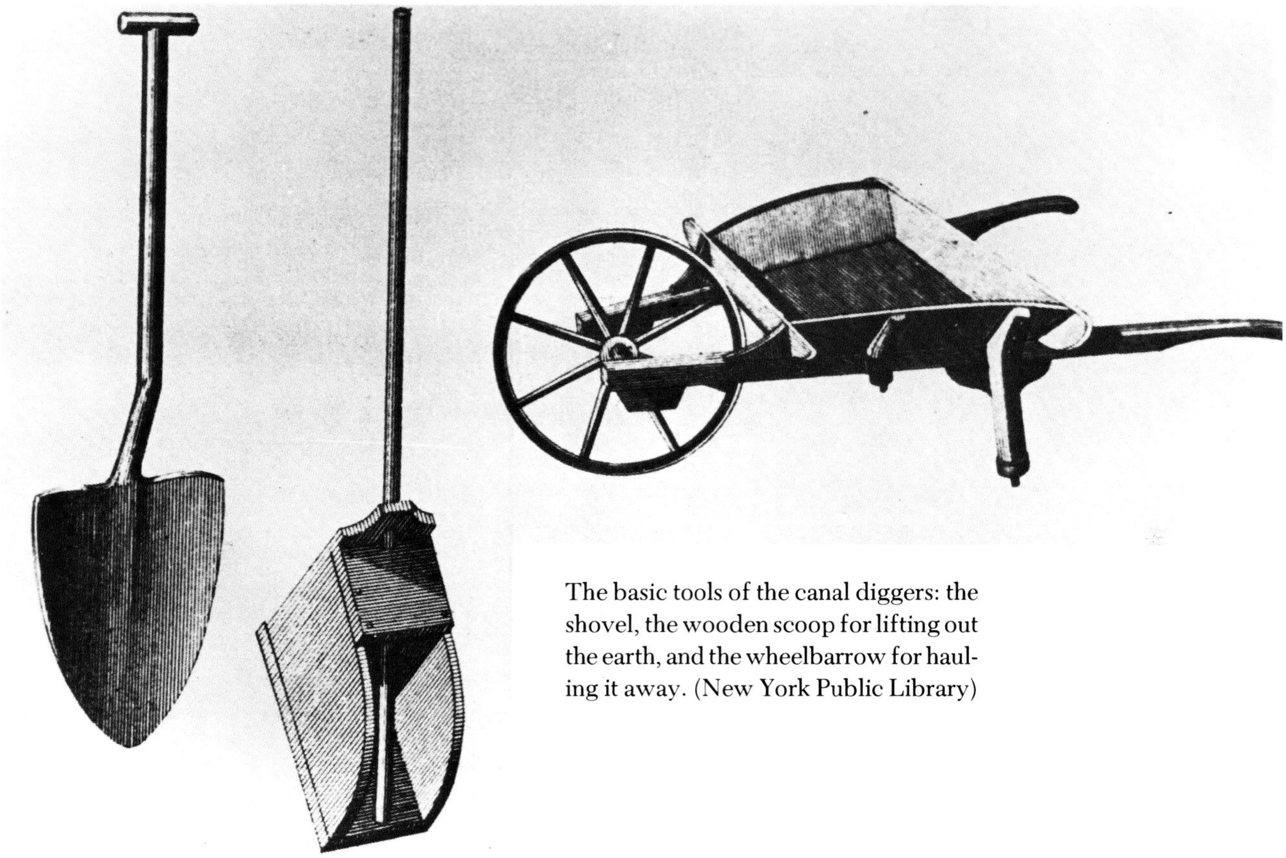

The basic tools of the canal diggers: the shovel, the wooden scoop for lifting out the earth, and the wheelbarrow for hauling it away. (New York Public Library)

The canals were often built a few feet above and parallel to rivers, with a berm separating the two waterways. As specified, the canal channel was no less than 40 feet wide at the top but sometimes had a width up to 150 feet. A ten-foot towpath for the mules was provided on one side only. The banks and bottom of the channel were lined with clay for waterproofing. (James and Margot Jackson, SCHS, from a drawing by Cal Sacks.)

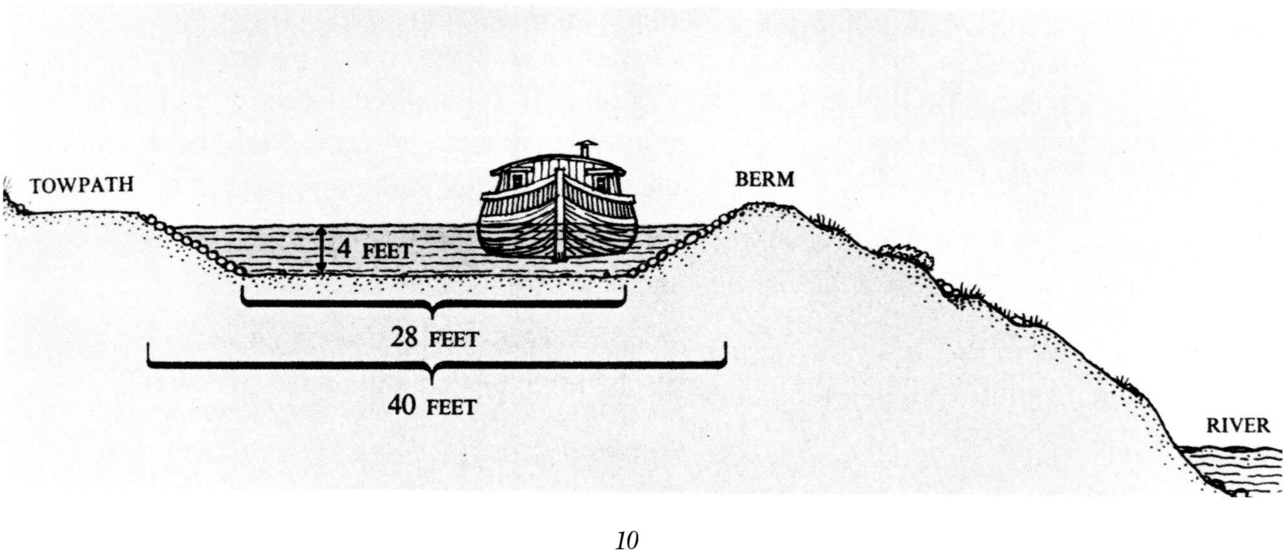

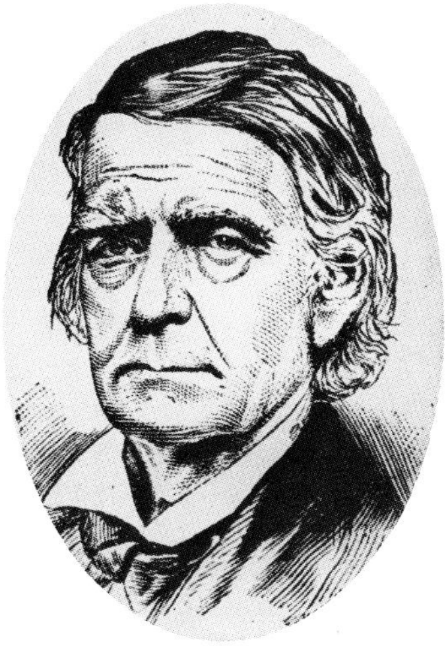

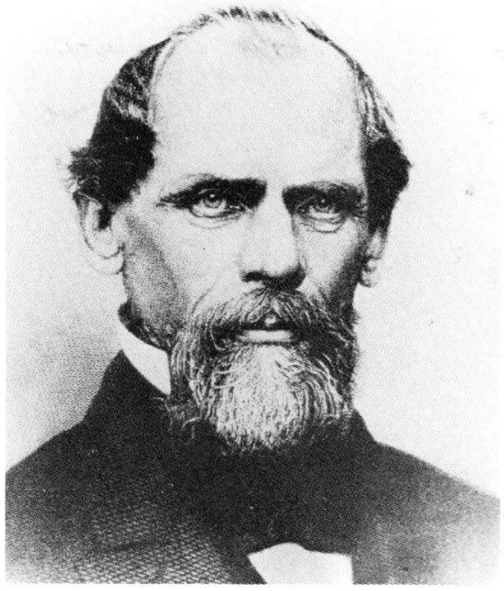

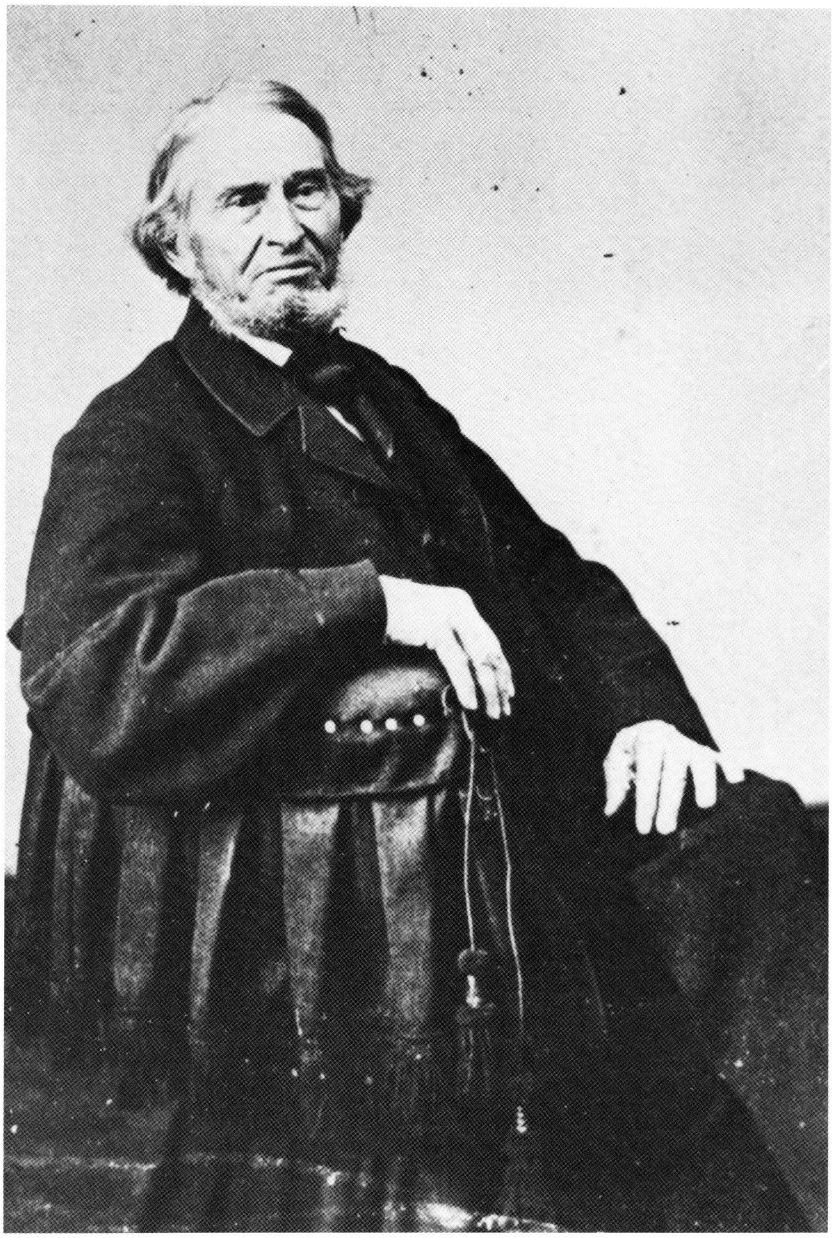

Above left, "Samuel Forrer holds the distinction of having had the longest association of any individual with the Ohio Canal System. For over fifty years, from the very beginning of Ohio's canals, he was variously engaged as rodman, surveyor, engineer, contractor and commissioner. Forrer, along with Alfred Kelley and Micajah Williams, rendered distinguished service to the State." Forrer was appointed deputy surveyor of Hamilton County in 1818, became resident engineer on the Miami Canal in 1826 and continued to work on aspects of the Miami & Erie until his retirement in 1873 (Frank W. Trevorrow, *Ohio's Canals,* 1973).

Above right, German engineer John Roebling. His first job in the United States was laying out portions of the Sandy & Beaver Canal. Roebling's preoccupation with the tensile strength of twisted iron-filament cable led to a new design for suspending the weight of an aqueduct's center span from a pair of towers. Late in life he designed a much more ambitious structure on the same principle—the Brooklyn Bridge. (CSO)

Left, Richard Howe, one of the engineers who built the northern section of the Ohio & Erie, was resident engineer on this canal from 1829 to 1850. His Akron residence west of the canal on Exchange Street is the oldest brick house in Akron; its front today is concealed by the facade of a retail store. (Nina Howe Stanford)

3

Hydraulics and Operation

Water is what the canals were all about. Instead of struggling to pull a few hundred pounds of goods in a wheeled vehicle over primitive, often muddy roads, water made it possible to float fifty to eighty tons of cargo in a single vessel. And water provides such a fluid, low-friction medium at moderate velocities that huge loads could be towed along at the speed of a brisk walk by no more than two or three mules. It wasn't swift transportation, but it was adequate—and it certainly beat the alternatives available at the time.

But "water seeks its own level," as the old aphorism says. It runs downhill when given the chance; and when contained, its surface is flat. But the land isn't. Even though canal engineers followed river valleys to achieve the flattest possible route, the canals somehow had to cross over the barriers between watersheds. For example, although the level of the Ohio River at Portsmouth was only a hundred feet lower than Lake Erie, the total amount of vertical travel that the Ohio & Erie would have to encompass was much greater. In the 116 miles between the river and Licking Summit, south of Newark, the canal would have to rise 413 feet. From this summit it would descend 160 feet to Dresden; it would then have to rise again 238 feet to the Portage Summit south of Akron, 78 feet higher than Licking Summit. Finally, in the 38 miles between Akron and Lake Erie, the canal would have to drop 395 feet.

For the most part, canal engineers overcame such obstacles with a system of locks—forming a kind of hydraulic stairway, whose operation is described below. On the Ohio & Erie, a total of 146 locks would be required to carry the boats up and down over a total vertical lockage of 1,206 feet. But locks greatly complicated water supply problems, as we shall see.

A lock was essentially a rectangular stone channel (although some Miami & Erie locks were constructed entirely of wood) ninety feet long, fifteen feet wide, and up to twenty feet deep. It served as a hydraulic elevator to raise or lower boats to a new level. The typical lift at each lock was eight to ten feet.

A pair of heavy wooden "whaler gates" closed the lock at each end. These could be opened and closed manually by means of long handles, called "sweeps," constructed of heavy timber. The gates were hinged so they opened upstream. When closed, their top edges formed a "V" with the point facing upstream so that the pressure of the "head" of water against the gates kept them closed.

Lock hydraulics worked by gravity. Near the base of each pair of gates was a "wicket," or "paddle"—a butterfly valve that was operated by means of an iron crank handle at the top of the gate (although some locks were later rebuilt with a sluice gate that controlled water flow through a passage in the wall of the lock).

In lifting a boat from one level to another, one of the boatmen (or the lockmaster if there was one) would first drain the lock, if it was filled, by opening the wicket in the lower gates. When the water level in the lock had equalized with the water level below the lock, the lower gates could be opened by shoving against the heavy wooden beams extending from the gates.

After the boat was drawn inside, the lower gates would be closed once more, as would the lower wicket, and the paddle in the upper gates would be opened—filling the lock to the level of the canal above. This permitted the upper gates to be opened, and the boat could then continue its journey. The time required to negotiate a lock was ten to fifteen minutes.

A boat would be lowered by reversing the process (as shown in the illustration). It is, perhaps, intuitively more obvious that lowering a boat would necessarily release to the lower canal level a volume of water equal to the "fall," or the difference between the two water levels. But the fact is that a single lockage cycle requires the same amount of water, whether the boat is going upstream or down.

For a lift of ten feet, this process would release more than 13,000 cubic feet of water—about 100,000 gallons irrevocably lost downstream. To provide for the passage of a hundred boats per day (generally considered the maximum a lockmaster could handle) it was calculated that an average flow of 1,500 cubic feet per minute from a source above the lock would have to be maintained. But, after taking into account evaporation and leakage, canal engineers figured that at least ten times this amount would be required. With a uniform flow of boat traffic both ways, it would be possible to double the number of boats passing through a lock using the same volume of water—providing a boat were waiting at the upper level to descend each time one was raised. But, as the Second Annual Report of Canal Commissioners observed in 1824, "It would be extremely vexatious to compel a boat to wait until one should arrive to pass in the opposite direction."

To begin with, then, it was necessary to provide a plentiful source of water above each of the canal summits. For the Ohio & Erie's southern summit, engineers dammed the great salt lick and surrounding swamp at the source of the Licking River south of Newark. The resulting canal reservoir became Buckeye Lake. At the northern summit they turned their attention to a chain of small, interconnecting natural lakes which were the headwaters of the Tuscarawas River. These lakes were also dammed and enlarged to form the Portage Lakes Reservoir, encompasssing about a thousand acres of water on three levels. In the western part of the state, the Miami & Erie was served by three reservoirs, of which Grand Lake St. Marys is the largest—the others being Loramie and Lewiston reservoirs. On both canals, a number of smaller feeder reservoirs were constructed.

Supplementing the reservoirs were feeders from creeks and rivers, which were dammed to raise the water level sufficiently to permit it to flow into the canal, usually controlled by a gate. Often, particularly on the Miami & Erie, these dams were the sites of "slackwater canals," which permitted the canal boats to negotiate the river channel for some distance to avoid the expense or difficulty of building a separate canal ditch (see following chapter). Some slackwaters were river crossings—in lieu of aqueducts. Such crossings had to provide a bridge over which the mules could pass to the towpath on the opposite bank.

At night, and during periods of less than peak traffic, excess water was shunted around each lock by means of a bypass race and spillway. Pioneer millers were quick to see the waterpower going to waste at these spillways, and a string of lock mills soon sprang up along the canals. These mills channeled the spillway water through their races and performed the service of grinding wheat and other grain into the more compact form of flour before it was loaded aboard the canal boats.

To avoid the cost of lock building, and to keep the canal channel low enough to take advantage of watersheds at higher elevations, a dramatic engineering alternative was the "deep cut." The longest of these is south of Spencerville, on the Miami & Erie. It is a continuous, single-level ditch from five to fifty-two feet deep, blasted through the blue-clay ridge that separates the Auglaize River valley from the St. Marys watershed. The Miami & Erie's Deep Cut, more than six thousand feet long, has been designated a National Historic Landmark. On the Ohio & Erie, a shorter deep cut lowered the water level of Licking Summit. The Pennsylvania & Ohio branch canal also had a deep cut at its summit, nearly five thousand feet long and over forty feet deep in places.

Even more daring was the construction of two tunnels on the Sandy & Beaver Canal—a privately financed, seventy-three-mile branch canal, or feeder, that connected with the Ohio & Erie Canal at Bolivar and extended to the Ohio River at the Pennsylvania line. The larger of its two tunnels was 1,060 yards long—more than half a mile—much of it through solid rock. The tunnels made it possible to shorten the canal by at least fifteen miles, as well as to lower the water level sufficiently to take advantage of feeders from higher-level streams.

The short Cincinnati & Whitewater Canal, which provided that Ohio River city with access to Indiana's Whitewater Canal, also boasted a tunnel near Cleves, Ohio. The Cincinnati & Whitewater tunnel was nearly a thousand feet long and was lined with brick.

Over the years, the canal system suffered from occasional washouts and other local hydraulic disasters, particularly during annual spring floods. During periods of drought, water supply itself became a problem; but for the most part it proved to be adequate. In an age when engineering calculations were done in longhand with a feather quill, early nineteenth-century canal planners had done a remarkable job.

Map detailing the hydraulics at Licking Summit, near Millersport. The combined Old and New Reservoirs are today known as Buckeye Lake (from *Hannum's Atlas of Fairfield County*, E. S. Hannum, Lancaster, Ohio, 1866, courtesy F. W. Trevorrow).

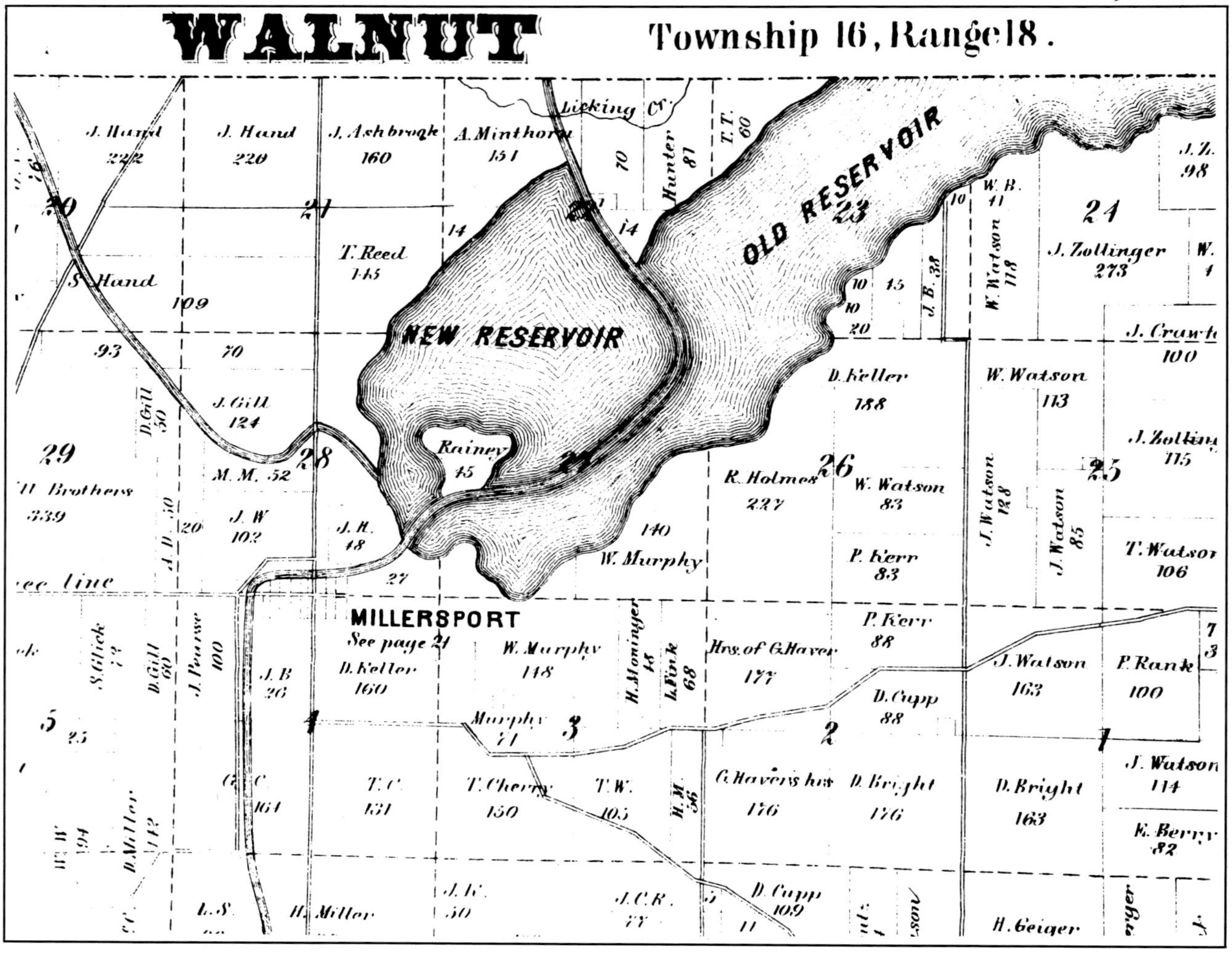

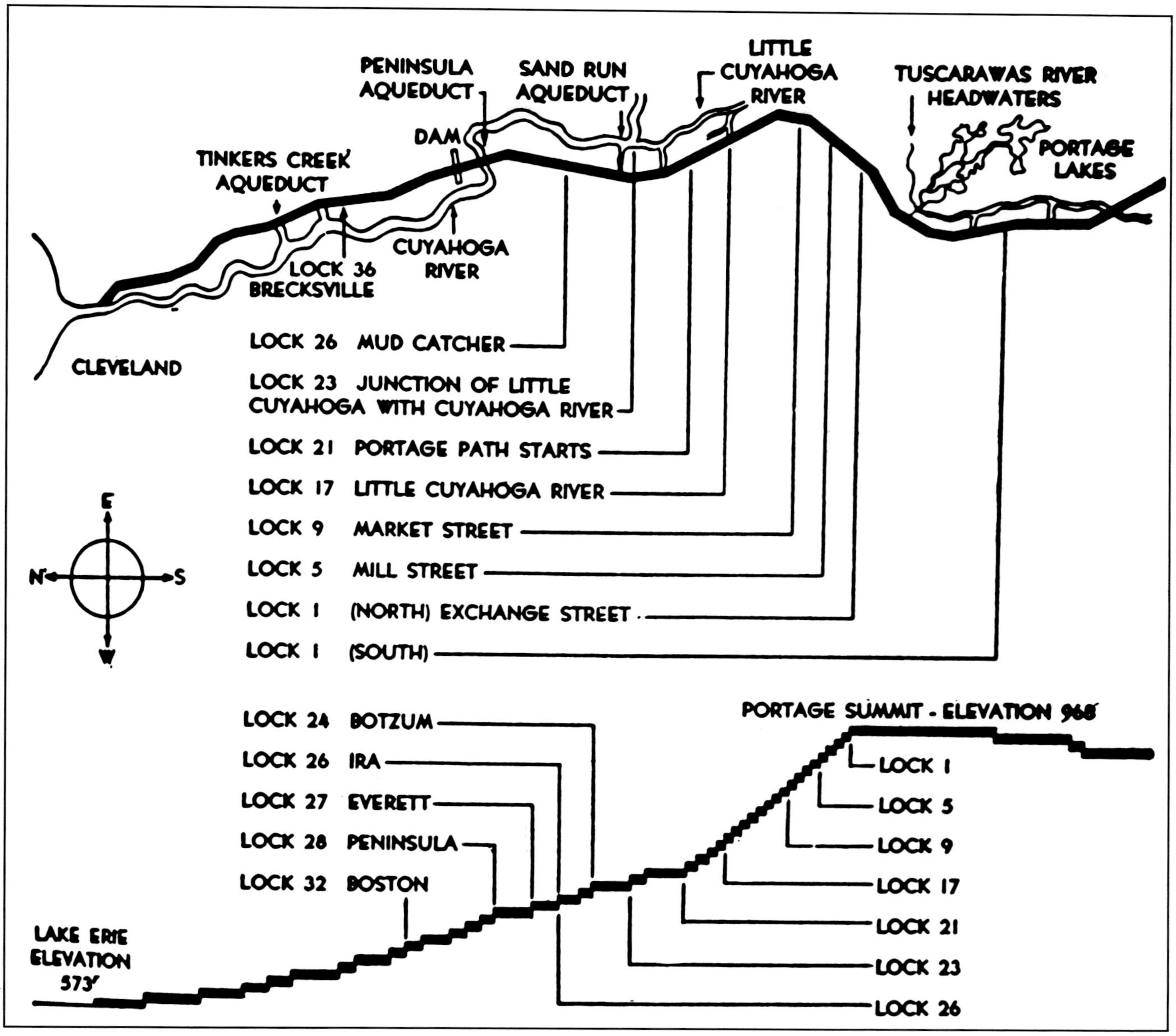

This chart shows the hydraulics and number of locks required in the first forty miles of the northern end of the Ohio & Erie Canal, which ascended 395 feet from Lake Erie's 573-foot elevation to the Portage Summit of 968 feet at Akron.

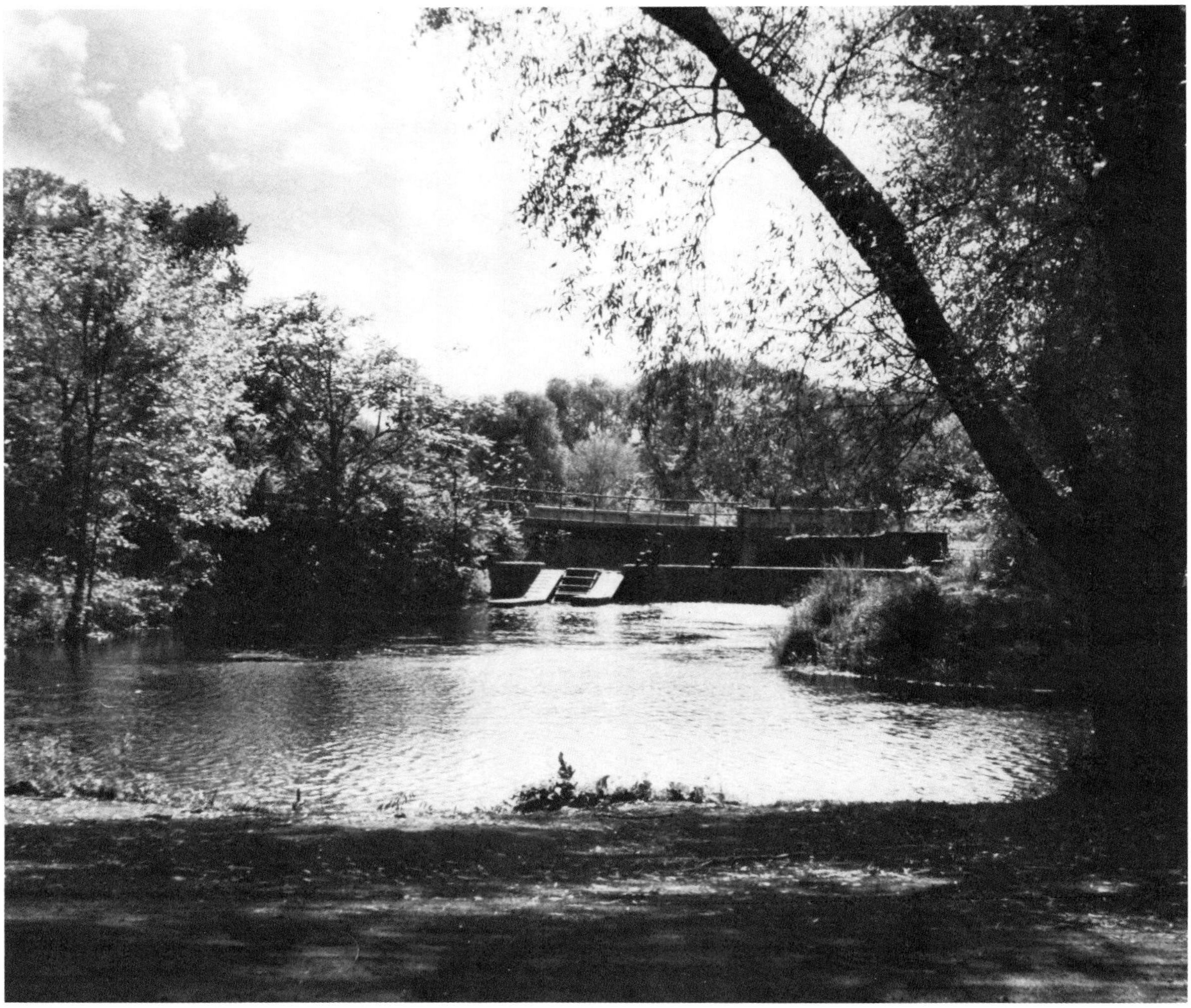

Above, Sluice gate and feeder from the Portage Lakes Reservoir as it enters the Summit Level of the Ohio & Erie Canal opposite Young's Tavern south of Akron. Water which flows to the left goes to the Cuyahoga River, into Lake Erie, and eventually down the St. Lawrence River to the North Atlantic Ocean. That which flows to the right finds its way into the Tuscarawas River, the Scioto, the Ohio River, and finally down the Mississippi into the Gulf of Mexico. (Photo by Jack Gieck, 1951.)

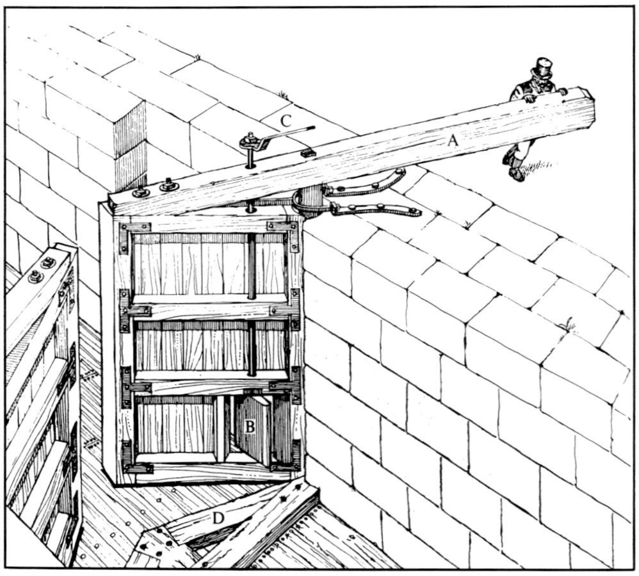

Right, Locks typically were built of sandstone blocks tuck-pointed with lime mortar, with a wooden floor. Set into the walls with iron hardware at each end were a pair of whaler gates constructed from white oak; the gates closed against the wooden lock sill (D), forming a "V" pointing upstream to resist the pressure of the head of water. Near the bottom of the gate is a butterfly sluice valve (B), called a paddle or wicket, operated by the iron control lever (C) which permitted water to flow into or out of the lock. The gates were opened and closed by means of long wooden balance beams (A) called sweeps. (James S. and Margot Jackson, "The Colorful Era of The Ohio Canal," from a drawing by Cal Sacks.)

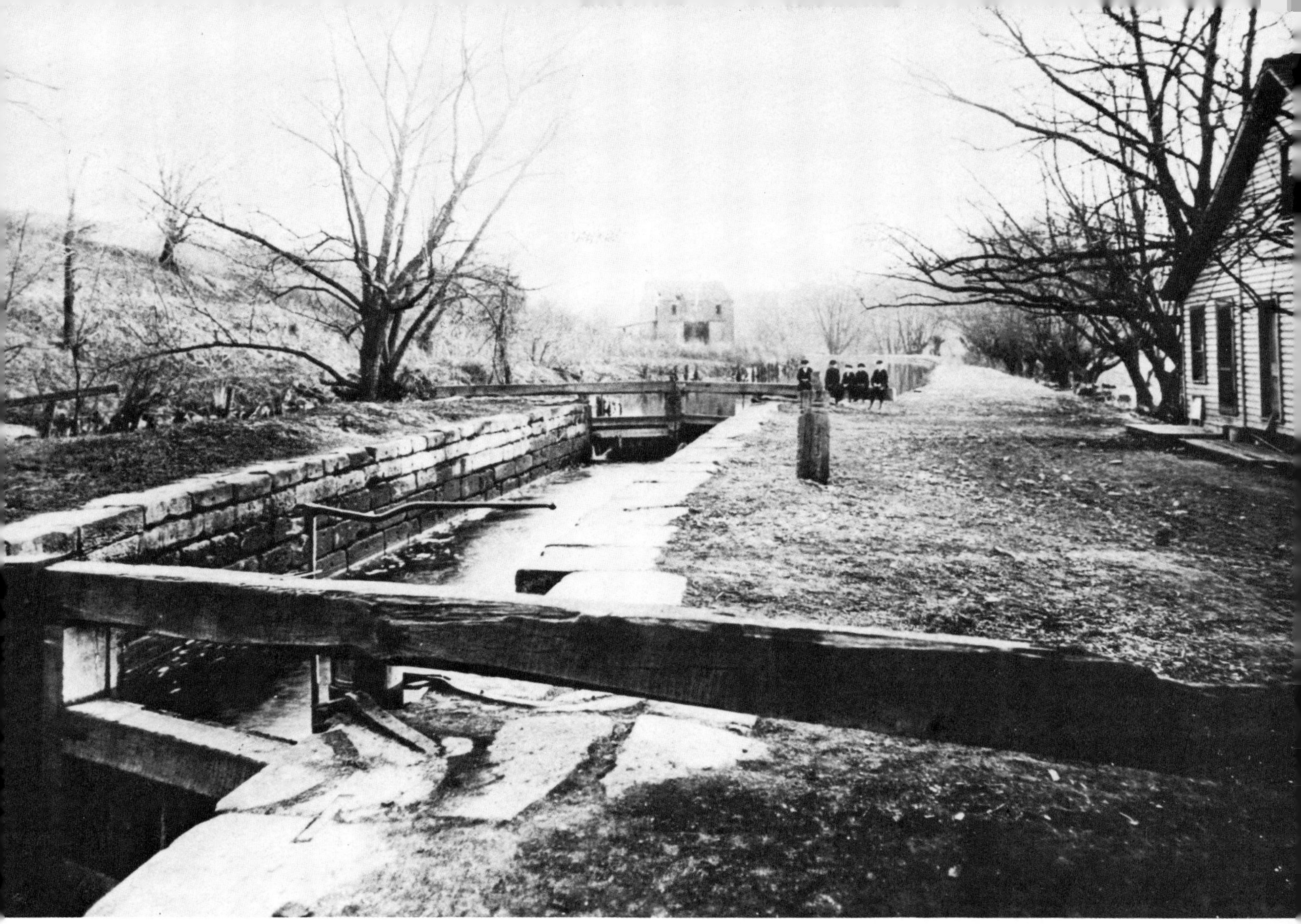

Lock components are clearly visible (above) in this scene at Five Mile Lock in Brecksville (on the Ohio & Erie south of Cleveland): sandstone walls, wooden gates with long balance beam sweeps, the control handle in the foreground for opening and closing the butterfly valve in the lower gate (foreground). Water is at upper level; 1894. (Baus Collection, RVF)

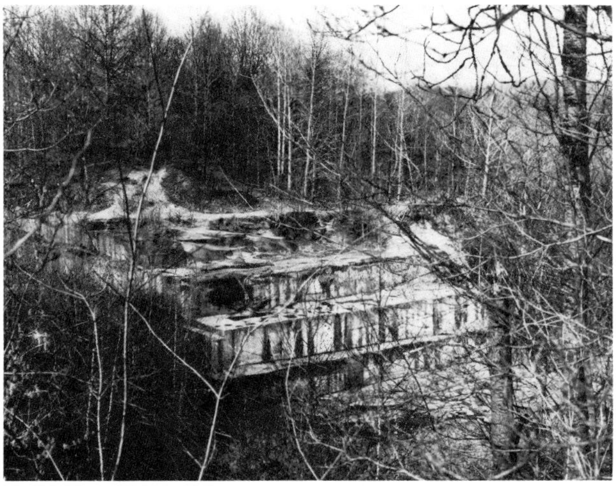

Deep Lock Quarry (above), where sandstone was quarried for many locks in the Ohio & Erie's northern sector. Imperfect lock stones were hauled north by canal boat to become Cleveland's breakwater. (Photo by Jack Gieck.)

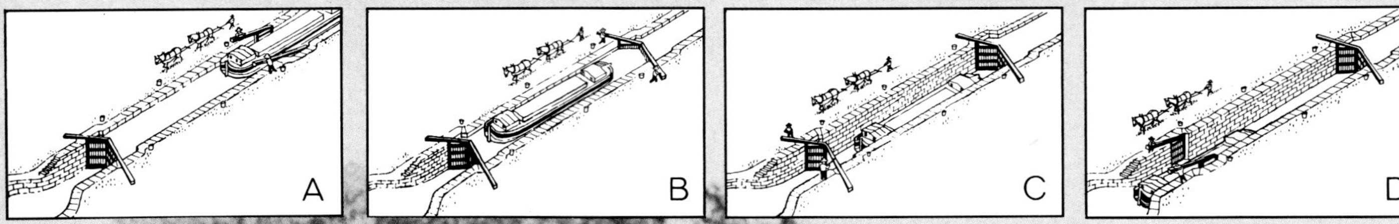
A
B
C
D

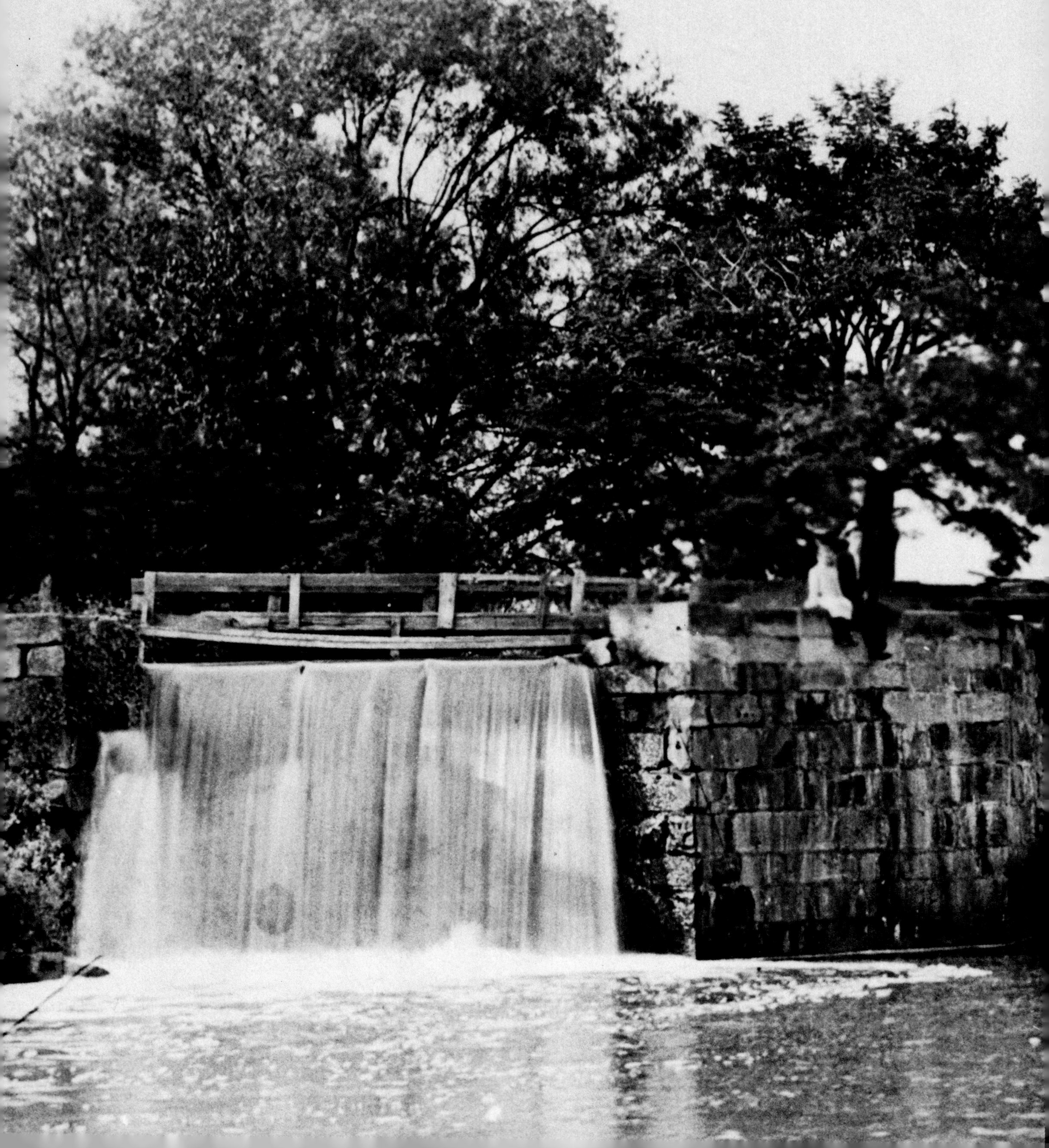

Left to right: Locks operated like hydraulic elevators, utilizing an engineering technology centuries old. To lower a boat from a higher level, the lock would first be filled by opening the butterfly valve, or wicket, in the upper gates until the water levels were equal. The upper gates could then be swung open to admit the boat (A). After the boat was locked in (B), the wicket in the lower gates would be opened to drain the lock until its water level matched that of the downstream elevation (C). The lower locks would then open easily to permit the boat to continue on its downstream voyage (D). Each lock cycle released about 100,000 gallons of upstream water. Providing an ample water supply for lock operation, especially during periods of peak traffic, was a major engineering consideration in designing Ohio's canal system.

At night, and during other periods of less-than-peak traffic, excess water was shunted around the locks through a bypass race and over a spillway—clearly in evidence (below) in this 1890 scene at Trenton. (Pockrandt Collection, SCHS)

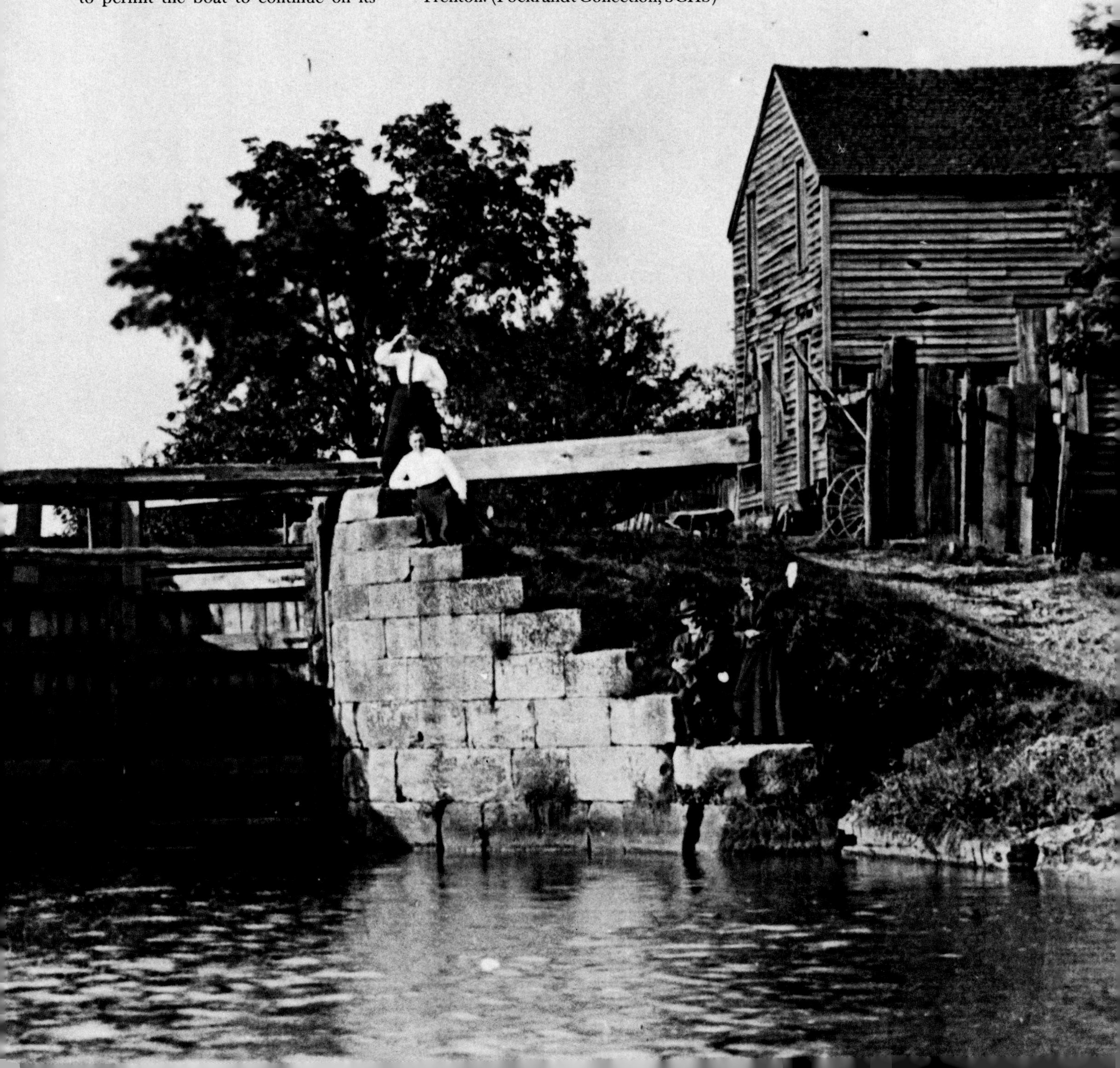

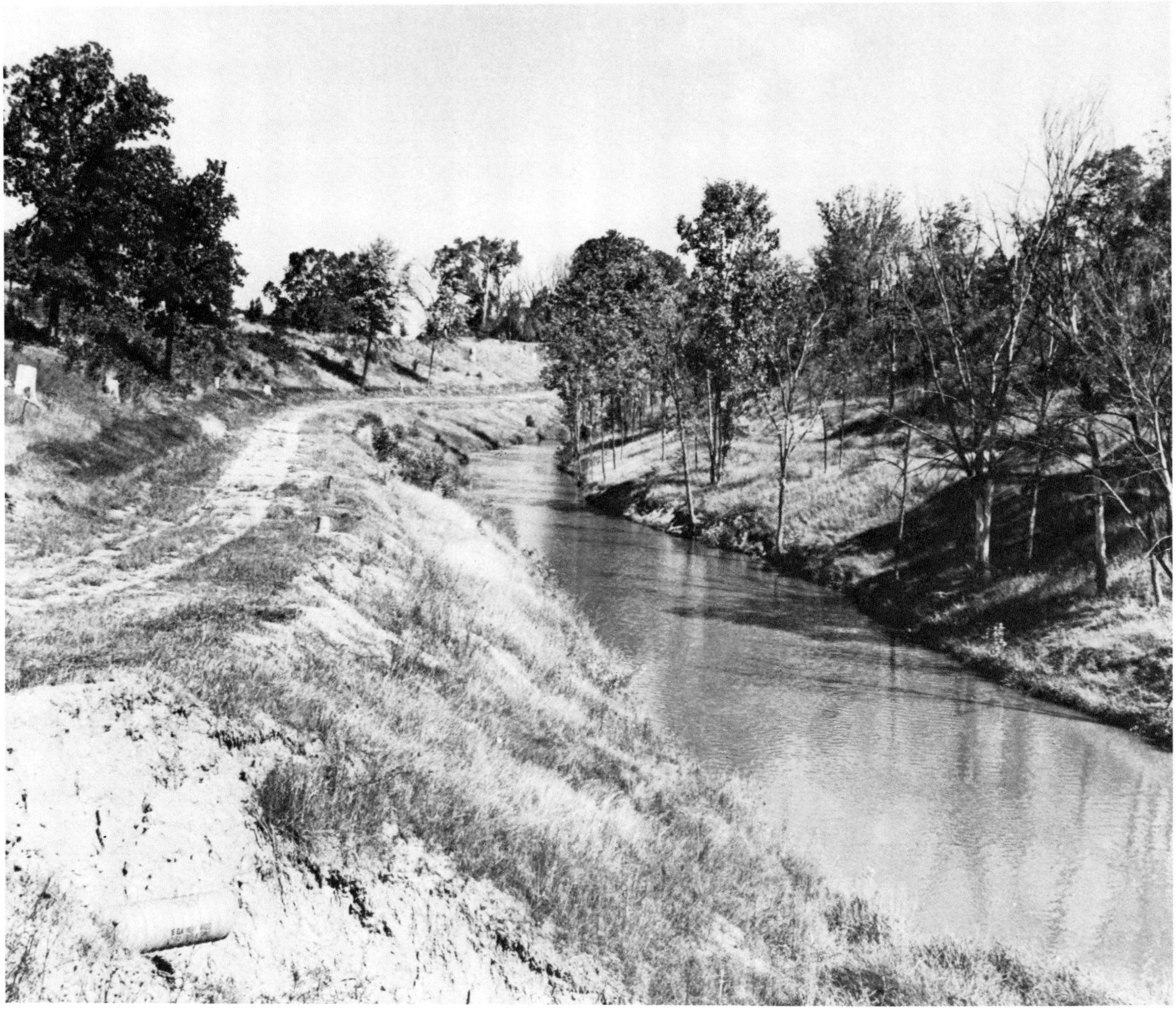

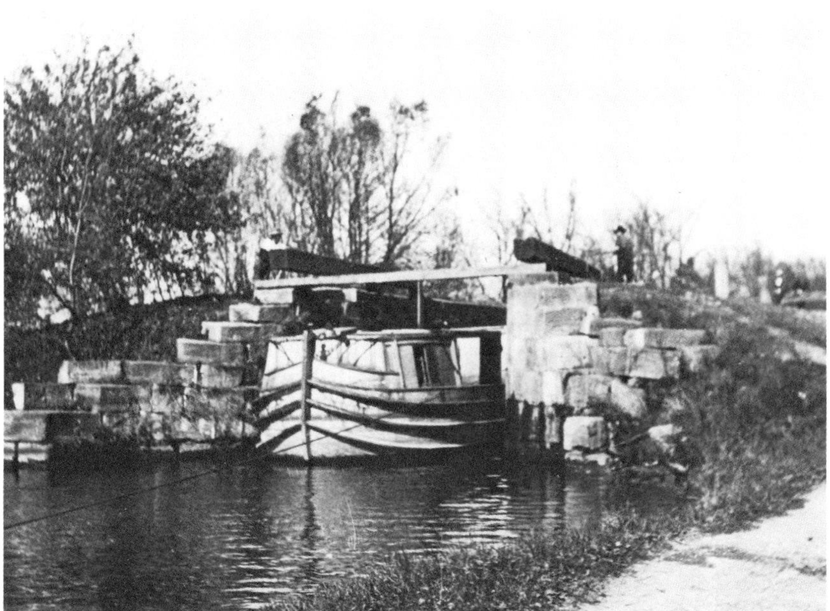

At left, a boat passes out of lock at Clinton. Above is Deep Cut, south of Spencerville, in 1962. From five to as much as fifty-two feet deep, this 6,000-foot trench was dug through the blue-clay ridge that separates the Auglaize River valley from the St. Marys watershed, significantly lowering the Miami & Erie's summit level in order to improve its water supply hydraulics. Deep Cut has been designated a National Historic Landmark. (Pockrandt Collection, SCHS)

4

Aqueducts, Culverts, and Slackwaters

Following river valleys as they often did, the canals frequently encountered tributary streams which had to be crossed. Indeed, the canals also had occasion to cross the major rivers they paralleled, sometimes having to switch sides as the riverbank became a vertical rock cliff. The usual expedient to solve such problems was either an aqueduct or a culvert—both based on canal technologies developed centuries earlier during the construction of European canal systems.

An *aqueduct* was a wooden flume, or rectangular trough, that constituted a hydraulic bridge. It fit into the channel-shaped tops of stone abutments on both banks of the stream to be crossed; if it had a long span, the aqueduct was supported on additional stone piers set into the river bottom. Many aqueducts were built with roofs, like covered bridges, partly to protect the construction from the weather, but also to provide additional bending strength in the cross section to support the heavy burden of water. Such structures required constant maintenance, especially after a ravaging winter. Nearly every spring saw a major overhaul, removing rotted lumber and rusted bolts and nails, and especially repairing leaks—which otherwise added to water-supply problems. Even so, all aqueducts leaked some—and when cold weather descended, many became winter wonderlands.

•

For relatively small rivers and creeks, *culverts* were a much more waterproof solution; although seldom used on the Ohio & Erie Canal, they abounded on the Miami & Erie. A culvert was built as a stone arch spanning the stream; the arch was then covered with earth, so that it resembled one of today's highway culverts that are formed by burying a metal pipe. In a standard clay-lined channel through this supporting bed of earth, the canal traversed the length of the culvert. And some were long, indeed, with multiple arches. The magnificent Turtle Creek Culvert in Shelby County was longer and taller—and certainly more graceful—than many aqueducts.

That at least one contractor took great pride in such structures is evidenced by the monument James Durbin erected to himself above the site of his Bad Creek Culvert near Texas, Ohio (see chapter 7 on the Miami & Erie Canal).

When the level of the canal was not significantly different from that of an intersecting waterway, it could sometimes be crossed by a "*slackwater navigation*," in which the boat was simply poled or towed across the river—ideally with a bridge provided for the mules. Usually, a dam was built in the river to assure reliable water depth. Occasionally, slackwaters were built to provide a river channel for several miles, as an alternative route to avoid the expense of constructing a parallel canal channel—or when rocky cliffs or rugged terrain on both sides of the river made canal construction impossible. In such cases, navigation was diverted into the river by means of a series of locks and dams. A dam at the lower end raised the water level in the river to a sufficient depth to maintain navigation. Adjacent to the end of the dam a lock was located, permitting adjustment of water level to that in the canal below the dam. If re-entry to "slackwater" was made further downstream, another lock provided adjustment of the level in the canal to that in the river.

Slackwater navigation was employed on the Miami & Erie Canal between Defiance and Maumee, through the Black Hand Gorge on the Ohio & Erie, and at several points on the Hocking Canal. The Muskingum River Improvement was entirely slackwater navigation. At some points on the Hocking, and in Black Hand Gorge, the towpath sometimes had to be cut from a rock face, or had to be built up from the rocky river bottom at the base of a cliff.

Having to pole a boat across a slackwater crossing near a dam could be a dangerous undertaking, as witness the happening described by Terry Woods in his book on the Walhonding Canal:

"The slackwater crossing [at Six Mile Dam] . . . was a hardship to the boatmen. Lock #5 was on the north side of the river, right at the edge of the slackwater pool. A captain heading toward Roscoe would take in his towline and put his mules aboard. Then, with the careful use of tiller and pikepole he would drift and pole across the river to the south bank and enter the canal again. The return was even more hazardous as the angle was slightly upstream—against the current. During periods of high water, a boat, its cargo and crew could, and sometimes did, go over the dam into the river."[1]

Woods goes on to recount how Captain Edmond Hicks, with his wife aboard her namesake, the *Liza Jane,* got in serious trouble trying to make this crossing during high water. Hicks ultimately lost control of the boat, which headed for the dam. Resourcefully, if not gallantly, the captain managed to leap from the deck and frantically swam to shore, just in time to see both Liza Janes go over the dam.

Somehow the boat did not break up in the process. Pluckily, Mrs. Hicks managed to get to the tiller and steer the boat down the river, through the Ohio Canal Feeder into Roscoe Basin, and then (no doubt with the help of the mules and remaining crew) back into the Walhonding Canal. There is no record of how the captain's abandoning ship affected the Hickses' marriage.

There is evidence that belatedly, several years later, a rope and pulley arrangement was strung across the river at this point, allowing boats to tack their way between Locks Four and Five with a significant reduction in the crew's adrenaline.

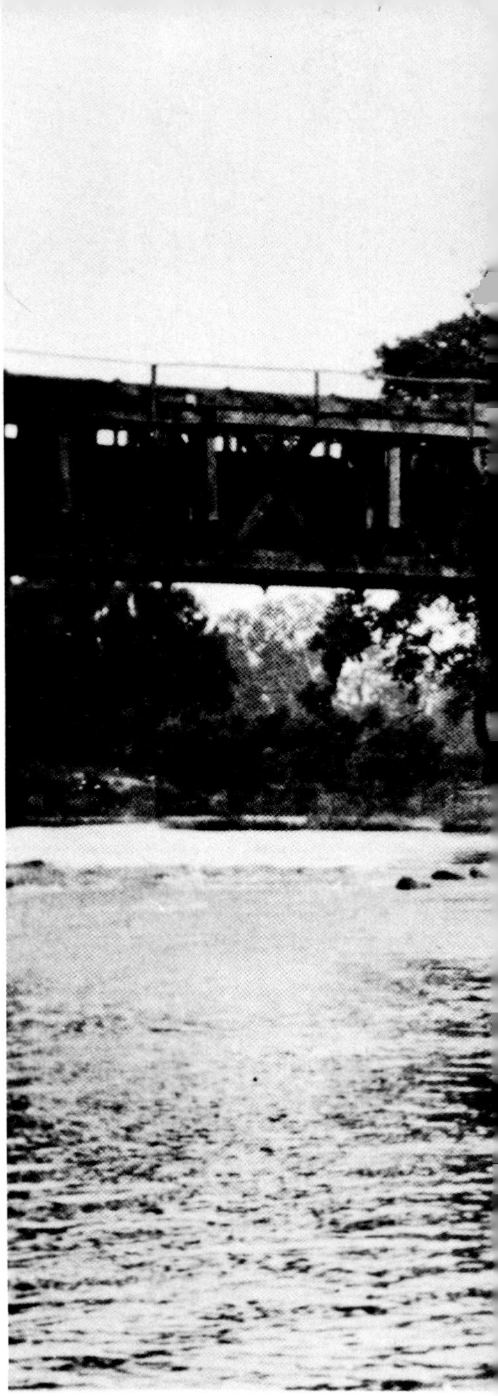

1. Terry K. Woods, *Twenty-five Miles to Nowhere: The Story of the Walhonding Canal, with Canal Guide* (Coshocton, Ohio: Roscoe Village Foundation, 1978).

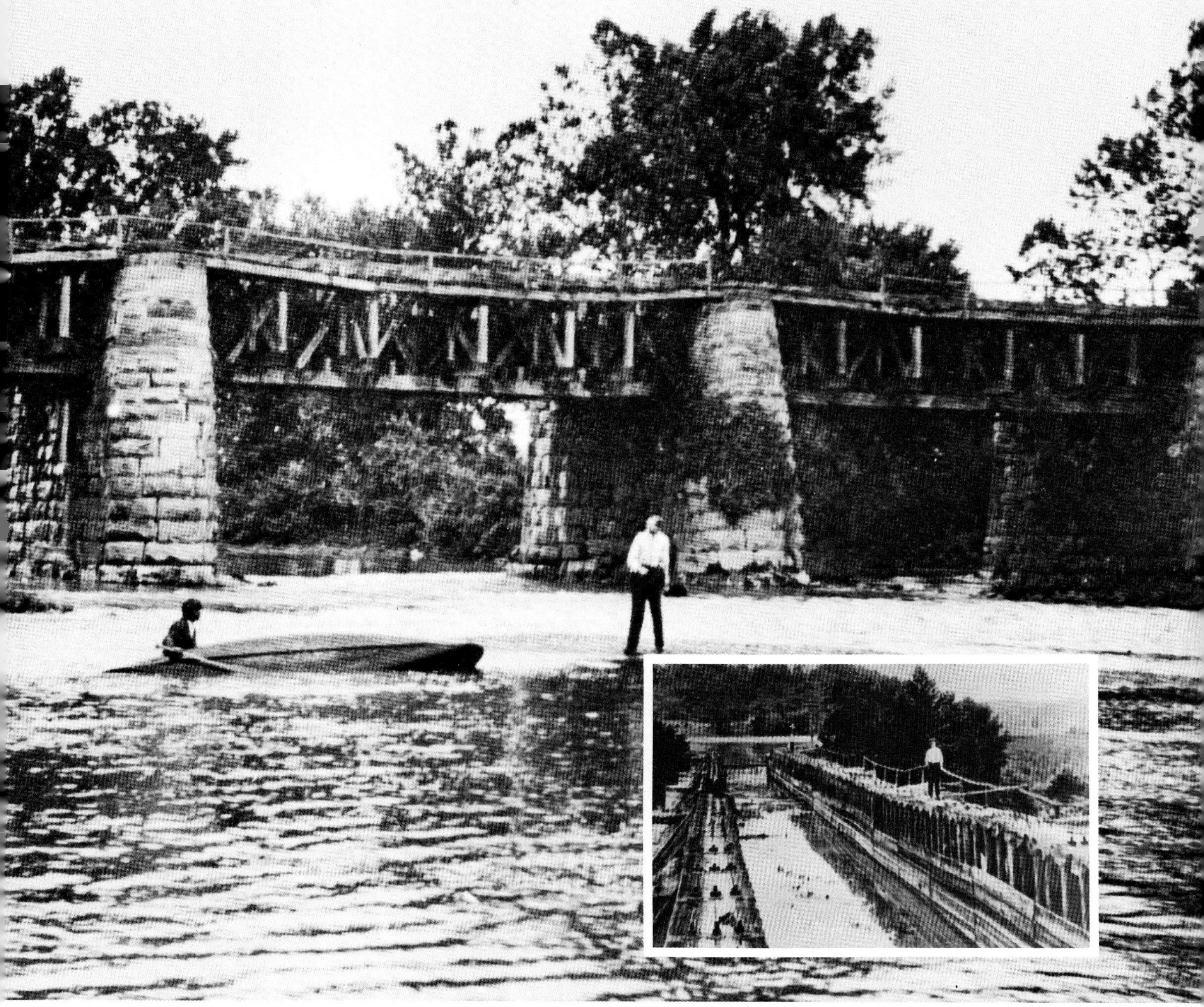

This Ohio & Erie aqueduct, which carried the canal over the Walhonding River near Coshocton, is a classic structure: a wooden flume, or rectangular trough, supported on stone piers set in the river, with ends attached to stone abutments on each bank. The fence at the top indicates the edge of a narrow plank towpath. Although its bending strength is improved by wooden trusses containing "X" members on each side, there is significant sag in the flume between the piers due to the weight of the water. As with many canal photos, the photographer's motivation for the picture was probably the overturned canoe in the foreground. Inset is a view of the water in the flume. The pedestrian is on the towpath, the plank floor of which is not quite visible. (RVF)

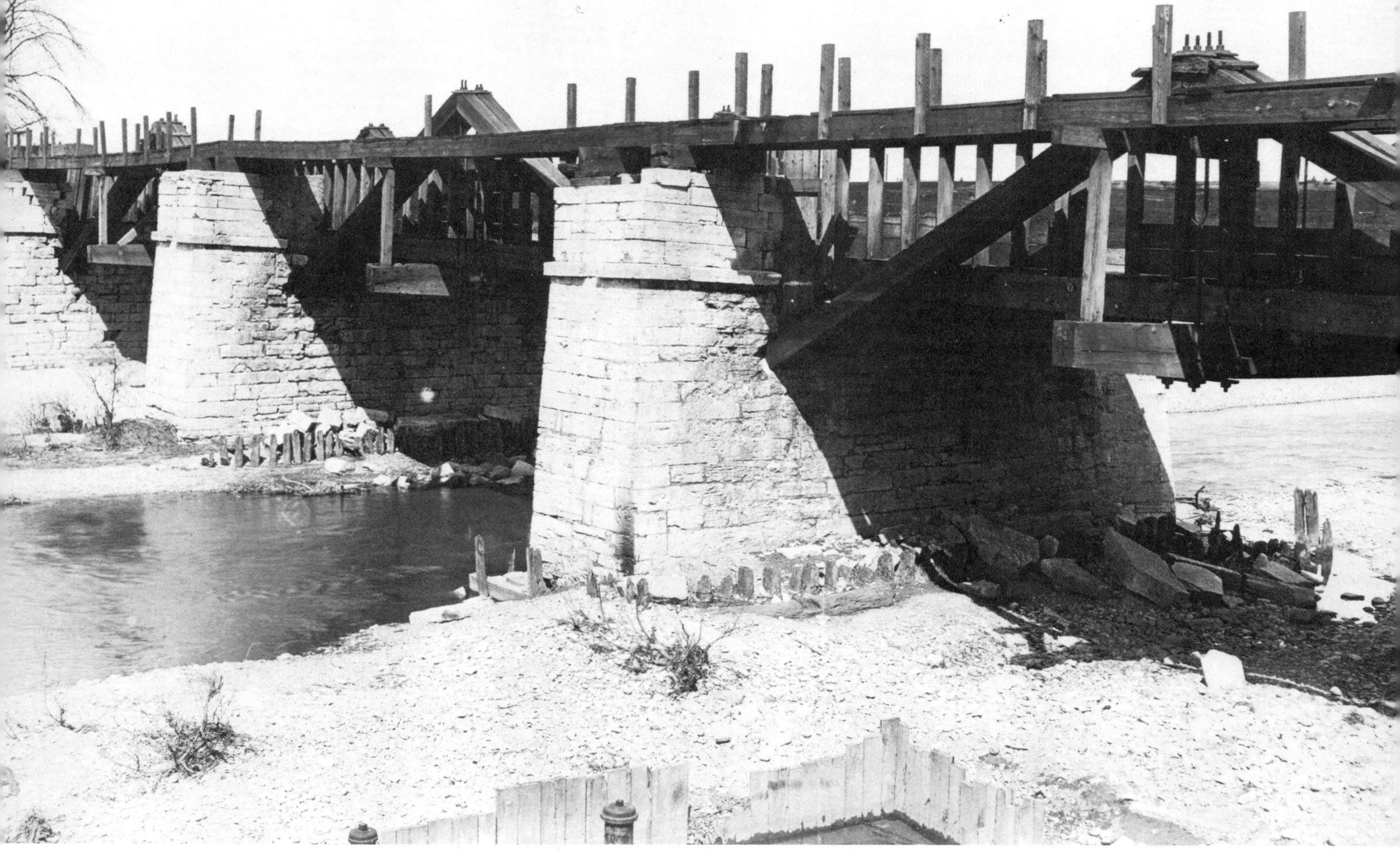

Above, This design of the Mad River Aqueduct on the Miami & Erie at Dayton is more sophisticated than the one below. The center of each span hangs from steel rods bolted to the apexes of triangular trusses, whose bases are anchored in the many stone piers, which also provide direct support. (RVF)

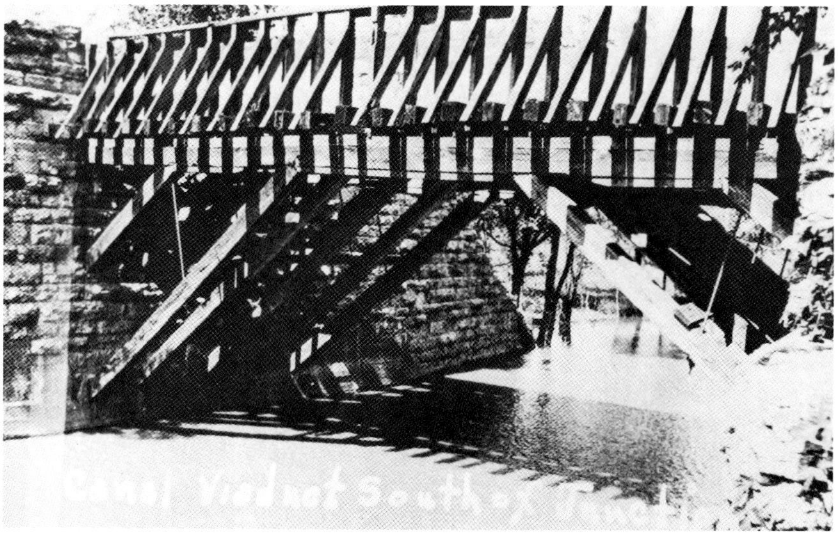

Right, This little aqueduct on the Miami & Erie south of Junction looks strong enough to support a battleship. A row of triangular buttresses reinforces the sides of the flume, while multiple tiers of slanting wooden compression members set into the abutments also form strong triangles as they support its underside. (CSO)

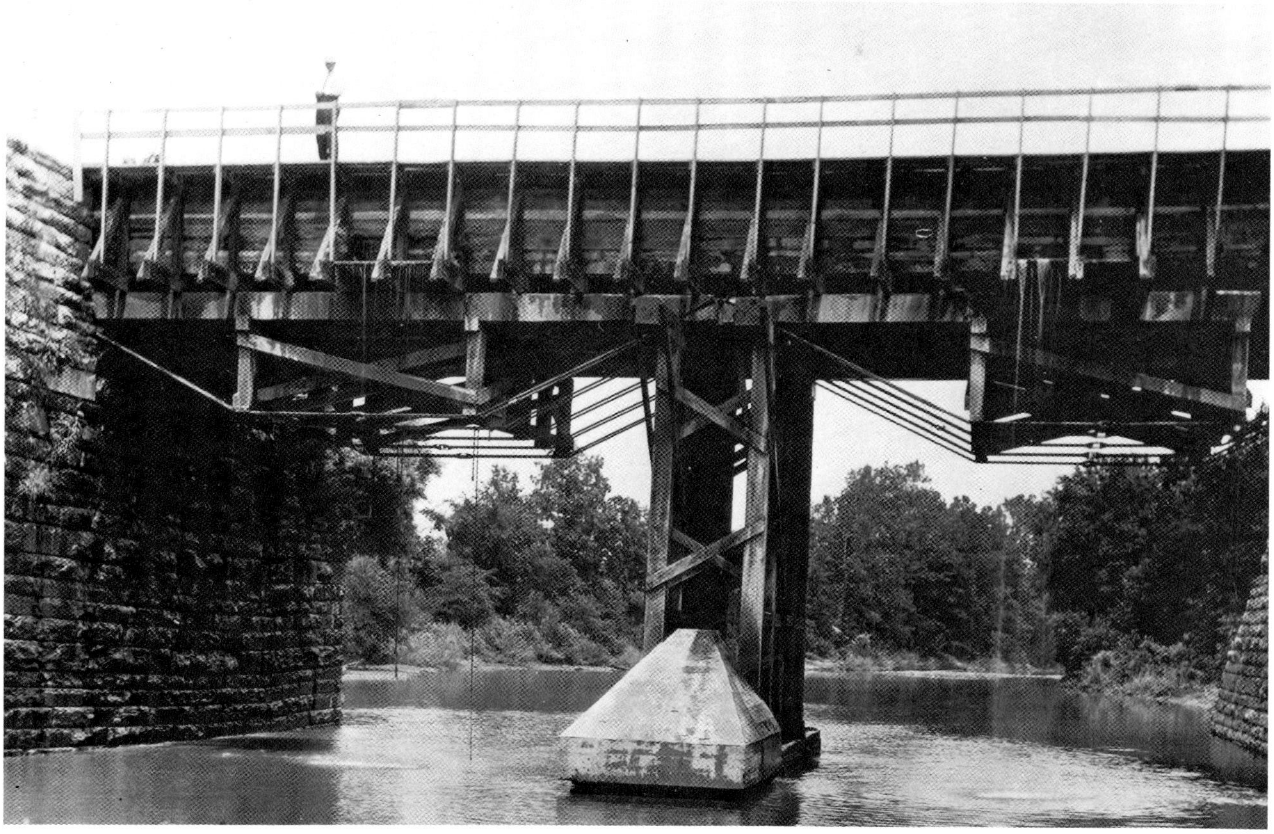

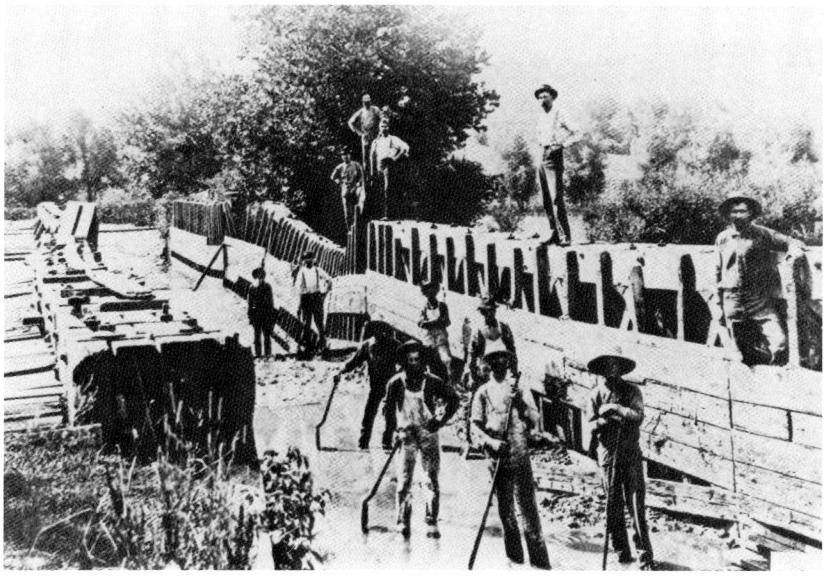

Above, This aqueduct spanning the Little Auglaize River on the Miami & Erie near Melrose utilizes a design that provides maximum strength with minimum weight of construction materials. Steel rods in tension effectively support the center of the trough midway between the wooden pier and the abutments. There is no perceptible sag in the structure. A spectator stands on the towpath; 1916. (OHS)

Left, Each spring brought with it the heavy chores associated with plugging leaks and replacing rotted timbers, rusty bolts and nails. The photographer is obviously not contributing to getting the job done in record time. Wooden towpath is visible at left. (RVF)

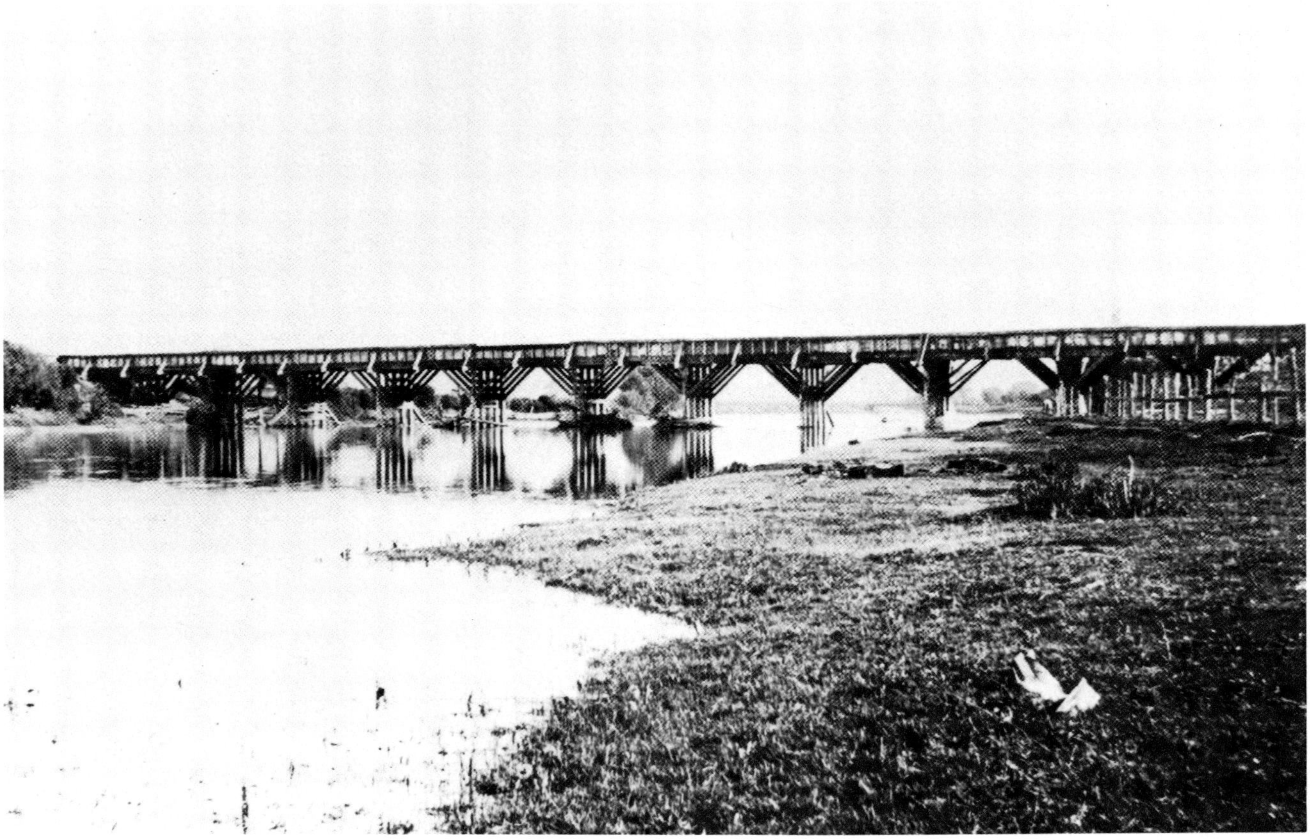

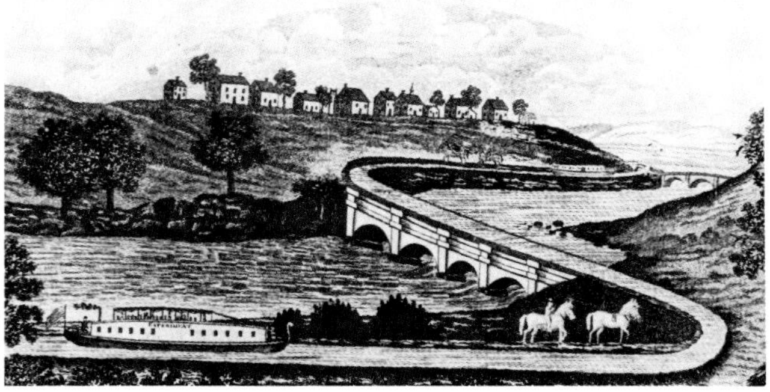

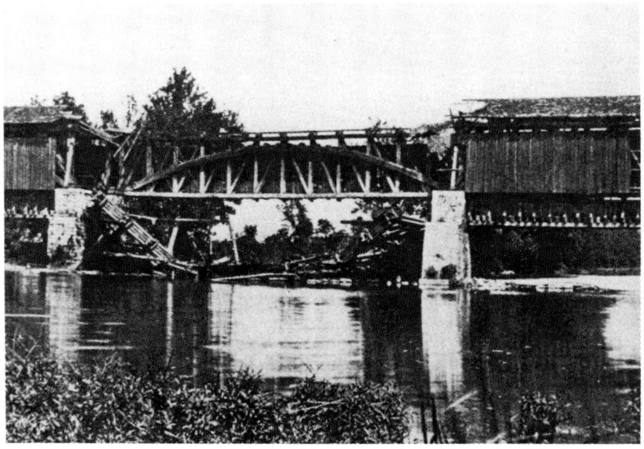

Top, The Sandy & Beaver Aqueduct at Bolivar was of a unique, all-wood construction. Banks of parallel, Y-shaped members anchored to the wooden piers form nearly a hundred strong triangular shapes visible in the picture. The construction does a remarkable job of support without detectable sag. (CSO)

Above left, A fanciful early engraving by an unknown but obviously imaginative artist depicts the aqueduct over the Great Miami River on the Miami Canal. In fact, the aqueduct was much shorter, ran straight across the river, and was covered. But the drawing says something about canal enthusiasm, which was rampant at the time. *Right,* The real Great Miami Aqueduct after its 1903 collapse. With its roof and protective cover partially stripped away, the aqueduct's arched "Howe truss" construction is clearly visible. A number of the system's larger aqueducts utilized Howe's design in order to increase the span between piers. (CSO)

The greatest of all Ohio aqueducts spanned the Scioto River at Circleville. This aqueduct was 425 feet long, and 20 feet wide. Its roof structure, similar to that of a covered bridge, not only provided protection from the weather to boat crews and the structure itself but also improved its bending strength by creating a box-beam cross section. After the aqueduct was abandoned in 1911, the town acquiesced to its becoming a local headquarters for games of chance. On April 17, 1915, it burned and fell into the Scioto River. The fire is said to have been set by a sore loser of the night before. (Pockrandt Collection, SCHS)

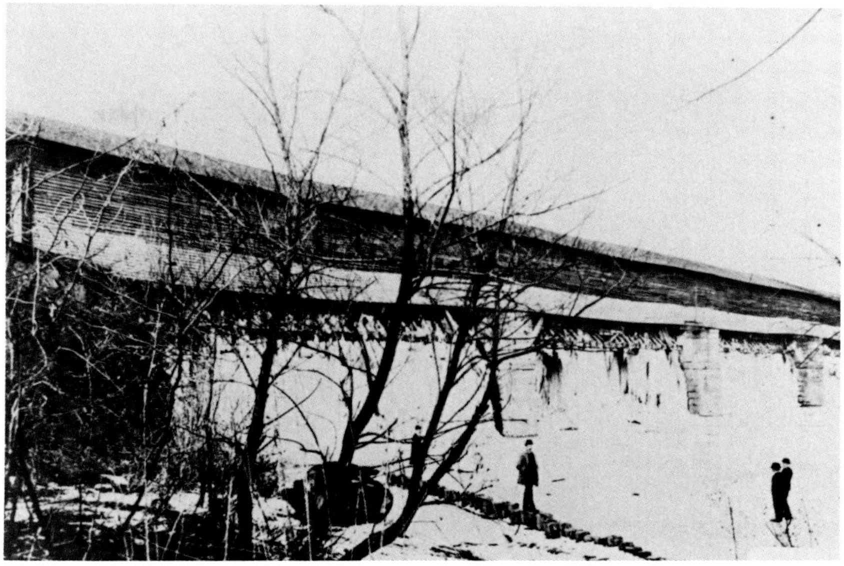

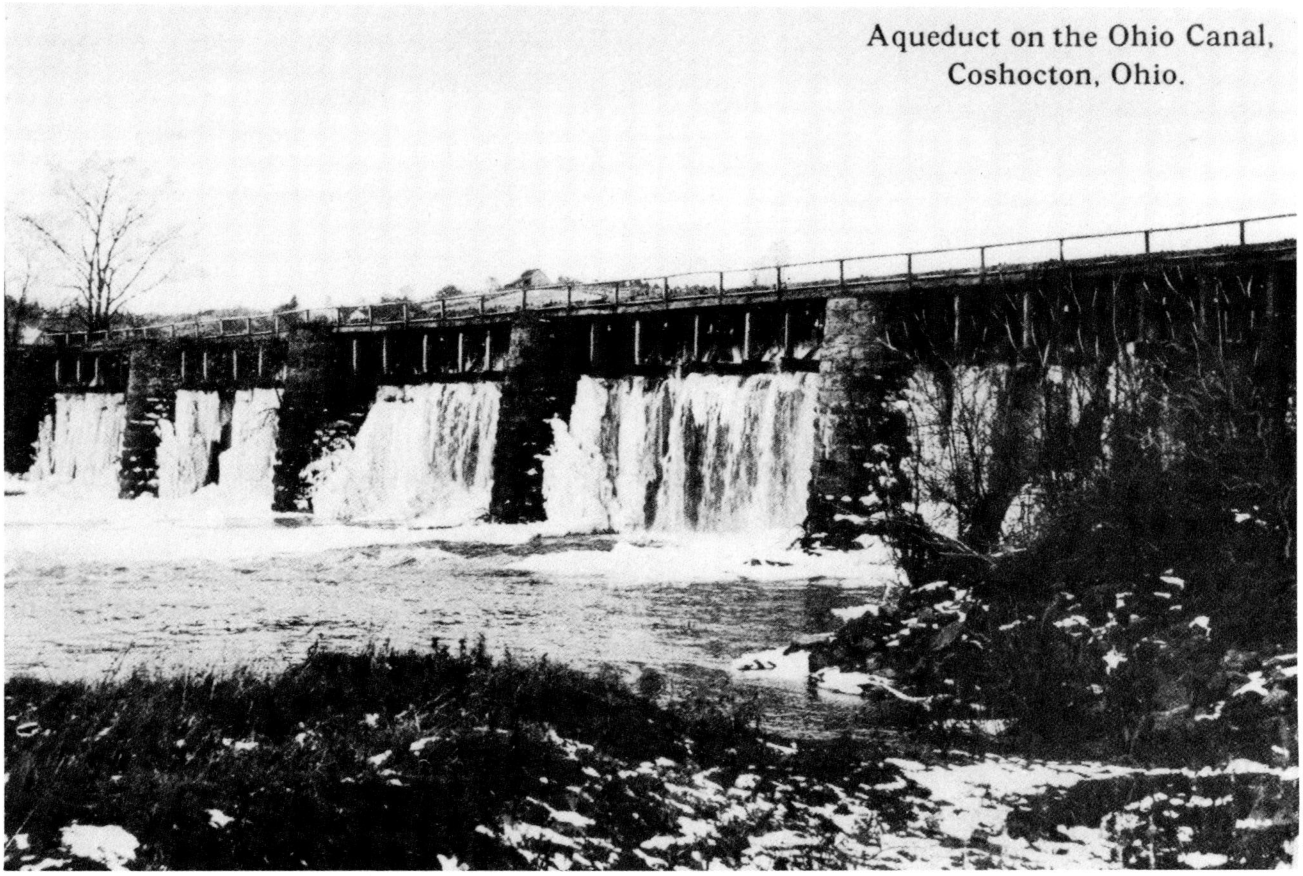

Aqueducts were a popular subject for nineteenth-century postcards. This one of the Ohio & Erie's Walhonding Aqueduct at Coshocton pictures torrential leaks which say nothing for the efforts of the repair crew pictured earlier. In fairness, this scene may have been recorded during high water, before spring reconstruction. (CSO)

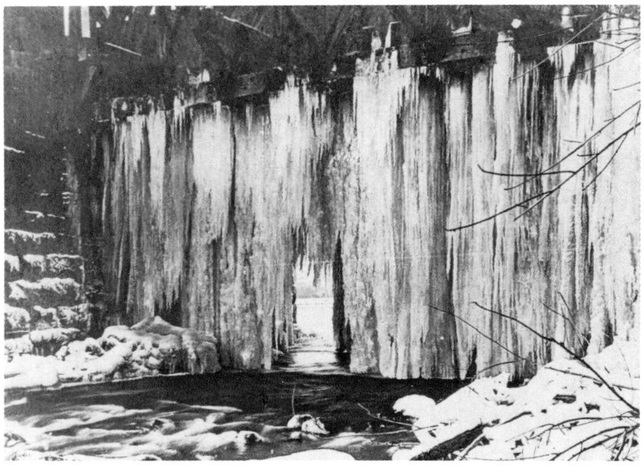

Left, This winter wonderland presented by the Walhonding Aqueduct at Coshocton was captured by another postcard photographer. (RVF)

Below, Tinkers Creek Aqueduct near Valley View, one of two surviving aqueducts on the Ohio & Erie Canal (a smaller Mill Creek Aqueduct is located several miles north), has been maintained in recent years for industrial water supply feeding a Cleveland steel mill. The ironwork superstructure in the photograph is part of a parallel highway bridge on Canal Road. The inset photo shows the wooden trough of water crossing Tinkers Creek. The steel beams that would obviously block boat passage have recently been installed to strengthen the flume. Wooden towpath appears at right. (Photos by Jack Gieck.)

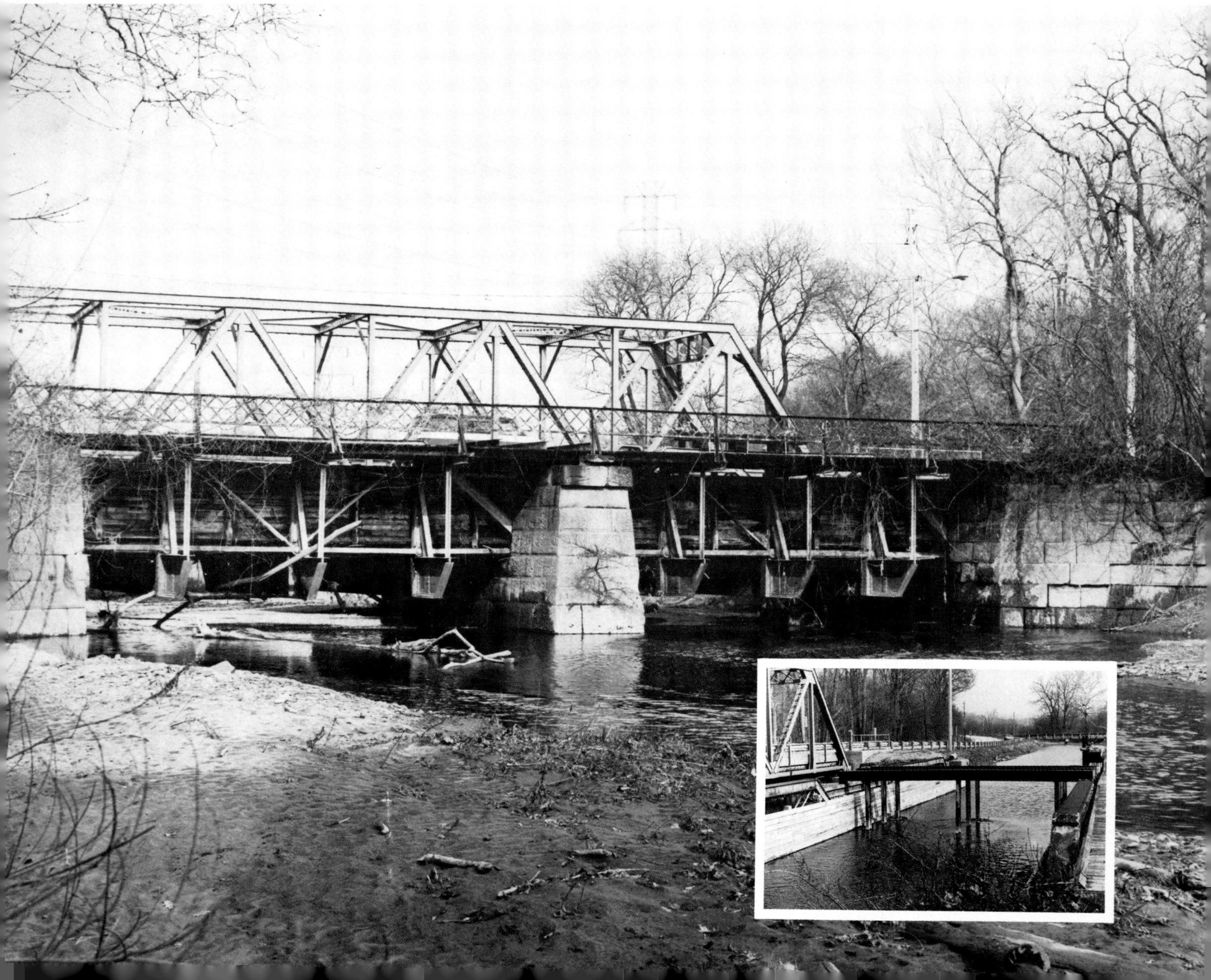

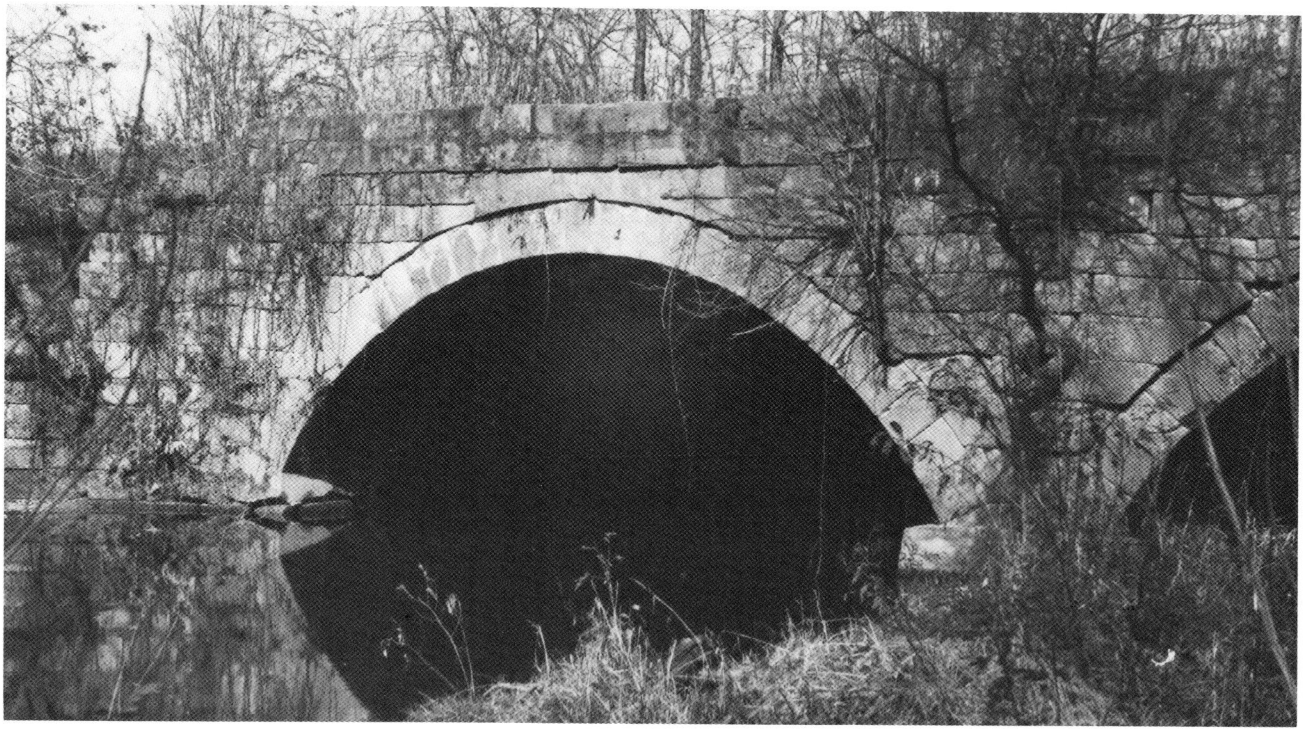

Above, Rocky Fork Culvert, on the Ohio & Erie east of Newark. An alternative to the aqueduct, especially for spanning relatively small waterways, was the stone culvert. Culverts were constructed by building a stone arch over the stream and erecting parallel walls to contain a fill of earth. The top mortared course constitutes the "coping." The clay-lined canal channel, supported by the bed of earth, traversed the length of the culvert. Such structures avoided the leakage problems characteristic of aqueducts; 1974. (Photo by F. W. Trevorrow.)

Below, This massive culvert, over Six Mile Creek south of Junction, carried the channel of the Miami & Erie Canal high above the level of the creek. (CSO)

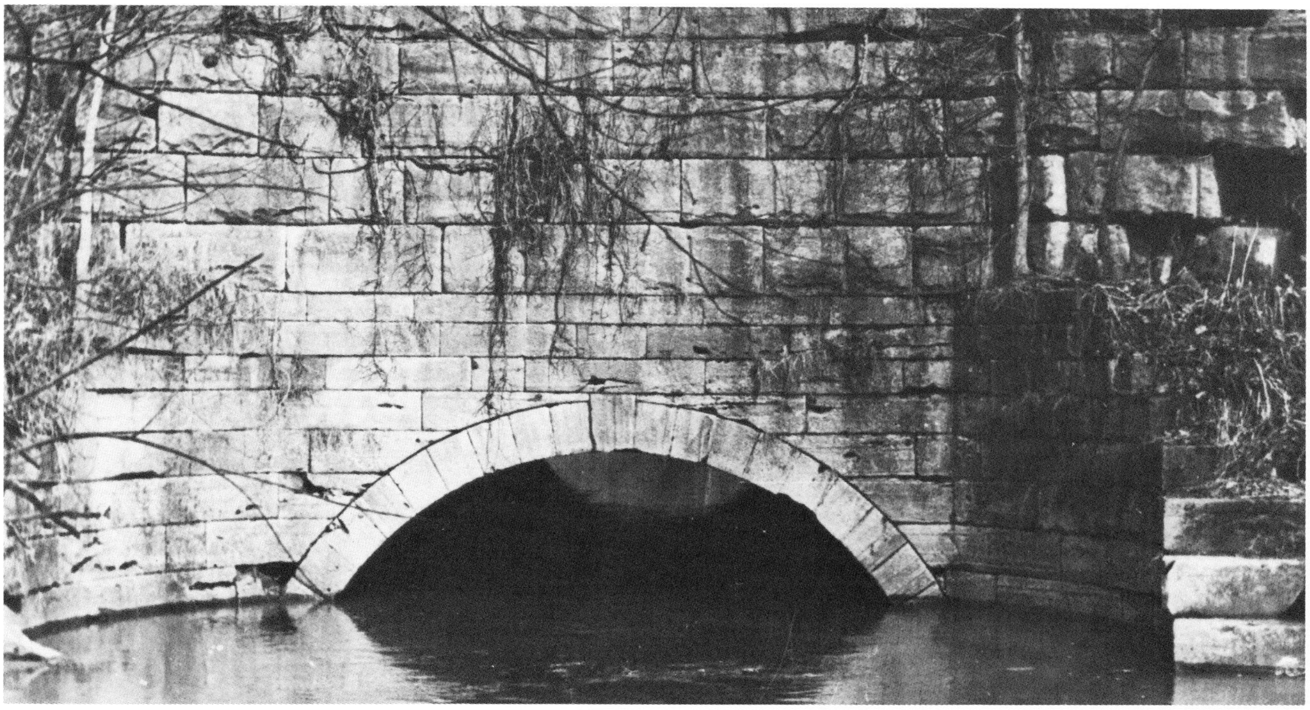

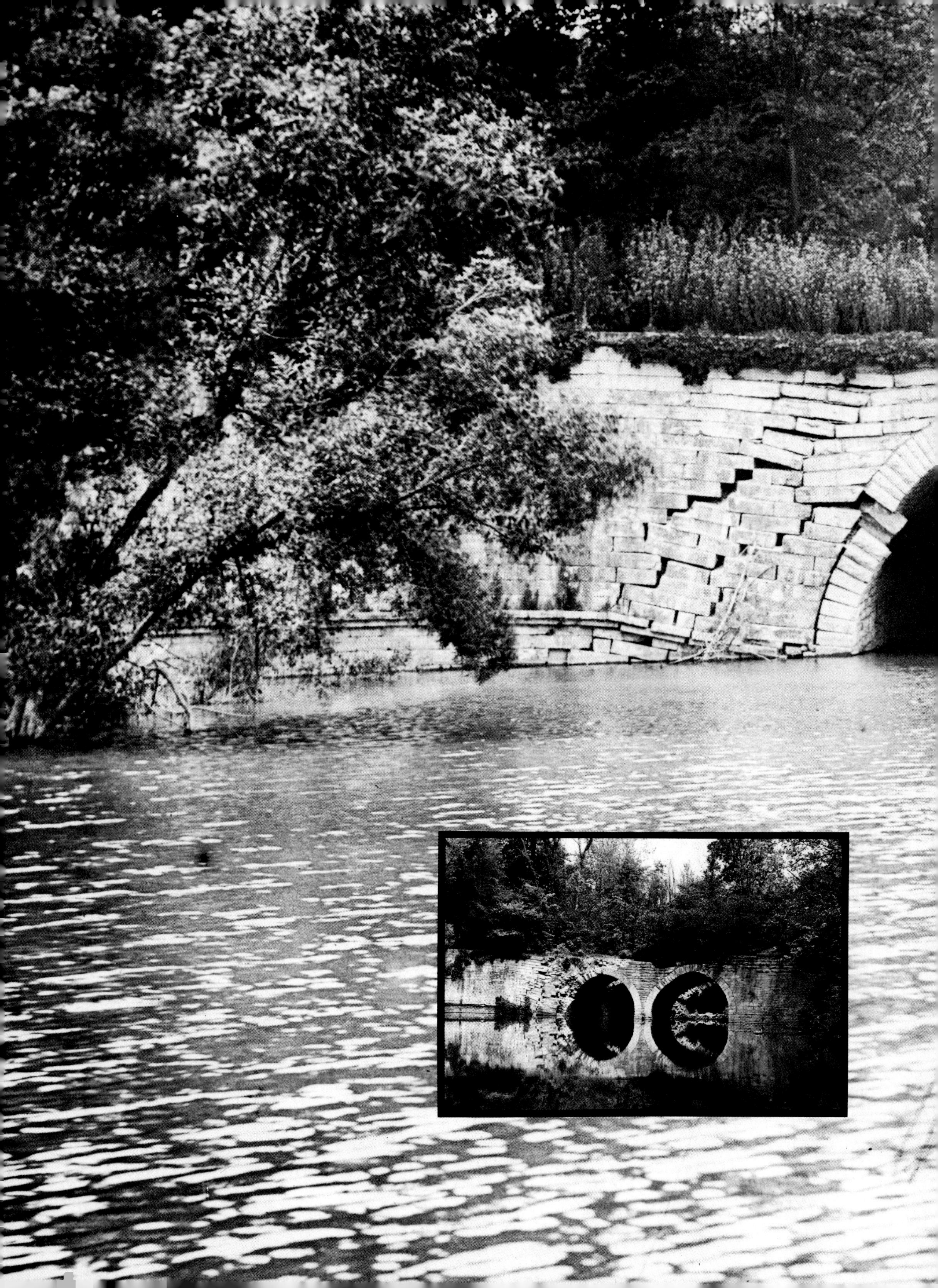

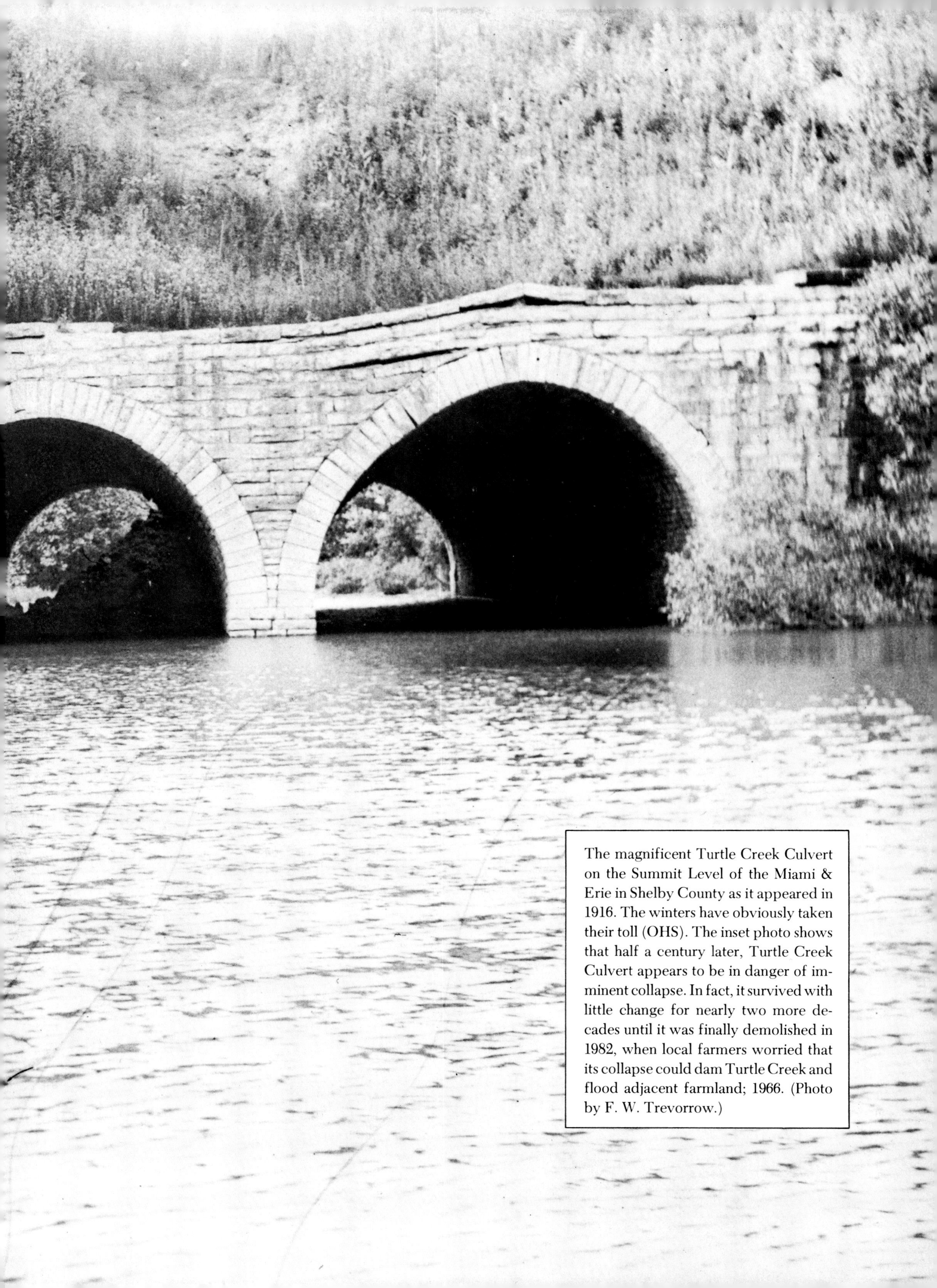

The magnificent Turtle Creek Culvert on the Summit Level of the Miami & Erie in Shelby County as it appeared in 1916. The winters have obviously taken their toll (OHS). The inset photo shows that half a century later, Turtle Creek Culvert appears to be in danger of imminent collapse. In fact, it survived with little change for nearly two more decades until it was finally demolished in 1982, when local farmers worried that its collapse could dam Turtle Creek and flood adjacent farmland; 1966. (Photo by F. W. Trevorrow.)

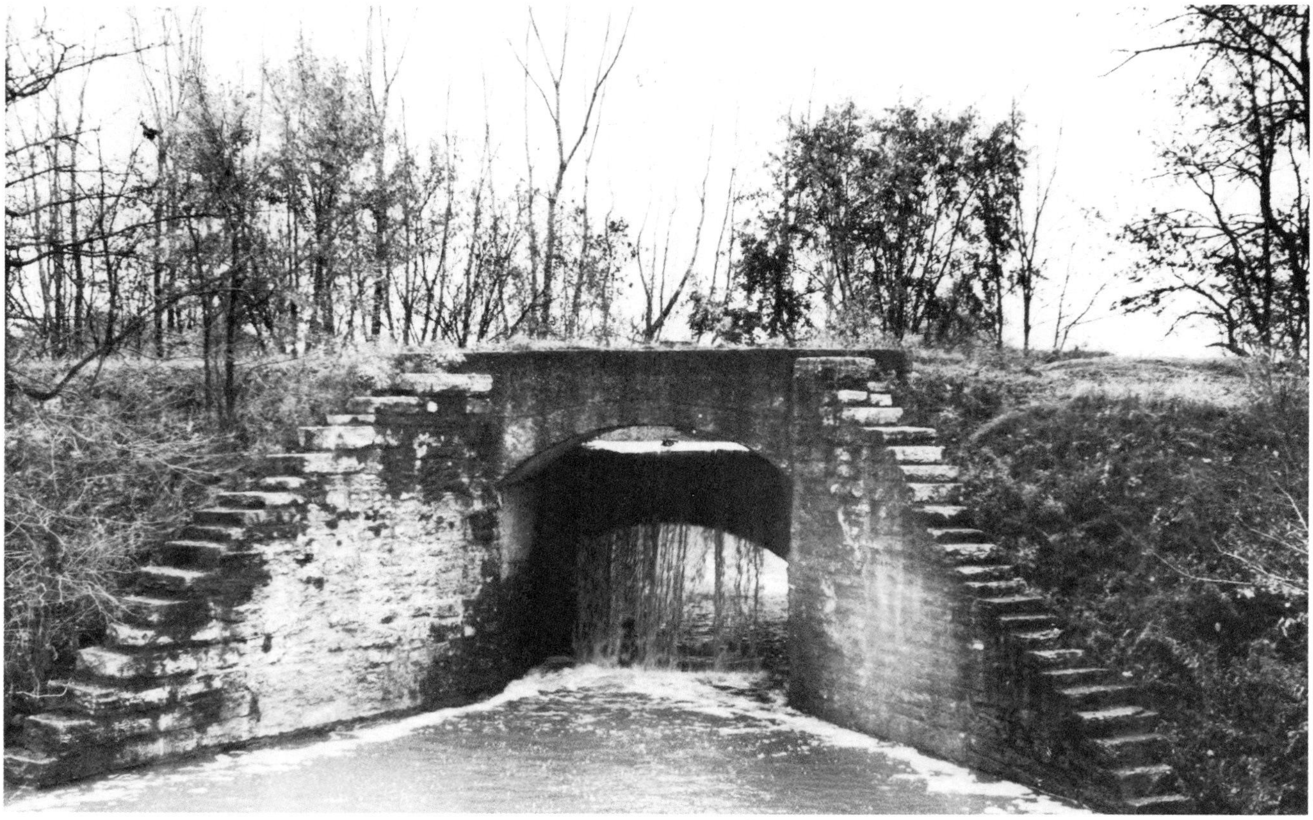

This unique bilevel structure is another Six Mile Creek culvert on the Miami & Erie south of Deep Cut. The upper level carried the towpath; the canal (overflowing in the picture) traversed the lower level. Near here, a monument commemorates "Bloody Bridge," which crossed the same creek. Here, the monument says, Bill Jones killed Jack Billings over the love of Minnie Warren, in the roaring canal days of the 1850s; 1975. (Photo by F. W. Trevorrow.)

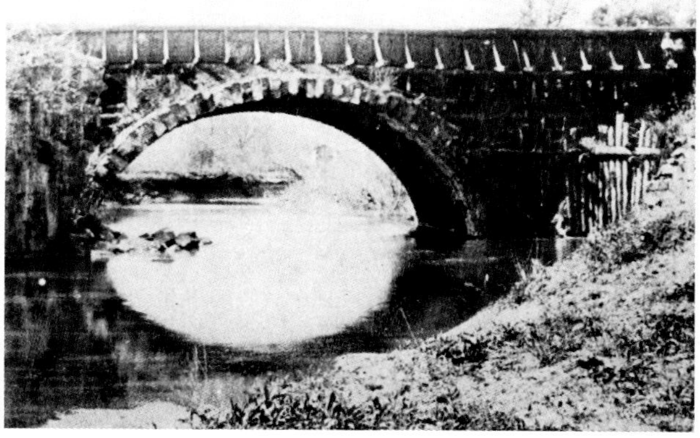

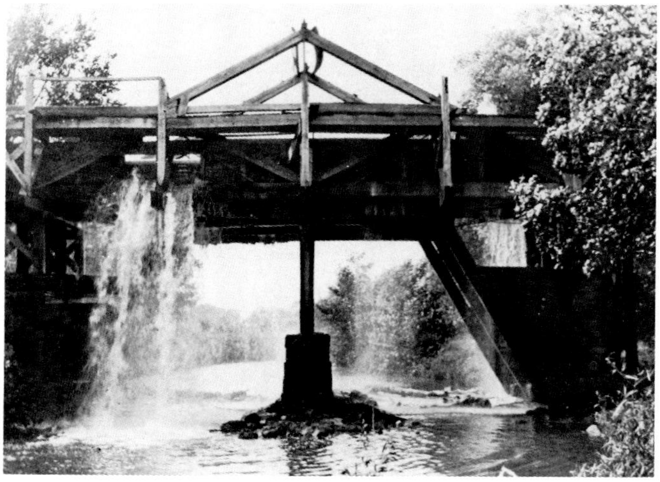

Longest Culvert (left), on the Miami & Erie, was actually a hybrid. The stone arch supports a wooden aqueduct constructed on top. For all its innovation, the structure collapsed before the end of the century; 1885. The aqueduct (right) which replaced Longest Culvert after its collapse looks like it could stand some repair itself; c. 1900. (Baus Collection, RVF)

The built-up towpath (right) at Black Hand Gorge on the Ohio & Erie rests on the rocky bottom of the Licking River. The cliff precluded digging a canal ditch parallel to the river at Licking Narrows. Instead, a slackwater dam was constructed and the famous "black hand" rock formation was blasted from the cliff to make room for the towpath. Water level in the slackwater channel here was much higher before demolition of the dam. (Photo by F. W. Trevorrow, 1966.)

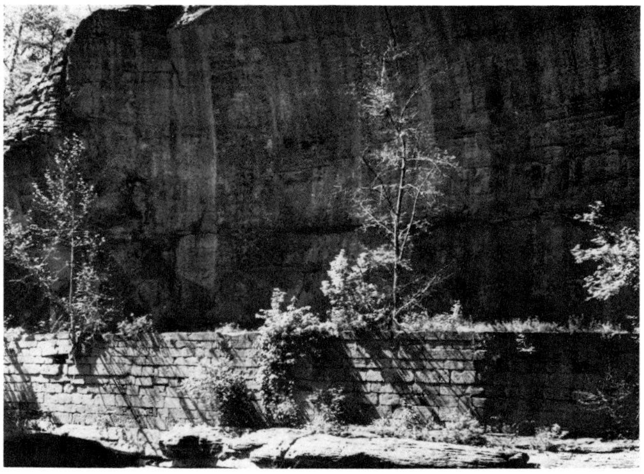

Lock 13 (below), renumbered later as Lock 40, was both a guard and lift lock at the lower end of the slackwater. It prevented flooding of the canal below State Dam and dropped the boats to the level of the canal below this point. Independence Park is now located here; 1957 (CSO).

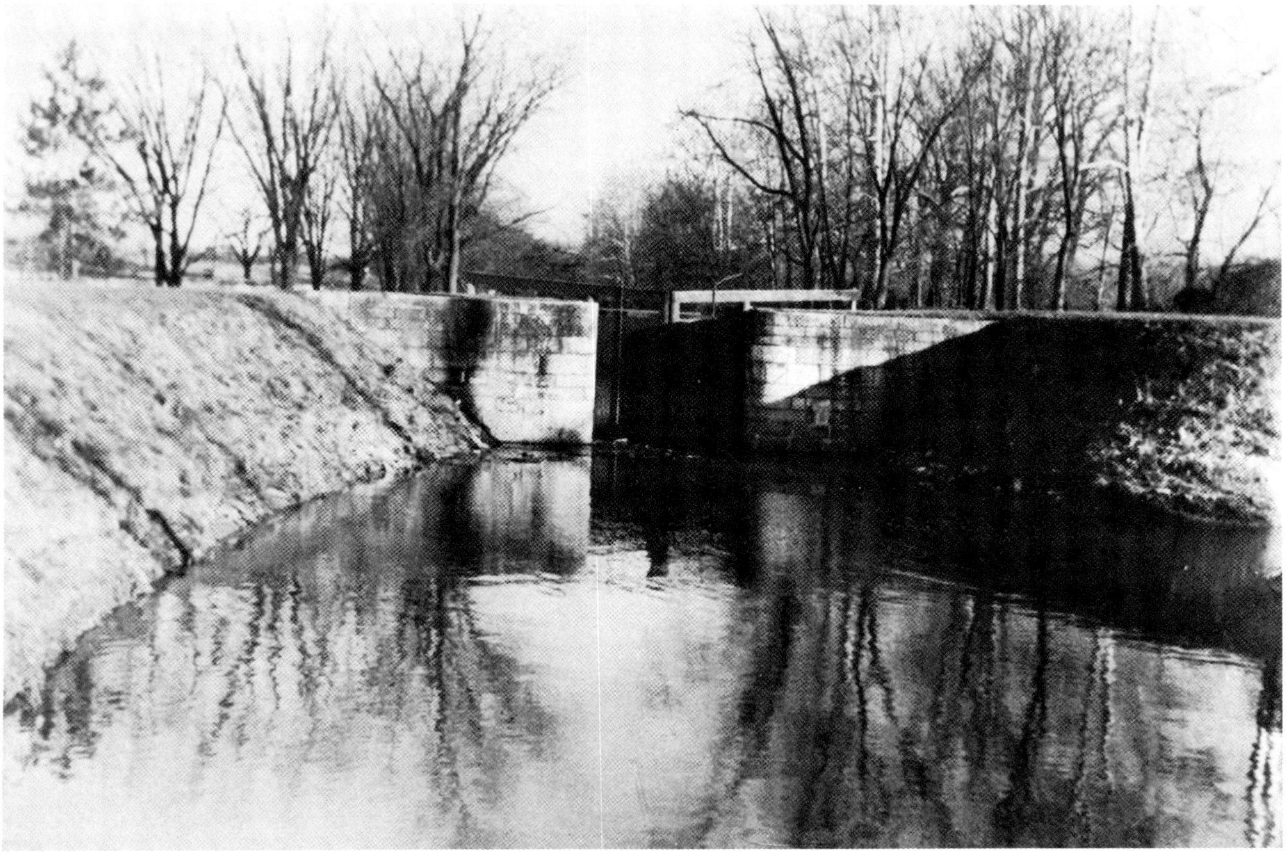

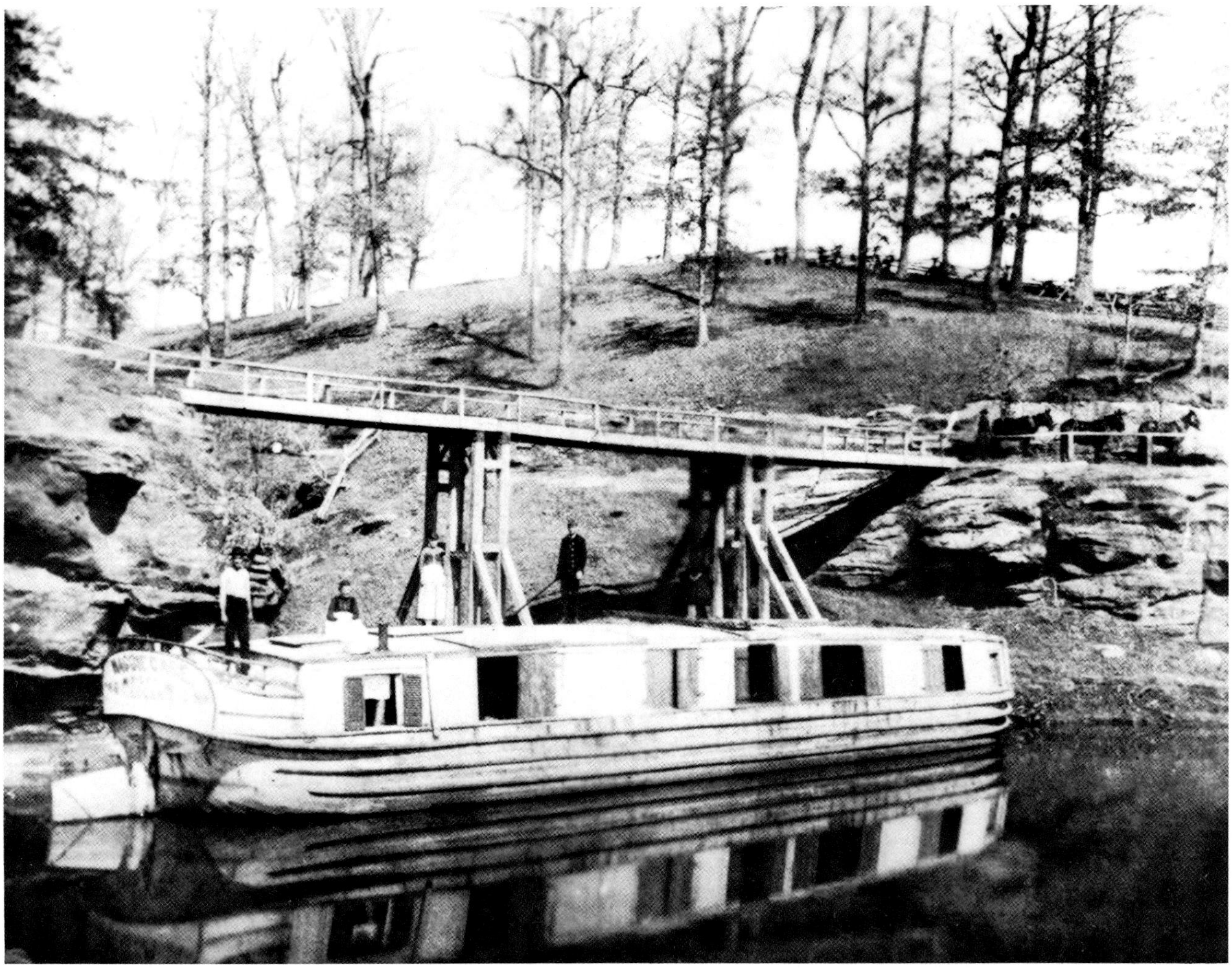

Above, The *Maggie Case* negotiates a slackwater on the Hocking River two and one-half miles west of Logan, on the route of the Hocking Canal. Since it would have been extremely difficult to build a canal parallel to the river through this terrain, a State Dam was built at the falls of the Hocking River, in Logan, to back up the river and deepen its channel for more than six miles. Mules at the upper right have just crossed a bridge over the gorge cut by Monday Creek to reach the towpath on the high ledge cut from solid rock on the far side of the bridge. The captain is wearing his dress uniform for the occasion; the steersman is manning the tiller to compensate for the sideward pull of the towline. Their wives are in attendance for the photograph, which appears to be a contact print from the original glassplate negative. (CSO)

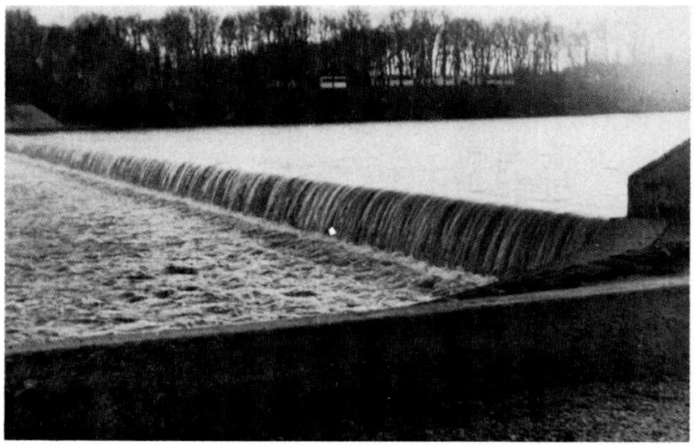

Right, State Dam at Independence, located three miles east of Defiance on the Maumee River, permitted boats to travel in the river channel for several miles and ultimately make a turn back into the Miami & Erie Canal ditch at Defiance; 1958. (CSO)

5 The Boats: Naval Architecture of Another Kind

Mingled smells of freshly planed wood and turpentine permeated the new boatyard near Akron's Exchange Street as a crowd began to gather on June 27, 1827. Many of those present were Irish and German laborers who had recently finished building the chain of fifteen locks here in the single mile north of Lock One. They gathered around the drydock to examine the creation of Alexander and Edward Wheeler, who might have felt a little like Noah. Suddenly, cued by the foreman's shout, the crowd grew expectantly silent. The sound of sledges echoed back from West Hill as wooden blocks were knocked out from beneath the hull of the *State of Ohio*. With a grinding, scudding cry from its keel, Ohio's first canal boat slid sideways down the slanting ways and splashed broadside into the canal—innundating the opposite bank with a small tidal wave. A roar of approval went up from the crowd gathered for the occasion; some had questioned whether it would float. An impromptu band struck up a lively tune as businessmen congratulated each other.

The boat was finished just in time. Only a few days later, the state's Governor Allen Trimble would arrive by stagecoach from Columbus, together with the secretary of state, the canal commissioners, and other distinguished guests. Their arrival would attract an even bigger crowd on the morning of July 3.

Practically all of Akron's 250 mostly male citizens gathered to watch the distinguished party board the brightly decorated *Ohio* to participate in ceremonies that would open the first segment of the Ohio & Erie Canal from Akron to Cleveland. On signal from Captain Henry Richards, Job Harrington's carefully groomed black horses took the slack out of the towline, and the boat glided away from the landing with John Sterns manning the tiller—as the throng lining the banks drank toasts to the new era at hand. And it was, indeed, a new era. For the *Ohio* was the flagship of a humble fleet of mule-drawn vessels that would transform Ohio from a frontier wilderness into the third most populous state in the Union.

That afternoon in the Cuyahoga Valley at Boston, another boatyard would contribute the *Allen Trimble* to a flotilla that would be joined the following morning by the *Pioneer*, loaded to capacity with Cleveland celebrities in the Flats. A small cannon aboard the *Pioneer* would salute the governor's arrival that Independence Day morning—answered by cannon volleys from the shore. The successful launching of the three boats signalled the start of a new industry in Ohio, as boatbuilding facilities began to spring up next to lumberyards along the hundreds of miles of canal routes.

Canal boats were not designed for speed. In fact, a state speed limit of four miles per hour was imposed on the canals to prevent erosion of the clay banks. The boats were designed instead to haul maximum tonnage with a minimum of tractive effort. Their motive power was, at most, about three horsepower—less than the smallest outboard trolling motor today. Yet they hauled enormous payloads.

The boats were seventy to eighty feet long, fourteen feet wide, and rode low in the water when loaded with fifty to eighty tons of cargo; yet they drew less than three feet of water. With a pointed prow and gracefully shaped curved hull (albeit with a flat bottom) the boats took advantage of the remarkably low drag characteristics of water at slow speeds when it slips over the surface of a floating object in laminar, or viscous (non-turbulent) flow. The sides of the hull were protected from abrasion against the stone locks by horizontal fenders—heavy wood strips running the length of the boat.

The boats were usually towed by three mules in

tandem, hitched to a single towline. There were several standard boat configurations: *Freight boats* had three small cabins with two open cargo holds between. The captain and his family lived in this rear cabin, which contained a galley with a small cook stove, and bunk beds—some of which extended underneath the short afterdeck below the tiller. A smaller cabin at the bow was occupied by the rest of the crew. The third cabin, amidships, was the stable for the spare mule team, and it was stocked with hay. This middle cabin was eliminated on *state boats* used for canal maintenance, and on some short-haul freight boats—frequently those hauling lumber.

Typical freight manifests included salt pork, lard, iron ore, coal, pig iron, whiskey, flour, lumber, food staples, and miscellaneous merchandise. It is apparent from the list that farmers often processed their harvest into a form that preserved well (flour, whiskey, salt pork, lard) and brought higher dollar per unit toll (2/10 to 7/10 of a cent per ton mile) than they would realize by shipping their raw produce.

On some freight boats, a walkway covered the cargo holds, forming a narrow roof. When the sides of these holds were boarded up, usually with shuttered windows, the style was called a "line boat." The enclosed spaces were often used for hauling grain, and occasionally they were converted for carrying passengers. Boat names changed frequently—almost always with a new owner and often with a new coat of paint.

A full crew consisted of the captain, a steersman who handled the rudder, a bowsman who assisted the lockmaster (not all locks had one), and a mule driver—frequently a teenage boy with a sophisticated vocabulary calculated to motivate mules and a switch or whip to punctuate his commands. Some "muleskinners" rode the rear animal, but most walked. The steersman had to constantly compensate for the sideward pull of the towline.

Since the towpath was on one side only, specified procedures theoretically avoided the potential havoc created by intersecting towlines of boats passing in opposite directions. According to regulations, the boat yielding the right-of-way was required to heave to and sink its towline to the bottom of the canal, permitting the towline of the other boat to pass over the towline lying on the bottom. The competitive personality profile of canalers being what it was, however, such courtesies were rarely extended (see chapter 9 on Canalers and Their Families).

That some line boats carried both freight and passengers on the same trip is apparent from the following account by seventeen-year-old Cyrus Parker Bradley, a student at Dartmouth who kept a journal during his Ohio & Erie passage in June 1835:

"June 7. Sunday . . . The forward part of the boat is the gentlemen's cabin, about 10 by 5; then the main part of the boat for the freight, where are also stowed the midship passengers; then, in the after part of the boat, the dining room, perhaps 10 feet square, with a kitchen closet adjoining.

"Tuesday 9. Our little cabin was fitted most ingeniously with berths for ten persons, but so contracted were the limits appropriated for each that he could only draw himself onto his shelf lengthwise and lie, without stirring for fear of a fall or a broken head.

"Awoke. Looked out of my window upon a block of stores, which it seemed I could almost touch. . . . We were in Chillicothe . . . and the canal runs directly through it and is lined on either side with shops and stores."[1]

Passenger packets, designed solely for the purpose, were a later development. And with them came another innovation: women passengers. Riding even lower in the water, many of the packets presented a sleek, even streamlined appearance. A long, low cabin, usually having a sun deck on the roof, ran the entire length of the boat with a row of curtained(!) rectangular windows on each side. Mostly for aesthetic reasons the packets often were drawn by horses instead of mules, and they could travel up to five or six miles an hour (in violation of the state speed limit). Some lines advertised schedules from Cleveland to Portsmouth, 308 miles, in eighty hours. Unlike freight boats, which usually tied up at night, packets frequently travelled around the clock.

The packets carried fifty to seventy-five passengers and, because they rarely tied up, carried a larger crew than the freighters. In addition to the captain, and a matron for the lady passengers, they usually carried two drivers, two steersmen, and a cook. The boy driver usually rode the rear horse instead of walking along behind the team; but, as on freight boats, spare horses rode aboard with the pas-

1. *Ohio Archaeological and Historical Quarterly*, 1906.

sengers. Passenger fares were less than two cents a mile, half a cent of which went for tolls. This charge included meals and lodging while aboard. Accommodations were less than lavish, however. Husbands and wives travelling together, for example, slept in separate dormitories.

The cabin was generally divided into three compartments. The largest of these served as a combination assembly room, lounge, and diner. Folded against the walls of this room were shelves three feet wide; at night these were swung down to provide sleeping quarters for the male passengers and the crew. Men who were fortunate slept on straw mattresses and, occasionally, even sheets (which doubled as tablecloths during the day). When the boat was crowded some passengers had to settle for the floor, which was less desirable since some boats shipped water.

The second largest compartment was the private compartment for women passengers. The small third room served as the ladies' washroom. At least one lady passenger seemed to find these accommodations quite acceptable, as indicated in a letter written by a "Malvina" to a friend on October 27, 1847:

". . . left Dayton at 8 o'clock Monday morning on board the packet "Banner," Capt. McCarty . . . not a cloud to intercept the cheerful 23 miles and 7 locks. Paid $5 and a half from Dayton to Toledo . . . Tuesday at 8 o'clock [after a storm and a collision!] arrived at St. Mary's distant 65 miles from Dayton . . . Wednesday found ourselves at Maumee . . . We got to Toledo at one o'clock."[2]

But there are some less happy accounts. In November 1843, former President John Quincy Adams was invited by the Astronomical Society of Cincinnati to lay the cornerstone for a new observatory to be built in that city. Travelling from Boston to Buffalo by train, and then by lake steamer to Cleveland, he and his small party boarded the Ohio & Erie canal boat *Rob Roy*, under the command of a Captain Phillips, to take them south as far as Hebron—a stagecoach stop on the National Road (now U.S. 40). The account of this adventure in President Adams's diary sounds downright surly:

"It is divided into six compartments, the first in the bow, with two settee beds for the ladies, separated by a curtain from a parlor bed-chamber, with an iron stove in the center, and side settees on which four of us slept, feet to feet; then a bulging stable for four horses, two and two by turns, and a narrow passage with a side settee for one passenger to sleep on, leading to the third compartment, a dining hall and dormitory for thirty persons; and, lastly a kitchen and cooking apparatus, with sleeping room for cook, steward and crew, and necessary conveniences.

"So much humanity crowded into such a compass was a trial such as I had never before experienced, and my heart sunk within me when squeezing into this pillory, I reflected that I am to pass three or four days in it. . . . The most uncomfortable part of our navigation is caused by the careless and unskillful steering of the boat in and through the locks, which seem to be numberless, upwards of two hundred of them on the canal. The boat scarcely escapes a heavy thump on entering every one of them. She strikes and grazes against their sides, and staggers along like a stumbling nag."[3]

2. Reprinted in Canal Society of Ohio's *Towpaths* (1965), from *Allen County Reporter*, Allen County Historical Society.

3. *The Memoirs of John Quincy Adams*, 1876.

Tourists yearning for a taste of what it was like to travel by canal boat have a choice of three that operate with a considerable degree of authenticity, and a fourth that looks the part, even if compliance with U.S. Coast Guard regulations dictated radically different materials and motive power. Most of the boats operate on a daily schedule (weather permitting) during the summer months, and each excursion takes about an hour.

The pioneer effort in constructing an authentic nineteenth-century canal boat was led by a group of firemen, Rotarians, and other civic leaders in Canal Fulton in 1967. Two and a half years later, after researching boat plans, including those of a State Boat (which proved unusable) and contemporary construction methods, they had completed the *St. Helena II*, a beautiful if modest-sized freight boat having the three-cabin configuration characteristic of freighters, with two cargo holds between the cabins. Designed by Canal Society member Caroll Gantz, the boat was actually constructed utilizing nineteenth-century methods. Trees were felled and rough-sawed into planks, then steamed to make them sufficiently pliable to conform to the curvature of the hull.

The captain's cabin at the stern has been fitted with a cookstove and bunks and looks genuinely cozy as the residence of an entire family. Operated by the Canal Fulton Heritage Society, the boat travels south down a short segment of the Ohio & Erie Canal from a dock at the edge of town to Lock Four Park, where a replica of a locktender's house has been constructed. The lock has been refurbished and is operated on special occasions when water availability permits.

The *St. Helena II* is towed by two mules—an adequate team for its modest passenger load of perhaps nine thousand pounds. But the Canal Fulton mules never did get the hang of working in tandem. They let it be known that they work as a team, in parallel, or not at all. And the drivers have gone along—authentic or not—mules being how they are.

Also on the Ohio & Erie, near the city's reconstructed Roscoe Village, the Park Board of Coshocton funded the construction of the *Monticello II*. Built by the late Marshall Jacobs, the boat is patterned from Jacobs's memory of one excavated from the bottom of Coshocton's Middle Basin in 1959, with modifications contributed by material received from the Ohio Historical Center and the shared experience of the Canal Fulton boat builders. To provide a steady supply of water for the restored section of canal, a dedicated, eighty-foot well was dug for the purpose. The *Monticello II* is drawn by draft horses, who, like Canal Fulton's mules, prefer to work side by side.

In the western part of the state the *General Harrison* was built by the Ohio Historical Society for the Piqua Historical Area on the Miami & Erie Canal north of the city of Piqua in Miami County, where

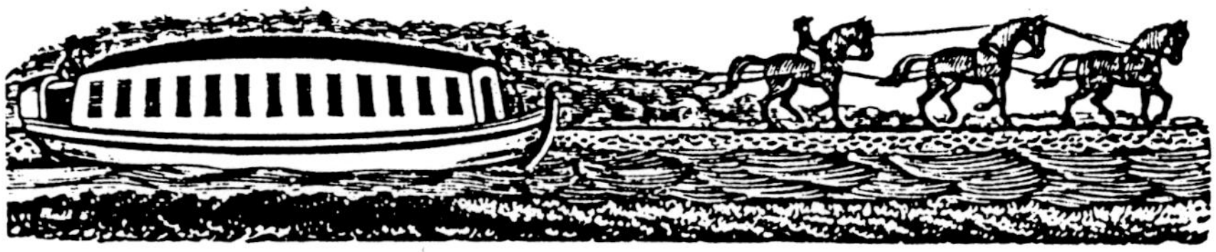

the canal bordered the property of federal Indian agent John Johnston. Johnston's farm residence and barn, dating from 1808–1815, have also been restored. Now shallower than when it was in commercial operation, the canal channel is fed by water which flows from the Johnston spring house.

The length of the *General Harrison* was based on the 77-foot lock north of the Johnston farm. Like the *St. Helena II*, planks were steamed to permit the wood to flex to the curved hull shape; but unlike the Canal Fulton builders, the Piqua designers elected redwood to resist decay better than native oak. The boat was completed in 1970 and began operation a year later, after several teams of mules were actually trained to work in tandem by pulling logs. A team of two tandemized mules now tows the boat. As Ohio Historical Society associate curator Donald Hutslar observed in 1978: "Aside from the initial pull, the back mule does most of the work, and the mules are obviously aware of the advantageous lead position. The boat handles with surprising ease. The effectiveness of the rudder depends on the point of attachment of the tow rope—the further amidships, the more sensitive the rudder. Load weight also affects the steering.

"Various boat crews have become adept at turning the boat in the basins by dropping the tow, starting the turn with poles, then picking up the tow on the opposite side; done correctly, the boat never stops moving. The biggest steering problem is a strong crosswind, which makes handling very difficult." (Donald A. Hutslar, "Canal Boat *General Harrison* and the Piqua Historical Area," *Towpaths,* no. 3, 1978.)

From Promenade Park visitors can see the chubby little *Sandpiper* bucking the currents of the broad Maumee River in downtown Toledo. Short for its width, the *Sandpiper* looks every inch a canal boat—except that it has no tow line. Powered by a 130-horsepower, 6-cylinder Isuzu diesel engine, the boat has a steel hull an inch thick at the bow. With the engine operating at 3,000 r.p.m., the boat can achieve a top speed of 13 miles per hour. The *Sandpiper* was christened by the late canal historian John Vanderlip in 1984 with water from a section of the Miami & Erie Canal, bottled and bonded at Florida, Ohio.

Canal Society members Vanderlip and Caroll Gantz (designer of the *St. Helena II*) provided line drawings of typical canal boat hull configurations, but the Marine Safety Office of the U.S. Coast Guard at Toledo had the last word, requiring that the boat be built entirely of steel. The *Sandpiper* travels out from the mouth of Swan Creek, at the foot of Jefferson Avenue in downtown Toledo.

The following table compares the vital statistics of Ohio's current canal boats:

BOAT NAME:	*ST. HELENA II*	*MONTICELLO II*	*GENERAL HARRISON*	*SANDPIPER*
Location	Canal Fulton	Coshocton	Piqua	Toledo
Waterway	Ohio & Erie	Ohio & Erie	Miami & Erie	Maumee River
Completed	1970	1971	1970	1984
Length	60 ft	78 ft	77 ft	65 ft
Width	13 ft	14.5 ft	14.3 ft	16.3 ft
Displacement	22 tons	25 tons	23 tons	37 tons
Passengers	60	120	120	101
Motive Power	Parallel Mules	Parallel Horses	Tandem Mules	Diesel Engine
Season	Mid-May to Mid-Sept. Daily except Mon.	Memorial Day to Labor Day Daily	Memorial Day to Labor Day Daily	May 1 to Nov. 15 Daily

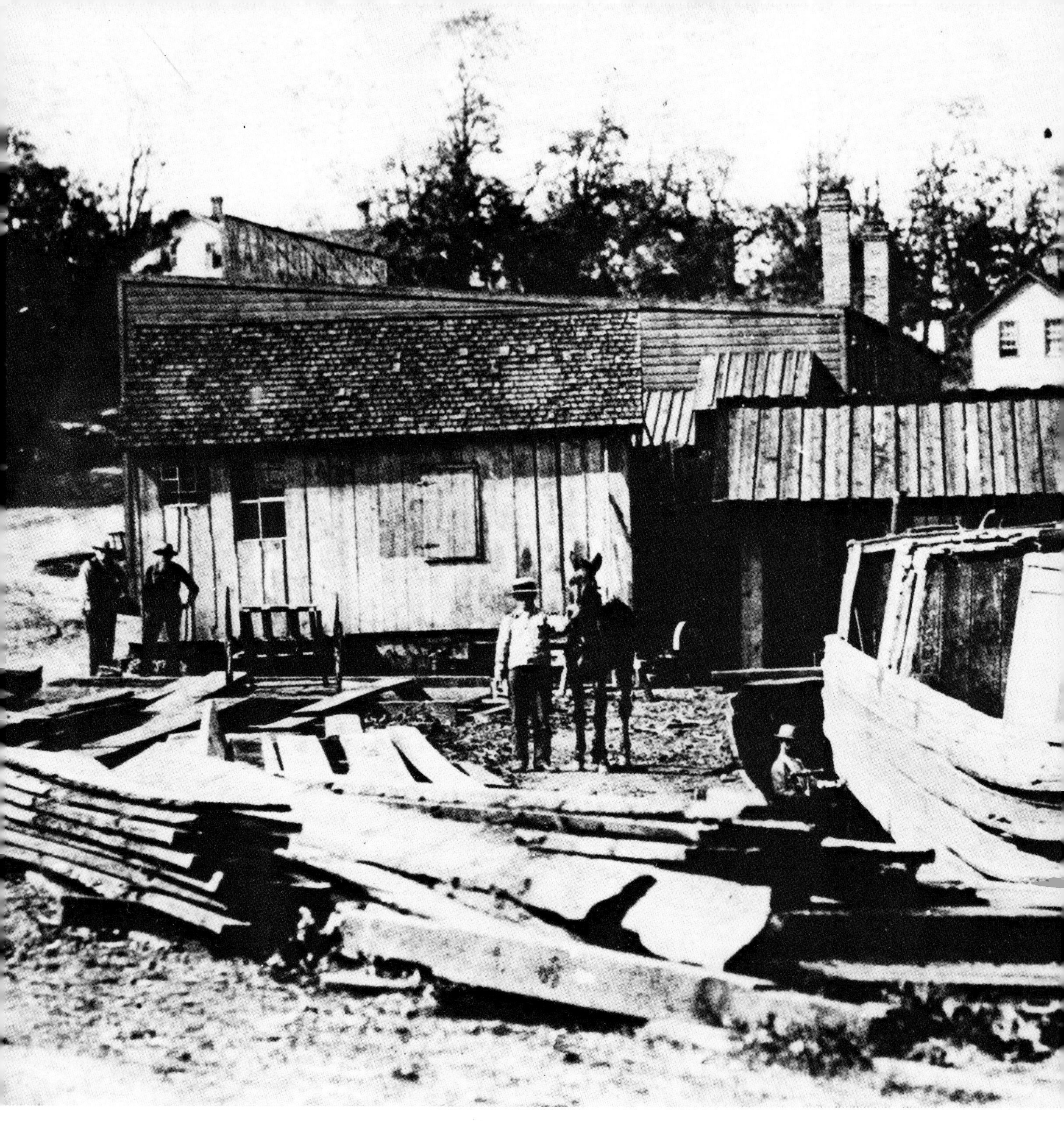

The *Narragansett* in drydock at W. J. Payne's boatyard at Lock 3, Akron, in 1891. It was near here that the *State of Ohio*, first boat on the canal system, was launched in 1827. Payne built a total of fifty-three canal boats in his facility. (Canal Visitor Center, CVNRA)

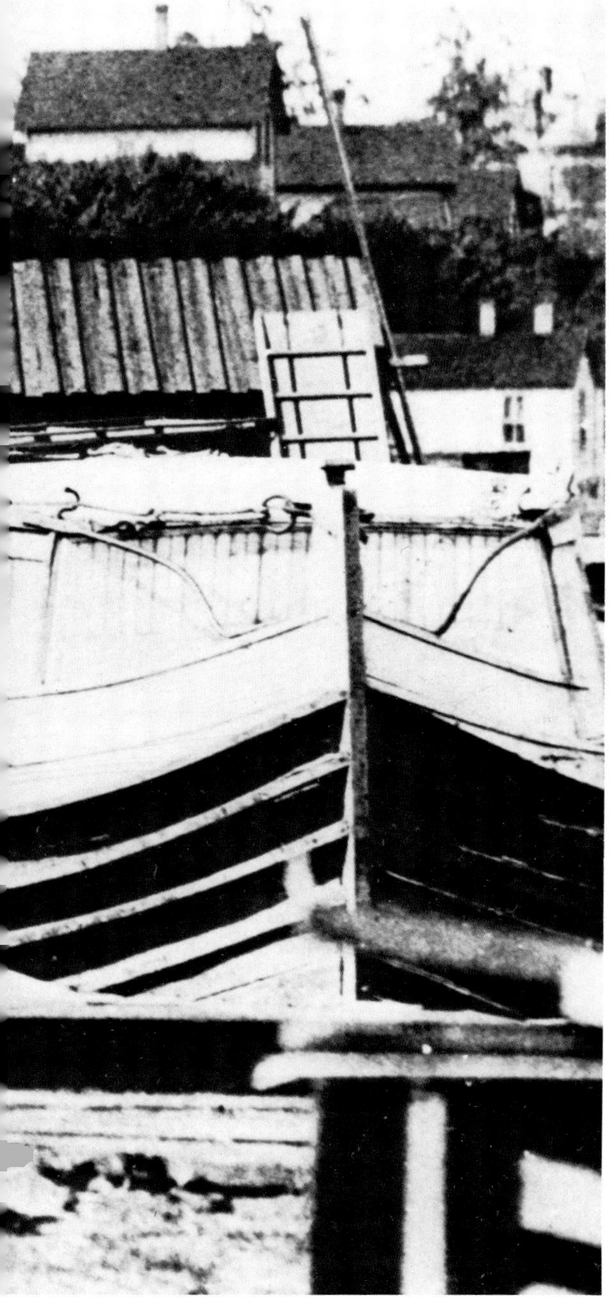

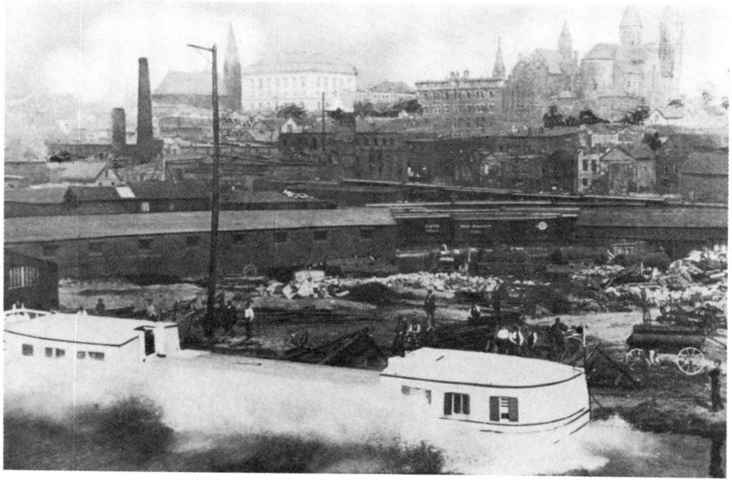

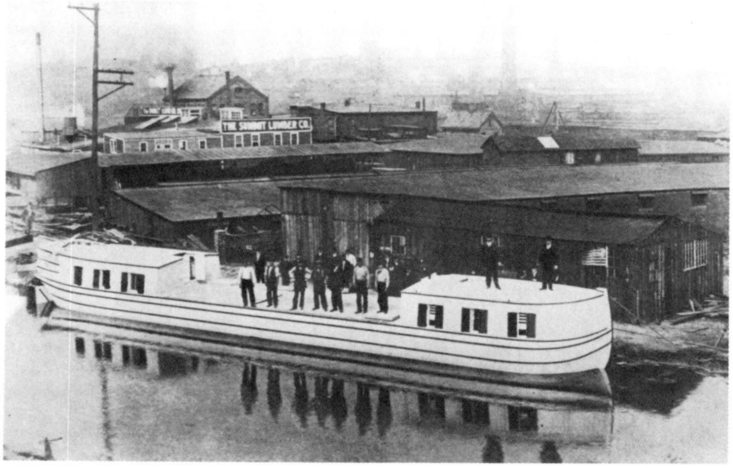

Sunset of a major nineteenth-century Ohio industry: The last state boat (top) is launched from the W. J. Payne boatyard near Lock 2 at the foot of Buchtel Avenue in Akron; 1907. Payne construction crew and dignitaries pose with completed state boat (bottom), having characteristic two-cabin configuration; 1907. More canal boats were built in this location than anywhere else in the country. (Pockrandt Collection, SCHS)

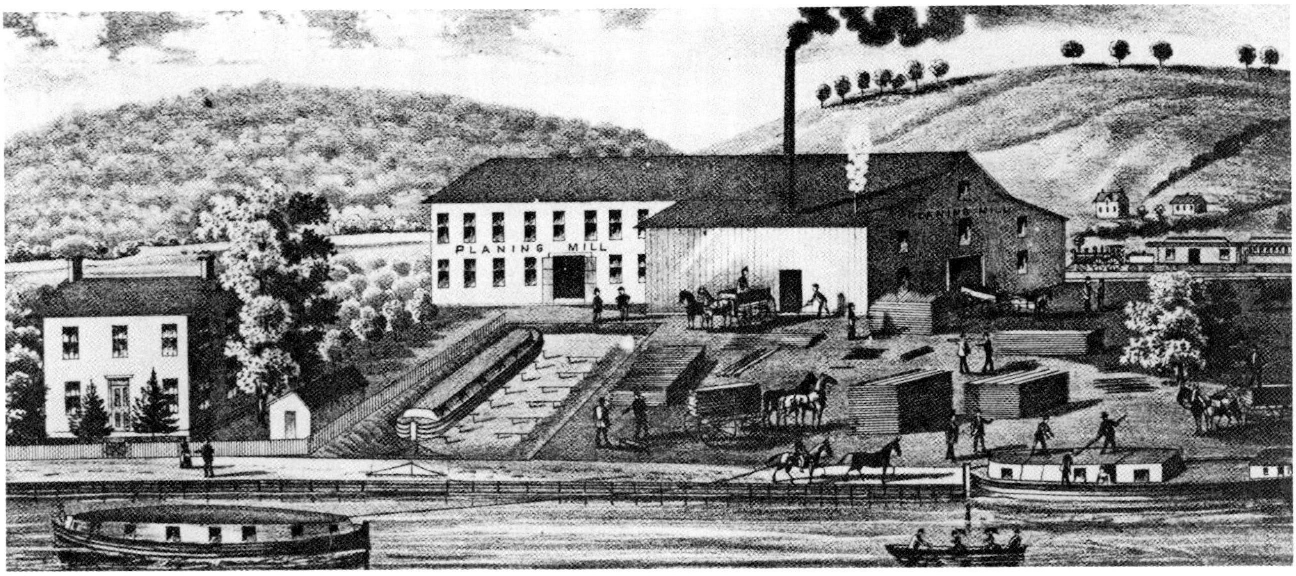

This drydock accommodates boat construction next to a lumber yard and planing mill on the Hocking Canal (foreground) at Nelsonville. (Engraving from 1875 *Athens County Atlas.*)

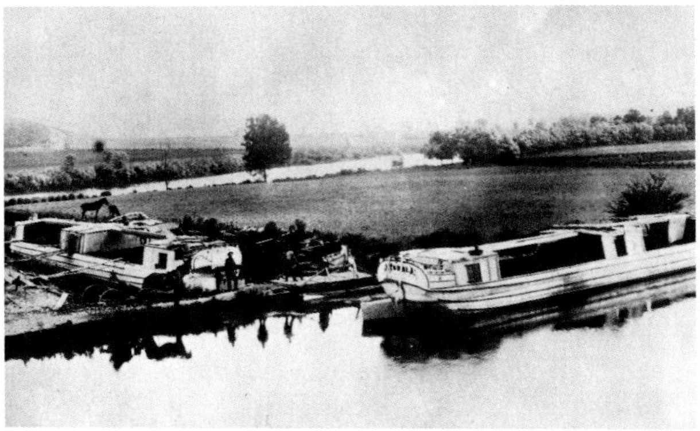

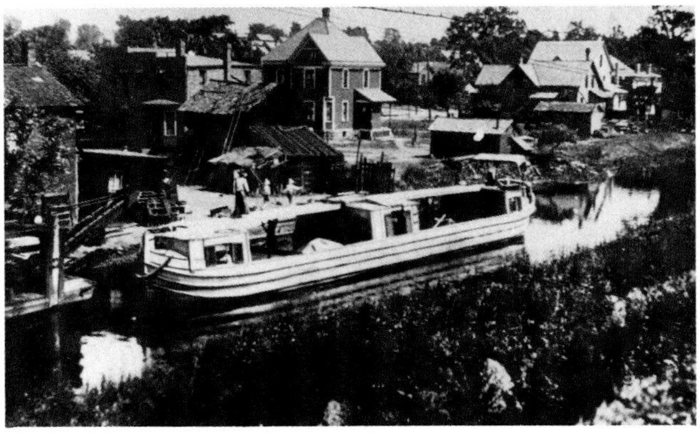

Left, The *Ohio* undergoes repairs at a rural drydock on the Ohio & Erie Canal, 1890. (OHS). *Right,* A three-cabin freight boat at Lock 5, Massillon. (CSO)

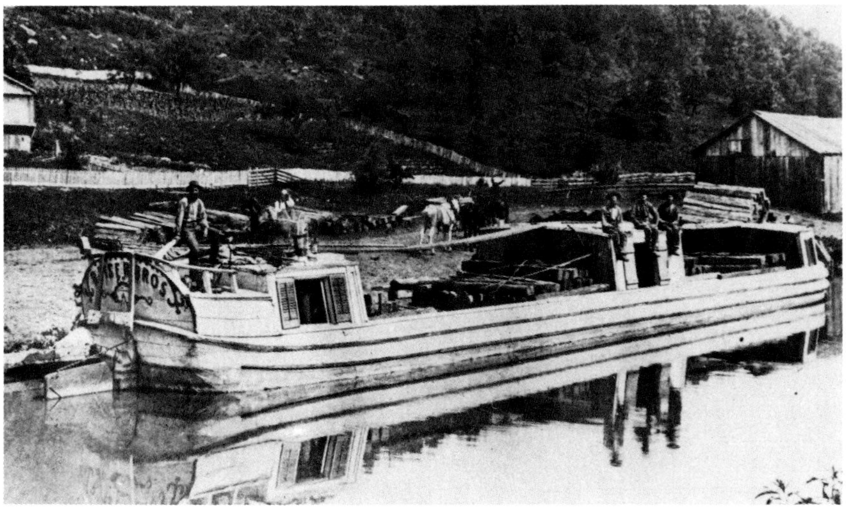

Captain Park Dalzell mans the tiller on his classic freight boat with three-cabin configuration. The rear cabin was the residence of the captain and his family; the forward cabin at the bow served as quarters for the remainder of the crew; the middle cabin (with three crewmen sitting on the roof) is the stable for the spare mule team. Both teams are on the canal bank at the moment, together with the boat's mascot, a goat. Ironically, the boat is hauling railroad ties from Jasper to Hebron—a cargo contributing to its own demise. A favorite photo from Pearl Nye's personal collection. (Canal Visitor Center, CVNRA, courtesy Cloea Thomas.)

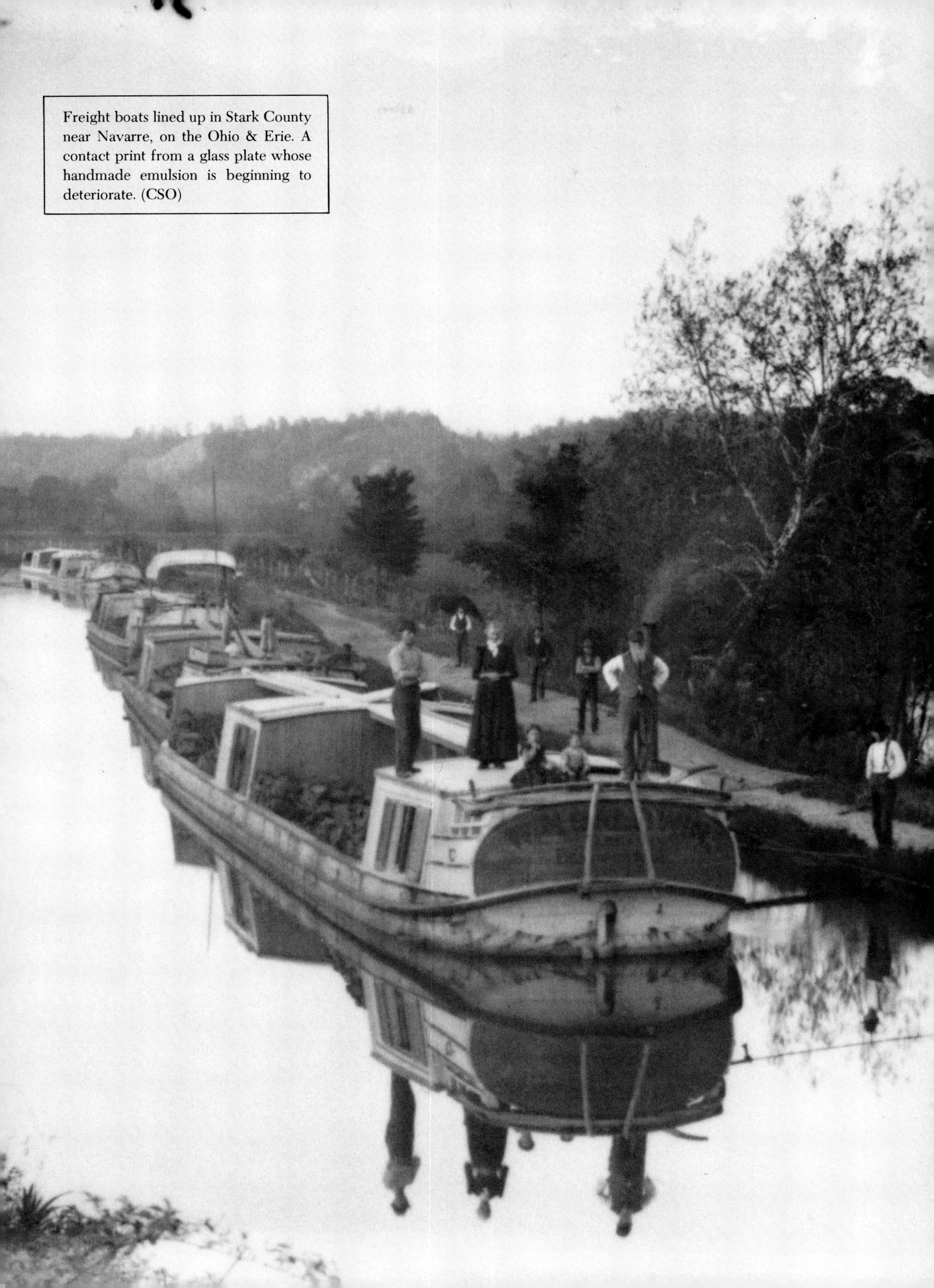

Freight boats lined up in Stark County near Navarre, on the Ohio & Erie. A contact print from a glass plate whose handmade emulsion is beginning to deteriorate. (CSO)

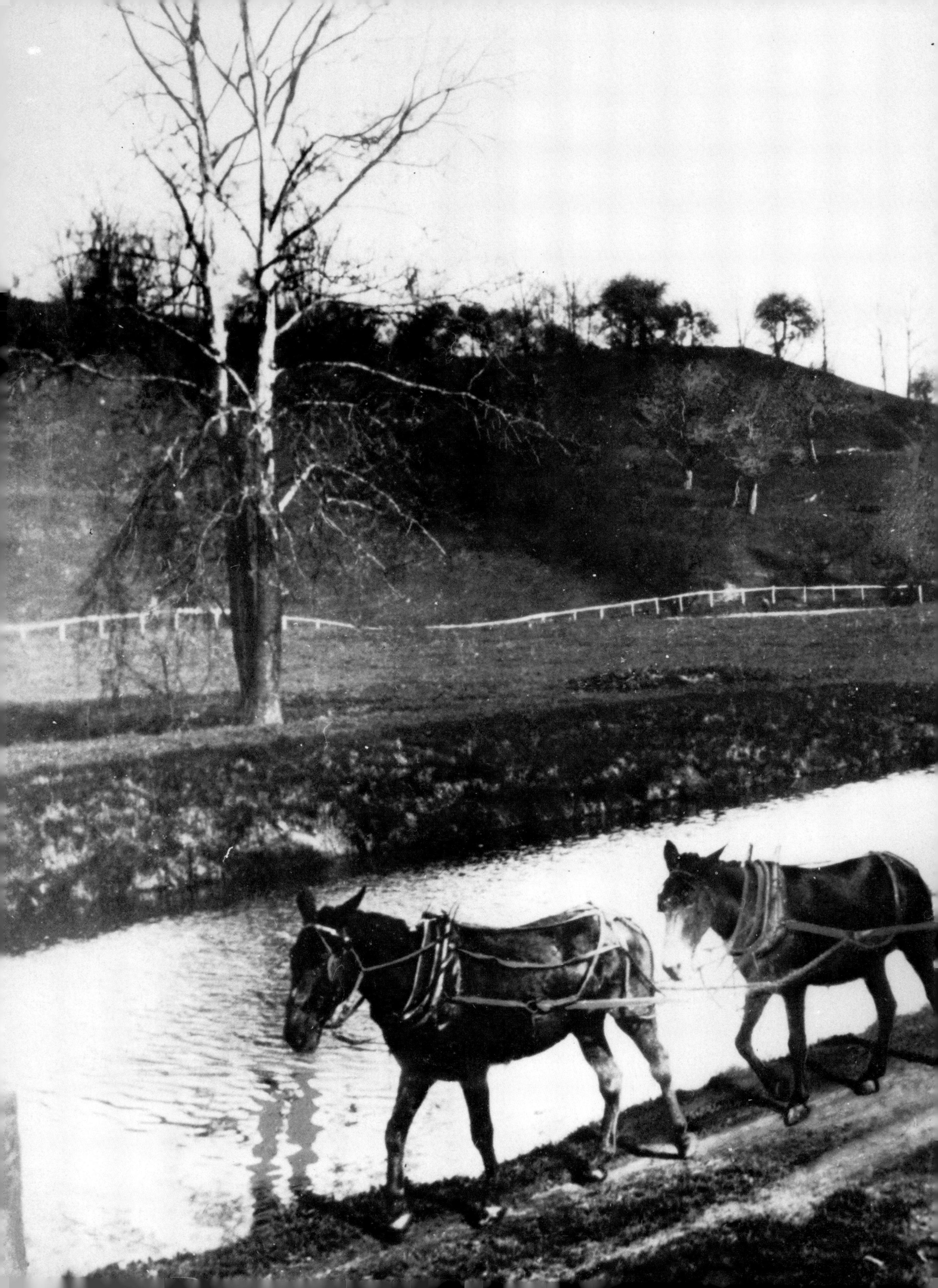

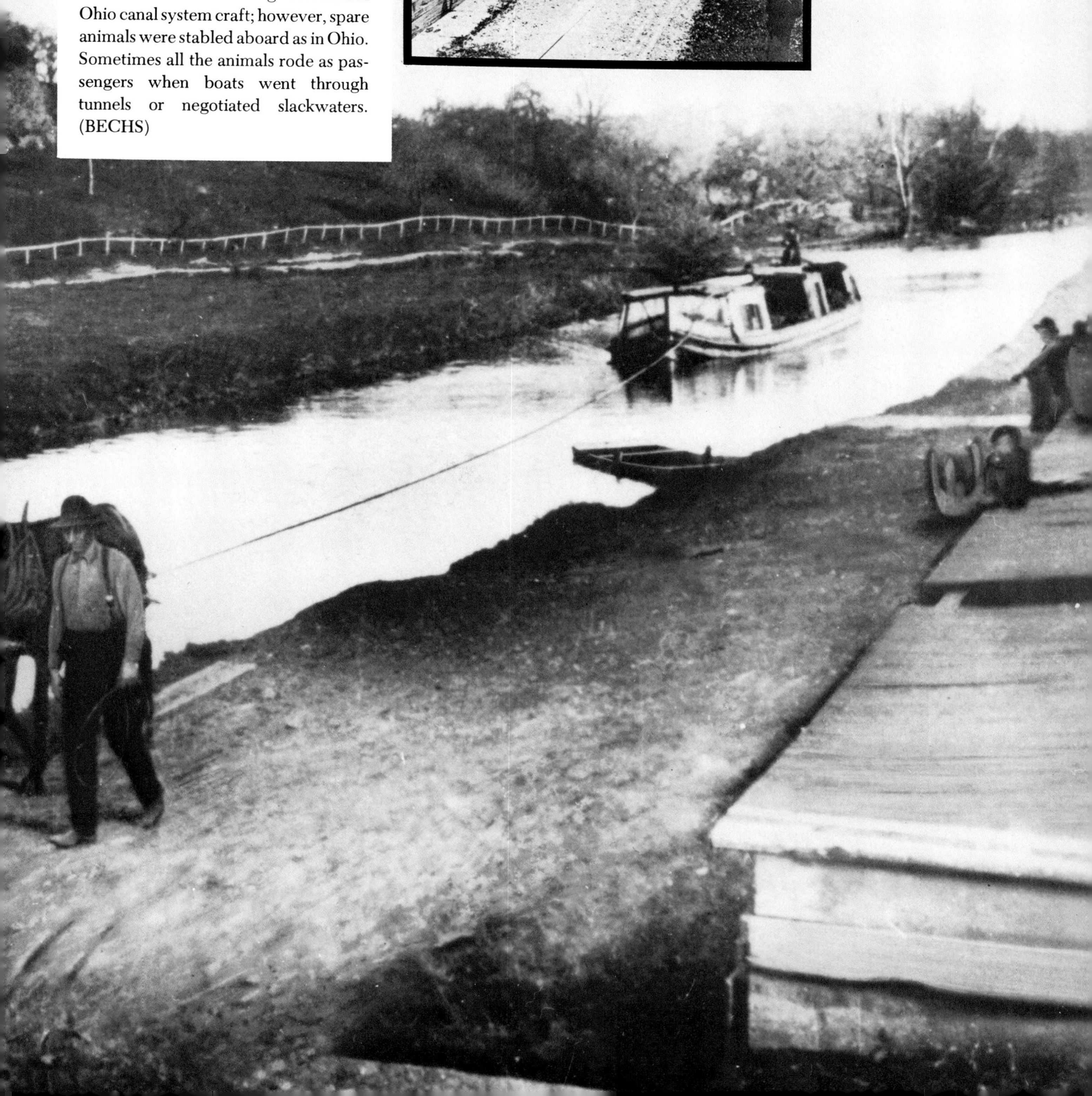

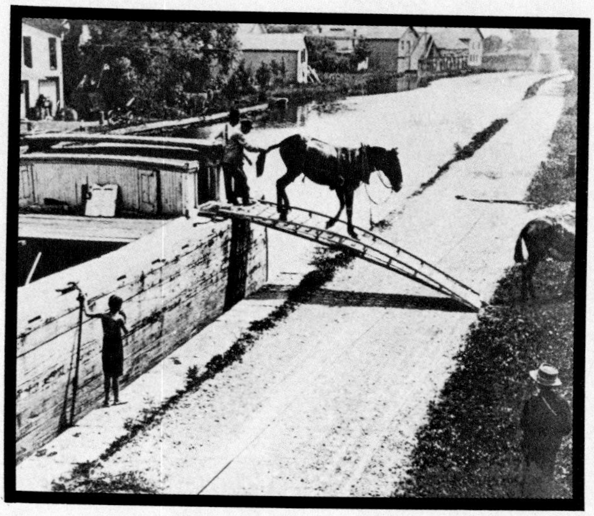

Boats were usually towed by three mules in tandem, hitched to a single towline and driven by a muleskinner. Since the towline is on one side only, steersmen had to compensate constantly for the sideward pull of the towline. The scene is on the Ohio & Erie near Independence (personal collection of Jack Gieck). *Inset*, The crew assists a spare horse down the gangplank of a New York Erie Canal boat, with its somewhat different configuration from Ohio canal system craft; however, spare animals were stabled aboard as in Ohio. Sometimes all the animals rode as passengers when boats went through tunnels or negotiated slackwaters. (BECHS)

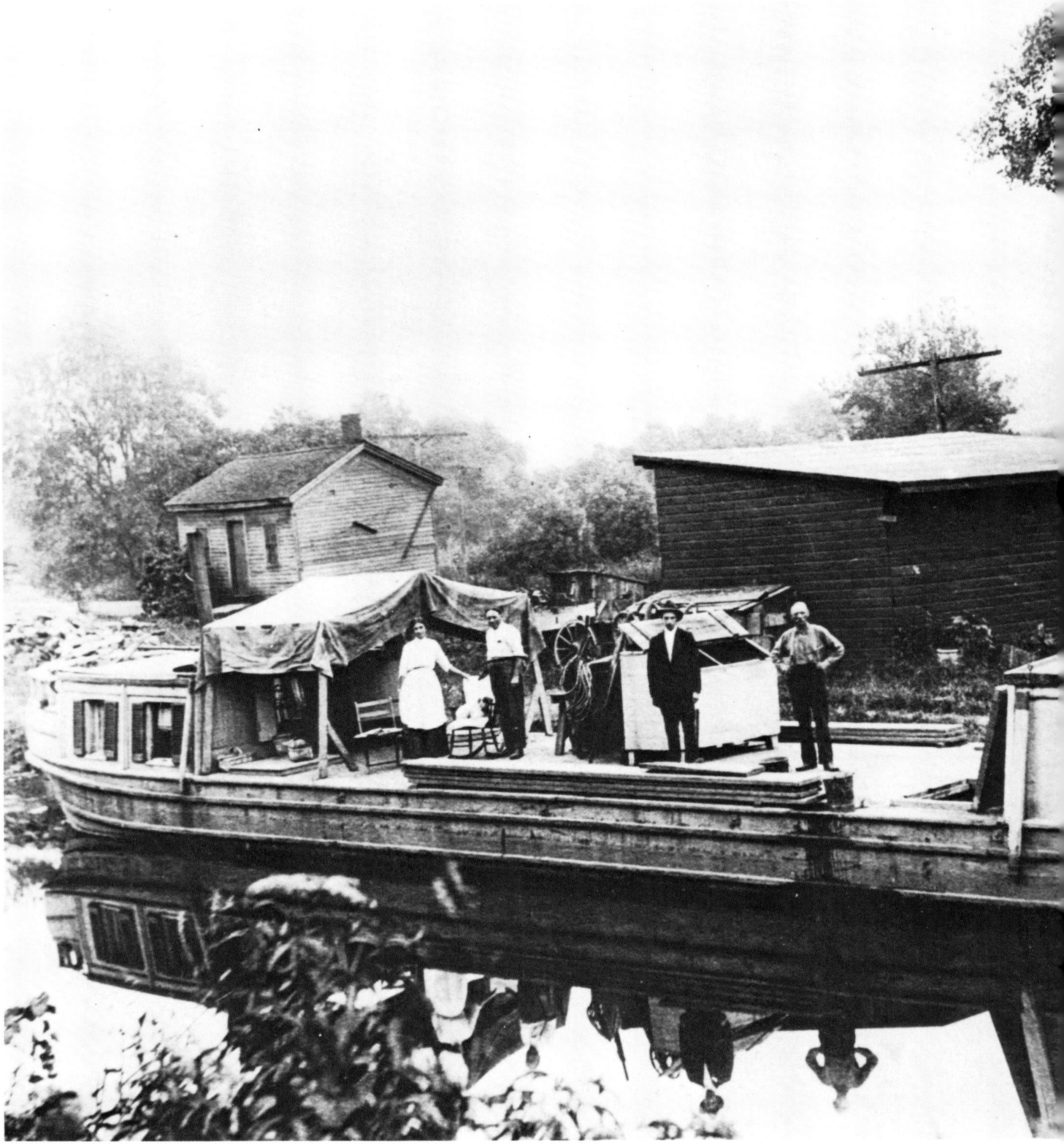

State Boat No. 1, near Stone Road Bridge south of Cleveland. Like all state boats (which did not require a spare mule team), the boat eliminates the center cabin, providing additional deck space for transport and storage of supplies and tools. The scene is apparently a weekend outing; 1900. (SCHS)

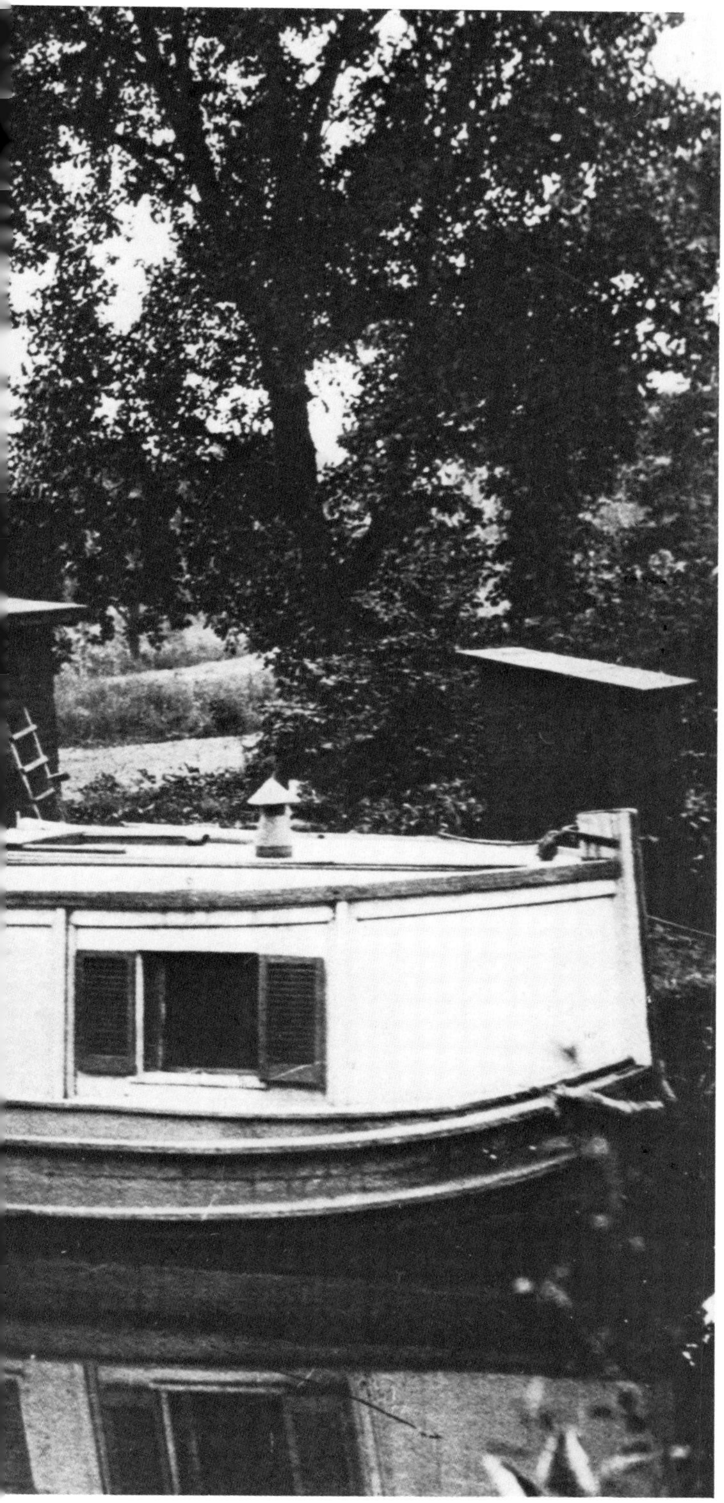

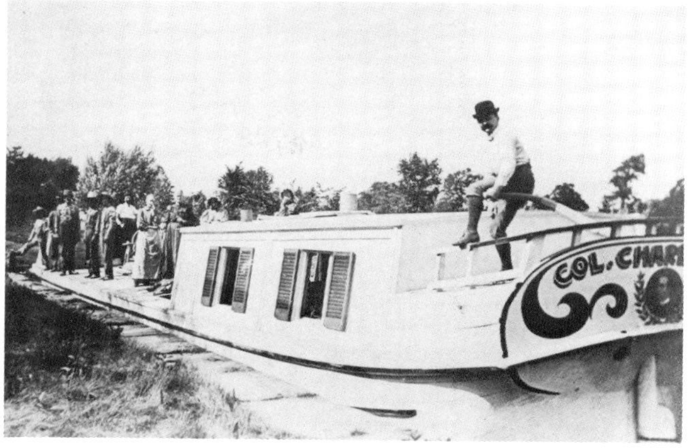

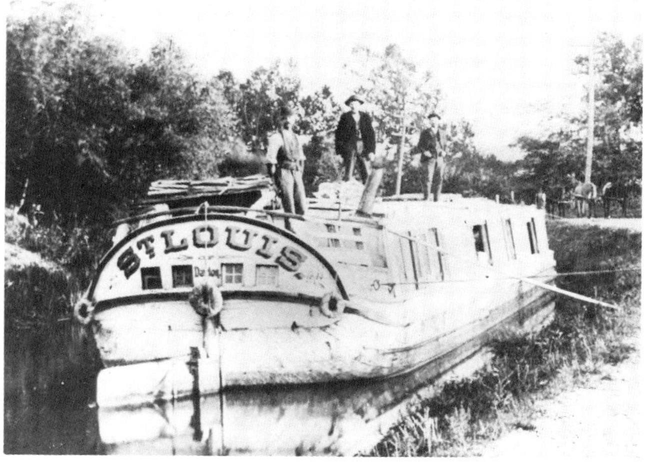

Superintendent Charles Harter (top), in derby hat, and his state boat, *Col. Charles Dick*, whose stern bears the likeness of a contemporary politician—or perhaps Colonel Dick himself; a not uncommon decoration on canal boats. (CSO)

The *St. Louis* (bottom), of Dayton: a roofed Miami & Erie freight boat with the sides enclosed, giving it a "line boat" configuration. Such boats were often used to haul grain and sometimes carried passengers. (RVF)

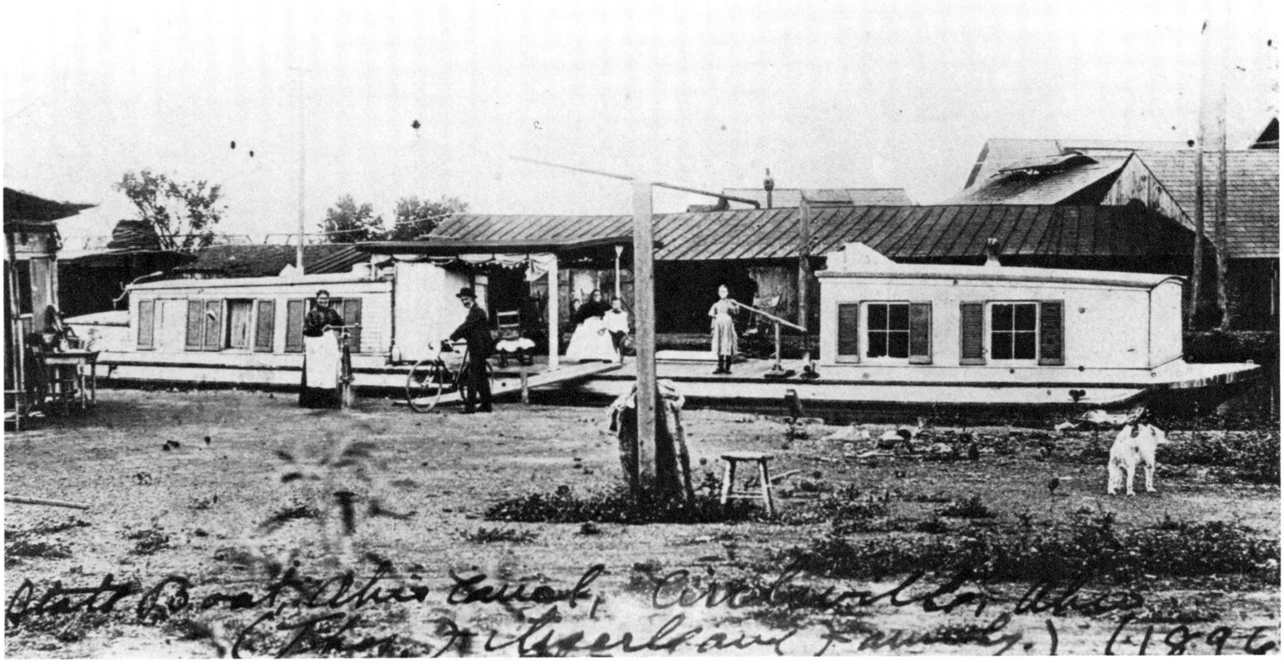

Above, This former state boat near Circleville was converted into a houseboat and excursion craft by the Tom Fitzgerald family, who apparently had recently purchased a pair of new "safety bicycles" (which had begun to supercede the high-wheelers available at the time)—the occasion for the picture; 1896 (SCHS). *Below left,* Many specialized boats were on the canal system. This one is a floating printworks and sign-maker's shop. Advertisements read "Rubber Stamps," "Signs—All Sorts," "Job Printing." The scene is on the Ohio & Erie at Bolivar. Aqueduct visible at upper left is on the Sandy & Beaver Canal; c. 1870 (SCHS). *Below right,* The *Sylph,* was a magnificent passenger packet representing the ultimate in canal travel luxury. Beginning its life on eastern canals, the *Sylph* is said to have operated on both the Ohio & Erie and the Miami & Erie after railroad competition began to affect eastern passenger packet revenues; c. 1860 (Pockrandt Collection, SCHS).

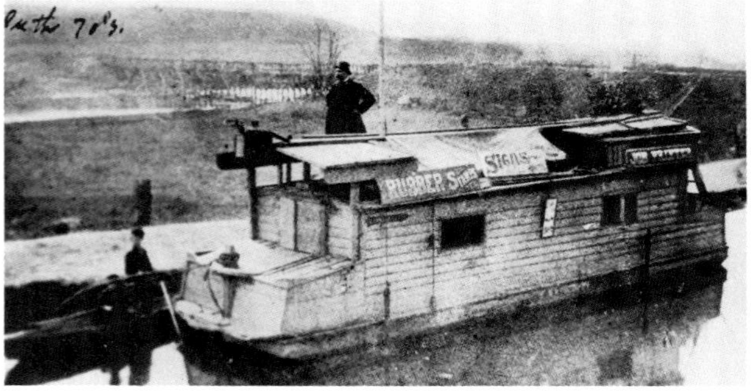

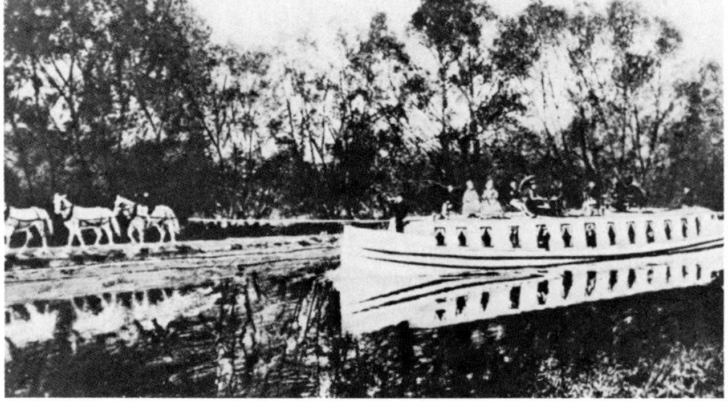

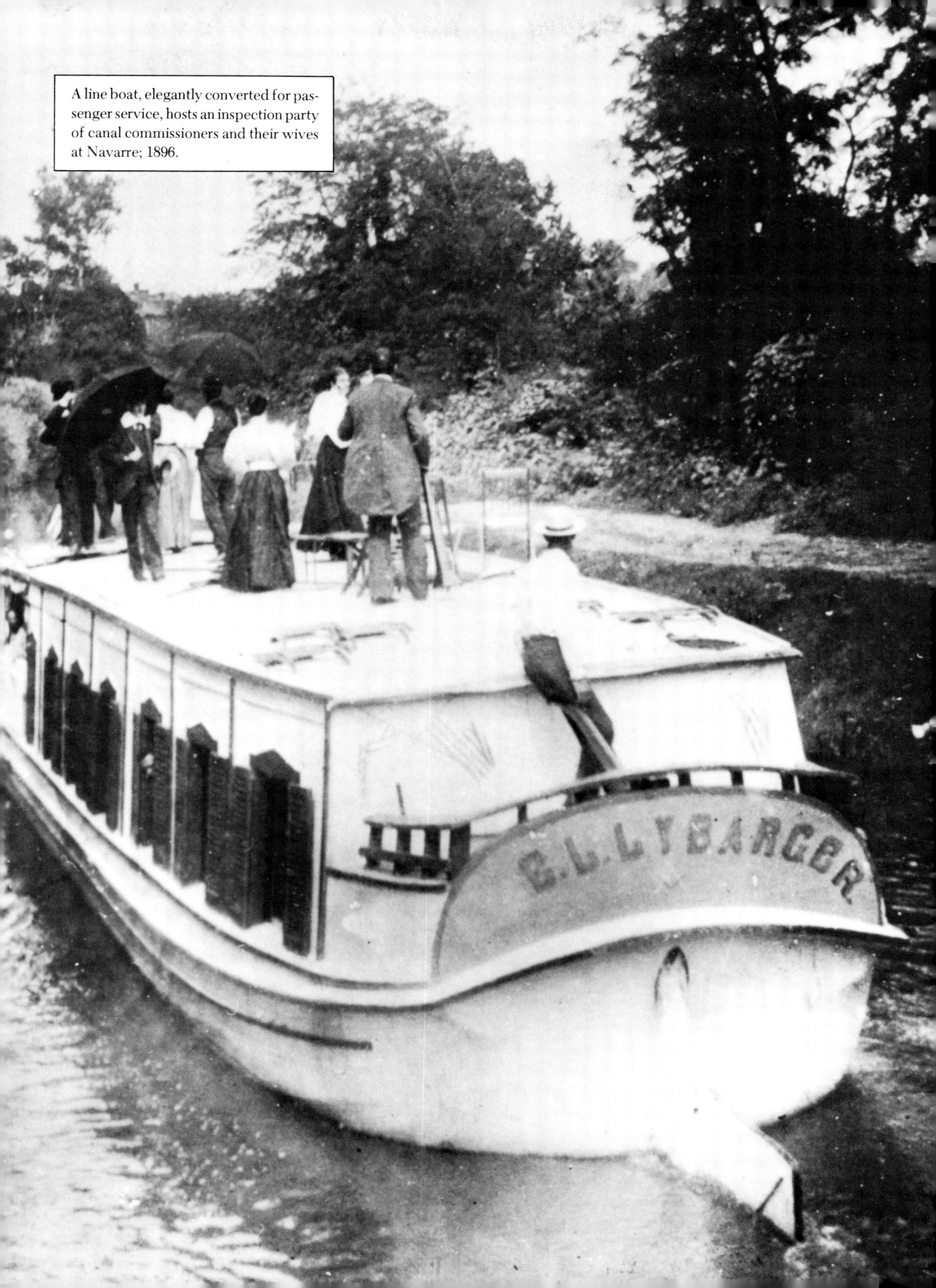

A line boat, elegantly converted for passenger service, hosts an inspection party of canal commissioners and their wives at Navarre; 1896.

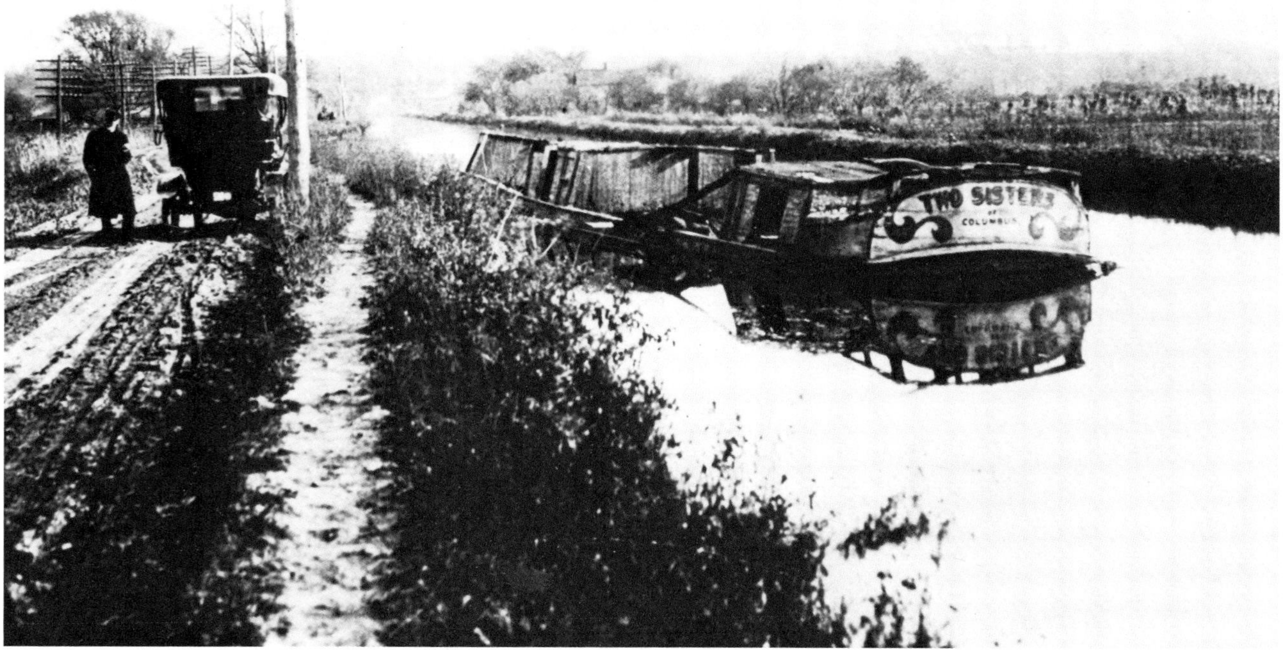

Above, Two Sisters of Columbus, a freight boat still operating on the Ohio & Erie in 1904, is visited by a gentleman who owns an example of a more recent transportation development.

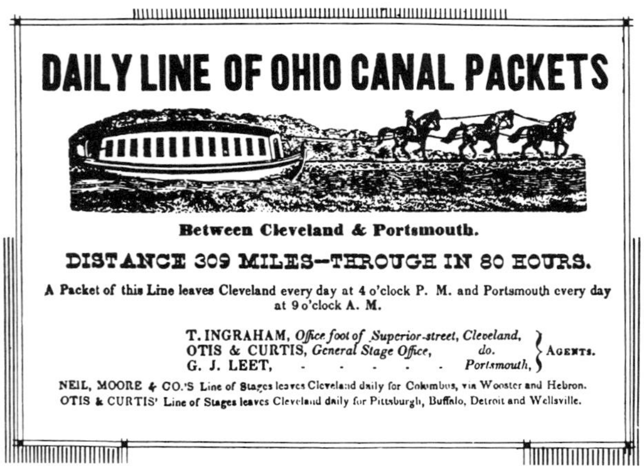

Above, A broadside (poster) advertising passenger travel on the Ohio & Erie Canal. (OHS)

Right, The last commercial boats operating on a surviving northern section of the Ohio & Erie Canal (1916–20), from Akron's Lock One, south through Summit Lake to the Portage Lakes, were Crescent Line steam launches with scheduled daily excursions. (CSO)

Below, An experimental self-propelled steamboat on the Miami & Erie in latter half of the nineteenth century. These experiments were discontinued because the excessive speed such boats were capable of proved damaging to the canal banks. (CSO)

Bottom, The pleasure launch *Dora,* operating excursions on the Hocking Canal, steams past the gas works at Lancaster. Small building above the pilot is a locktender's house. (Canal Visitor Center, CVNRA)

SEASON OF 1908

CRESCENT LAUNCH LINE

BOATS WILL LEAVE FOR LONG LAKE AS FOLLOWS

TICKETS GOOD ON BOATS No. 1, 2, 3 and 4

Leave Lock 1 W. Exchange St. For Long Lake	Leave Lakeside For Long Lake	RETURN Leave LongLake
5:00 A. M.	9:30 A. M.	6:30 A. M.
9:00 A. M.	1:30 P. M.	11:00 A. M.
1:00 P. M.	6:00 P. M.	5:00 P. M.
		8:30 P M.

Boats leaving for lake at 5 a. m. and 6 p. m. and returning at 6:30 a. m. and 8:30 p. m. commence July 3rd.

SUNDAYS

Leave Lock 1 For Long Lake	Leave Lakeside For Long Lake	Return Leave LongLake
9:00 A. M.	9:30 A. M.	11:00 A. M.
1:30 P. M.	2:00 P. M.	5:00 P. M.
	6:10 P. M.	6:00 P. M.
		7:30 P. M.
		8:30 P. M.

14 Miles Round Trip 25 cts.

Crescent Launches Connect With Boats for Turkeyfoot

FOR SPECIAL ENGAGEMENTS AND PRIVATE PARTIES CALL PEOPLES PHONE 2326 R 2

TICKET OFFICE AT LAKE SIDE

Office Hours 9 A. M. to 6 P. M.

J. W. HEIM, Residence Peoples Phone 4846 AKRON, OHIO

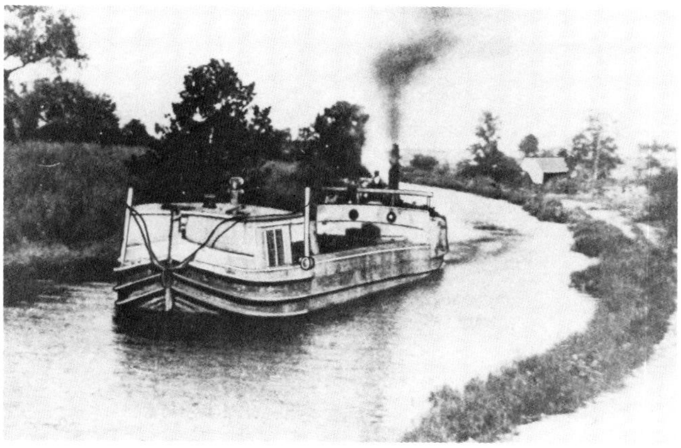

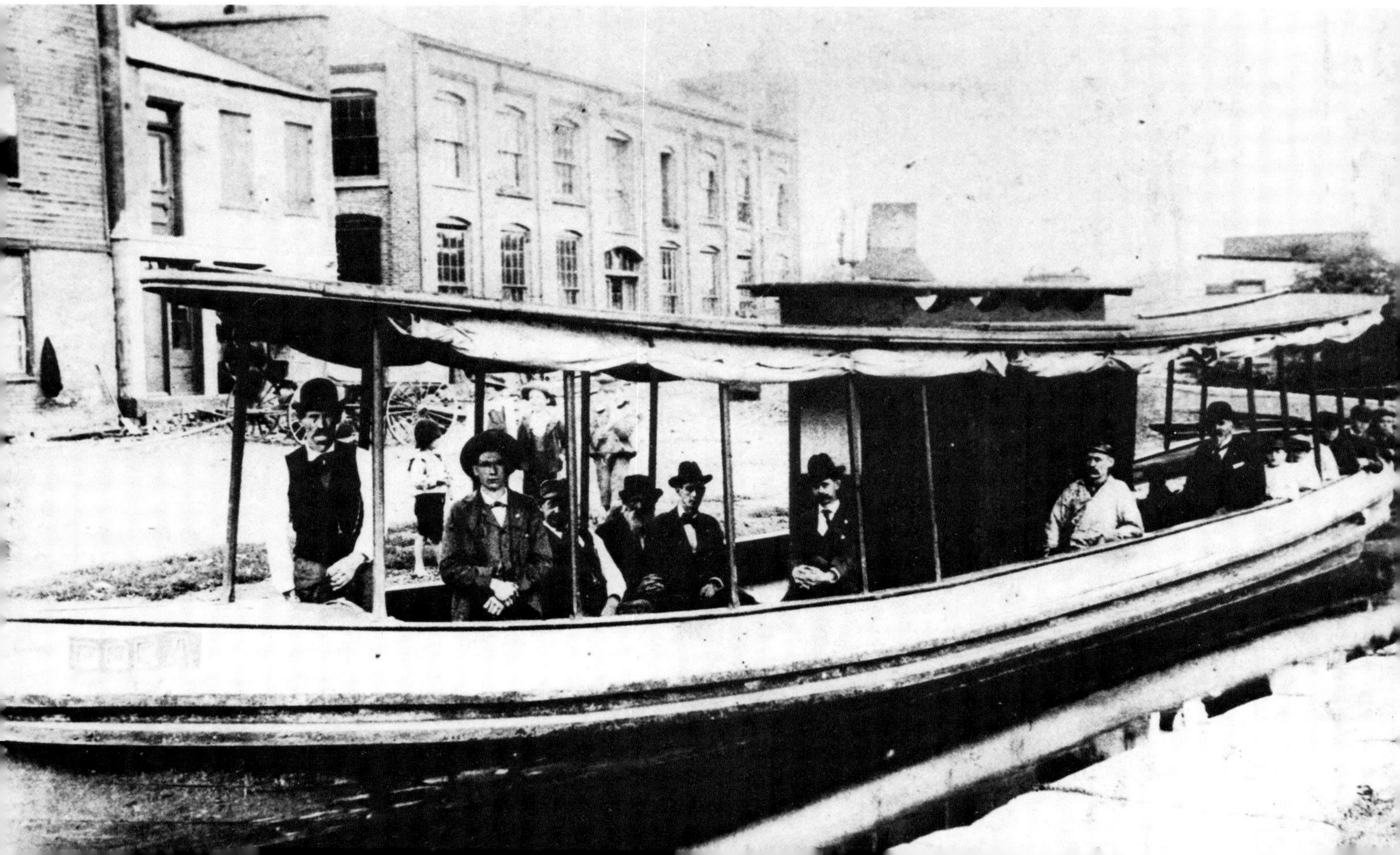

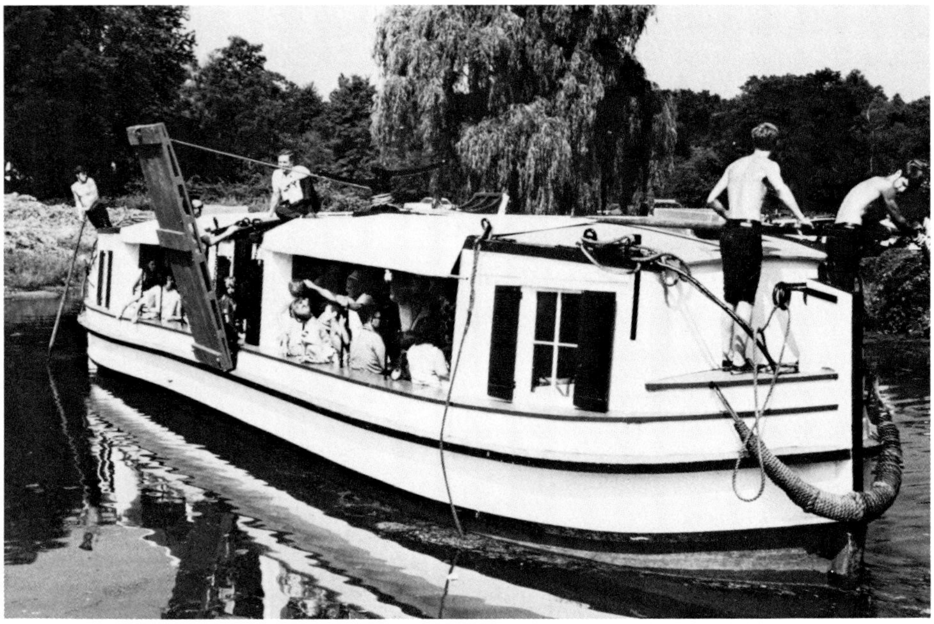

A number of civic groups at Canal Fulton cooperated to build the *St. Helena II,* seen above operating in 1970 on a preserved section of the Ohio & Erie Canal at Canal Fulton and hosting members of the Canal Society of Ohio. Nineteenth-century methods were used in 1967 to construct the boat with the greatest authenticity possible: e.g., trees were felled and rough-sawed into planks, then steamed to make them sufficiently pliable to conform to the curved shape of the hull. The craft is of freight-boat configuration, with the captain's cabin at the stern actually fitted out with a cookstove and bunks. Awnings provide cover to the cargo holds for the benefit of weekend passengers visiting the restoration. The gangplank at left would have been used to load the spare mule team into the center stable cabin. (Photo by F. W. Trevorrow.)

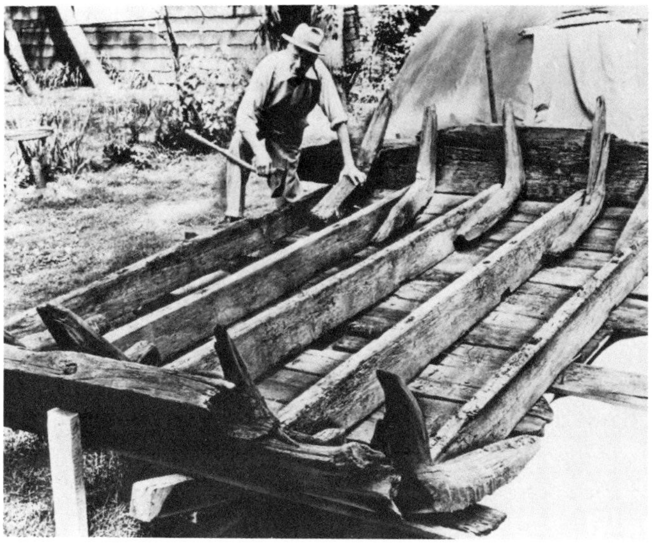

A century and a half after the launching of Ohio's first canal boat, these ribs and hull fragment (right) are thought by some to be from the *State of Ohio,* which sank in 1829, loaded with stoneware. In 1892, Akron historian Samuel Lane described "the hulk of the ancient craft now lying in the mud of a small cove in the berm bank of the canal near the residence of the late James Robinson" on the Summit Level of the Ohio & Erie in the Akron-Kenmore area ("Fifty Years and Over of Akron and Summit County" by Samuel A. Lane, Beacon Job Department, 1892). Indeed, the wreckage was observed in this location in 1896 by James T. Black, who, with the help of a former canal boat captain, pulled it out of the bank for safekeeping some thirty-four years later—in 1930. The relics eventually found a home at Hale farm; here, in 1970, they are examined by a Canal Society member. A credibility problem remains, however, since the *Herald* of October 1, 1829, reported that the *State of Ohio* sank nine miles south of Cleveland. (CSO/research by Terry K. Woods.)

The Ohio & Erie Canal 6

The Ohio & Erie Canal was not only longer than the Miami & Erie (308 miles vs. 250), but it was also the first to be completed between Lake Erie and the Ohio River. Launched with elaborate ground-breaking ceremonies at the Licking Summit on July 4, 1825, the Ohio & Erie's first thirty-eight-mile segment, between Akron and Cleveland, was completed in only two years—despite its large number of locks and other structures, including several aqueducts. This segment at first contained 44 locks; however, Locks 43 and 44, which permitted Great Lakes sailing vessels to enter Merwin's Basin in Cleveland, were eliminated in 1872. The entire length of the canal from Cleveland to Portsmouth was opened to traffic on October 15, 1832.

Beginning at an outlet lock on the Cuyahoga River in Cleveland, the canal ascended 395 feet through Boston, Peninsula, Everett, and Botzum, past Old Portage to Akron, where it climbed through fifteen locks in a single mile to Lock One. Here, a nearby basin provided access to the Pennsylvania & Ohio Canal after 1840. From the nine-mile Portage Summit, the Ohio & Erie passed through Canal Fulton, Massillon, and Bolivar (which, eventually, was the junction with the Sandy & Beaver), through Zoar farmland, through the towns of Canal Dover (now Dover), New Philadelphia, Trenton (now Tuscarawas), Lock Seventeen (now Seventeen), Port Washington, Newcomerstown, Canal Lewisville, and Roscoe. In Roscoe's Lower Basin, the Walhonding Canal descended through the Triple Locks to join the Ohio & Erie's main trunk.

From Akron, the canal dropped 238 feet to Webbsport, where the Dresden Sidecut came in from the canalized Muskingum River Improvement. From here it rose again 160 feet to another Lock One (Upper Taylor Lock) at the north end of the Licking

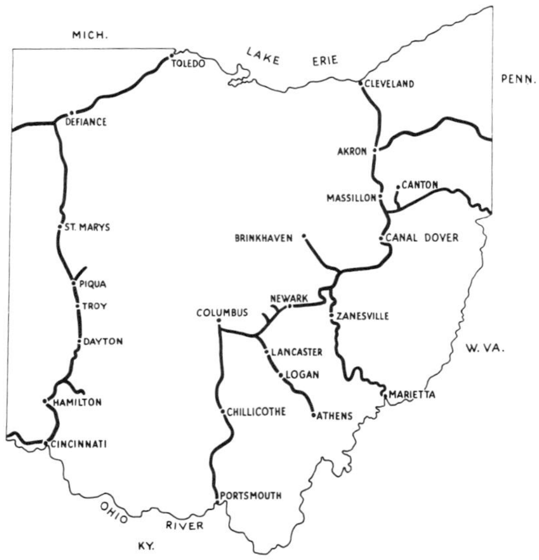

A simplified map of Ohio's canal system after both the Ohio & Erie (on the right) and the Miami & Erie (on the left) were completed, together with their feeders. (OHS)

Summit, having gone through Frazeysburg, Nashport, the Black Hand Gorge Slackwater at Licking Narrows, and Newark in the process.

At this secondary summit, the canal traversed two aqueducts and a culvert before entering the Licking Reservoir (now Buckeye Lake). After leaving the reservoir, it passed through the Deep Cut excavation, which extended the Licking Summit for a total of fourteen miles without a lock.

From Licking Summit, the canal locked down 202 feet through Baltimore, Carroll (the Hocking Canal Junction), Lockville, Canal Winchester, and Groveport, to Lockbourne, where the twelve-mile Columbus Feeder came in from Ohio's capital city. From Lockbourne it descended an additional 211 feet to the Ohio River at Portsmouth, crossing the Scioto River on a great covered aqueduct at Circleville, and passing through downtown Chillicothe along the way. It will be noted that the descent from Licking Summit to the Ohio River totaled 413 feet—greater than the 395-foot difference between Portage Summit and Lake Erie. Since, however, the Ohio River is substantially lower than Lake Erie, Portage Summit was the higher of the two summits (by 78 feet) in actual elevation above sea level; hence, the name of the county (Summit) where Akron is located.

In its 308 miles, the Ohio & Erie Canal climbed up and down a total of 1,206 feet, through 149 lift locks and 7 guard locks.

The pictures in this chapter are arranged geographically, from north to south, insofar as their location can be pinpointed. With the help of the remarkable table of distances in the Appendix prepared by Frank Trevorrow and Barnett Golding, the captions locate each picture in terms of estimated miles south of Cleveland along the canal route.

The greatest concentration of surviving vintage photographs for the Ohio & Erie turned up in the Cleveland-Akron area—and not just because the author and publisher happen to be located here. Not only was this thirty-eight-mile segment completed and operating well before any of the rest of Ohio's canal system, but it profoundly influenced the region's economy. The canal itself was a source of recreation, as evidenced by the many photographs of picnics and family outings—often on state boats; besides, this segment averaged more than one lock per mile in its 395-foot ascent from Lake Erie. Accordingly, it attracted many early photographers and influenced later generations of both amateur and professional archivists, historians, and canal enthusiasts to preserve images of the region's colorful era. Some of these early prints are contrasted with photographs picturing remains of the canal as they appear today throughout the Ohio & Erie Canal's 308 mile length.

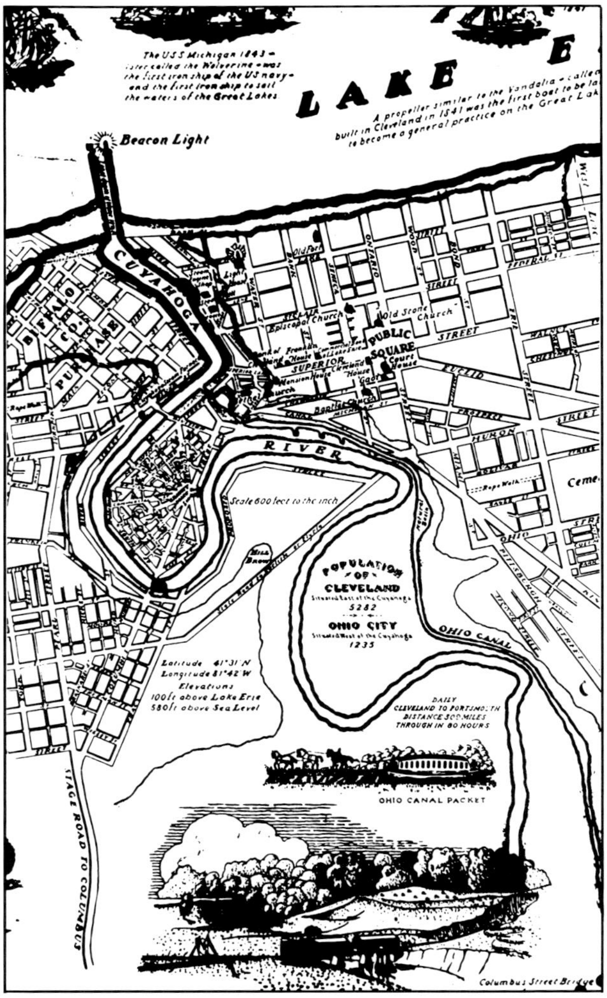

Left, An early map of Cleveland, showing the Ohio & Erie Canal entering from the lower right and paralleling the Cuyahoga River for nearly a mile before locking down to river level. It will be noted that Cleveland's population is listed as 5,282; Ohio City's inhabitants number 1,235. The map advertises daily passenger schedules to Portsmouth on Ohio Canal packets. (CSO)

Below, Announcement of the first freight boat to arrive in Cleveland, in the *Herald,* Cleveland, July 6, 1827. (WRHS)

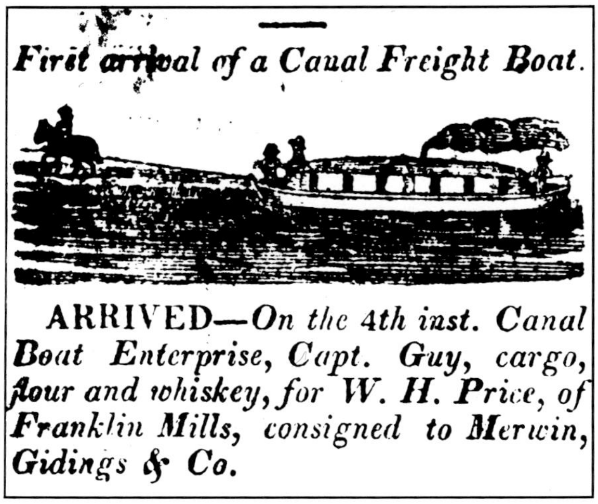

First arrival of a Canal Freight Boat.

ARRIVED—*On the 4th inst. Canal Boat Enterprise, Capt. Guy, cargo, flour and whiskey, for W. H. Price, of Franklin Mills, consigned to Merwin, Gidings & Co.*

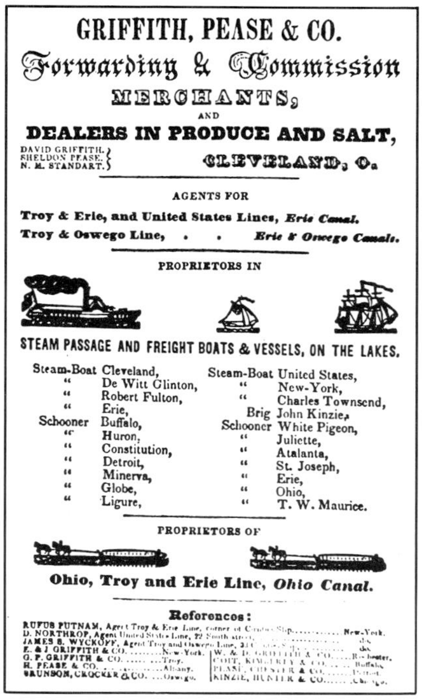

Left, Canal advertisement from the 1837 Cleveland Directory. (CSO)

Below, Mouth of the Cuyahoga River on Lake Erie, as seen from Cleveland's Terminal Tower in 1951. Horizontal straight line a short distance out into the lake is Cleveland's breakwater—originally constructed from rubble and imperfect lock stones brought here by canal boat from Deep Lock Quarry south of Peninsula. (Photo by Jack Gieck.)

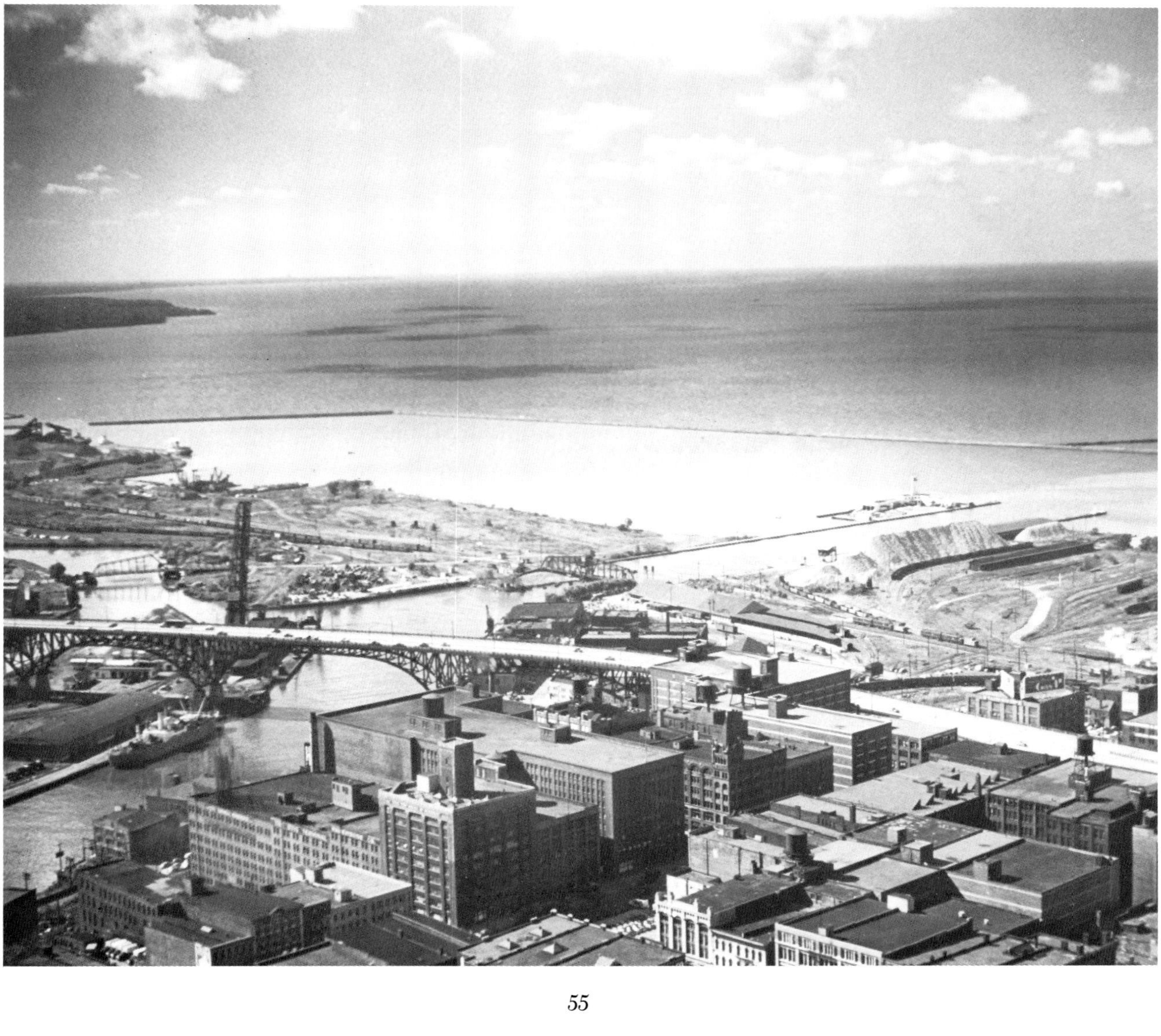

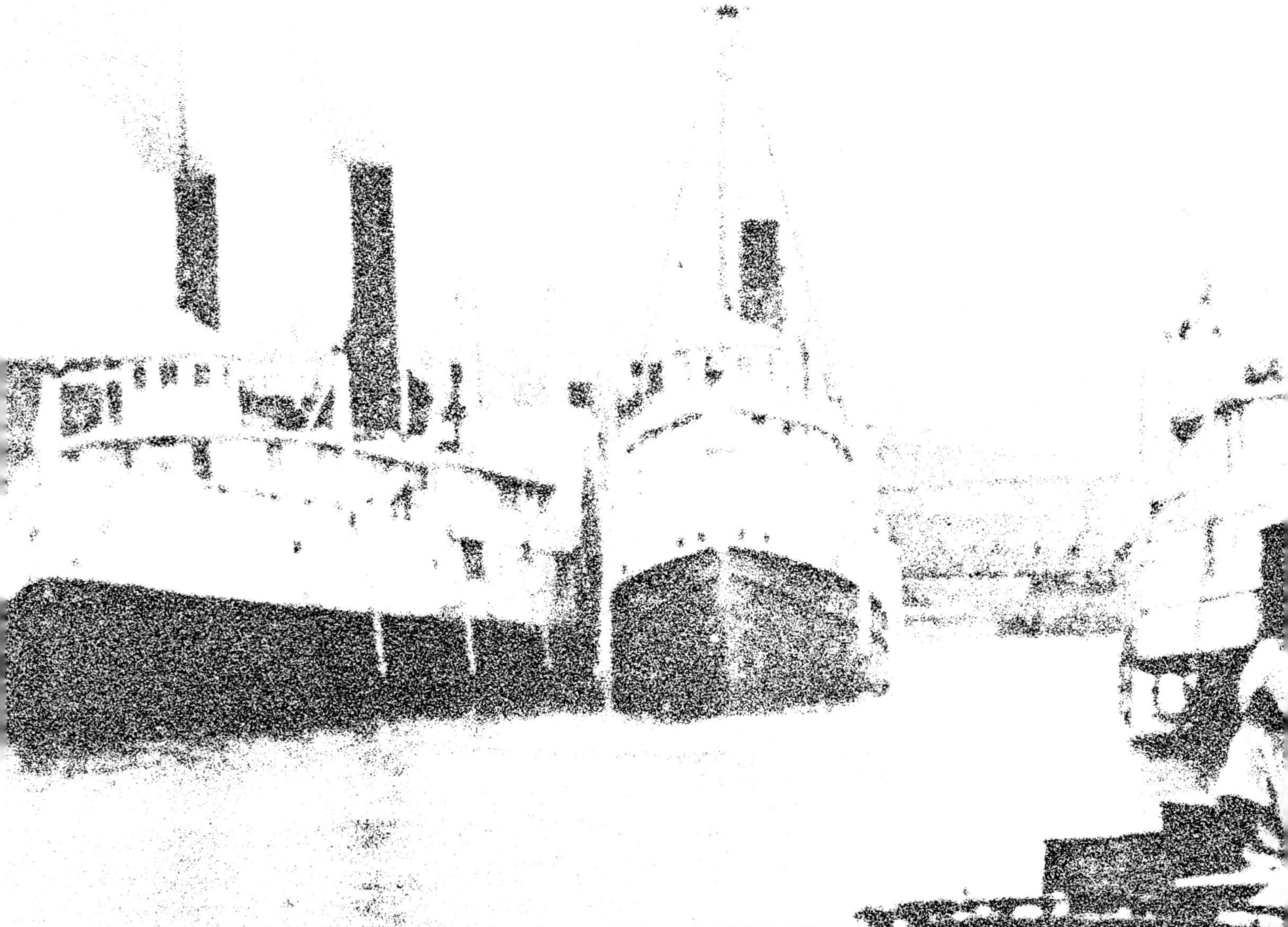

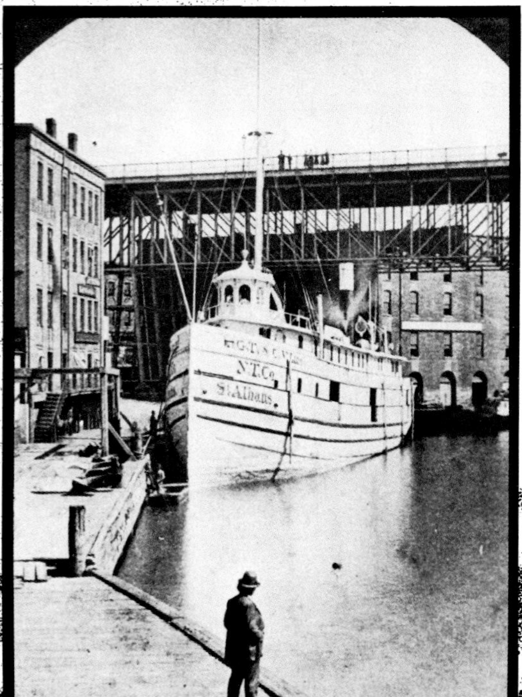
N.T.Co.
St. Albans

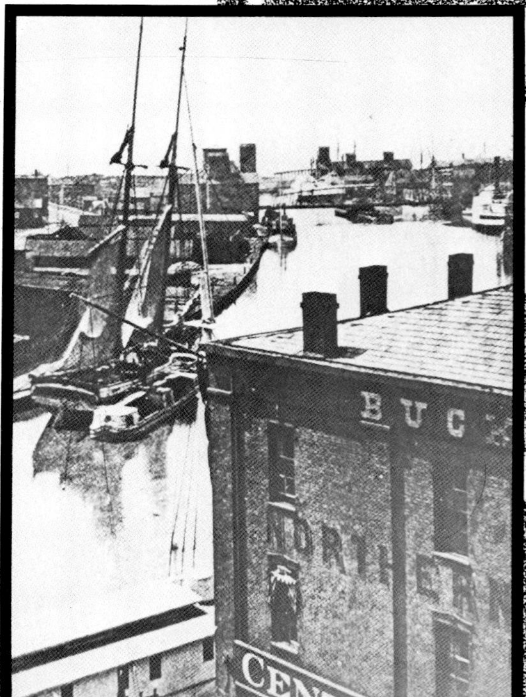

Canal boats with lake steamers in the background, in the Cuyahoga River at the foot of Cleveland's Superior Street, where the Ohio & Erie locked down to river level. Several crew members and a captain's wife are sitting on deck. Such encounters between vessels of different classes developed commercial, social, and even artistic significance. Fights between canal crews occurred as their boats nosed between each other for access to the steamers, whose crews made fun of the canalers and their boats. But when peace was restored and comradery ensued, the canalers learned stirring new music from the steamer crews: lake chanteys, which had wide musical intervals and dramatic sweeps, influenced subsequent canal songs; 1869 (Baus Collection, RVF). The left inset photo shows the five-story Canal Hotel, which also housed Cleveland's steamboat ticket office. A lake passenger steamer is docked beside the hotel on the Cuyahoga River. Behind the steamer, at extreme right, a canal boat may be identified; c. 1865 (Pockrandt Collection, SCHS). The right inset photo shows, in the Cuyahoga River channel, cargo being transferred from a canal boat to a Great Lakes sailing vessel at the foot of Superior Street opposite the entrance of the Ohio & Erie Canal; 1871. (Canal Visitor Center, CVNRA)

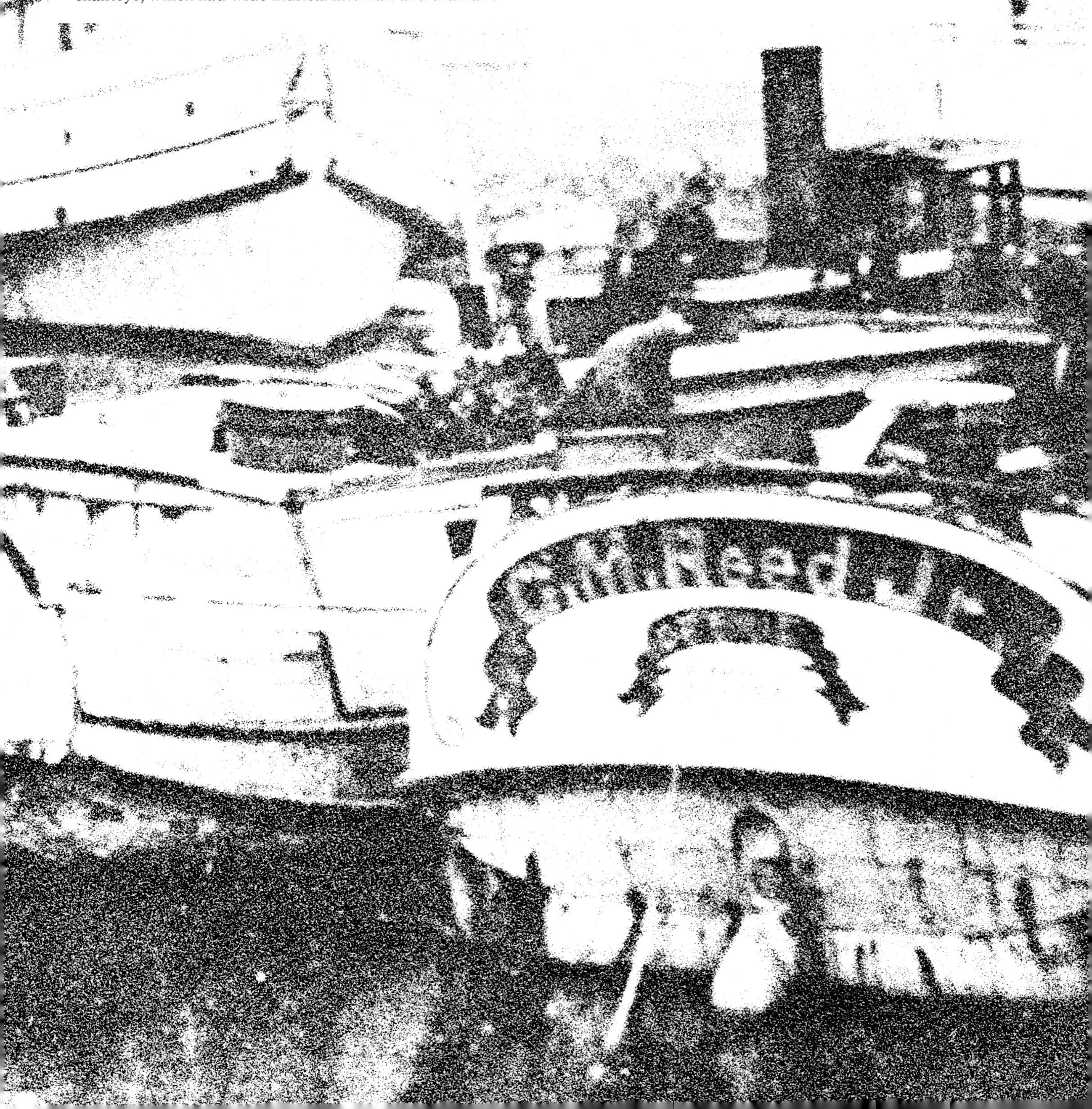

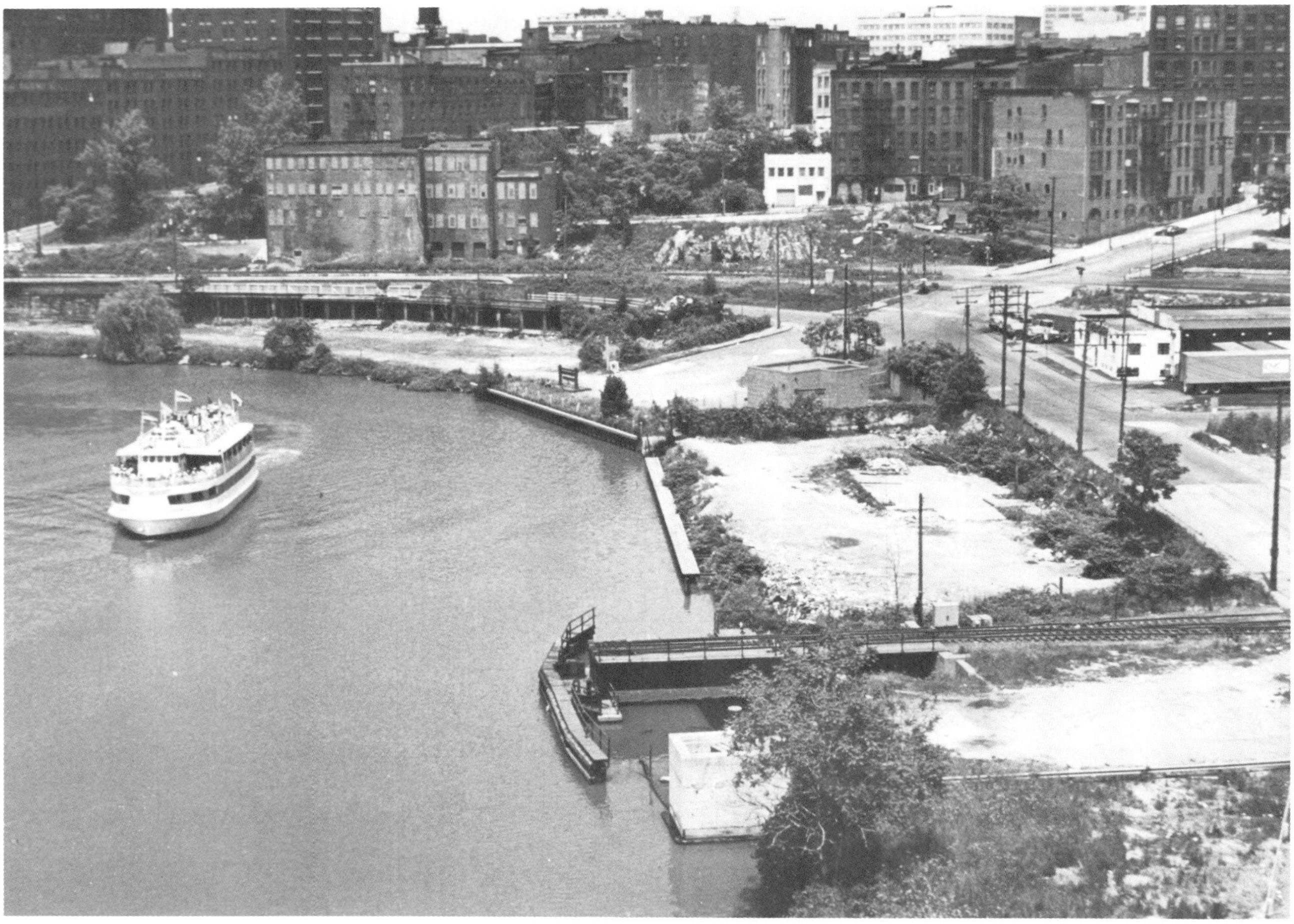

The former entrance (above) of the Ohio & Erie Canal into the Cuyahoga River is marked by the railroad dock in the foreground. In the background, at the foot of Superior Street, a small sign may be identified marking the landing site of General Moses Cleaveland and his Connecticut Land Company party in 1796. (From a 1978 color slide by Jack Gieck.)

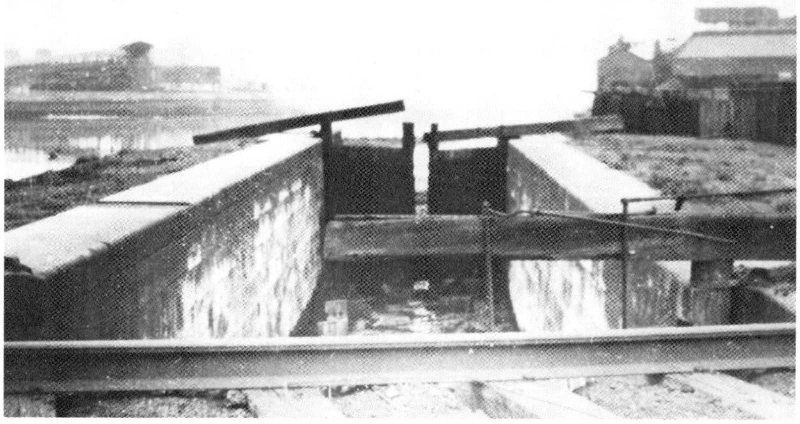

The outlet lock (left) of the Ohio & Erie Canal into the Cuyahoga River at the foot of Dille Street in Cleveland, after 1872. (The original River Lock was at Mile 0, at the foot of Merwin Street.) Mile: 3. (WRHS)

A rare light-sensitive "salt print" shows the canal in Cleveland in 1859. One of the businesses along the canal is the Star Candle Works. At extreme left, behind the bridge structure, is Cleveland's weigh lock. Mile: 3. (WRHS)

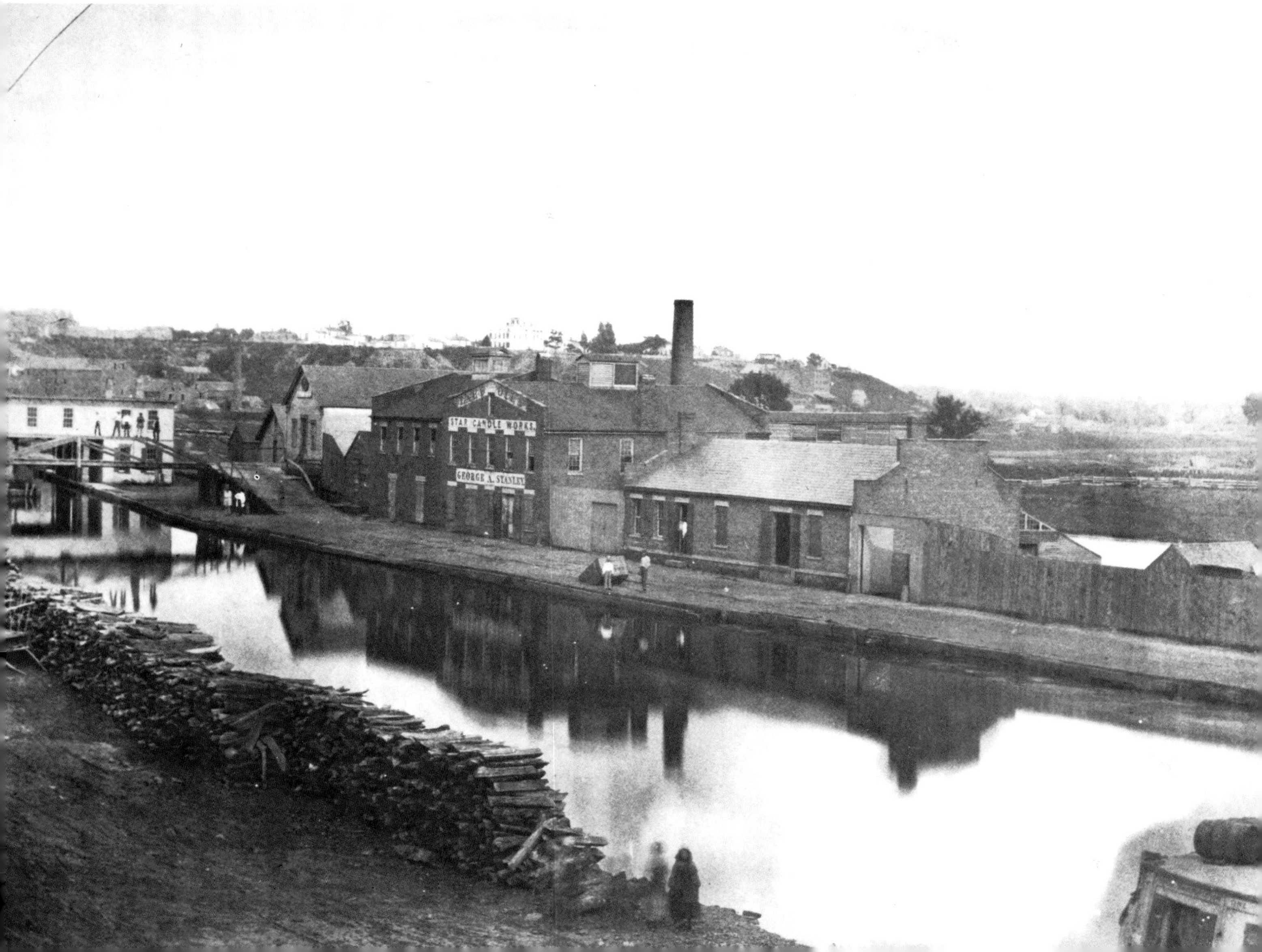

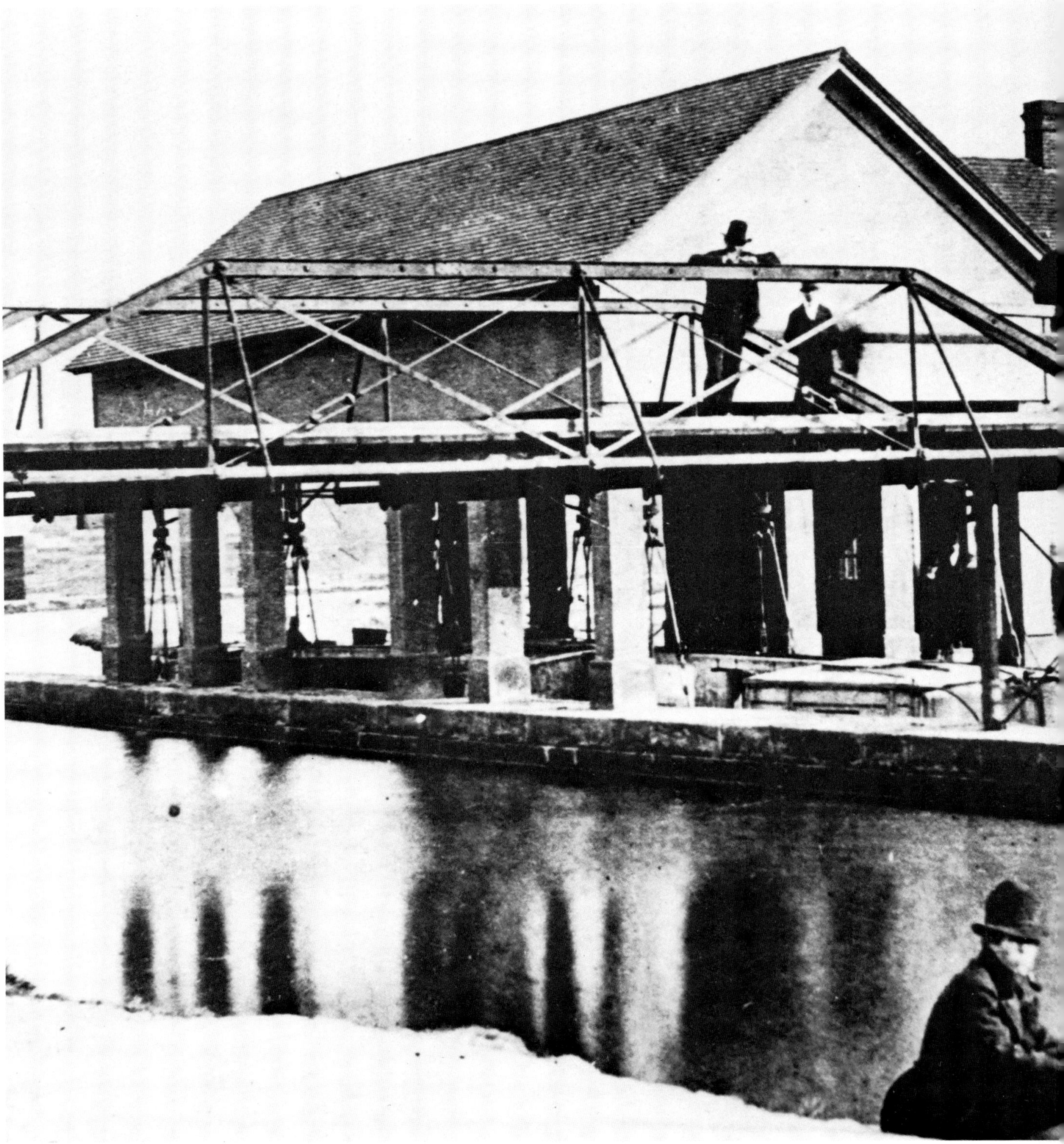

Above, Cleveland's weigh lock at the foot of Dille Street near Broadway around 1874. The lock has been emptied so that the boat is resting on the scale. One cabin may be seen protruding above the lock walls. (Baus Collection, RVF). *Top right,* This boat, raised back to canal level, leaves Cleveland's weigh lock around 1900. Mile: 3. (Courtesy F. W. Trevorrow)

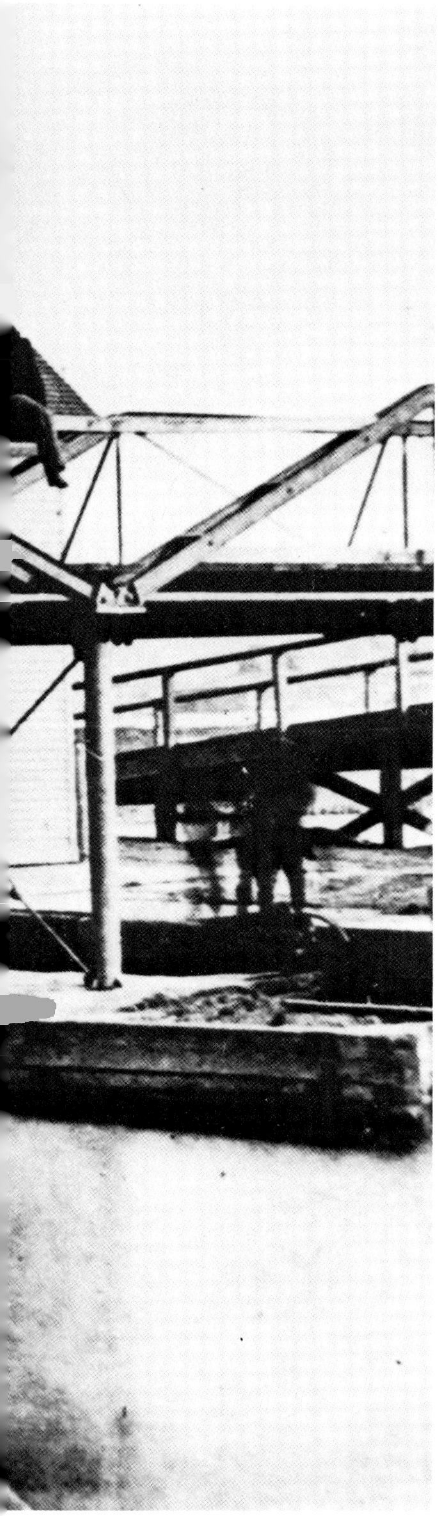

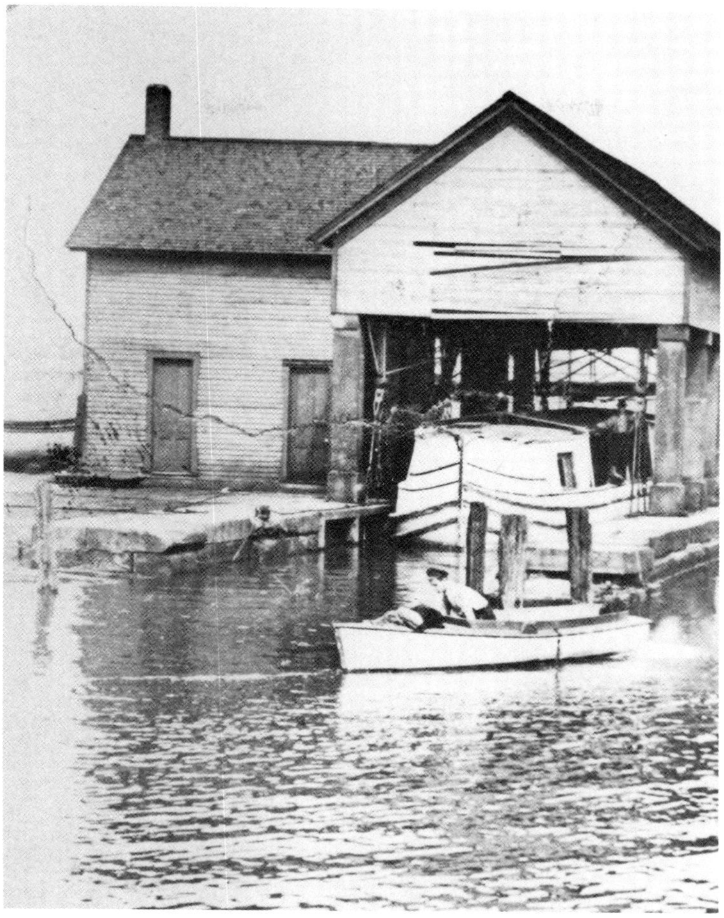

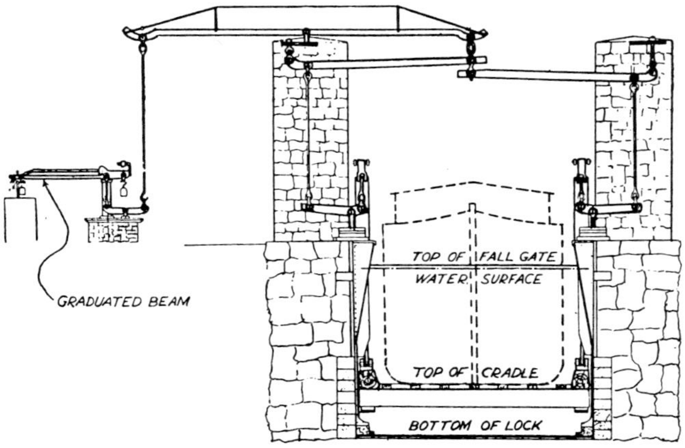

A cross section of a weigh lock shows how it operated. After a boat was drawn inside from the canal, the lock was drained so that the boat rested on top of the scale cradle and could be weighed directly. The weighmaster had a record of the empty weight of each boat on the canal; after this tare weight was subtracted, the correct toll could be levied. (Drawing by F. W. Trevorrow, courtesy the Scale Manufacturers Association.)

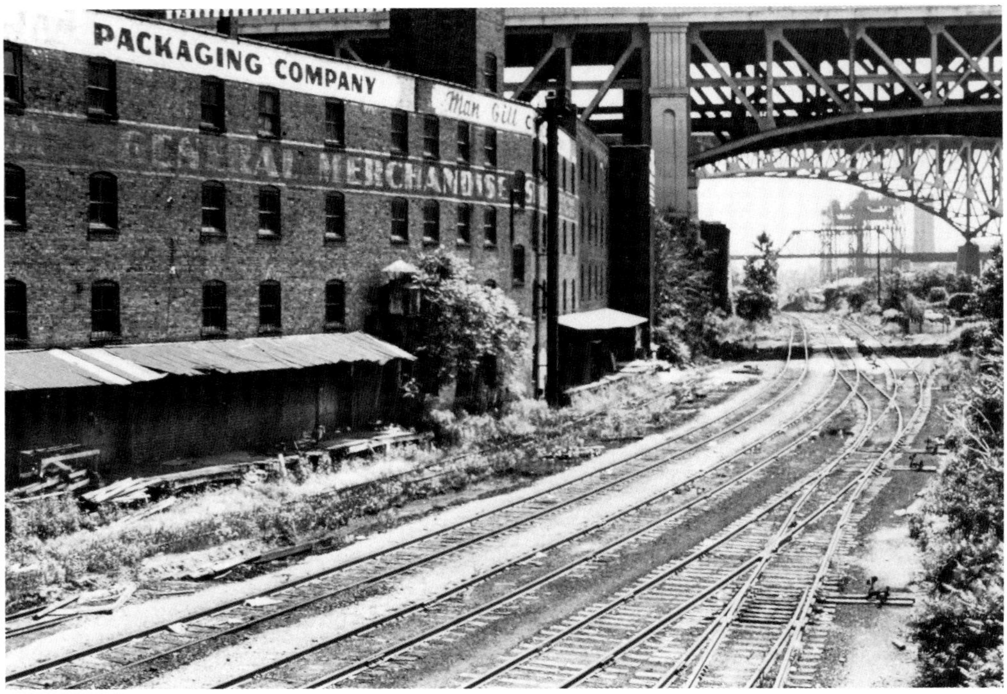

Railroad tracks run in the bed of the canal in Cleveland's Flats today. The buildings at left were originally canal warehouses; 1972. Mile: 4. (From a color slide by Jack Gieck.)

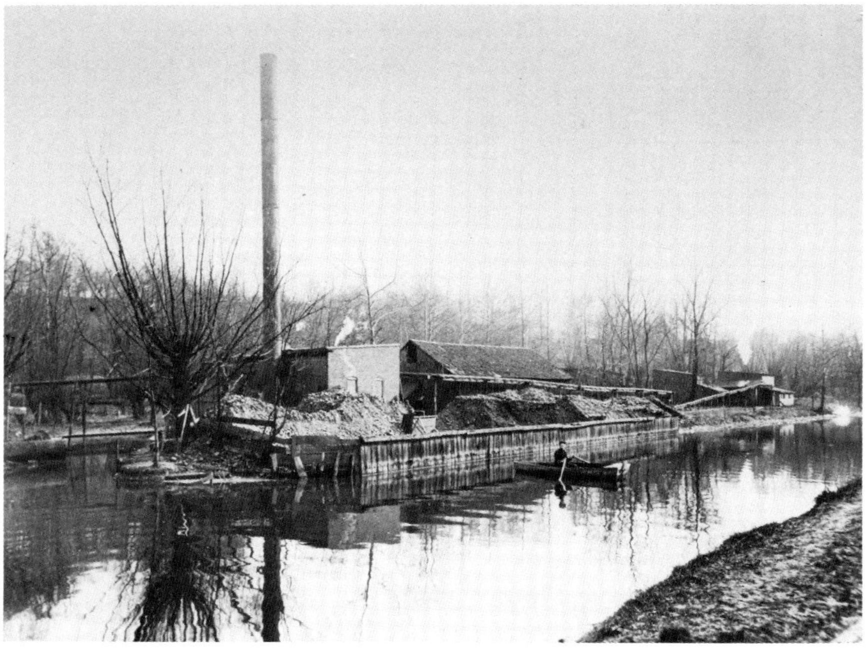

Austin Powder Company's charcoal plant near Five Mile Lock was located a discreet distance from the company's powder mill for safety reasons. The operation was a major supplier of gunpowder to the Union army during the Civil War; c. 1865. (CSO)

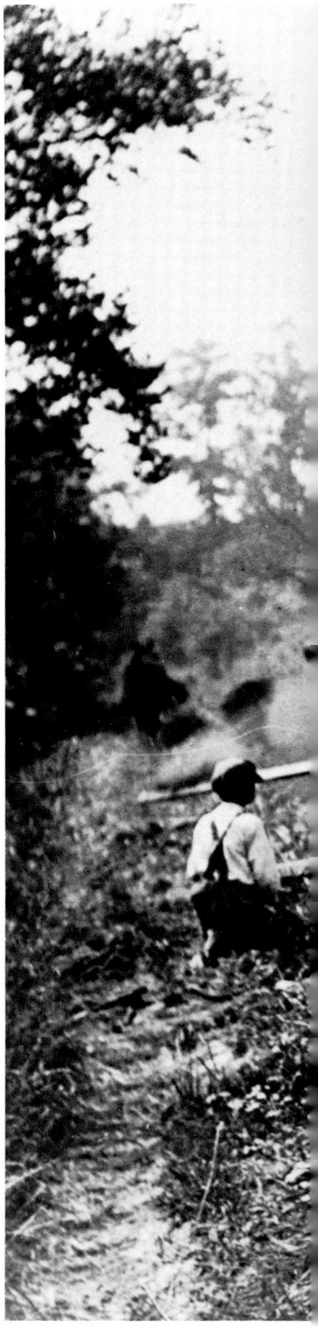

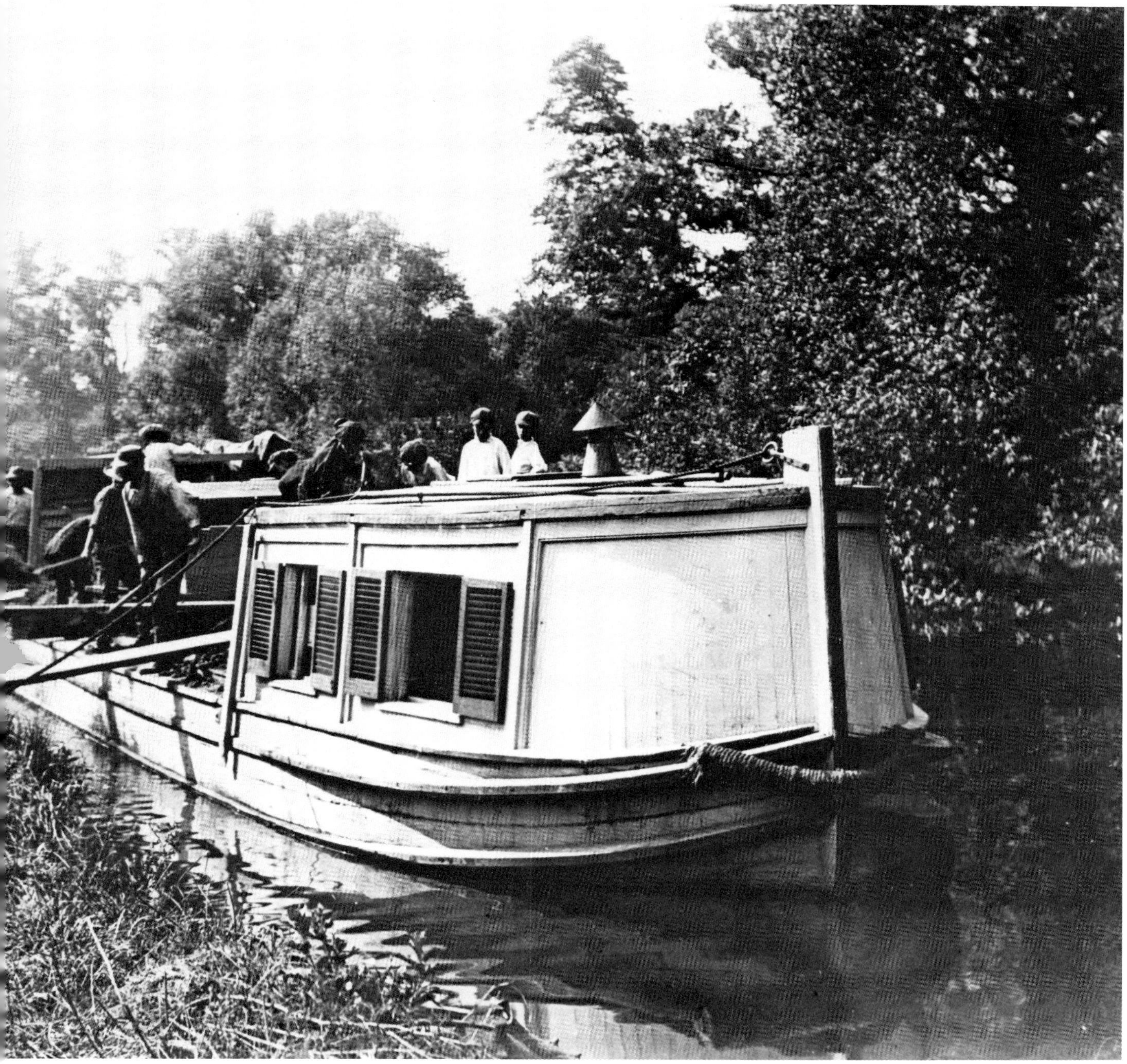

Above, A state boat crew prepares for work at Five Mile Lock; 1900. (Baus Collection, RVF)

Right, A line boat with visitors on board at Eight Mile Lock south of Cleveland. Mule team with driver riding rear animal is at upper right; 1881. (CSO)

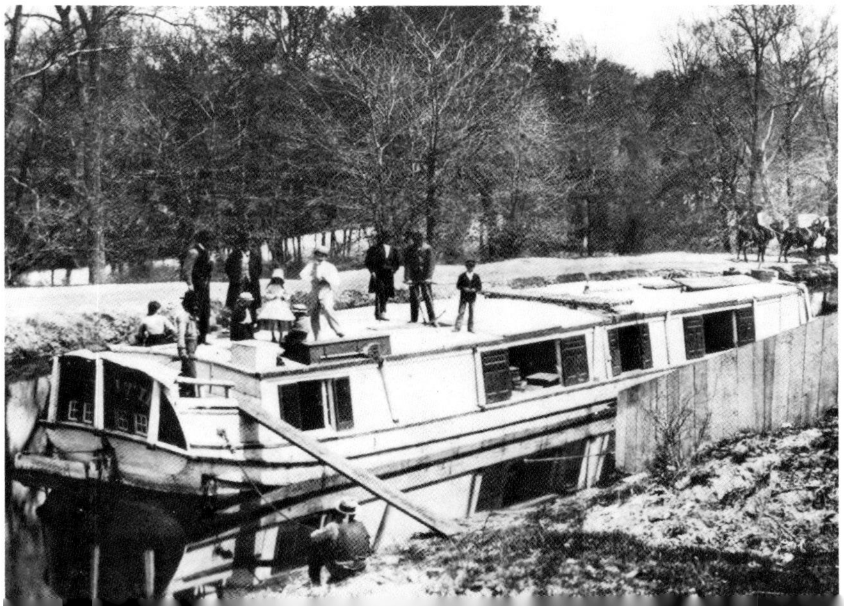

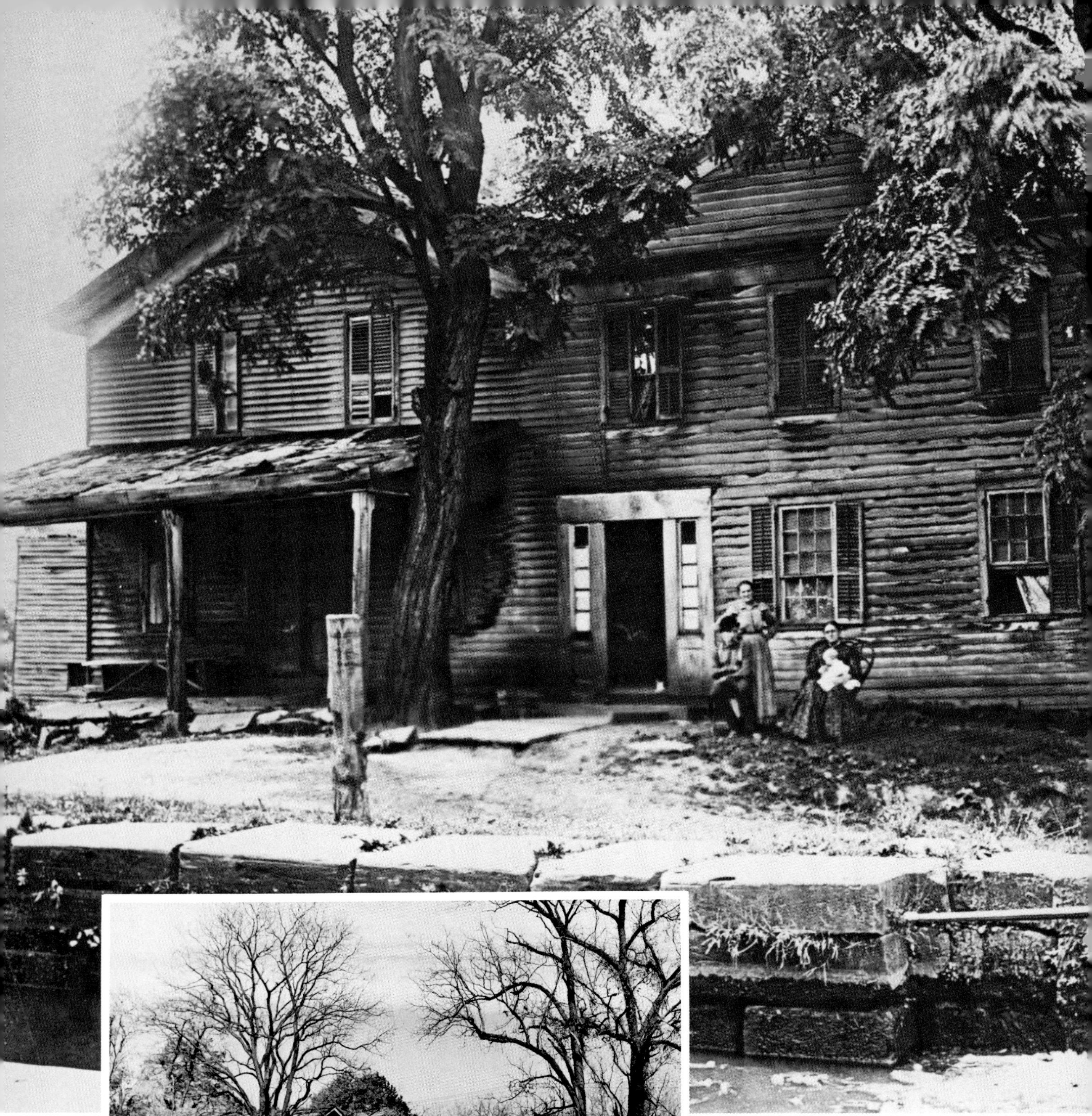

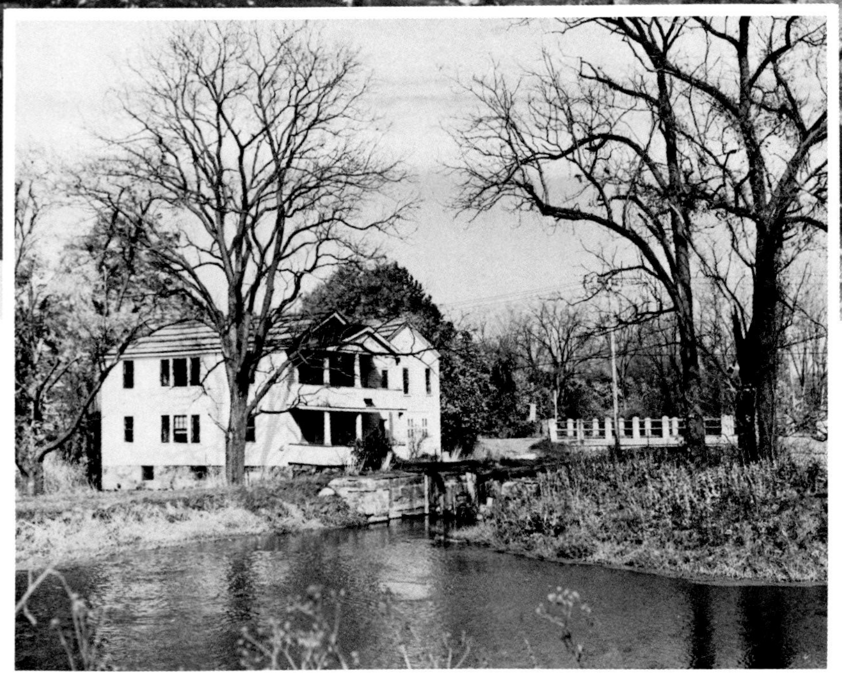

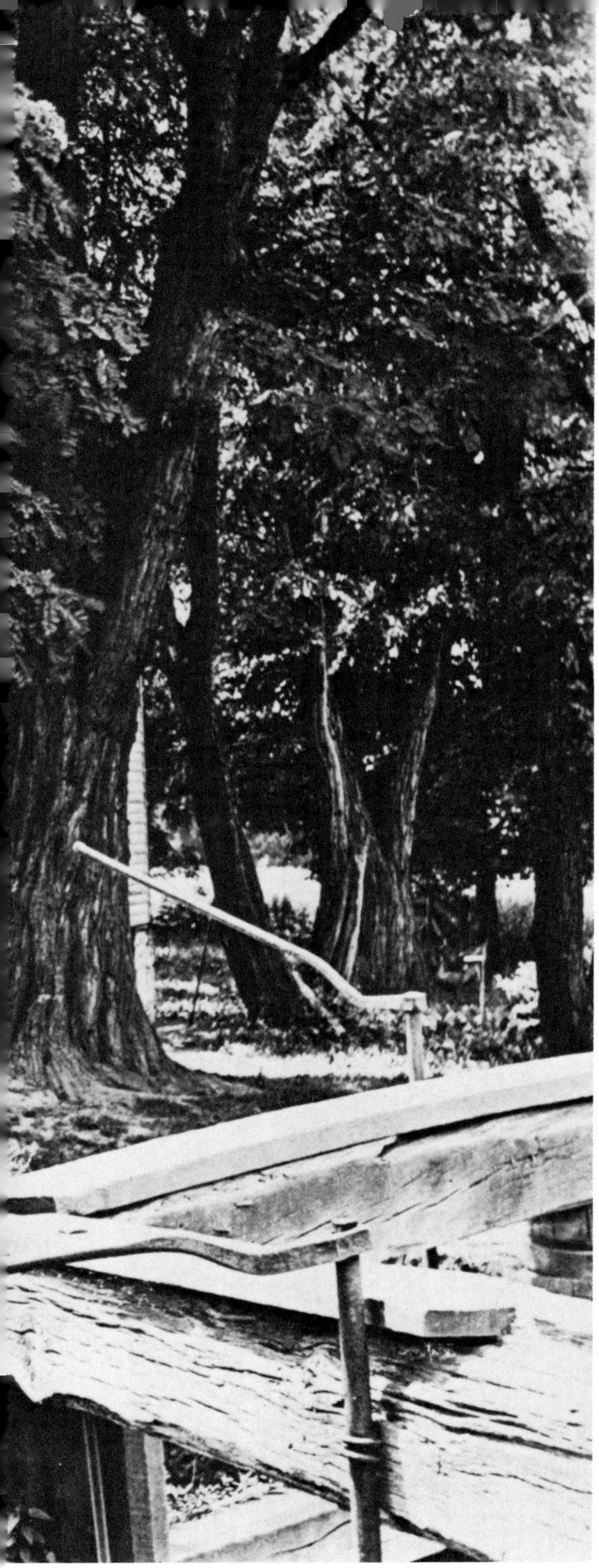

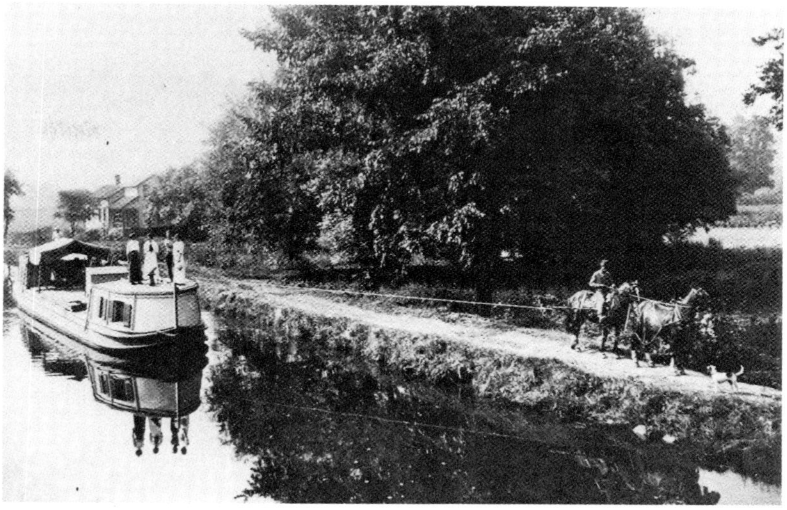

Wives and/or girl friends seem to have been invited along on this weekend outing of State Boat No. 1, south of Stone Road; 1902. Mile: 11. (Canal Visitor Center, CVNRA)

Often called the Locktender's House, canal records indicate that no lockmaster was ever stationed here. The buildings were, at various times, a saloon, an inn, a dance hall, and a general store; because of unspecified entertainment offered upstairs, the establishment at Lock 38 was also known among canalers as Hell's Half Acre. The lock in the foreground is filled to its upper level; lock gates, with wicket control handles, are clearly visible at the right (CSO). The inset photo shows the same house in 1981. The structure has since been restored by the National Park Service. Mile: 12. (Photo by Jack Gieck.)

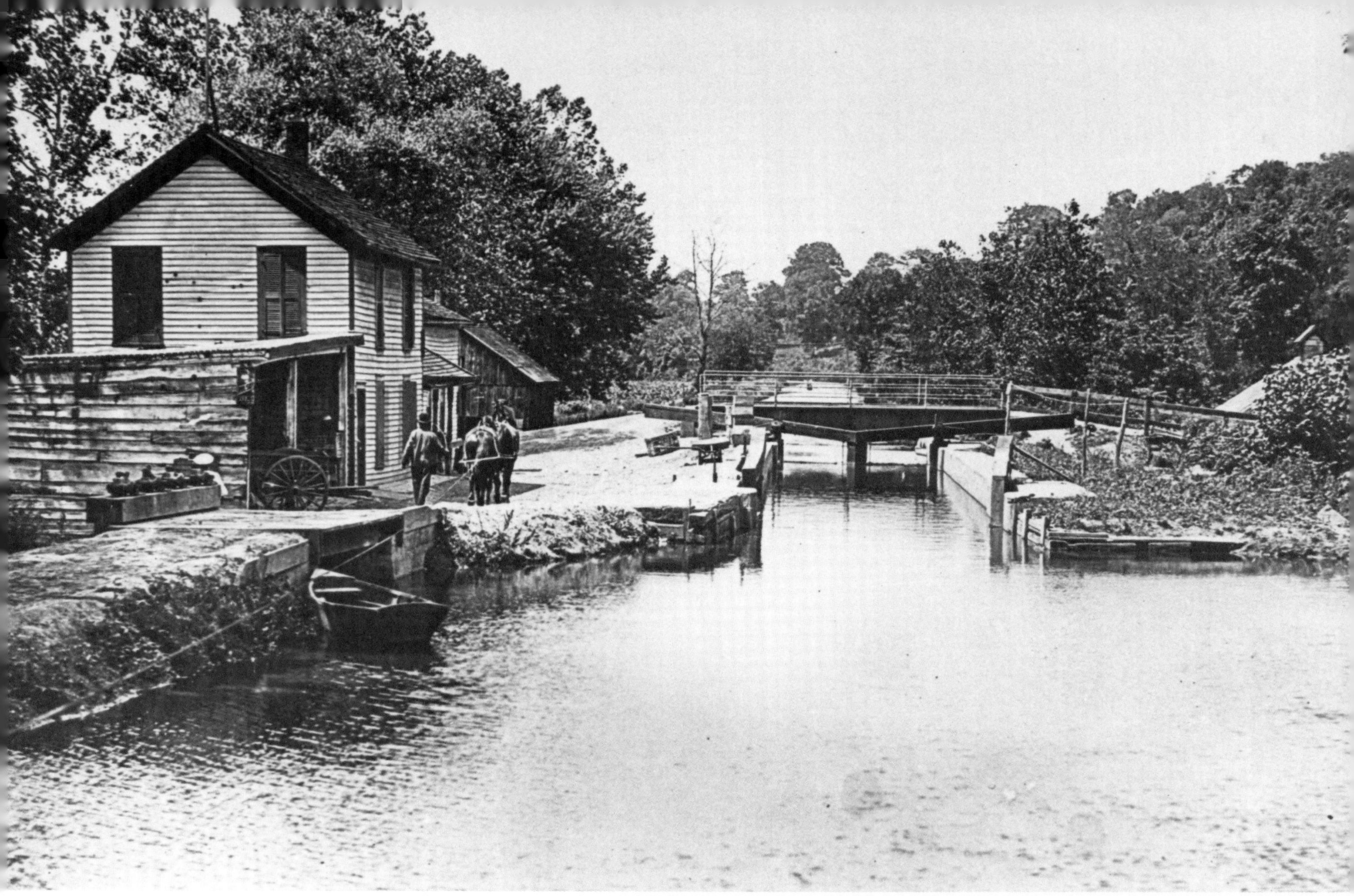

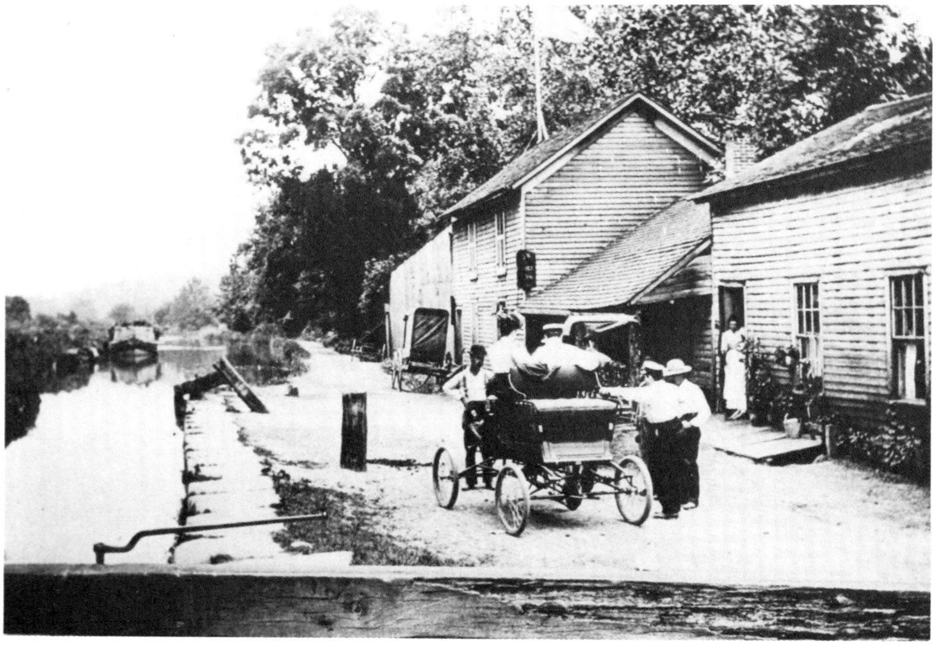

Lock 37 (above), at Fitzwater and Canal Roads, looking north from the deck of a canal boat about to negotiate the lock; the mule team is at left. The structures on the west bank are a general store and residence; across the canal on the east bank is Alexander's (now Wilson's) Mill. Mile: 14. (CSO)

As a canal boat (left) approaches from the south a newer mode of transportation captures the attention of spectators at Lock 37; 1898. Mile: 14. (Baus Collection, RVF)

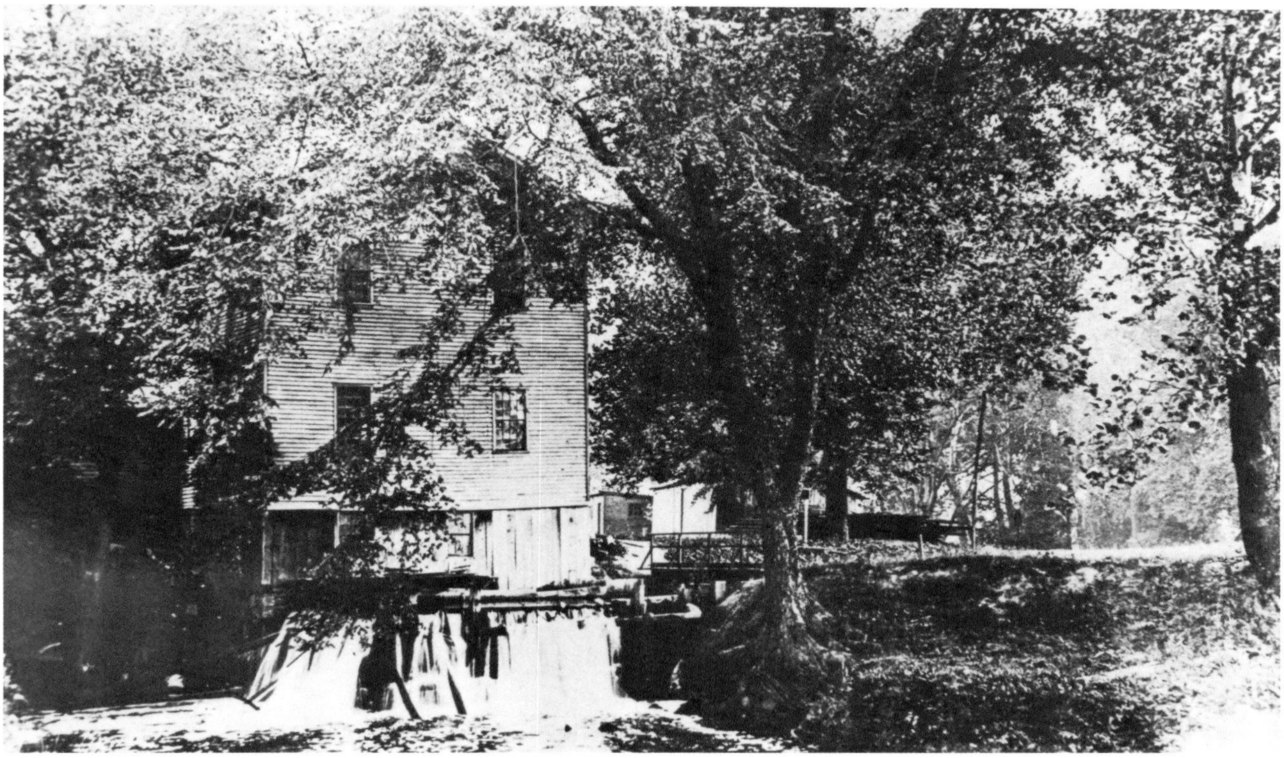

Alexander's Mill (above), built in 1853, as it appeared in 1915. Waterpower to operate the mill was produced by the substantial head of water, or fall, available at Lock 37 (Canal Visitor Center, CVNRA). The renamed Wilson's Mill (below) as it appeared in 1981. Serving as a lock mill for more than a century, it is one of the last surviving lock mills on the Ohio & Erie. Mile: 14. (Photo by Jack Gieck.)

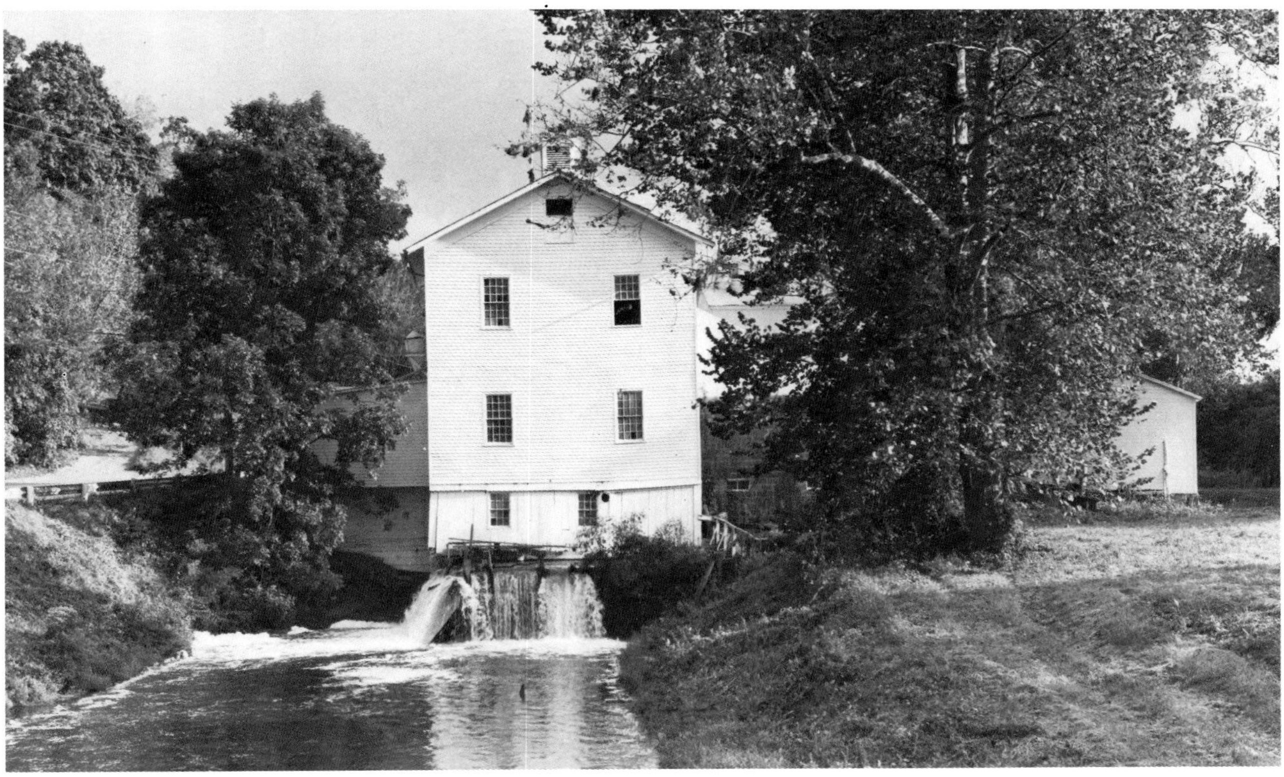

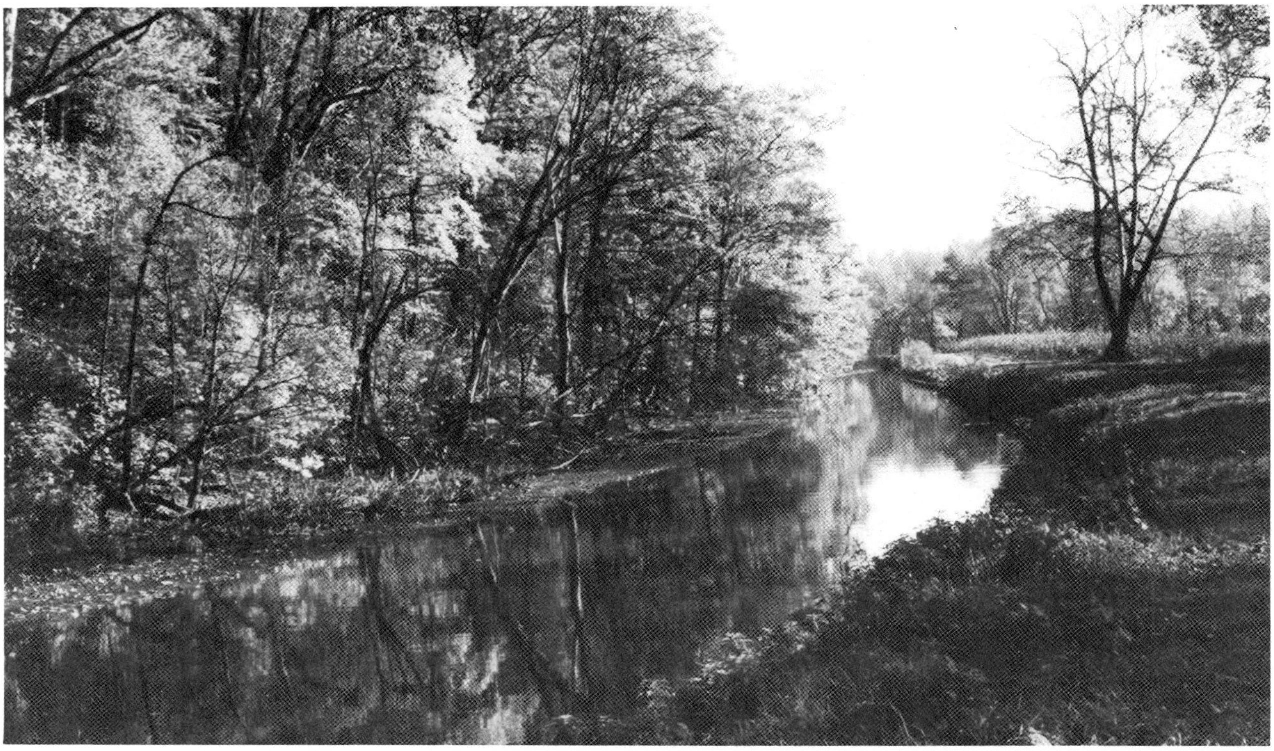

South of Wilson's Mill the canal meanders through the woods in a section known as the Pinery. The canal has been maintained in this area since 1913 to provide industrial cooling water for a Cleveland steel mill. This section of the canal is said to have been well known to teenager James A. Garfield, who worked as a driver, or muleskinner, on his uncle's boat one summer, many years before he had presidential ambitions. Garfield gave up his life as a canaler after contracting malaria. Mile: 15. (Photo by F. W. Trevorrow, 1970.)

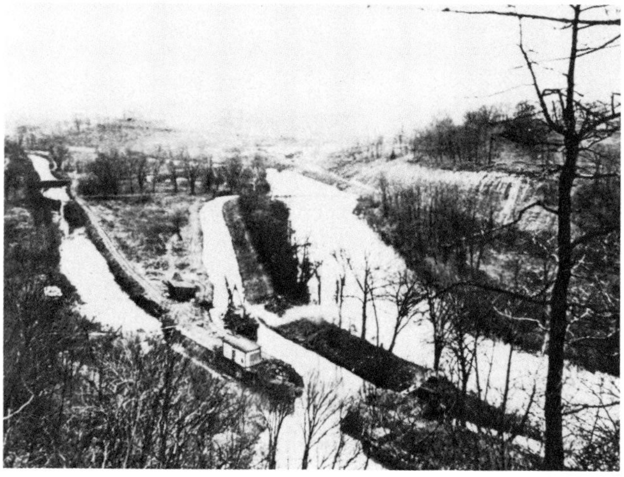

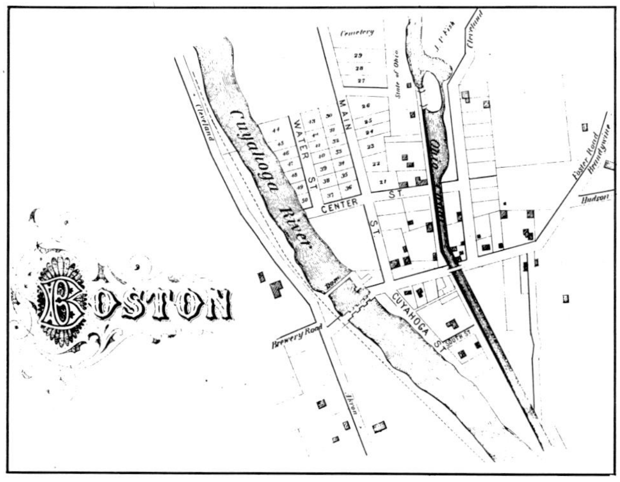

Left, Brecksville Dam and Feeder at Seventeen Mile Lock (Lock 36) during reconstruction by the state in 1907. The Cuyahoga River parallels the canal at right. The feeder may be seen entering the river in the distance to feed the canal in the foreground. (Canal Visitor Center, CVNRA, courtesy Cloea Thomas.)

Right, An antique map shows the canal at the village of Boston, originally a mill community, 21 miles south of Cleveland. (SCHS)

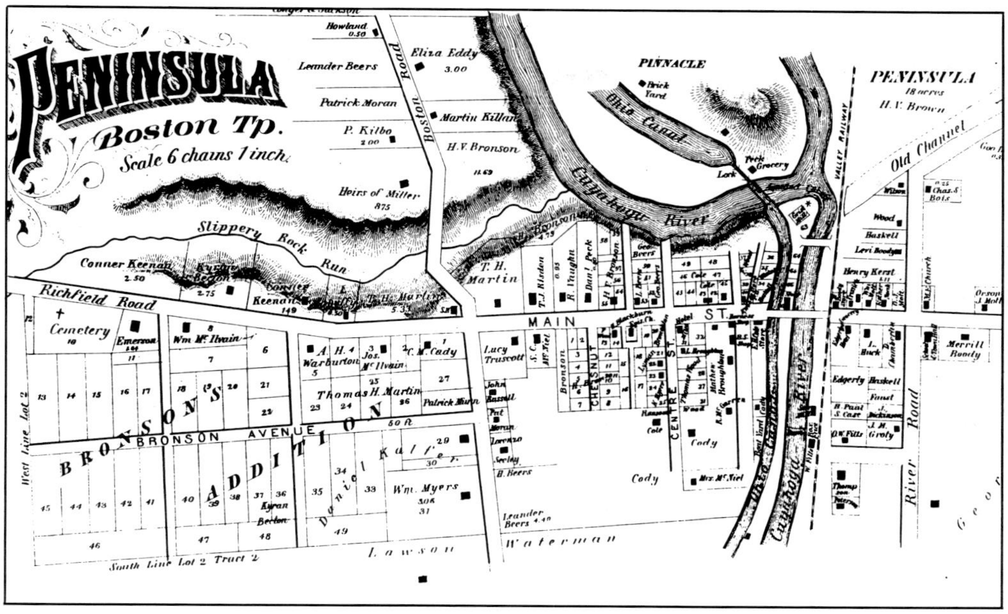

Left, An antique map shows the canal aqueduct crossing the Cuyahoga River at Peninsula. (SCHS)

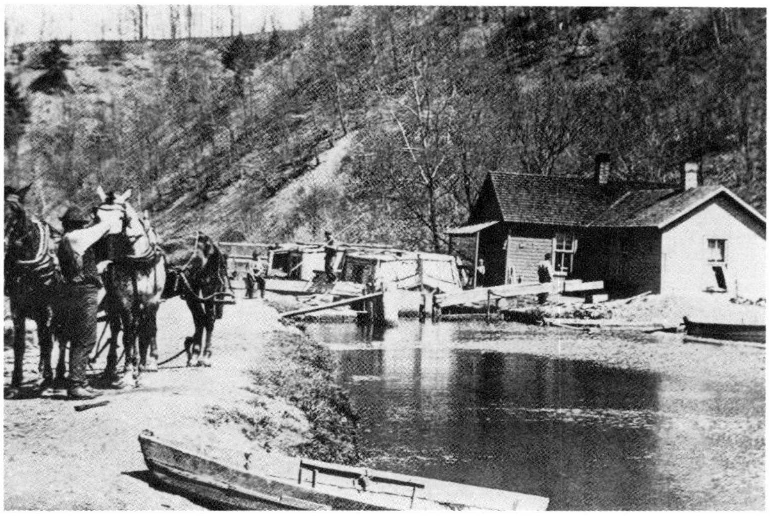

Canalers and mules negotiate a lock north of Peninsula. (Pockrandt Collection, SCHS)

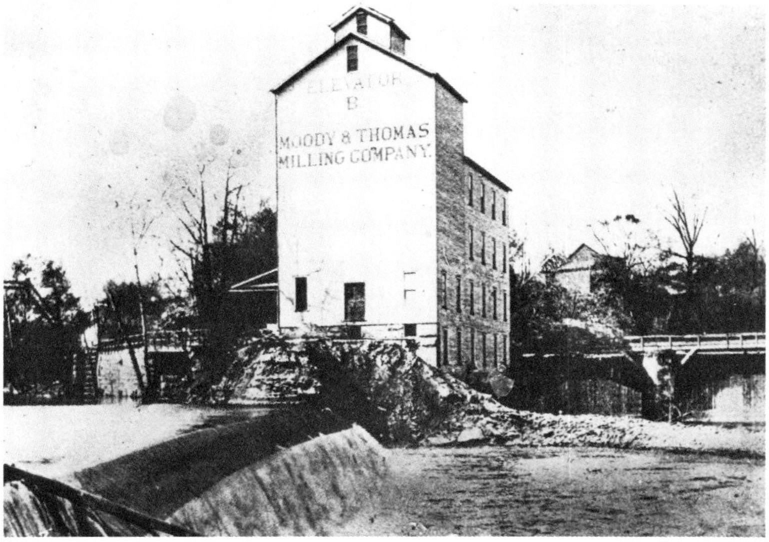

At Peninsula the Cuyahoga River was dammed to provide waterpower for the huge Moody and Thomas Mill, which stood between the river and the canal at the south end of the aqueduct, seen in the background at right. The mill was constructed so that it could unload flour directly into canal boats. Mile: 24. (Canal Visitor Center, CVNRA)

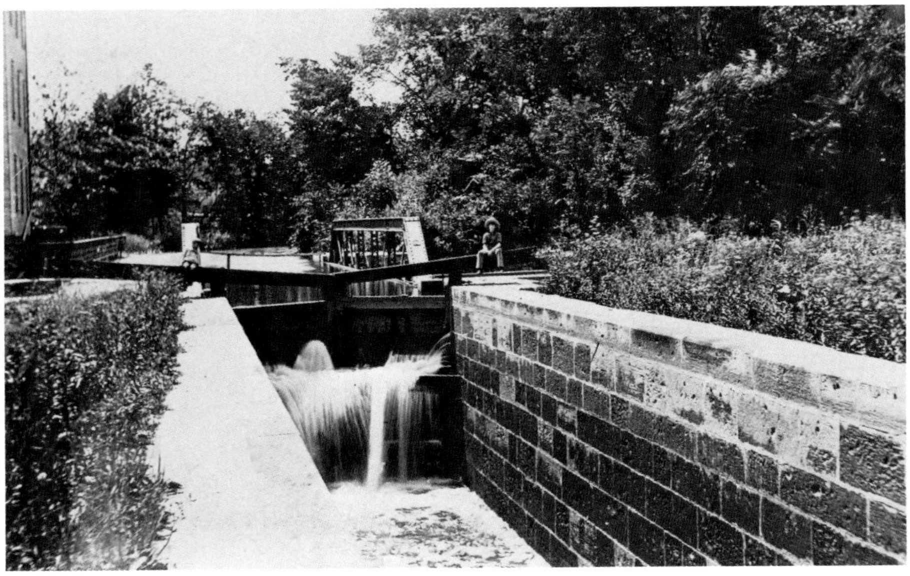

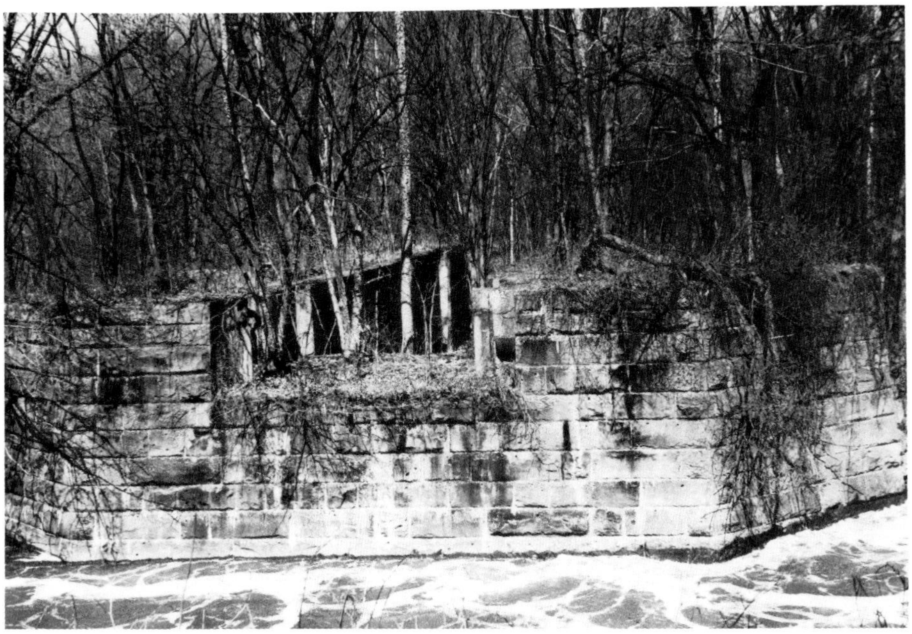

Top, Lock 29 (adjacent to mill at left) which raised boats to aqueduct level, looking south; an aqueduct is visible beyond the lock gate. A Huck Finn-like character lounges on the sweep of the gate at right; his sister sits on the other side. *Bottom,* The same lock, looking north, shows aqueduct abutment at the river's edge in 1967. (CSO)

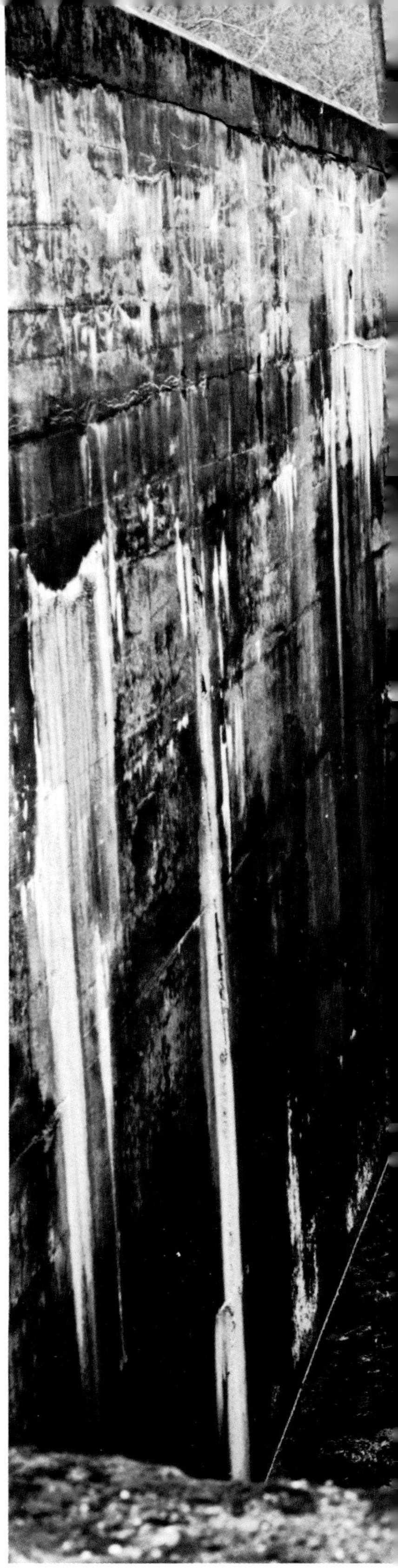

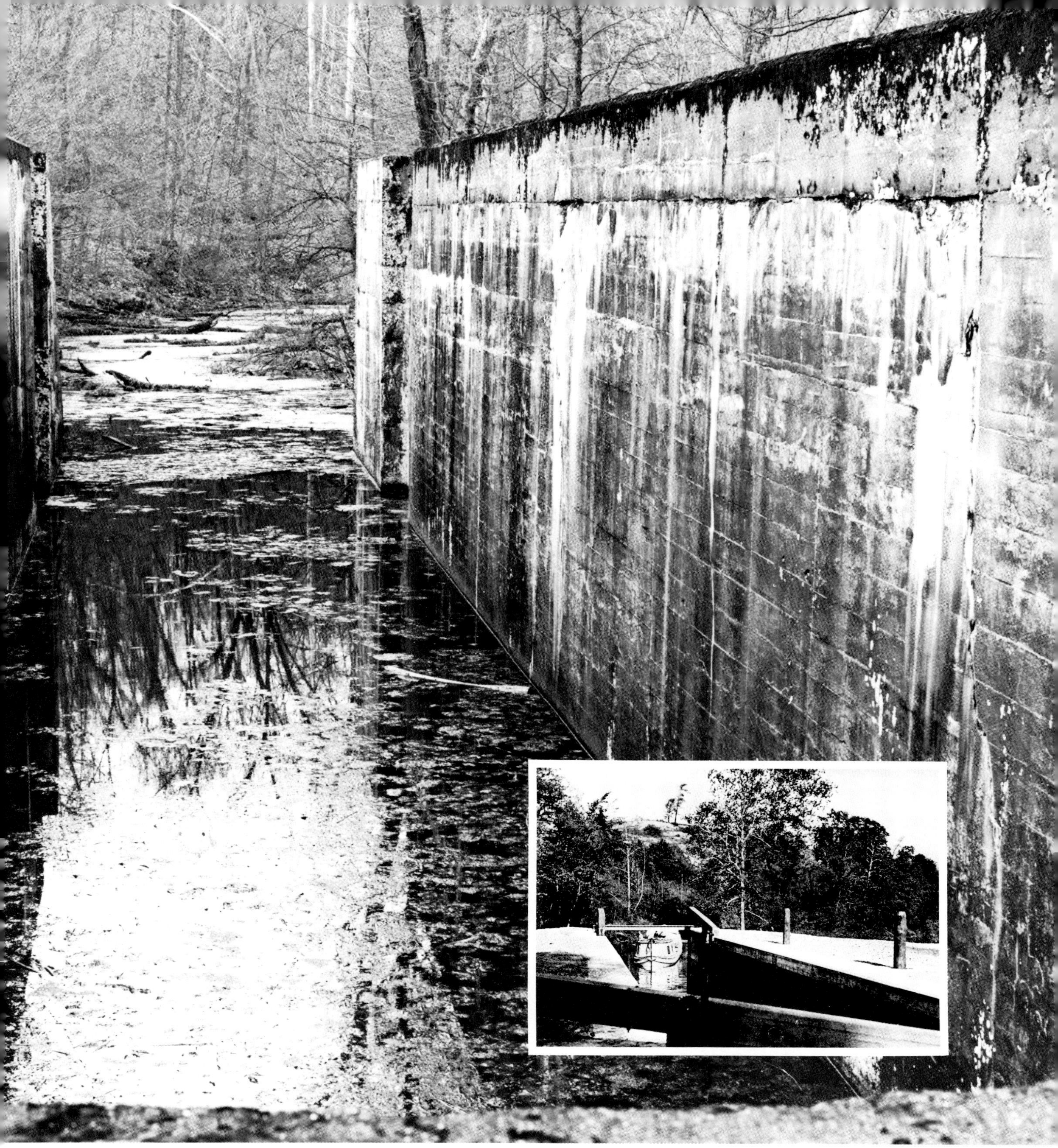

Deep Lock in 1986, now the centerpiece of Deep Lock Quarry Metropolitan Park, is presently maintained by the Akron Metropolitan Park District (although the property lies within the Cuyahoga Valley National Recreation Area). Several of the park's trails also lead to nearby Deep Lock Quarry (Photo by Jack Gieck). *Inset,* A boat enters Lock 28, Deep Lock, south of Peninsula. The lift at Deep Lock was seventeen feet, the deepest lock in the entire system. Mile: 25. (Pockrandt Collection, SCHS)

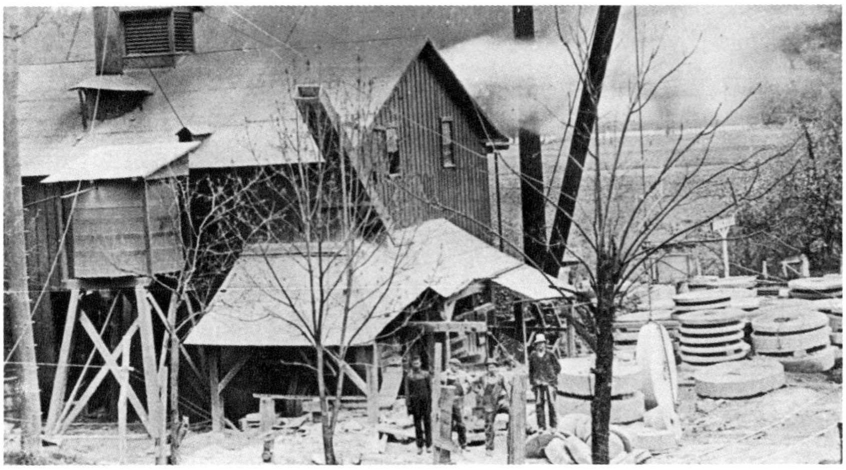

Deep Lock Quarry's main product was millstones for area grist mills; the site also supplied sandstone building blocks for lock walls and building foundations in the Cleveland-Akron area. (SCHS)

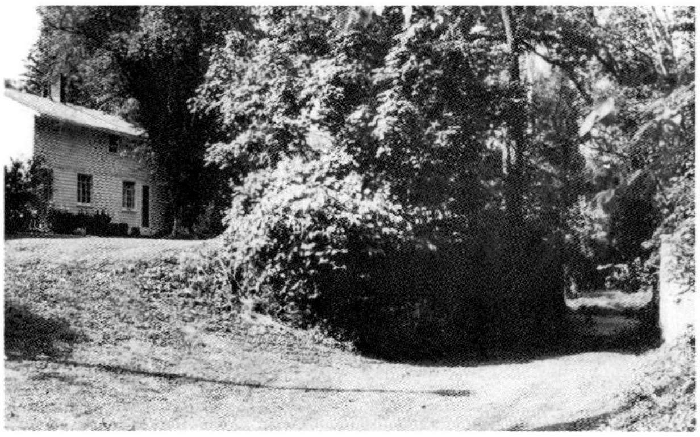

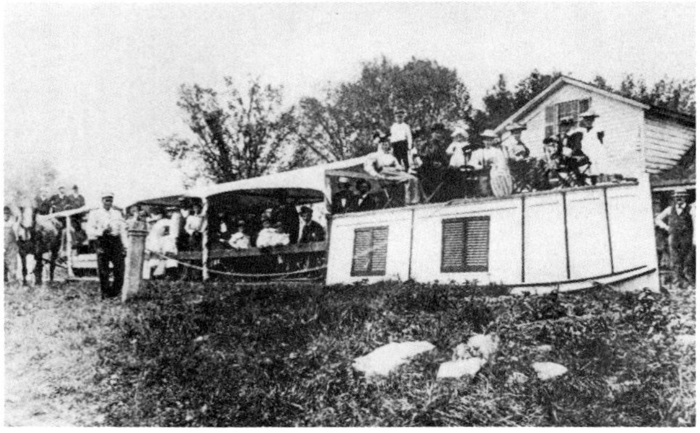

Above left, Johnny Cake Lock, so named because passengers subsisted on fried cornbread for several days here while their boat was stuck aground in mud. The house at left was a general store, an inn, and may have also been a lockmaster's house. (From a color slide by Jack Gieck, 1978.) *Above right,* A Sunday excursion boat tied up at Johnny Cake Lock (mule team at left); the same house appears in the background. Mile: 27 (CSO). *Below,* Captain Charles Hatch and crew on a state boat near Ira in the early 1880s. Mile: 29 (CSO).

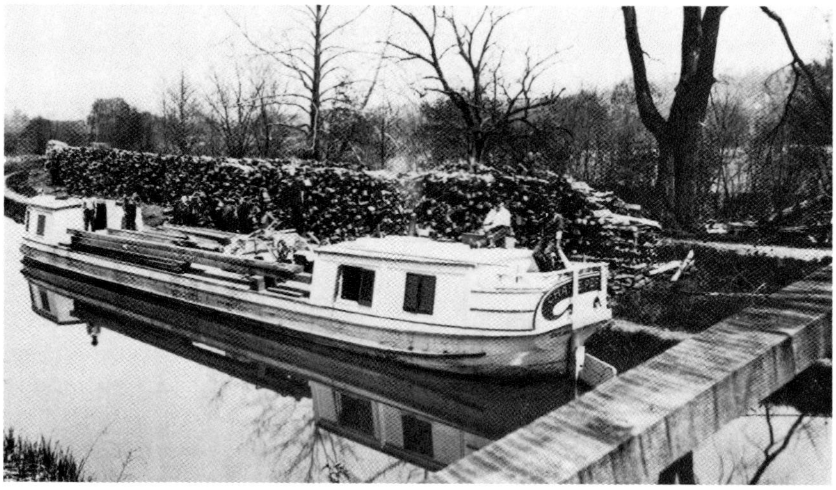

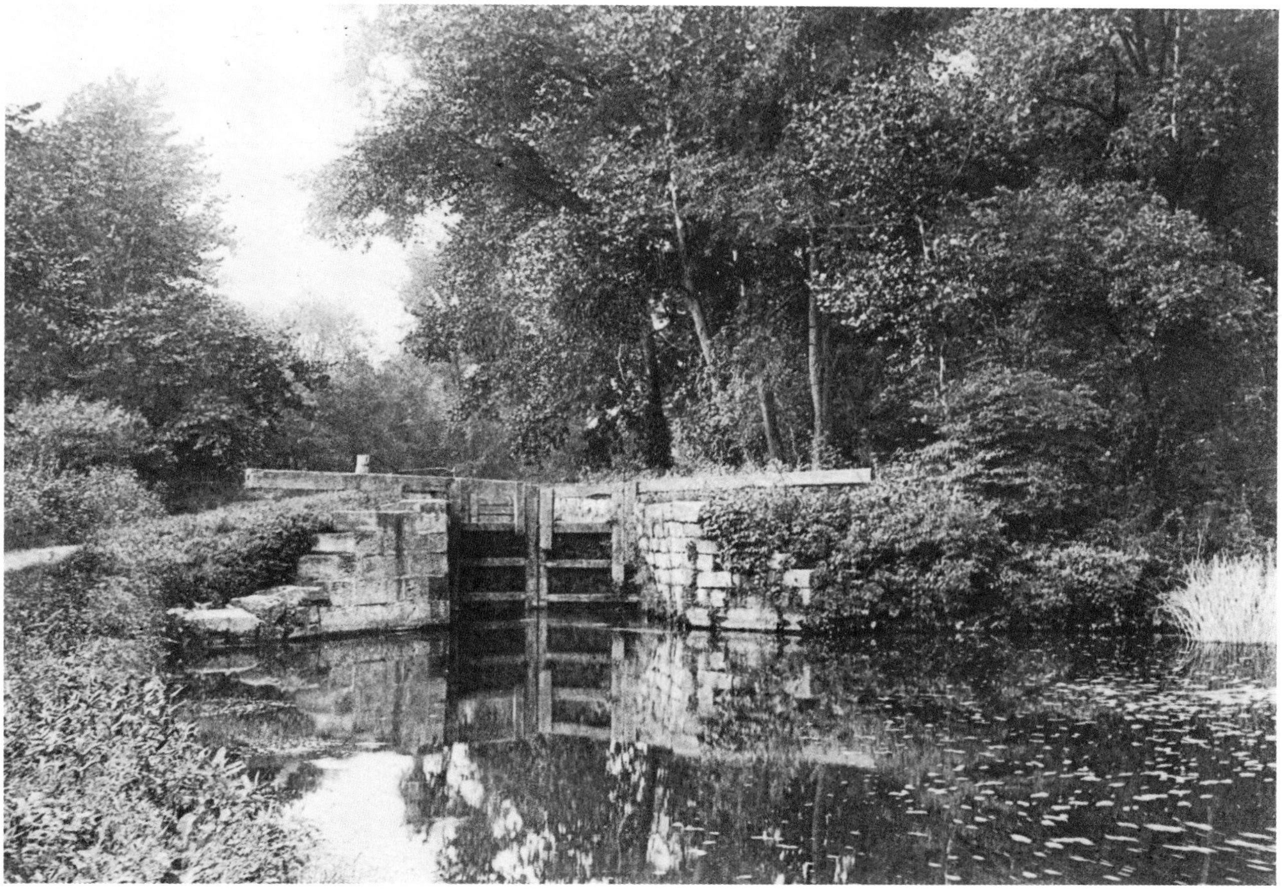

Lock 21, north of Akron, was last in the two-mile chain descending from the Summit. Here, as a young mule-skinner, James A. Garfield is said to have had his first fight. His horses got their lines tangled with those of the animals pulling another boat, and Garfield was thrown into the canal (or so the legend goes). The lock was near today's intersection of Portage Path and Riverview Road. Mile: 35+. (CSO)

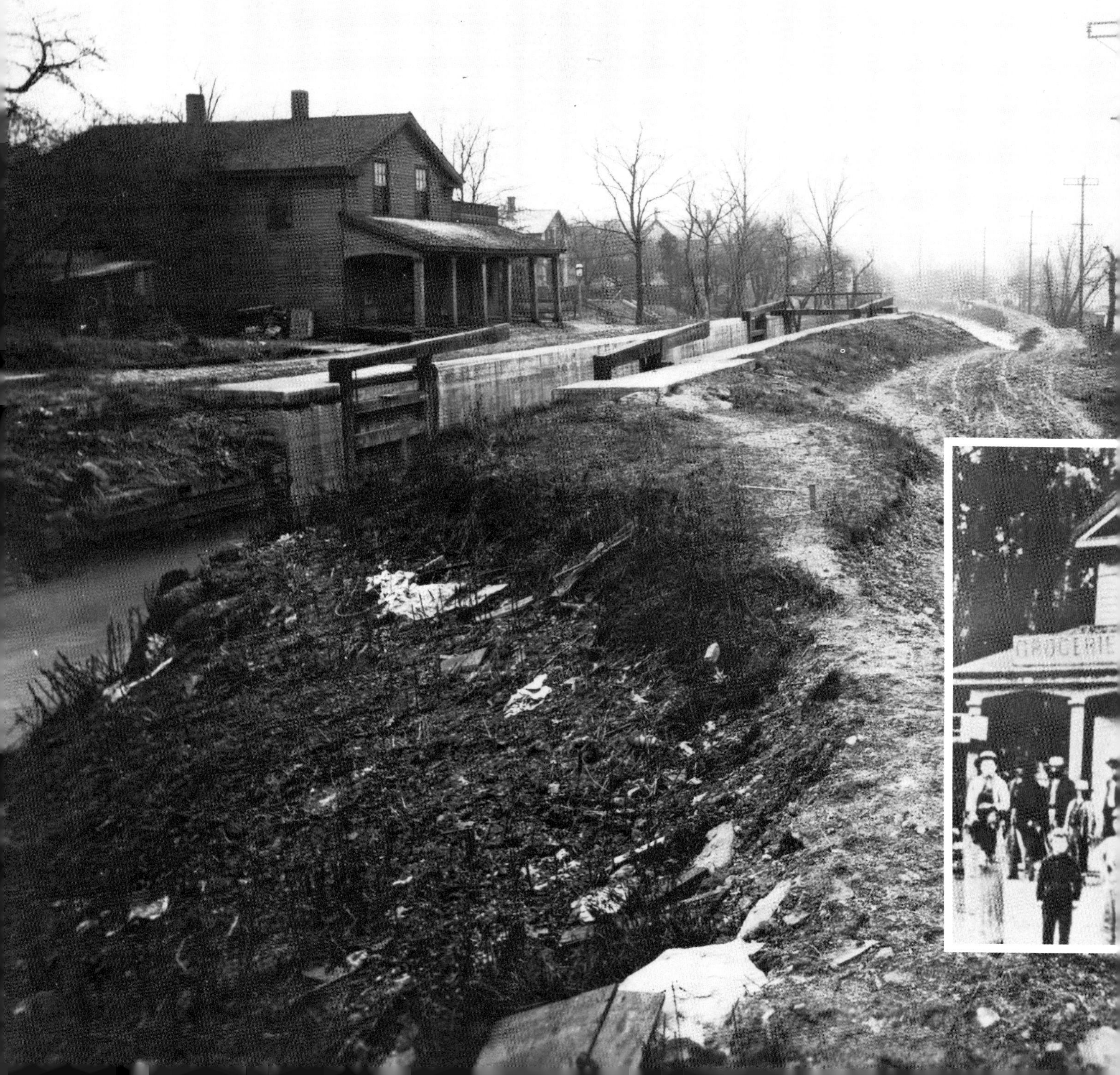

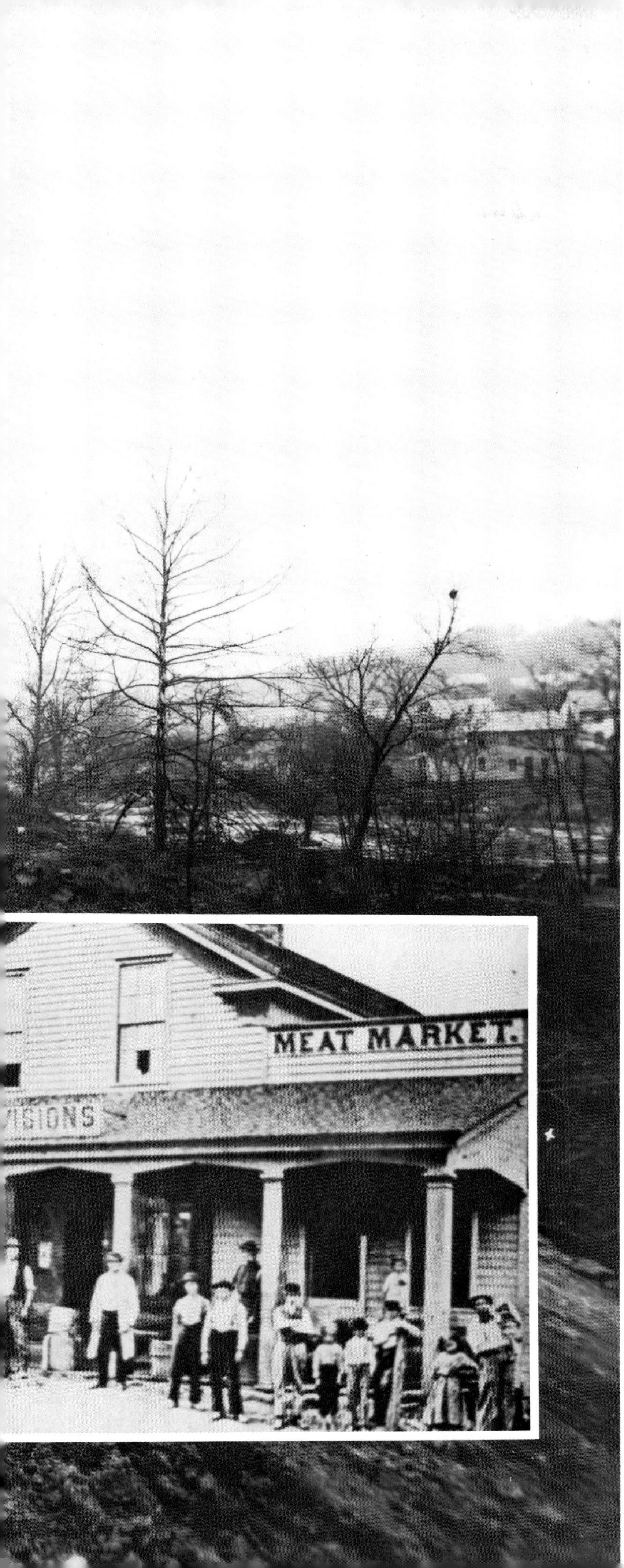

Mustil's store around 1907, after Lock 15 had been rebuilt by the Daley brothers. The building still stands today down a short trail off Akron's North Street (OHS). Mustil's store, shown in inset photo around the 1860s, was a popular, busy canal marketplace. Mile: 36+. (CSO)

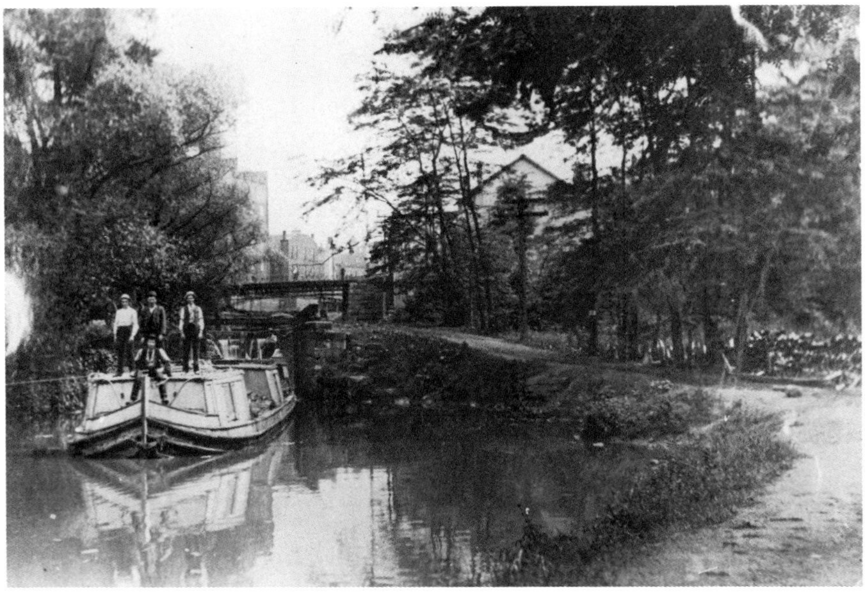

Above, A freight boat leaves an Akron lock near Ash Street, heading north on the canal. Mile: 37. (Pockrandt Collection, SCHS)

Below, The chain of locks below West Market Street in Akron's cascade of locks, before construction of the city's Inner Belt; 1966. Mile: 37. (Photo by F. W. Trevorrow.)

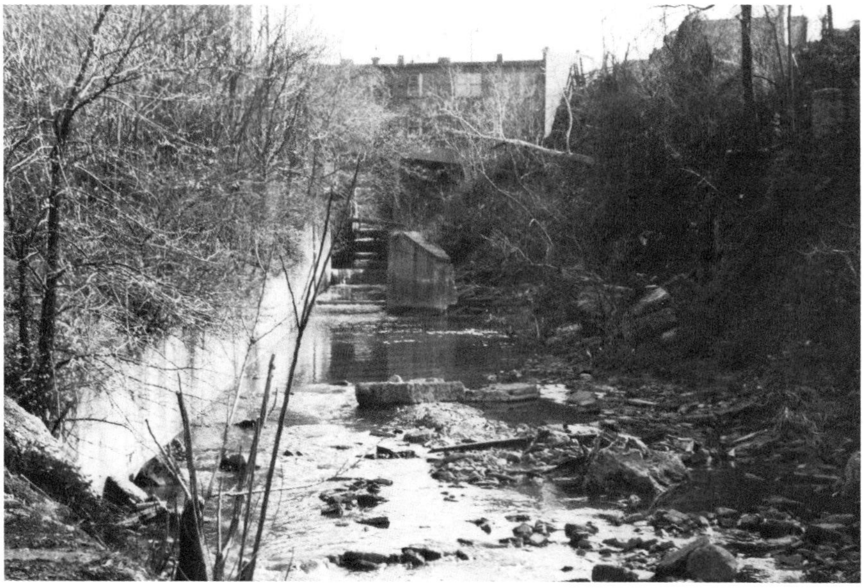

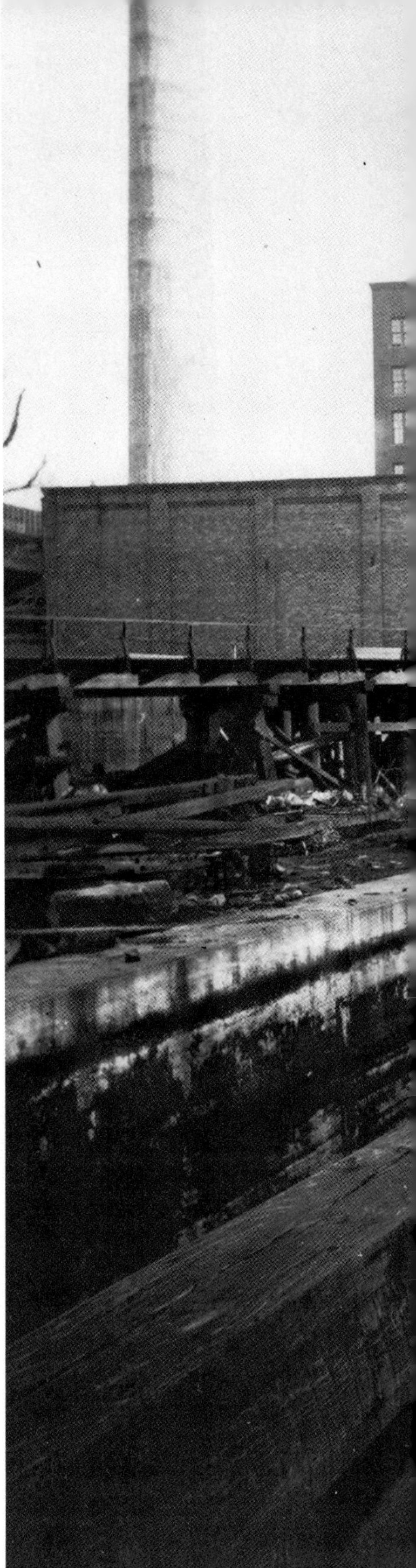

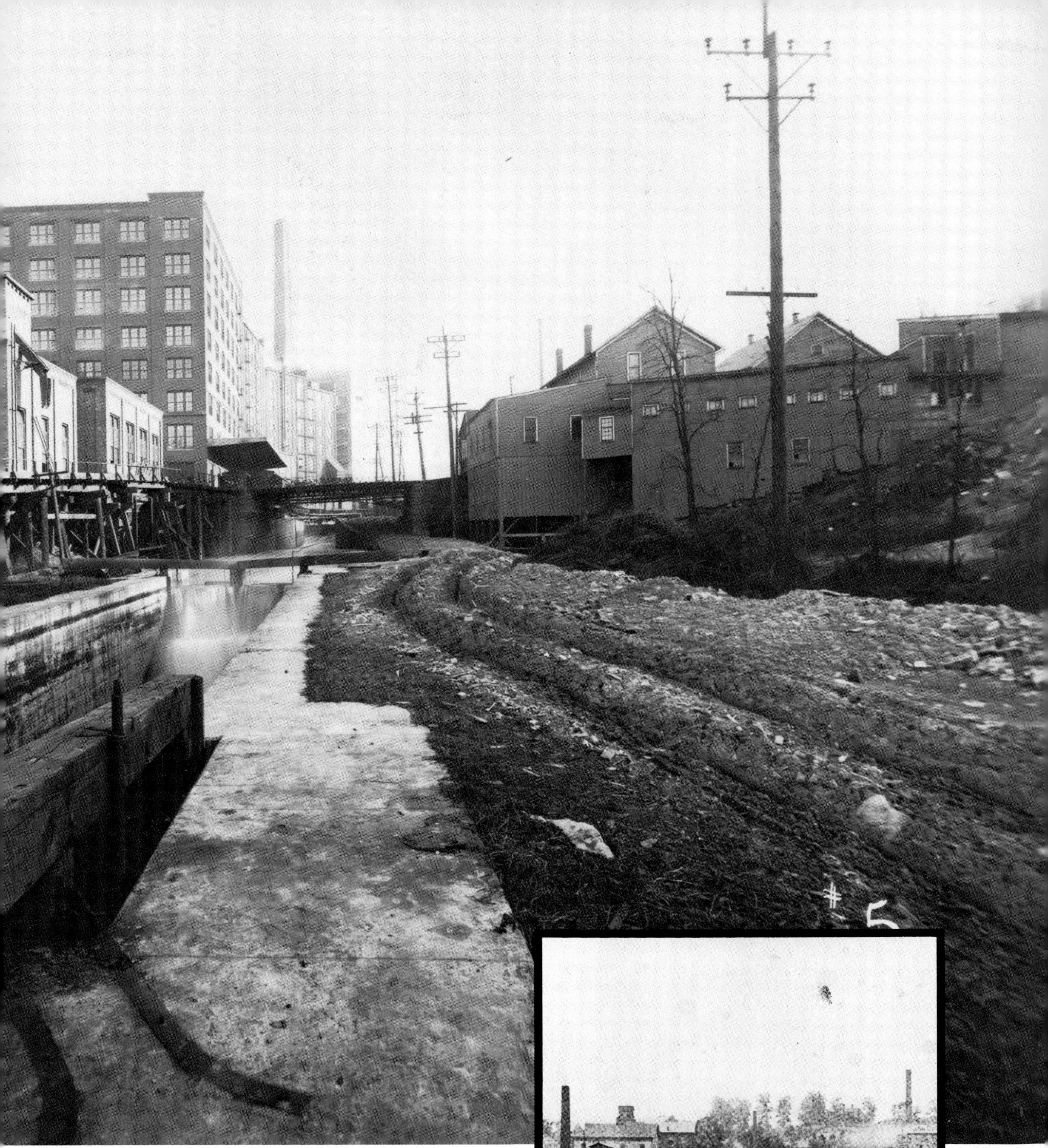

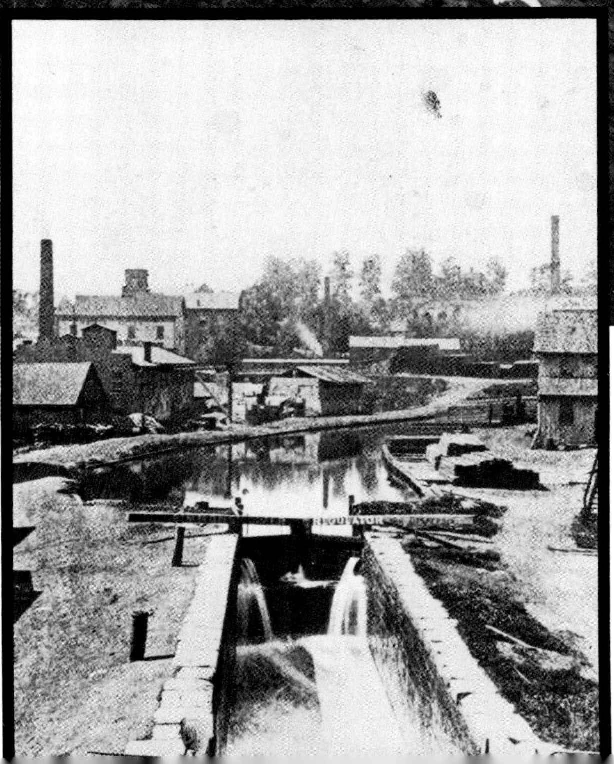

Lock 5, Akron, with Quaker Oats plant in background—legacy of Ferdinand Schumacher's string of nineteenth-century mills which founded the American breakfast cereal industry; c. 1908 (OHS). Inset is a vintage photo of Lock 8 in Akron, flanked by Cascade industries and canal warehouses. Mile: 37+. (CSO)

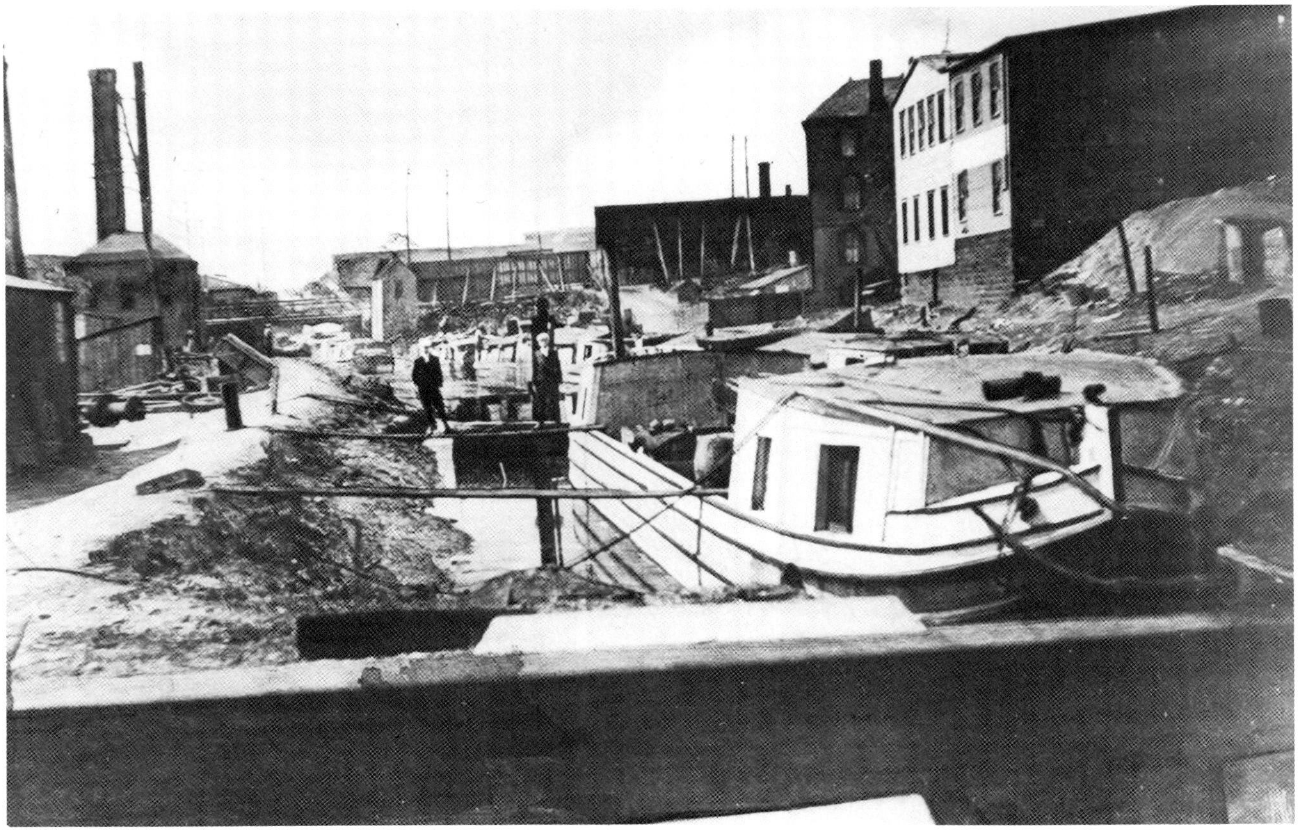

Above, A State Boat at Lock 2 leads a string of canal boats through Akron's chain of locks descending from Portage Summit; 1906 (Pockrandt Collection, SCHS). *Right,* Just over three-quarters of a century later, in 1983, Akron's Canal Park at Lock 2 was dedicated. A ghostly sculpture of a three-cabin freight boat stands on the left bank. (Photo by F. W. Trevorrow.)

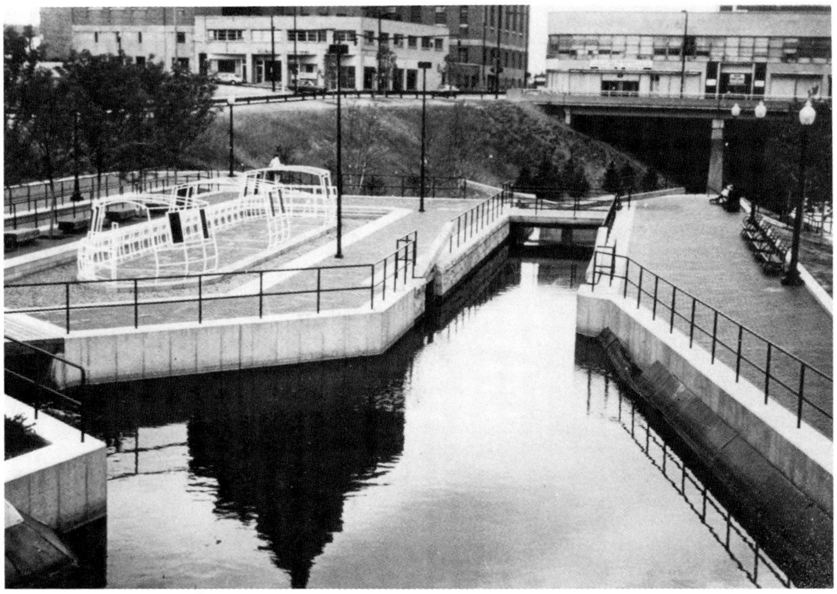

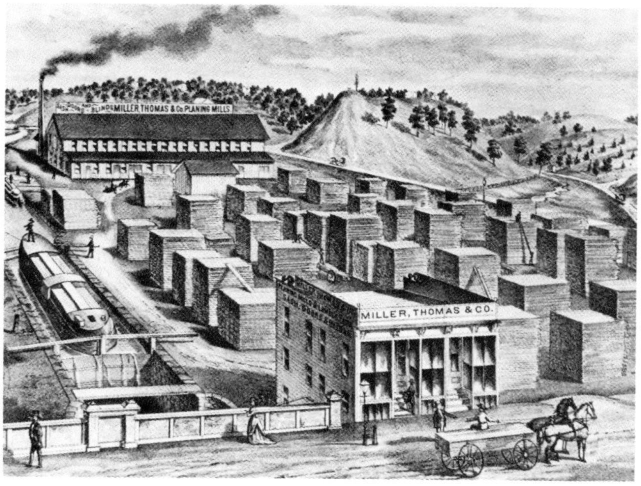

Industries flanking the canal near Akron's Exchange Street are advertised in this engraving from the 1874 Summit County Atlas. Mile: 38. (SCHS)

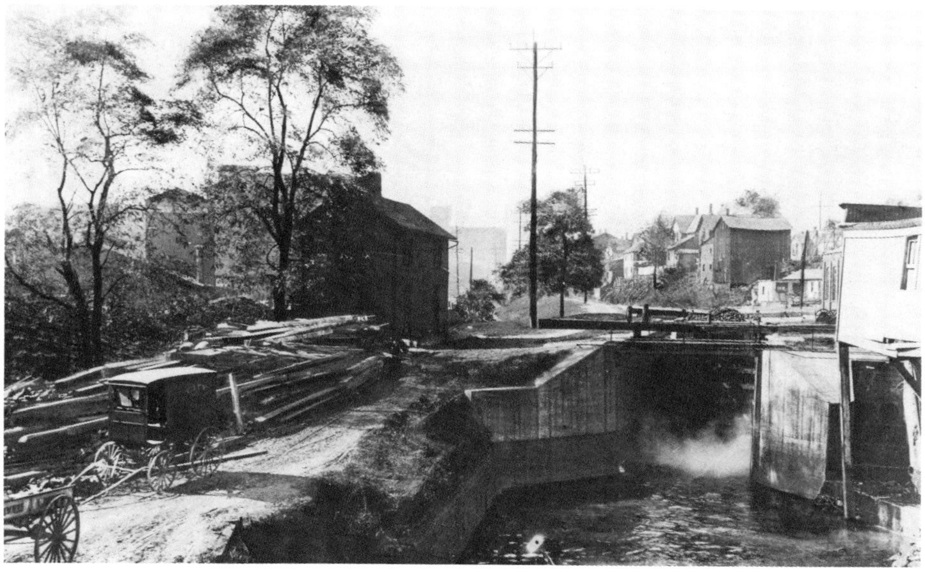

Lock One at Akron's Summit Level and State Canal Office in 1890. Mile: 38. (Pockrandt Collection, SCHS)

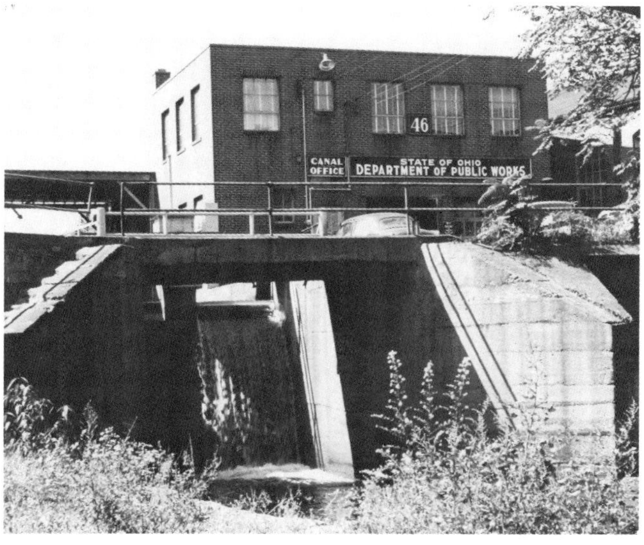

Lock One and Ohio Department of Public Works Canal Office in 1951. Mile: 38. (Photo by Jack Gieck.)

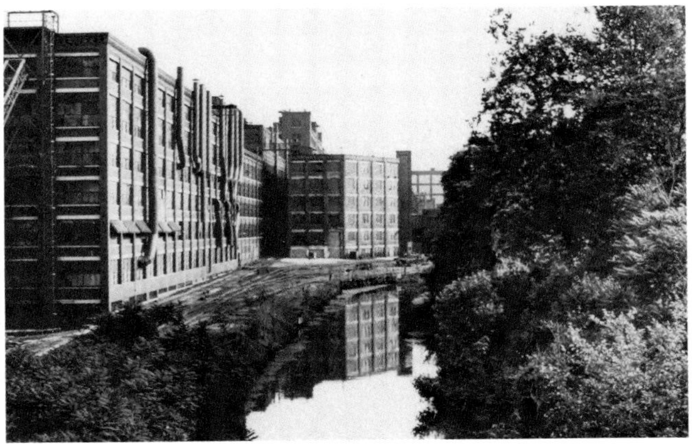

The rubber factories of B. F. Goodrich have lined the east bank of the Ohio & Erie Canal near Lock One since 1870, when Dr. Goodrich moved his firehose business there from New York. Akron's attractions included abundant cooling water (from the canal), inexpensive coal for generating vulcanizing steam, and a skilled labor force. Mile: 38+. (From a color slide by Jack Gieck, 1972.)

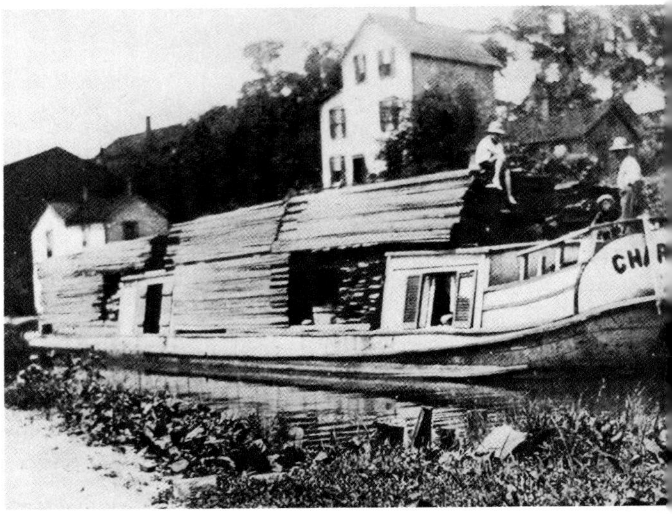

The Charlotta of Akron transports lumber south, along the Summit Level near Lock One; 1880. Mile: 38+. (Pockrandt Collection, SCHS)

"***One of these lakes,*** called Summit Lake, near three-fourths of a mile in length, forms part of the Canal. The towing path across it is constructed partly on floating bridges, made of light timber doweled together so as to form a perfect floor, secured in their proper positions by means of long poles driven into the bottom of the lake and rising above its surface; and partly by throwing up a bank along a projecting part of the shore between the bridges. The depth of the water and the marshy character of the shore and the bottom prevented the formation of a towing path of earth along those parts of the lake where the bridges are used." (Report of the Board of Canal Commissioners, January 22, 1833.)

Dropping the level of the Summit Lake by nine feet drastically reduced its size and uncovered land that had been submerged for thousands of years. Some 230,000 cubic yards of muck—each cubic yard weighing about a ton—had to be loaded into scows and carried away before the canal could be dug. Even then, some of the ground never firmed up sufficiently to build a towpath; thus, the need for the floating wooden footbridge for the mules.

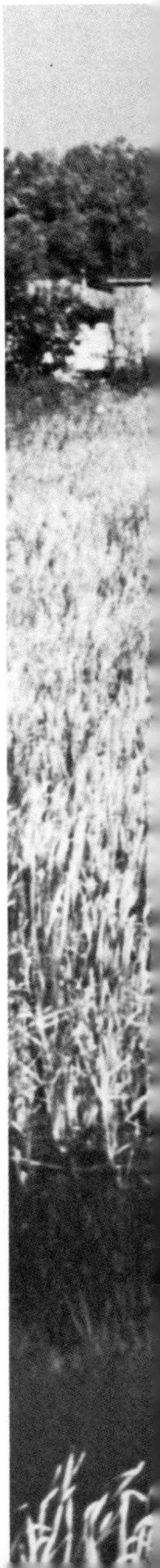

This marshy area in Summit Lake necessitated the floating footbridge; 1967 (from a color slide by Jack Gieck). *Inset,* Here, canal boat draft teams negotiated this floating wooden footbridge to cross part of Summit Lake. As years passed, the bridge timbers became waterlogged and less buoyant. Under the weight of the animals it sank locally to at least hoof deep and undulated as the teams crossed. The surviving handrail in the picture marks it as an early photo. Mile: 40. (SCHS)

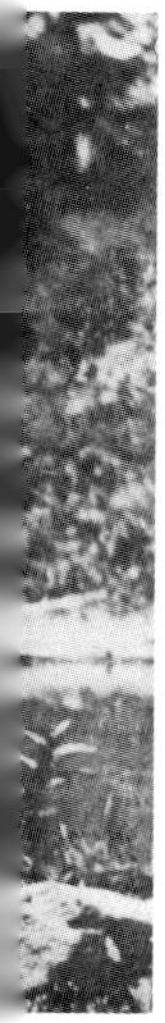

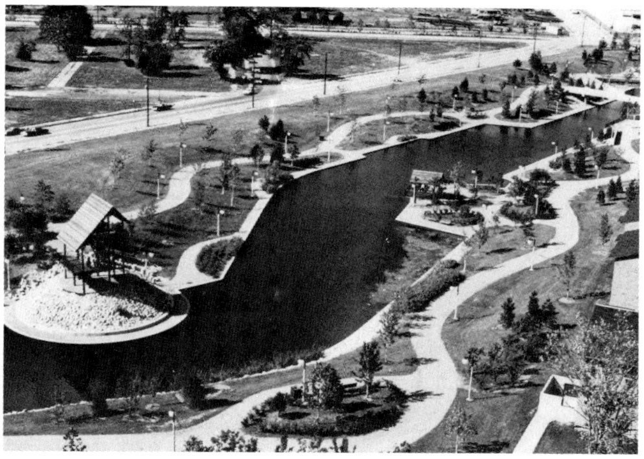

After architectural treatment complementing a public housing project, the canal channel in this area looks very different in Akron's Opportunity Park area in 1978. Mile: 39. (From a color slide by Jack Gieck.)

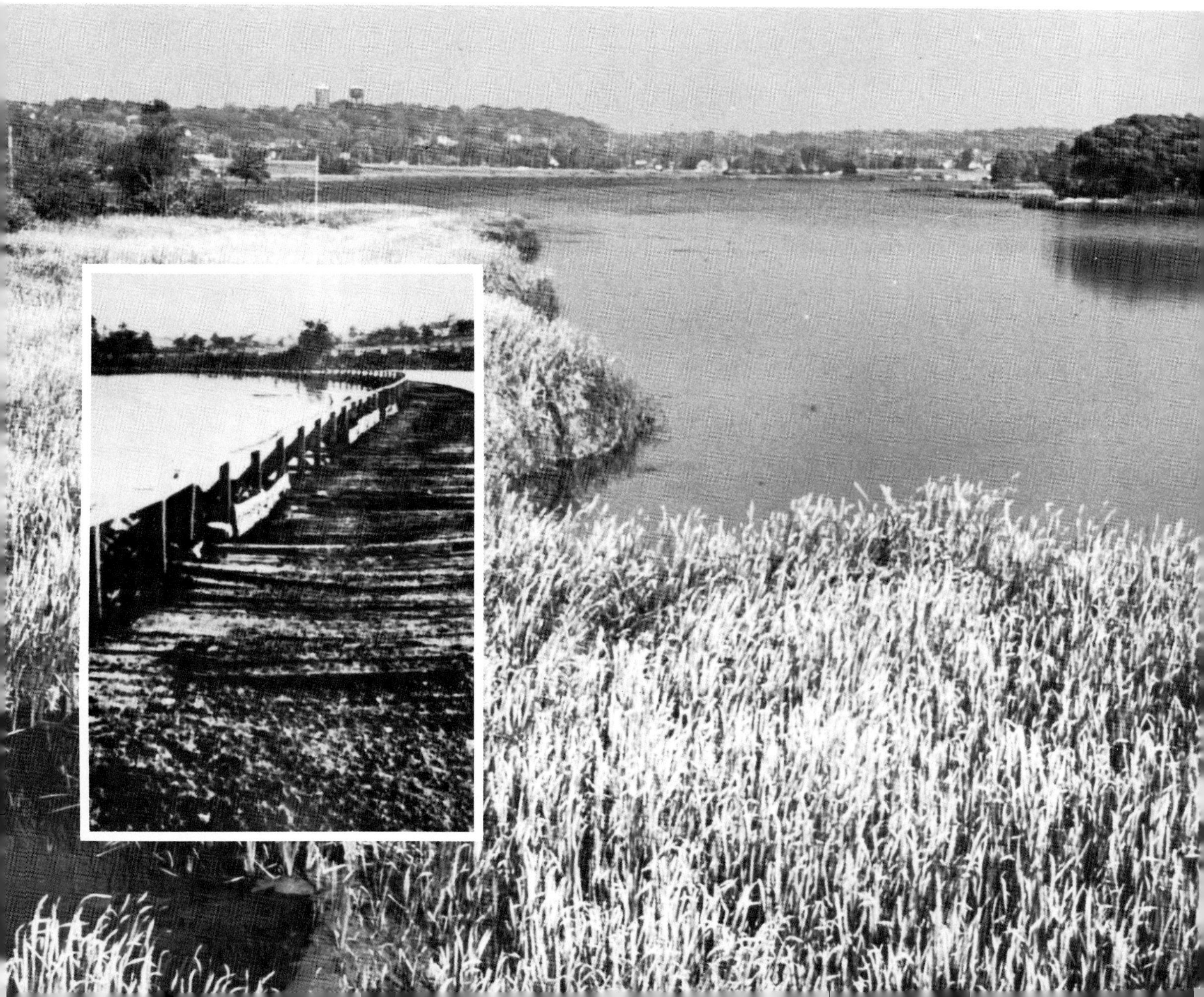

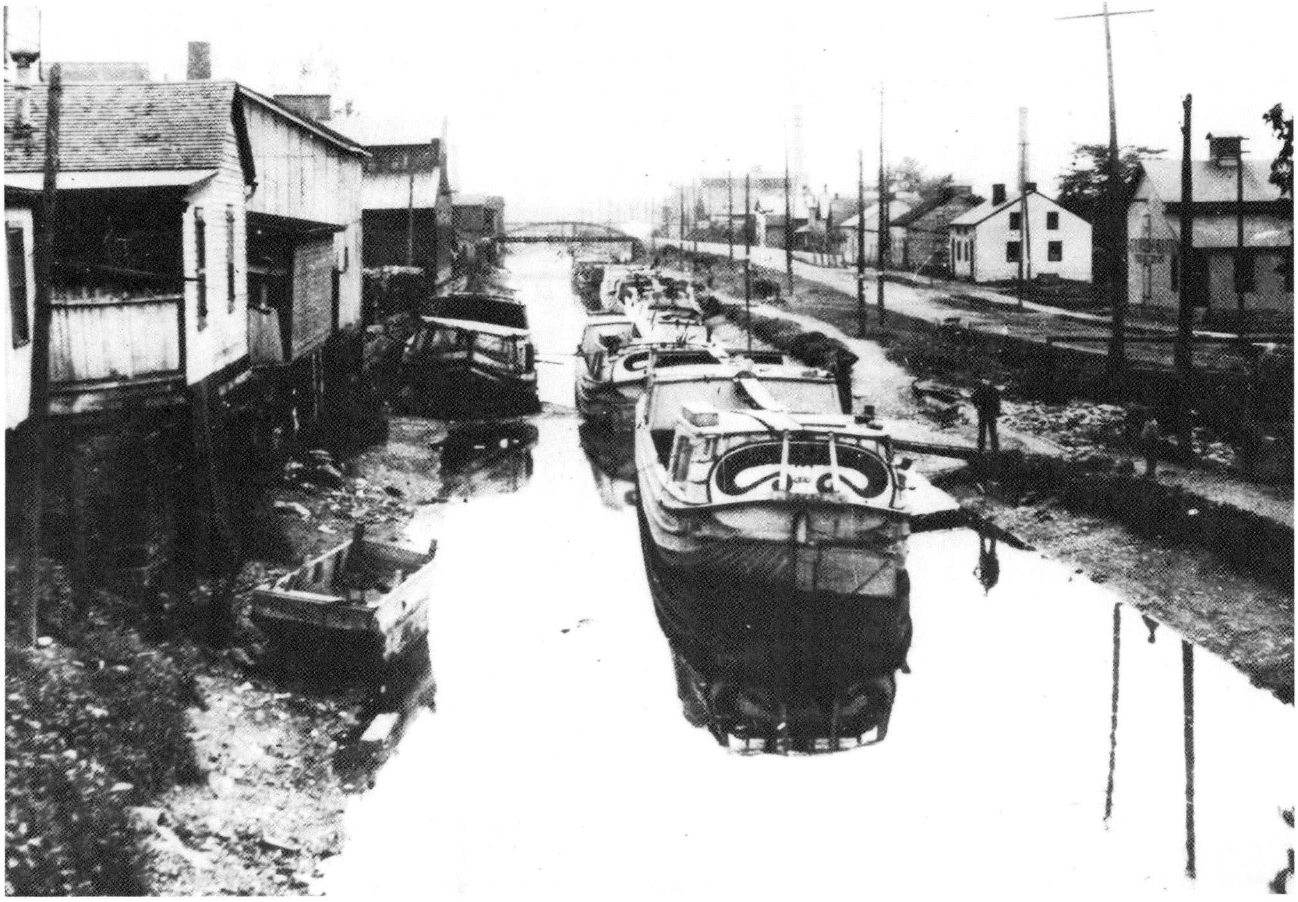

Boats are stranded at Massillon (above) due to a breach in the bank. (CSO)

The intersection of Massillon's Canal and Main Streets is shown (right) in background, with freight boats on the canal. Mile: 65. (Baus Collection, RVF)

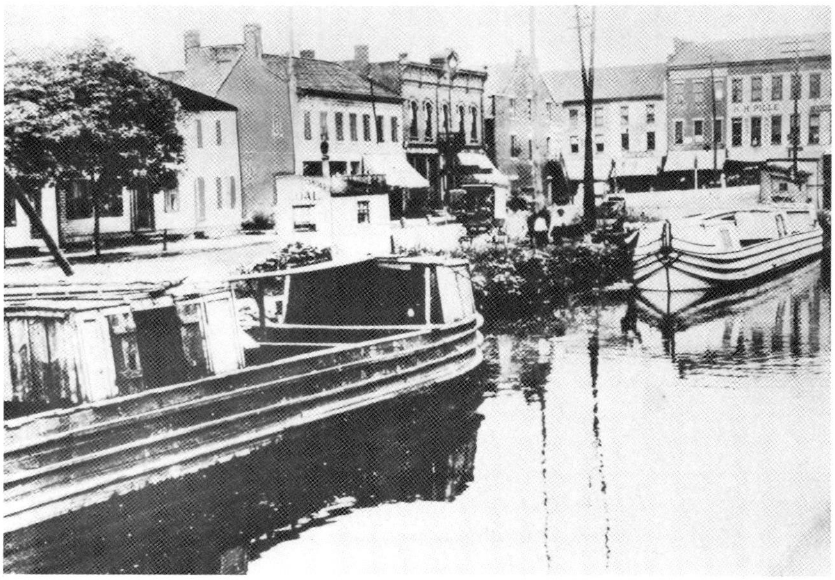

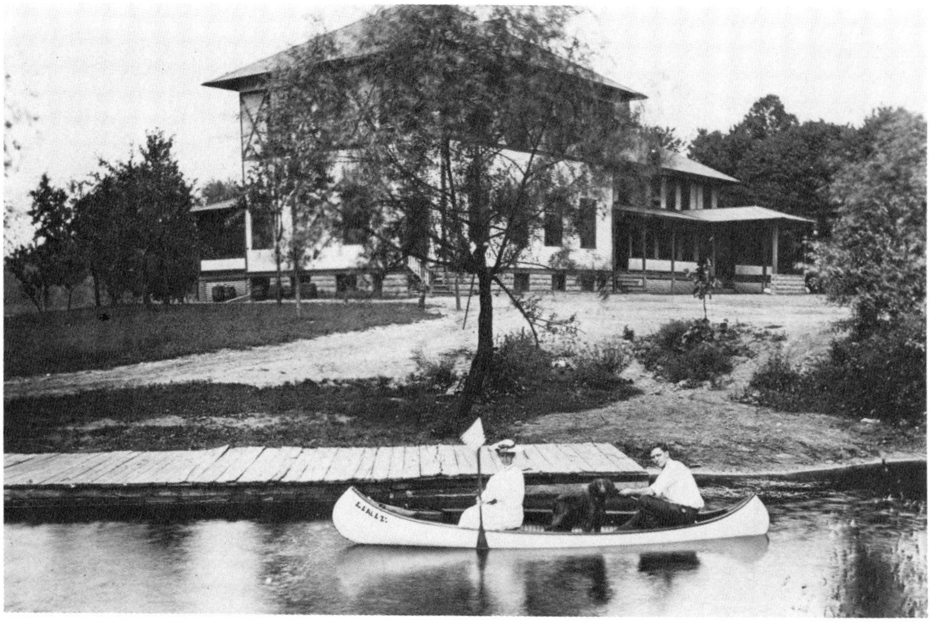

Young's Tavern is a canal hostelry dating back more than a century. It is opposite the Portage Lakes Summit Level feeder; 1910. Mile: 42. (Pockrandt Collection, SCHS)

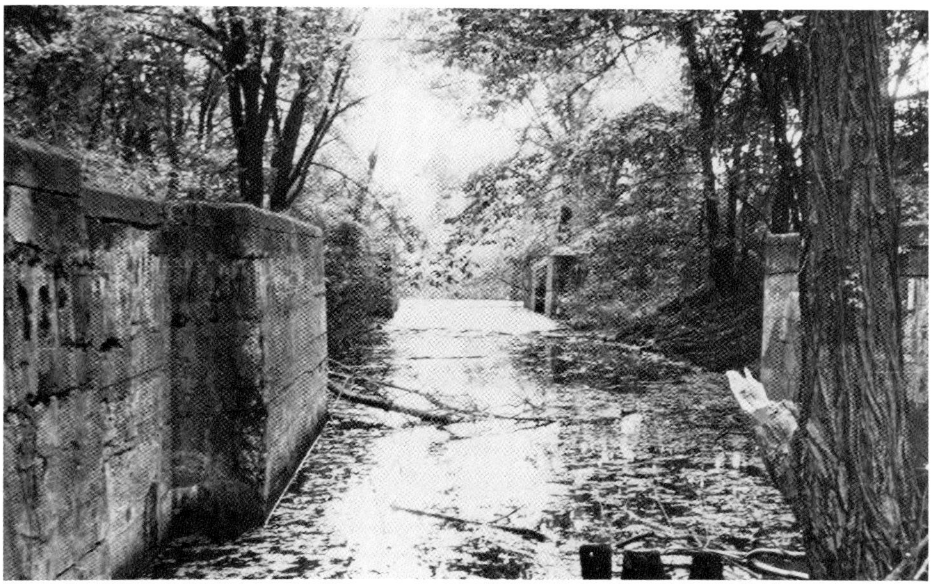

Guard lock at Clinton. Since guard locks were not lift locks, they suffered considerably less wear and often were built of stone only at each end in order to support the gates; "riprap" debris was used in between the stone ends. Mile: 52. (CSO)

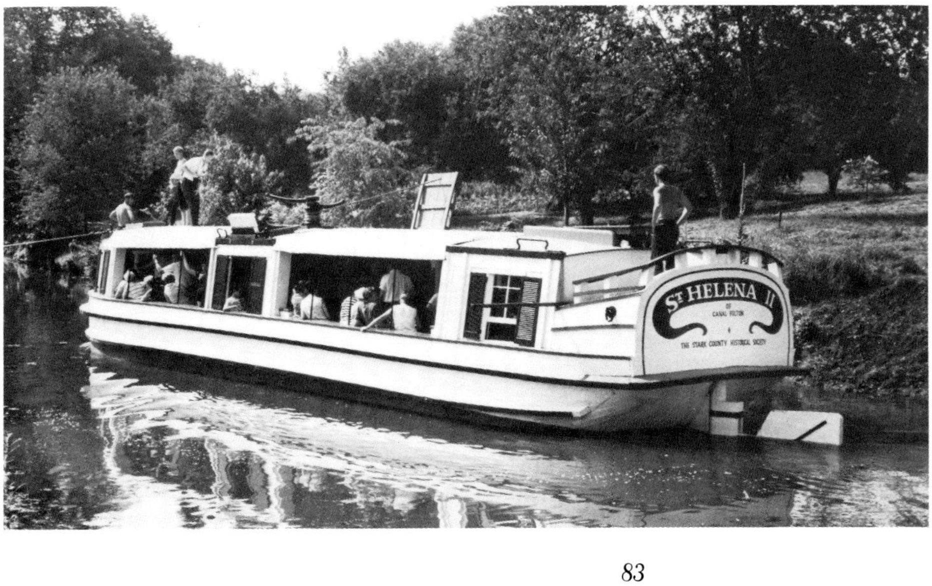

The *St. Helena II* is an authentic freight boat that was painstakingly constructed in 1967 by local Canal Fulton civic groups using nineteenth-century methods. A mule team tows the boat on a two and a half mile excursion for visitors. Mile: 56. (Photo by F. W. Trevorrow)

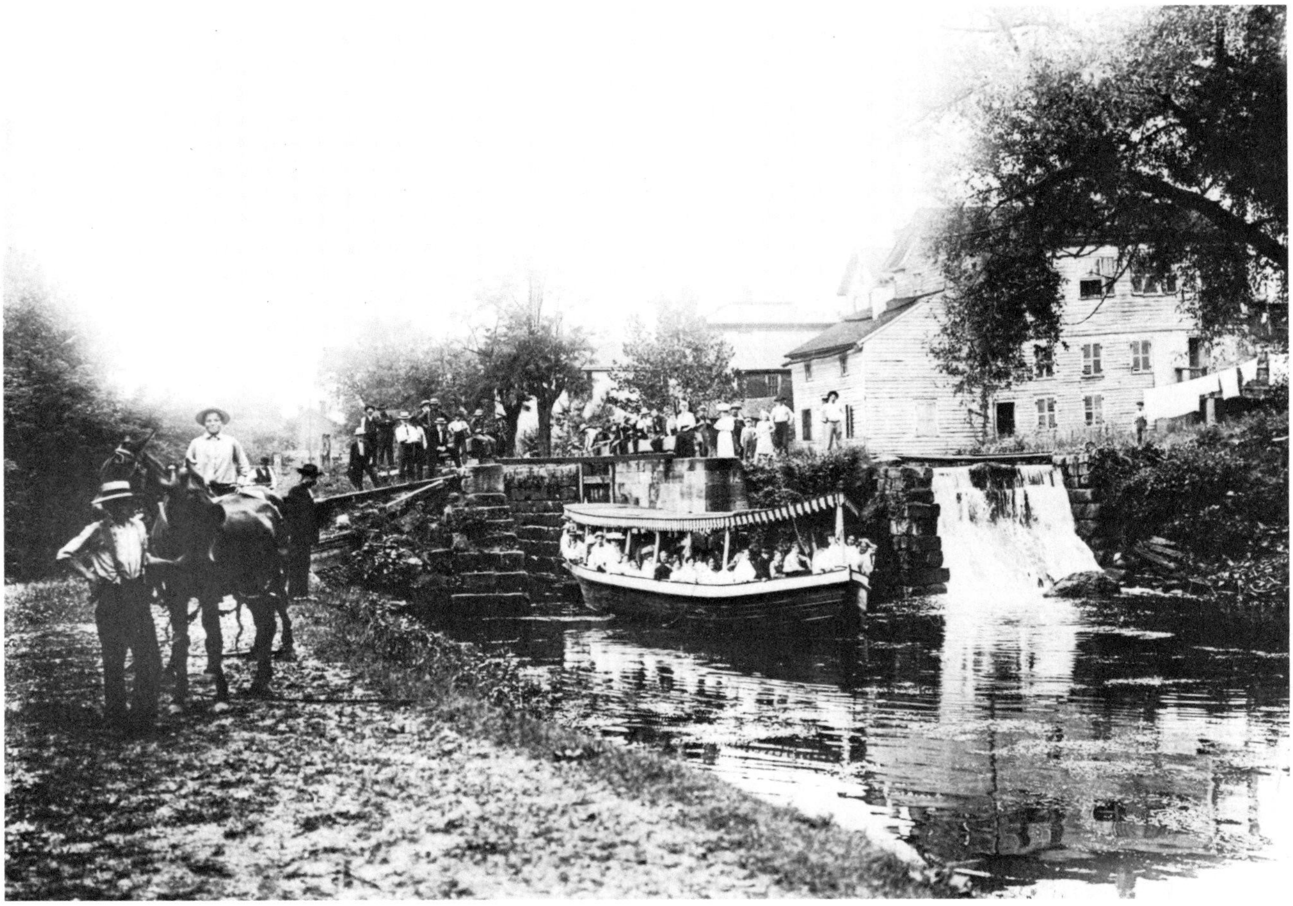

Repairs having been made, a pleasure boat begins a Sunday picnic as it emerges from Navarre's Lock 6 in 1896. Mile: 71. (Pockrandt Collection, SCHS)

The Little Italy hillside overlooks the canal at Navarre. The community was originally settled by canal workers. Mile: 71 (Pockrandt Collection, SCHS)

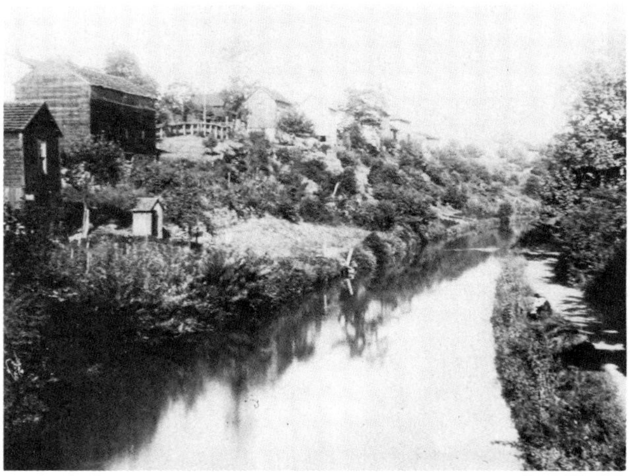

Photographed from Italy Hill, three coal boats line up to negotiate Lock 6. The canal elevation is a few feet above the Tuscarawas River in the background. (CSO)

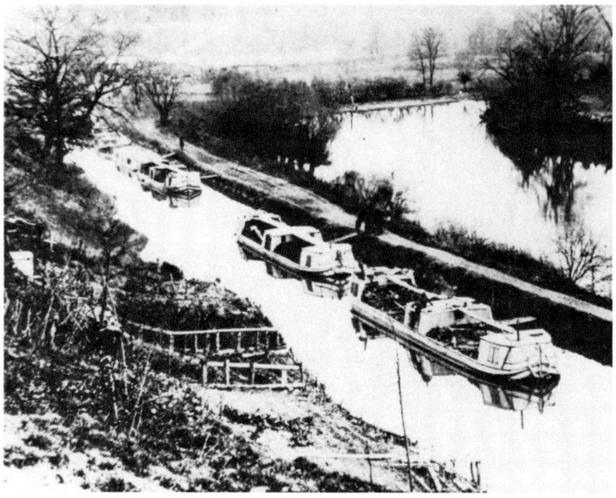

Right, Ohio's most recent ice age planed down the surface of the land to make coal veins readily accessable to shallow strip mining operations, like this one near the Ohio & Erie at Navarre; 1951. (From a color slide by Jack Gieck.)

Below, The E. L. Bennett Warehouse was constructed to load canal boats directly, an operation that will obviously be delayed until repairs are made, as these Navarre residents contemplate the situation; 1896. (Pockrandt Collection, SCHS)

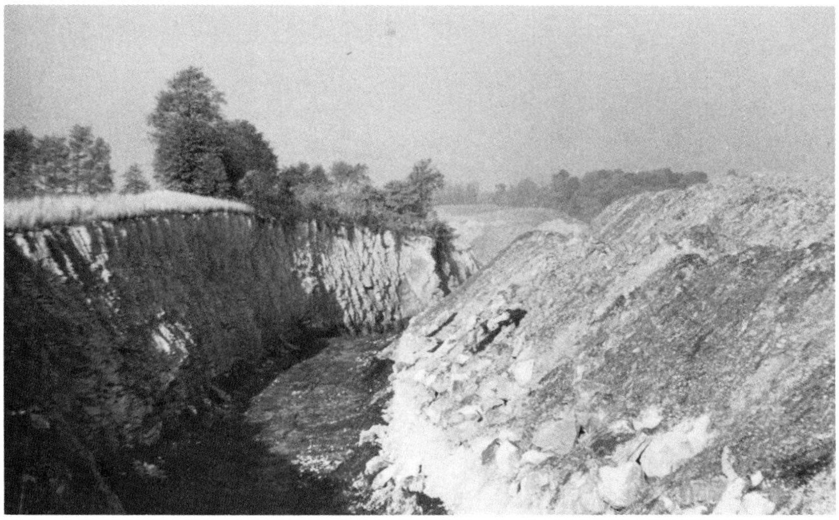

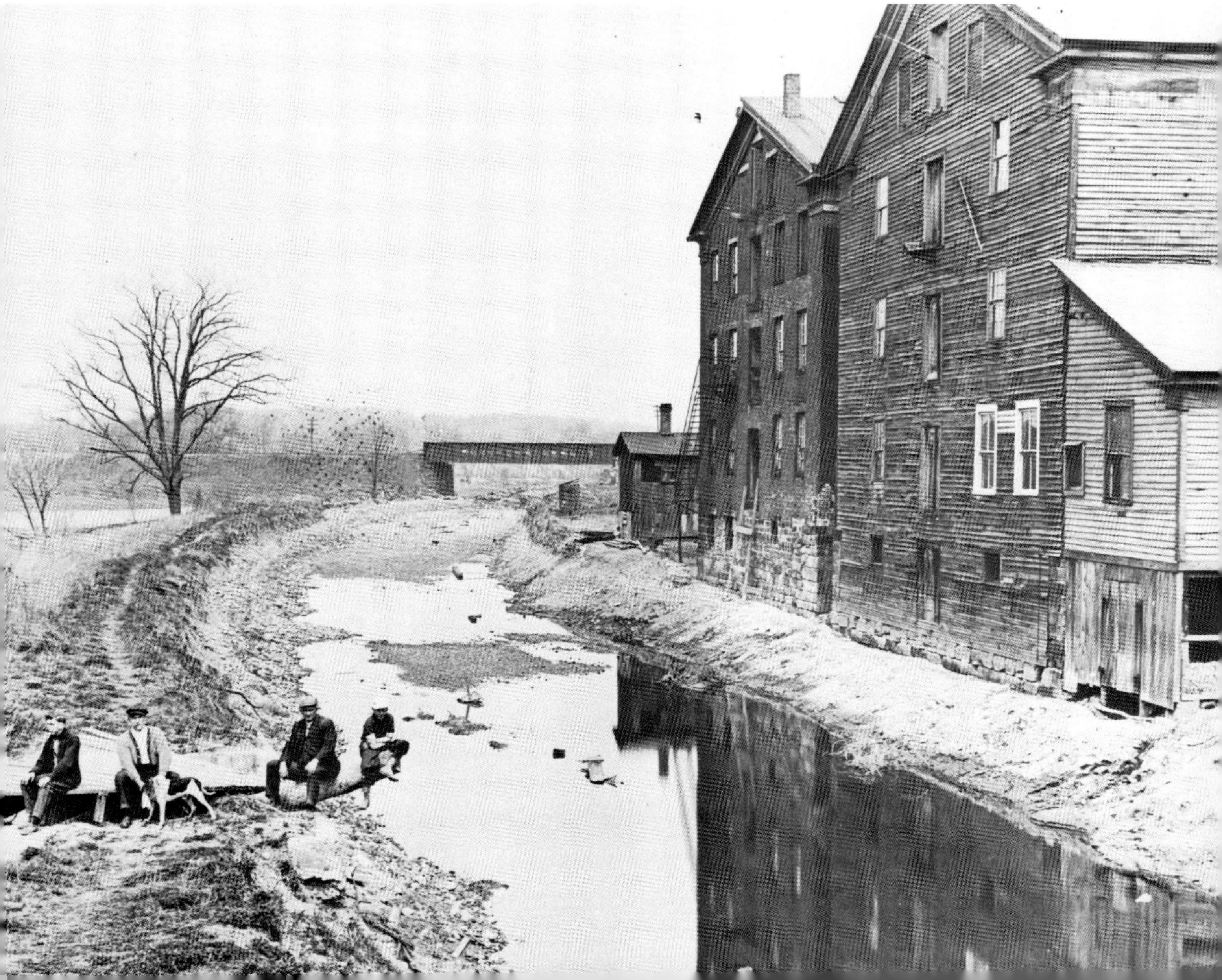

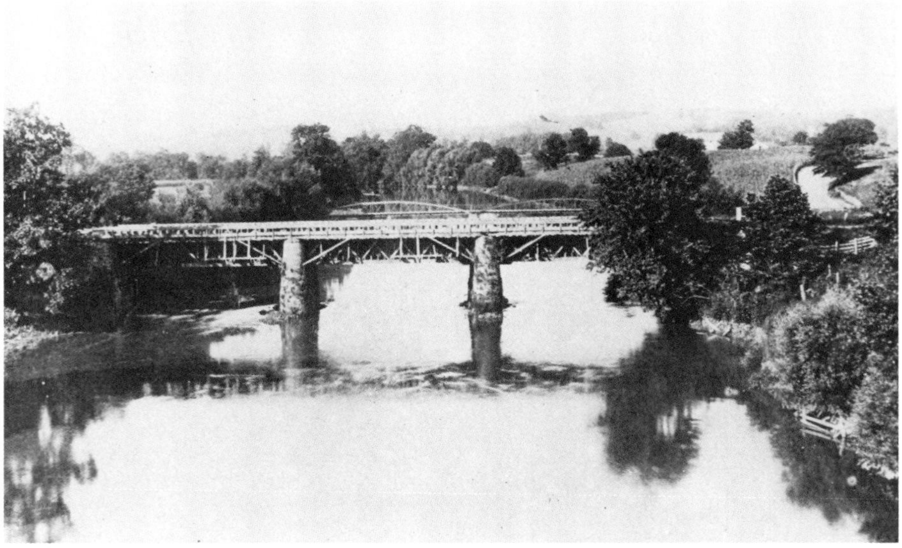

Above, The Sandy & Beaver Aqueduct crosses the Tuscarawas River before joining the Ohio & Erie Canal at Bolivar (Pockrandt Collection, SCHS). *Below,* This 1883 photo shows the aqueduct in a state of imminent collapse. Mile: 80 (CSO). *Bottom,* Only the two stone piers stand in the Tuscarawas River in 1951 to mark the crossing of the Bolivar Aqueduct; stone abutments are also visible on the banks. Ironically, a steam locomotive is crossing the parallel trestle. Mile: 80. (From a color slide by Jack Gieck.)

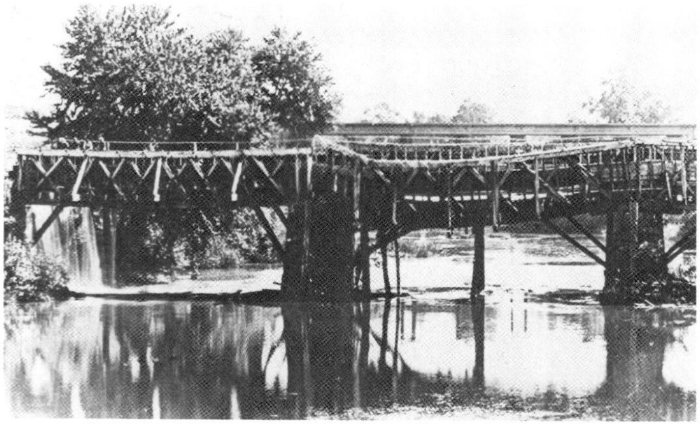

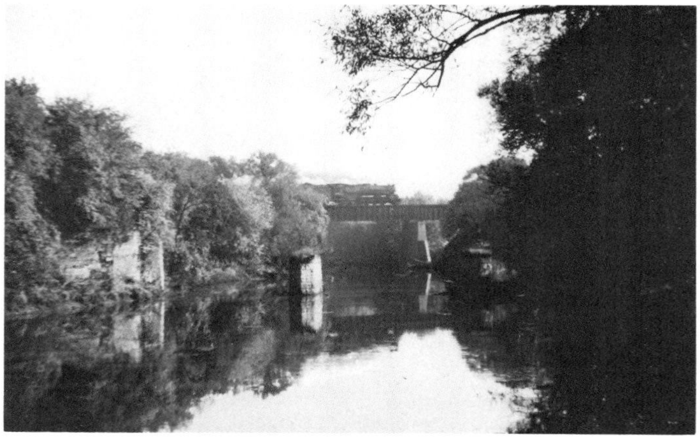

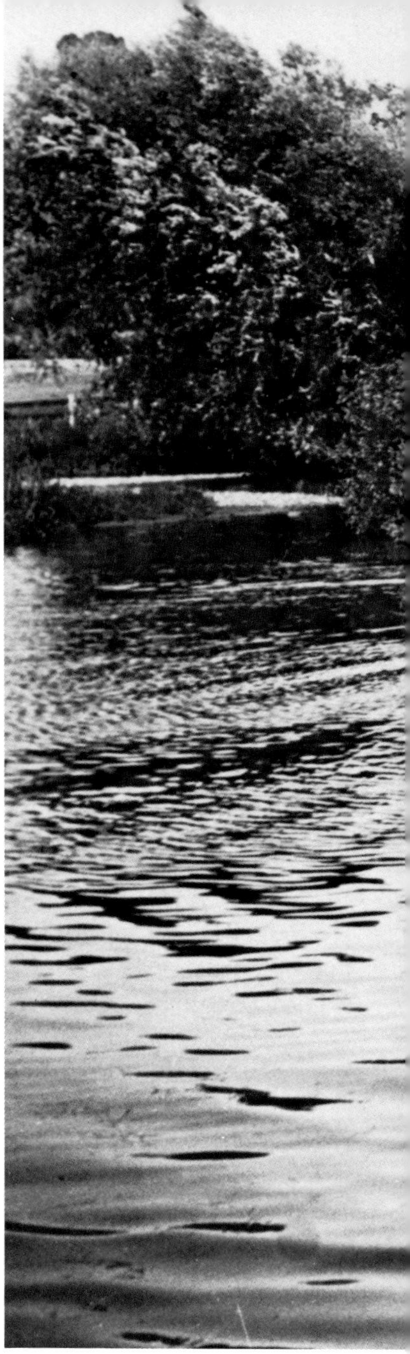

Above, Local boys enjoy Navarre's swimming hole on the canal in July 1896. Mile: 71+. *Inset,* A group poses on a passenger packet with shutters and ornate window treatment at the Zoar Feeder. Mile: 83. (Pockrandt Collection, SCHS)

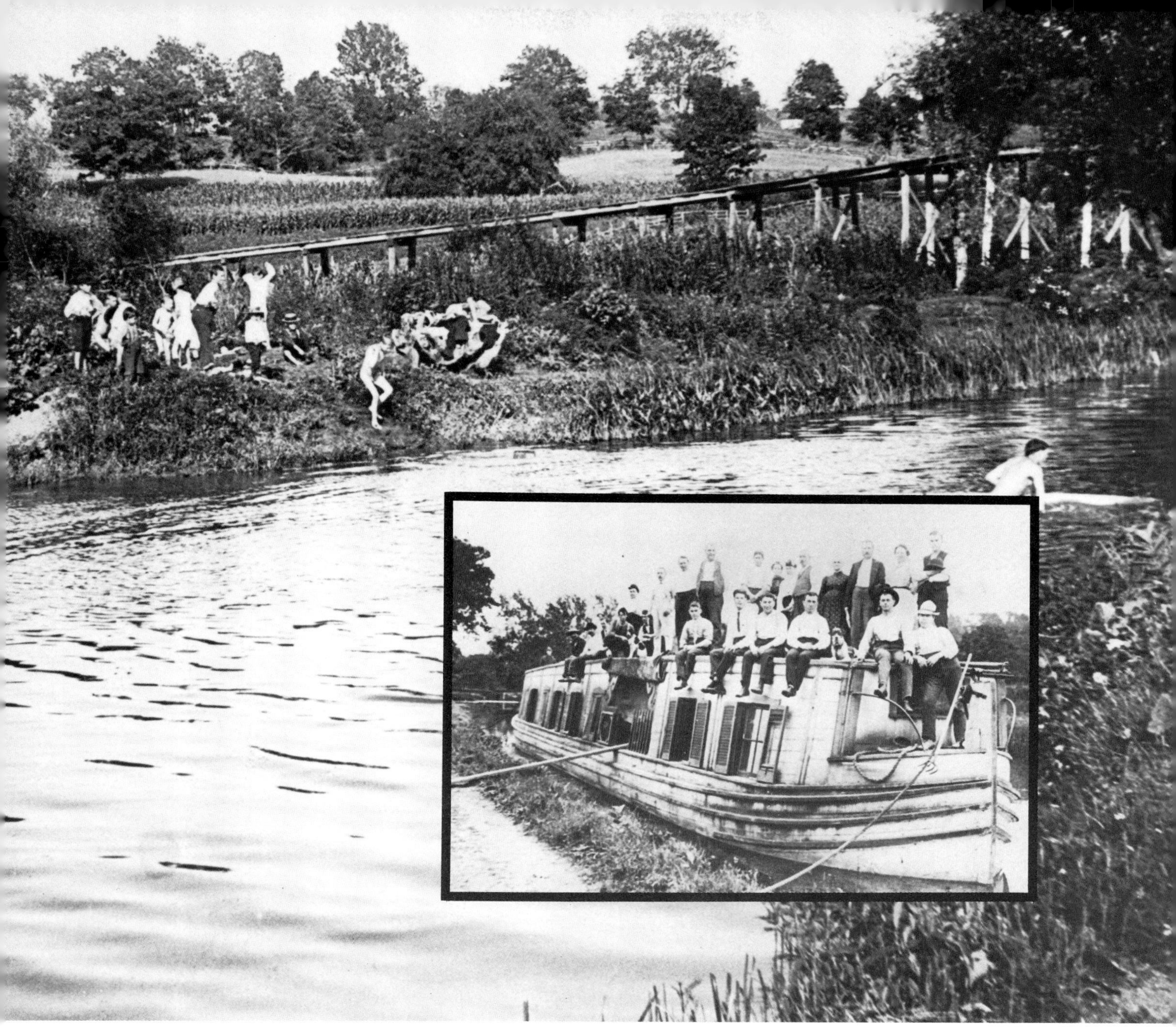

Some travelers kept a record of their expenses for later reimbursement. For the following expense sheet we are indebted to the Reverend B. W. Chidlaw, who made a trip from New York City to Cincinnati in 1836, taking about a month to arrive at his destination. It may be assumed that lodging and meals are included in the fares charged for boat travel:

LEG	*TRANSPORTATION*	*COST*	*MILES TRAVELED*
New York-Albany	Steamboat	$2.00	160
Albany-Utica	Stage/Canal	1.50	110
Utica-Buffalo	Canal	3.75	254
Buffalo-Cleveland	Steamboat	2.50	193
Cleveland-Portsmouth	Canal	4.00	309
Portsmouth-Cincinnati	Steamboat	1.00	100

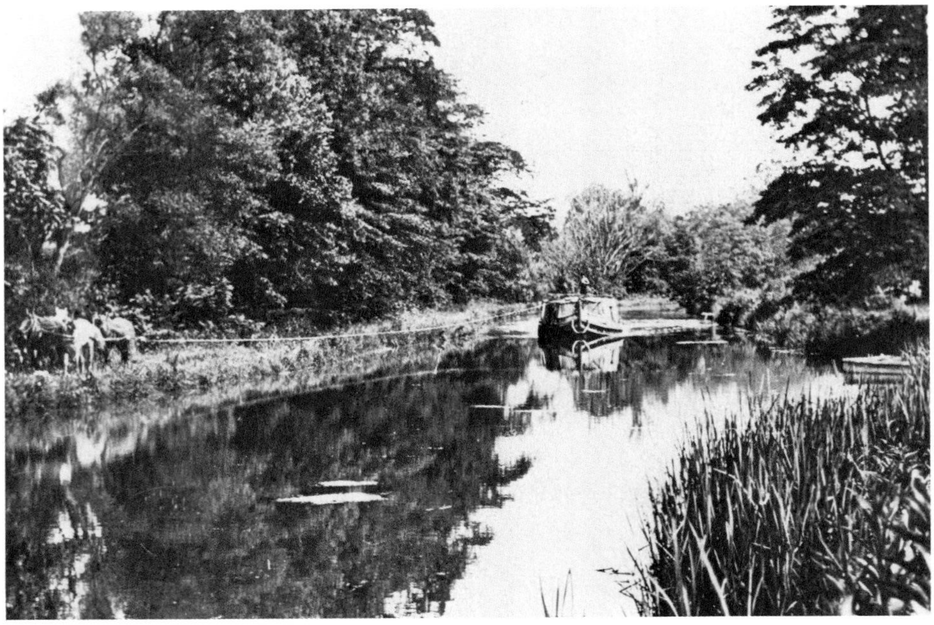

Above, A timeless canal scene above Canal Dover with a heavily laden freight boat towed by three mules who paused in a tightly knit group while the photograph was made. (Baus Collection, RVF)

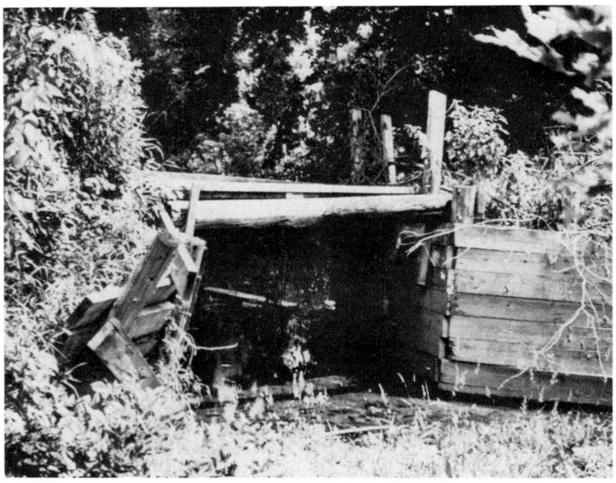

A towpath footbridge over a stream feeding the canal near Locks 8, 9, and 10, in the deep woods north of Zoar. Mile: 82. (From a color slide by Jack Gieck.)

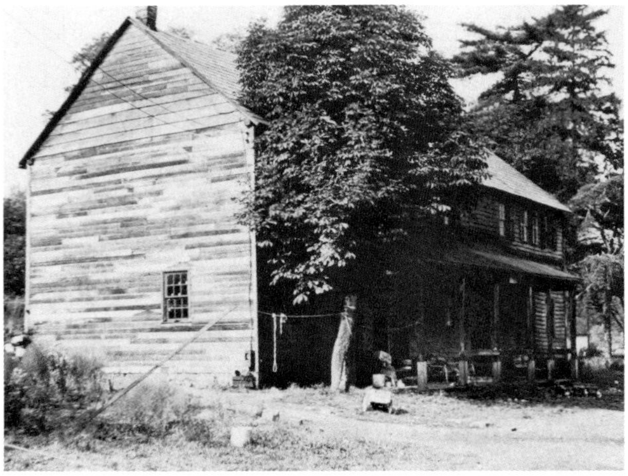

An old canal tavern south of Zoar. The fashionable clothes and freedom which canal passengers displayed here contributed to the dissolution of Zoar's communal society of German Separatists. Mile: 84. (From a color slide by Jack Gieck, 1951.)

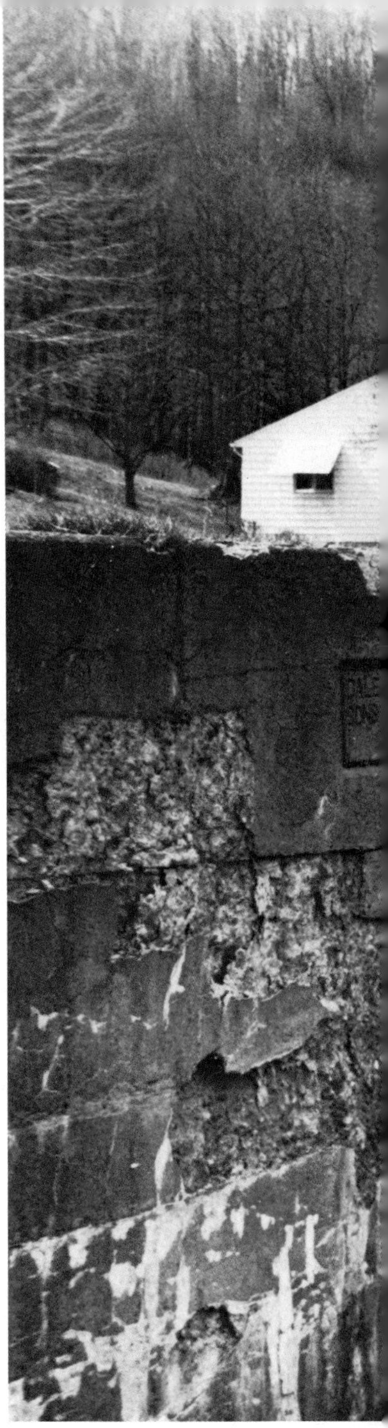

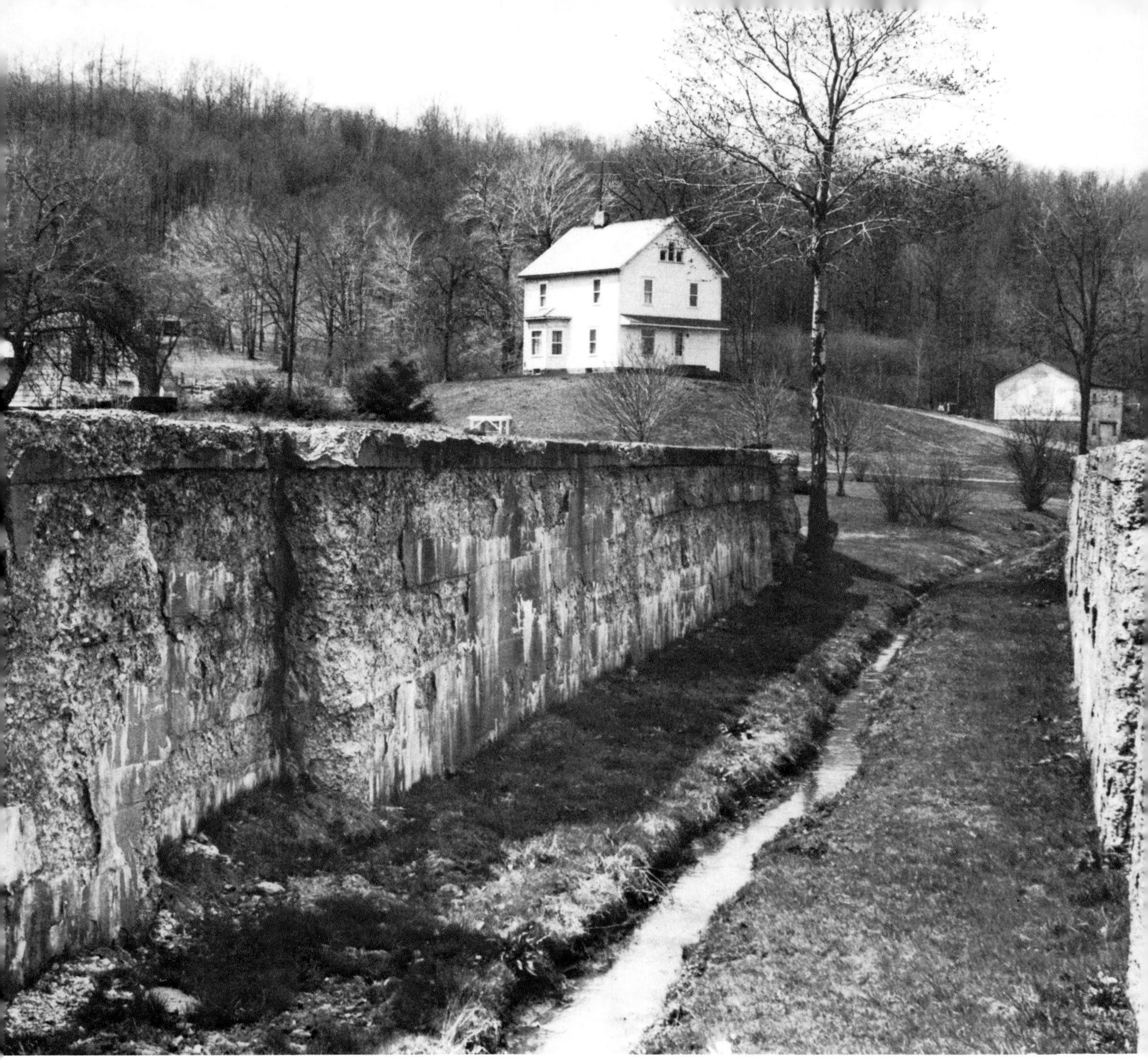

Lock 14, Blake's Lock (above) at New Philadelphia in 1968. An inscription indicates that the ubiquitous Daley brothers also refurbished this lock in 1907. Mile: 99. (Photo by F. W. Trevorrow.)

A map of Canal Dover (right) shows that the Ohio & Erie Canal paralleled the Tuscarawas River on the north bank. At the lower left is Sugar Creek, where a dam, originally built to serve a grist mill that predated the town, created a slackwater pool, which provided a crossing of the tributary. The backed-up water was also diverted into a hydraulic canal which crossed the river at the lower center for the benefit of the rolling mill and other industries on the south bank. Boats could enter the hydraulic canal from the Ohio & Erie to serve these local industries. (CSO)

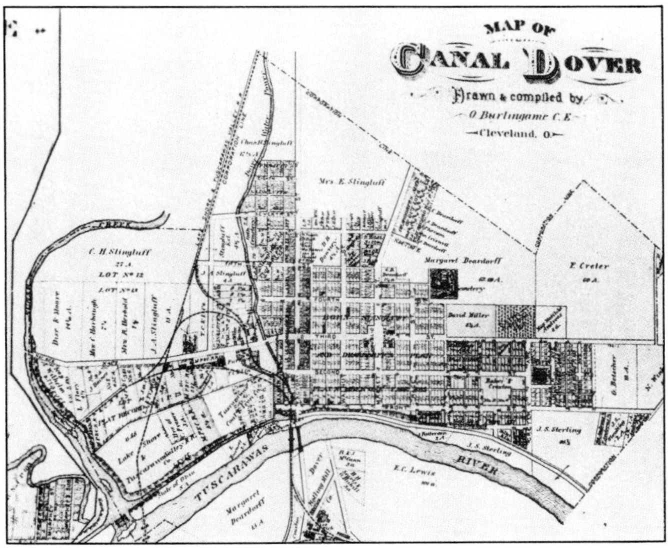

Upper Trenton Lock (below) awaits repairs, with the lower gates washed out by spring floods and the spillway denuded of water (Canal Visitor Center, CVNRA, courtesy Cloea Thomas). The *Valley Mills* (below left) emerges from the lock, with a band playing for the benefit of Sunday picnickers, who danced on deck and on the loading platforms (CSO). Ladies (below right) await the arrival of a boat approaching the lock, standing on the bridge across the spillway. Mile: 103. (Pockrandt Collection, SCHS)

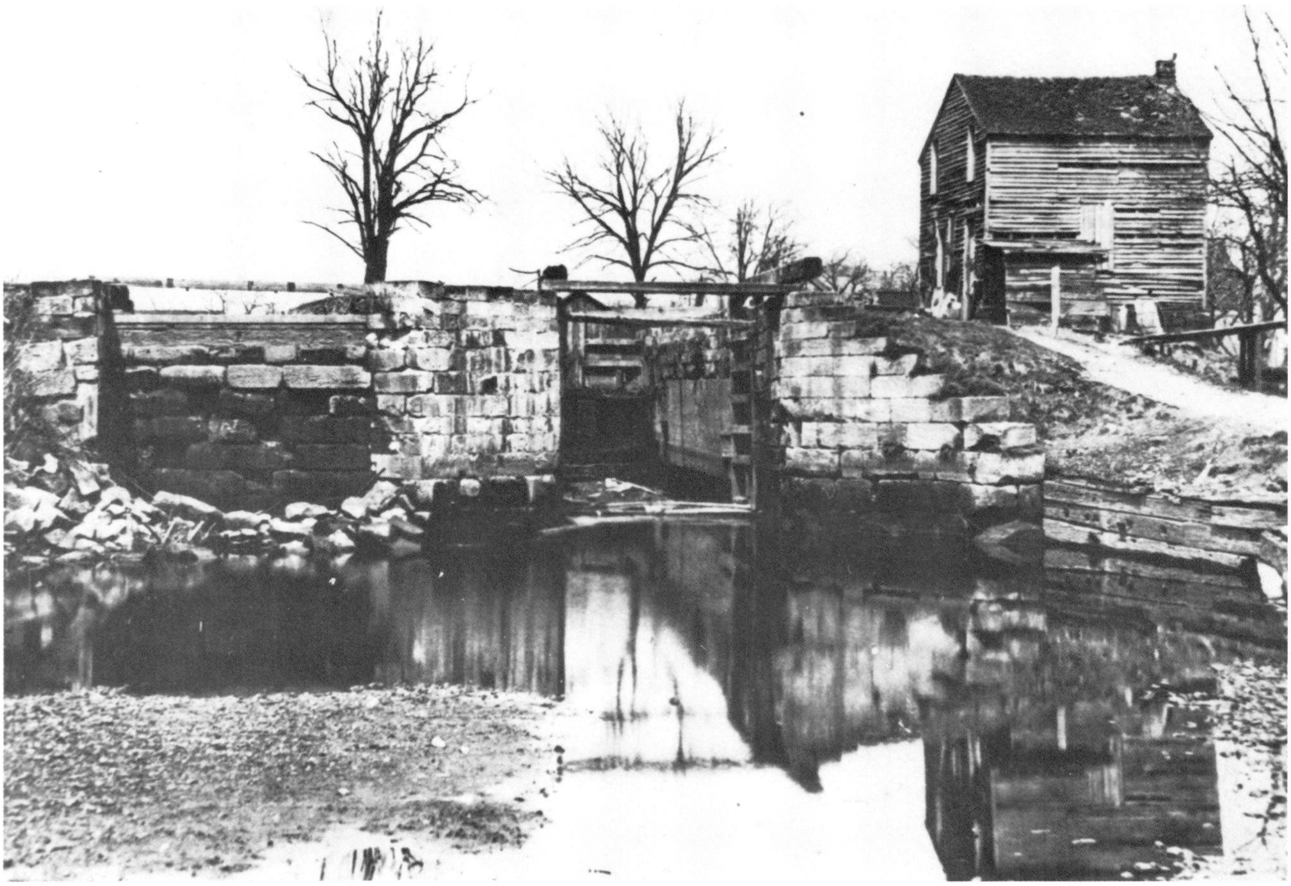

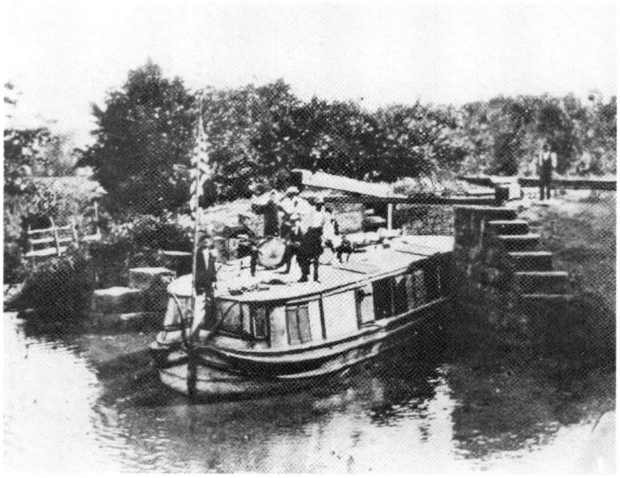

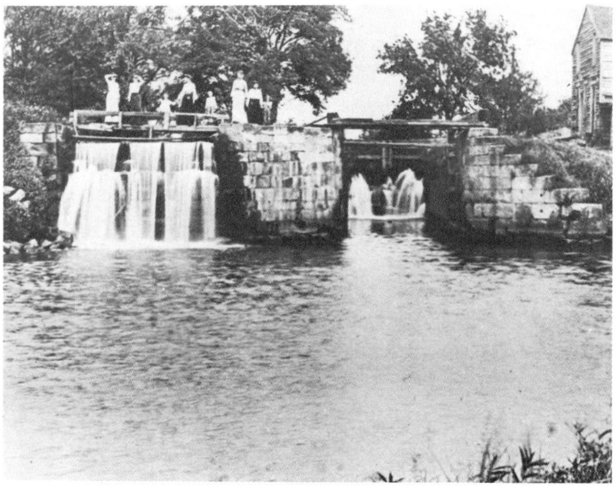

A scene above Lower Trenton Lock, in Tuscarawas County. Mile: 109. (Canal Visitor Center, CVNRA)

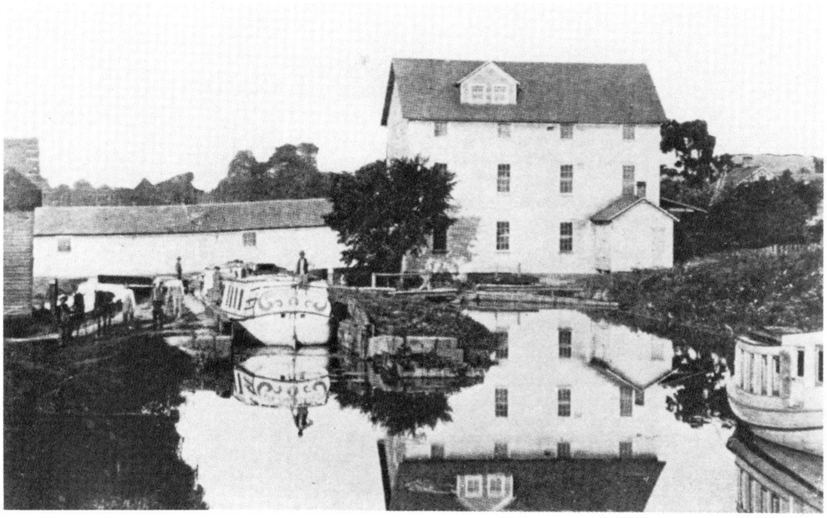

Towpath bridge below Lock 17, Lower Trenton Lock, to cross a feeder stream; 1896. Mile: 109. (Pockrandt Collection, SCHS)

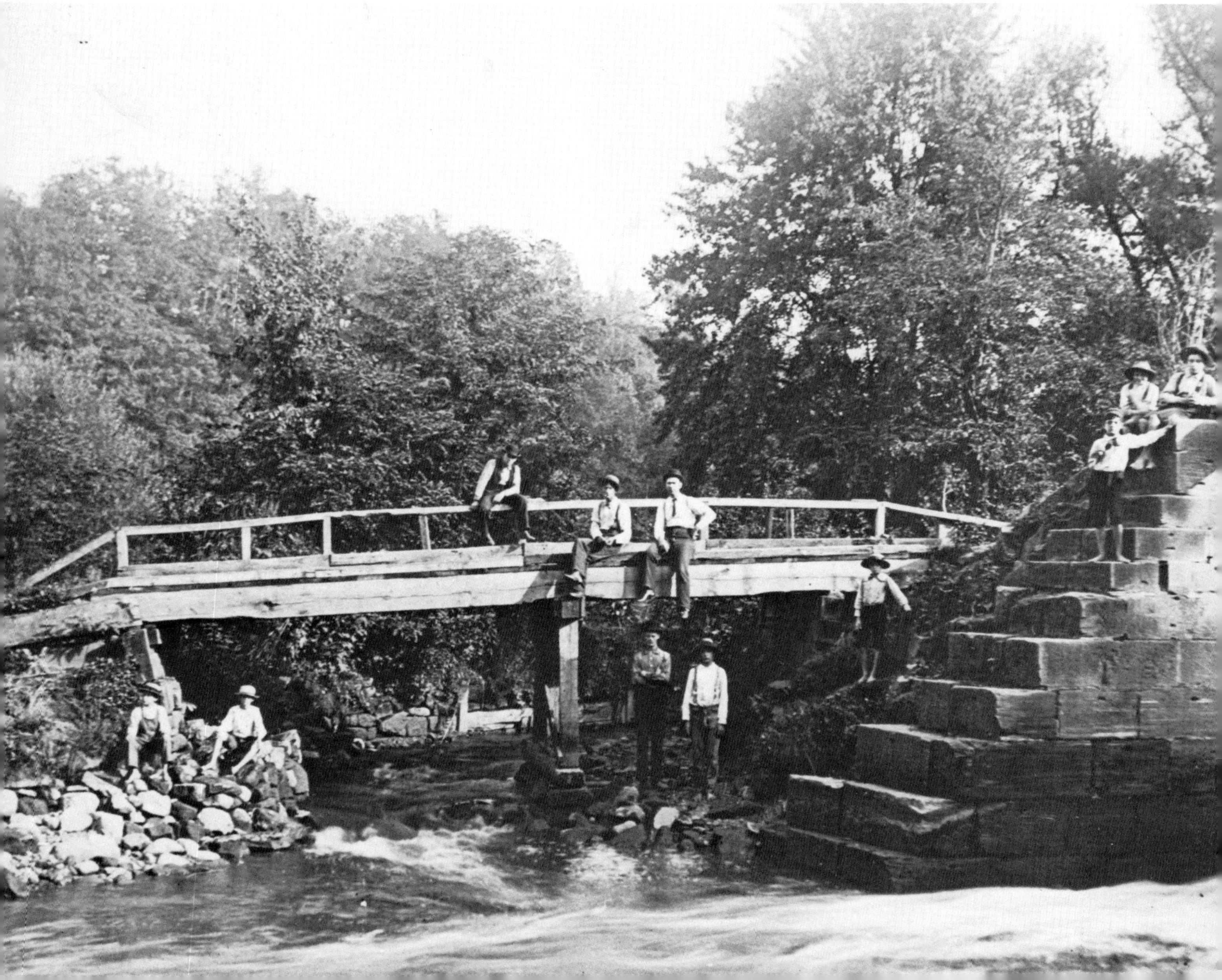

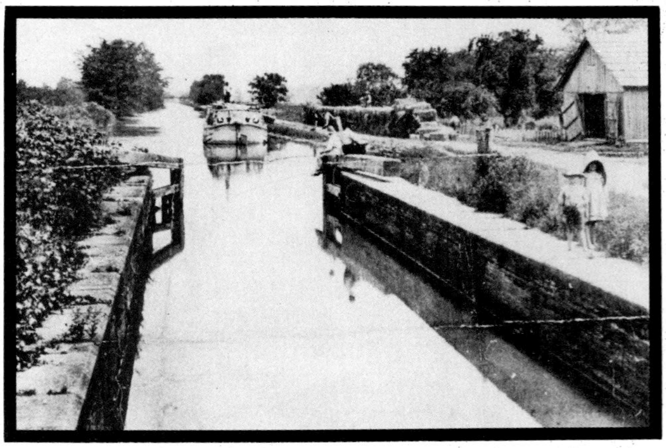

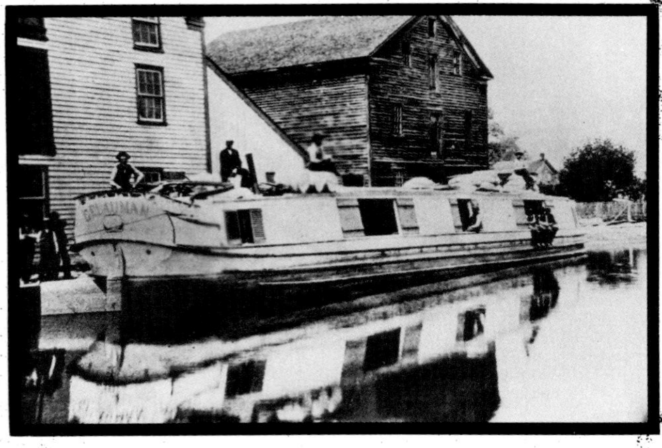

The *Marfield Mills* (left) enters Lock 18, east of Port Washington, as children watch the efforts of the photographer. Mile: 112. (CSO)

The line boat *G. R. Bauman* (right) at Port Washington. The left shutter on the window of the rear captain's cabin was evidently opened during the long exposure, resulting in its blurred image; c. 1880. (SCHS)

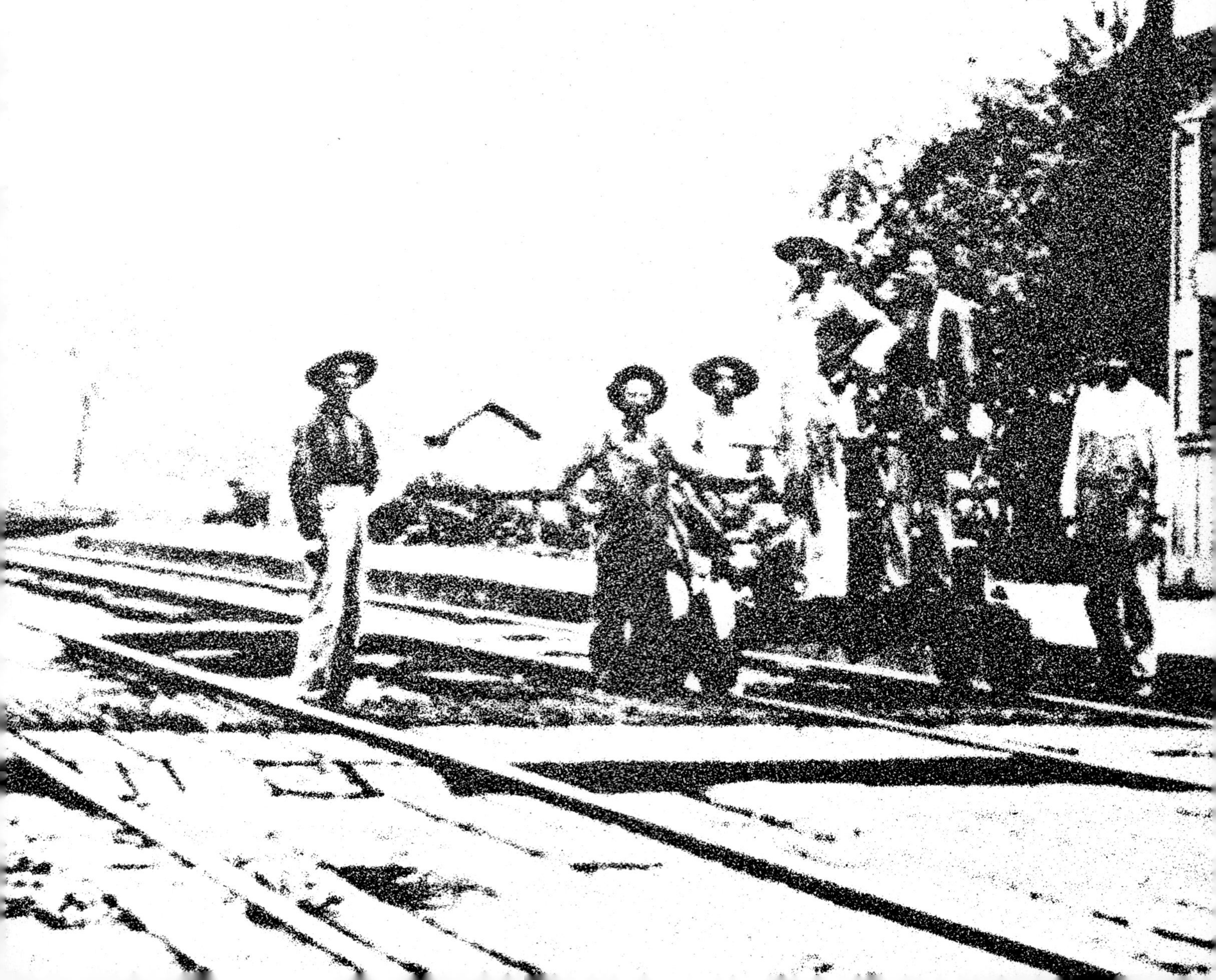

Lock 17 became the name of the post office and town adjacent to the lock, as seen here on the town's railroad station. The name of the village today is Seventeen, Ohio. (CSO)

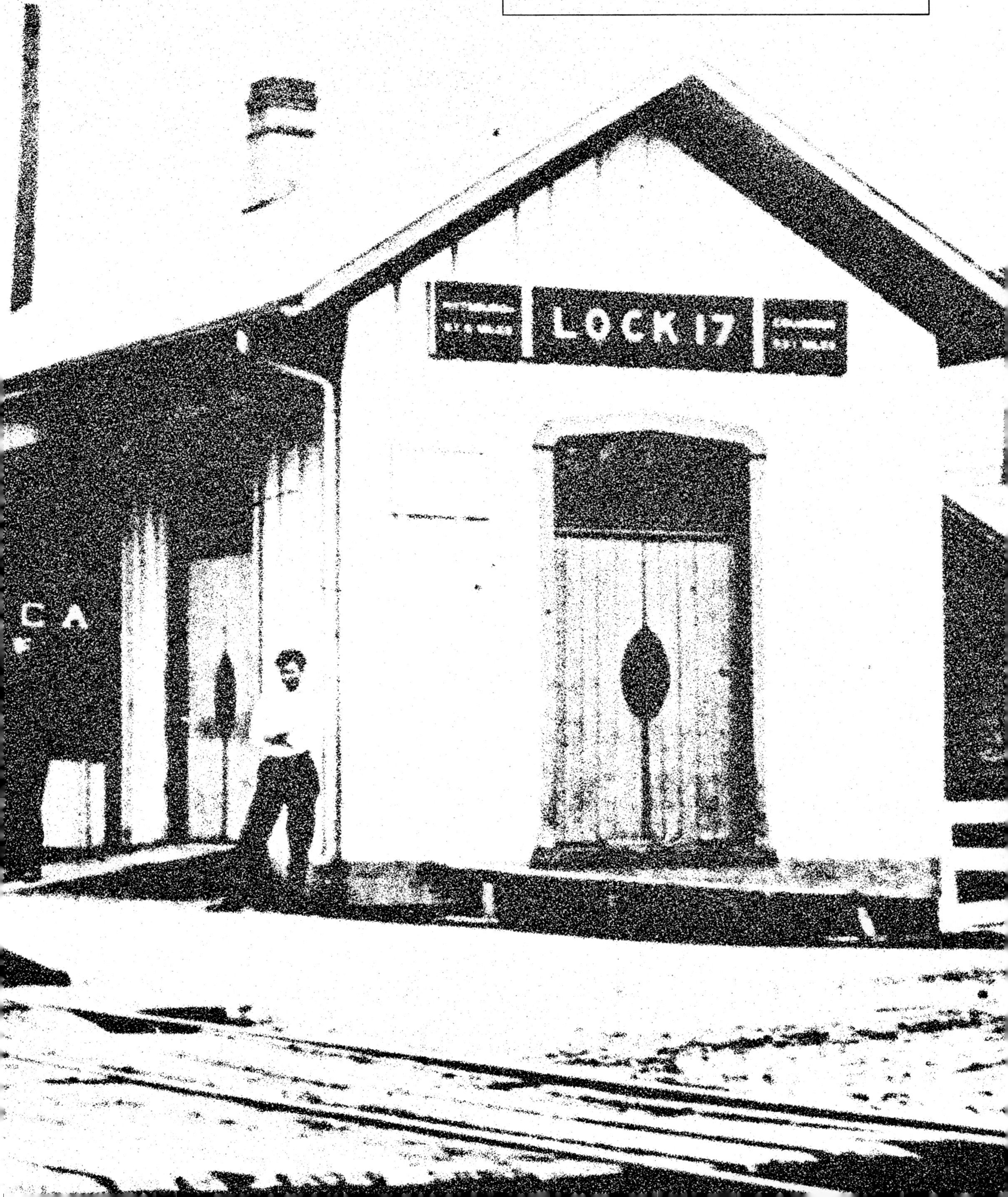

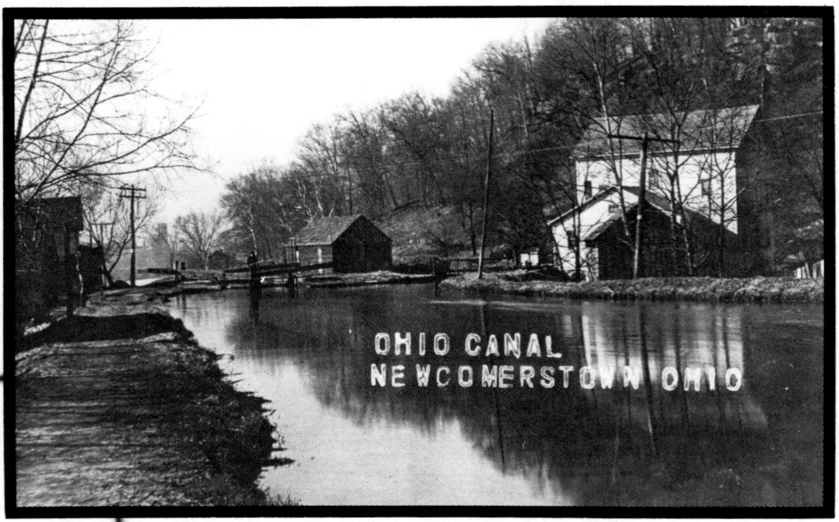

The photo at left shows the wide canal above Newcomerstown. The town was named for a captured white squaw, The Newcomer, in an Indian legend. Mary Harris was taken captive by the Delaware tribe as a young girl and became the second white wife of Chief Eagle Feathers. Shortly afterward, the chief was killed (very likely by his first wife) and The Newcomer fled. She was pursued and captured near this spot and returned to the Delaware Village on the Walhonding ("white woman" in Delaware) River at today's Warsaw, where she was executed by the tribe for the murder—probably unjustly. Mile: 117. (CSO; legend research courtesy RVF.)

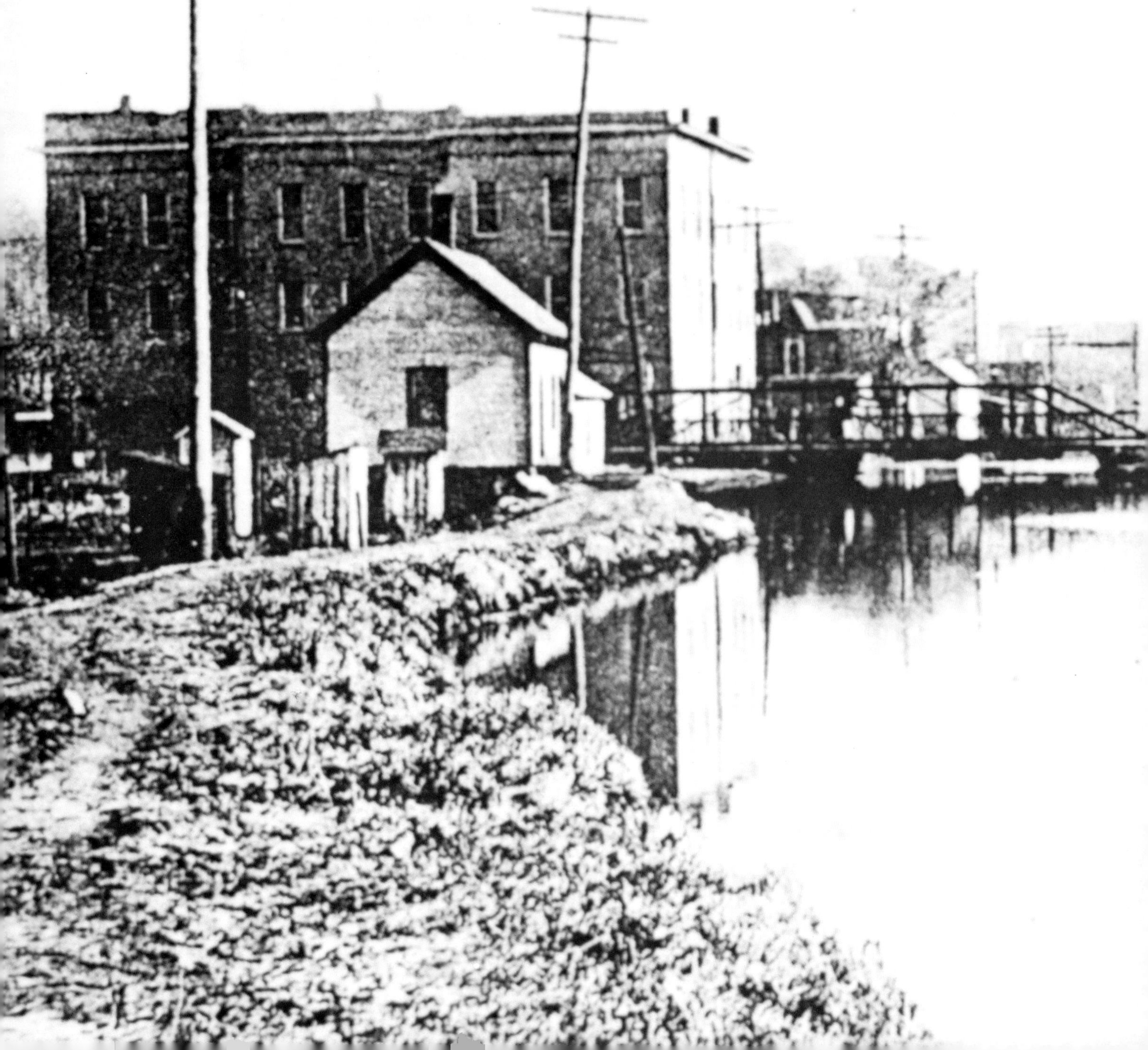

Below is the canal at Newcomerstown; the four-story structure with cupola is a grist mill (CSO, from an old postcard). At right is another view of Lock 21 and the mill at Newcomerstown. Mile: 118. (CSO)

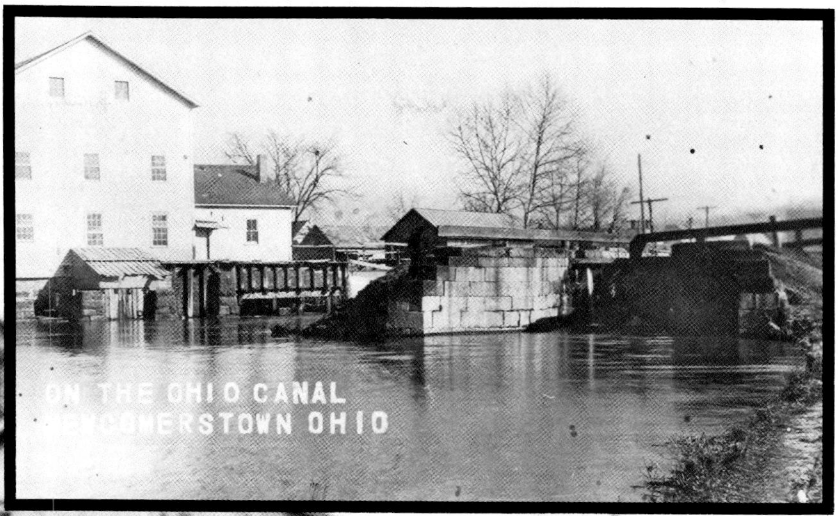

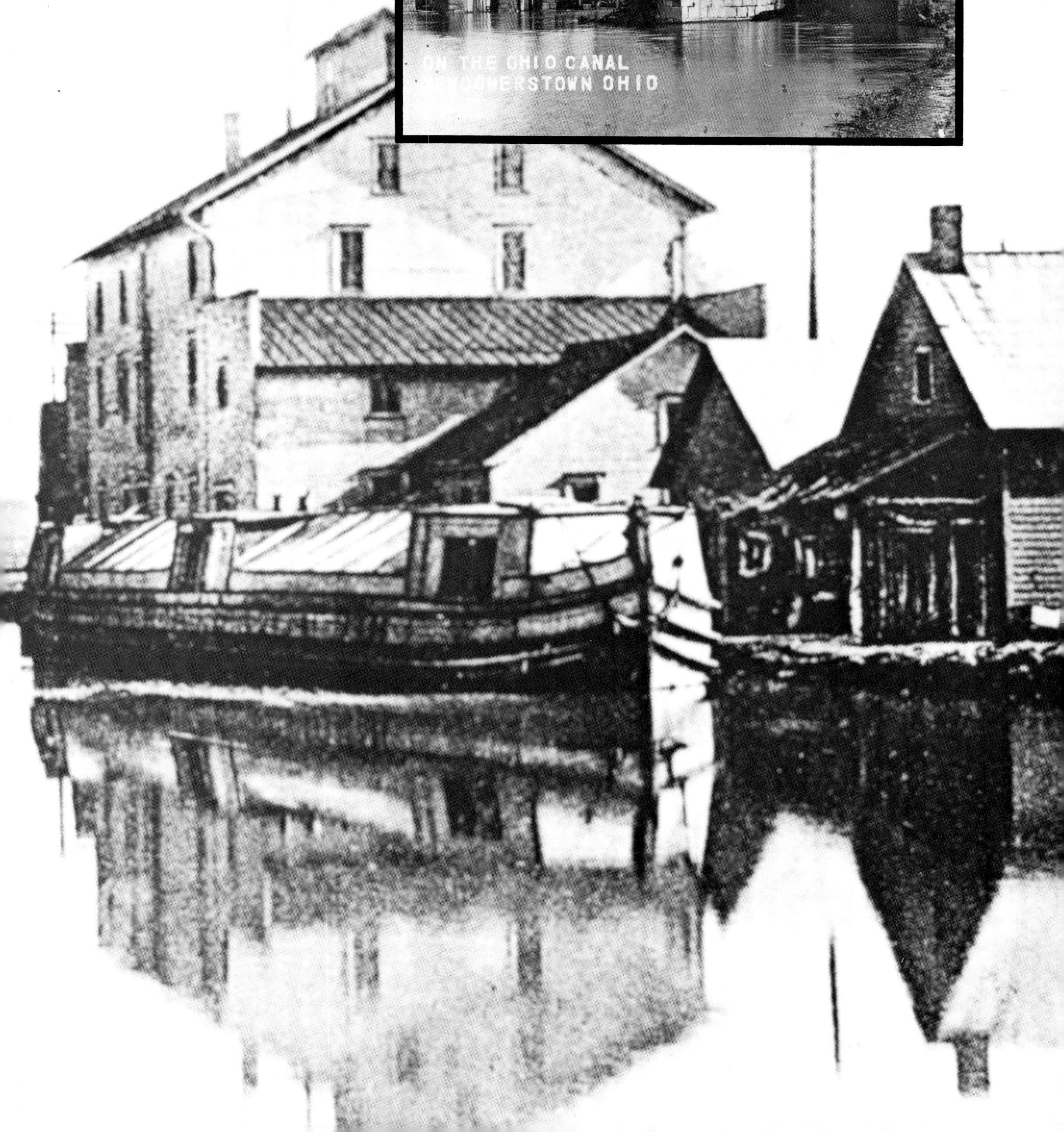

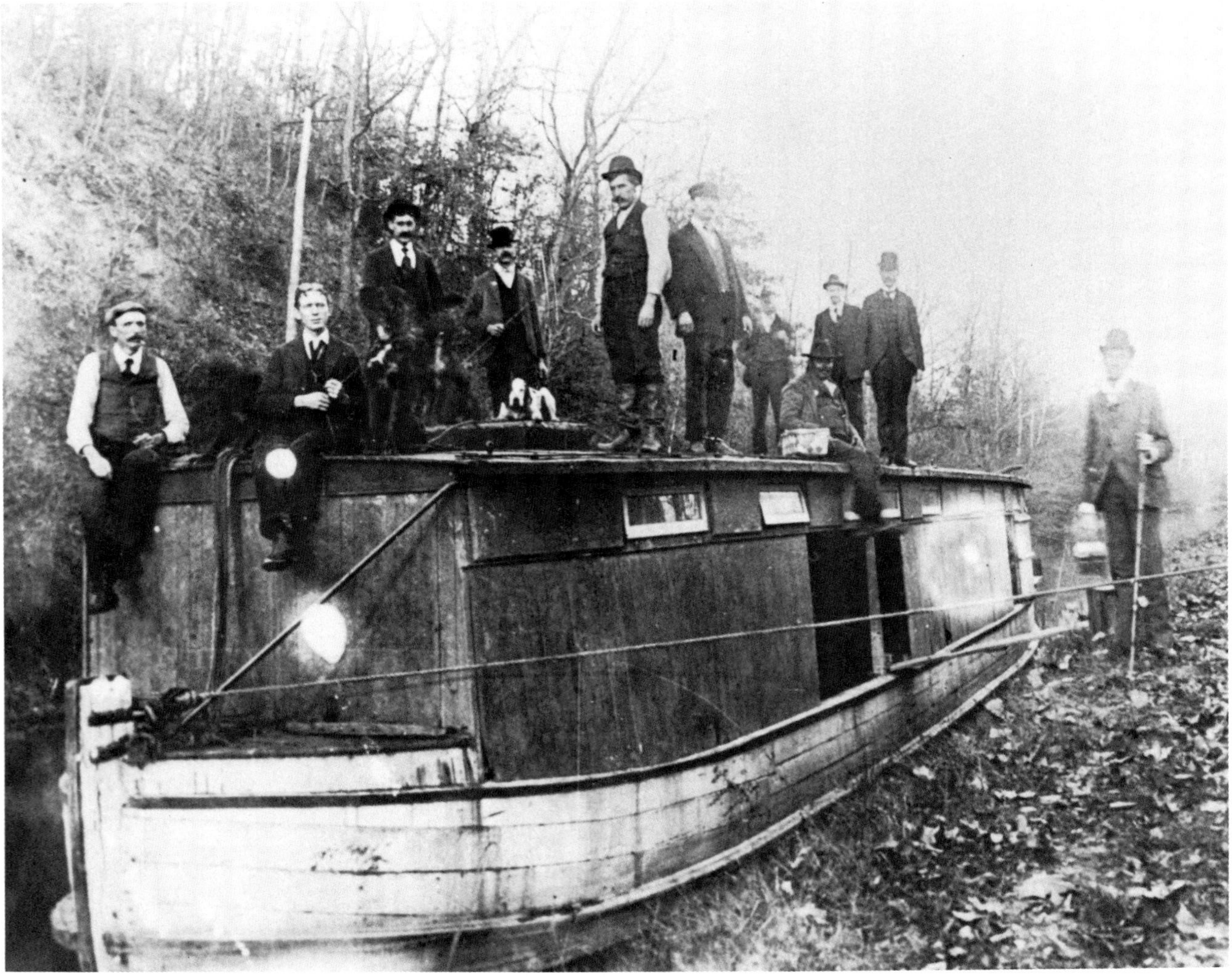

A hunting party aboard the Gleason Lumber Company boat near Roscoe. (CSO)

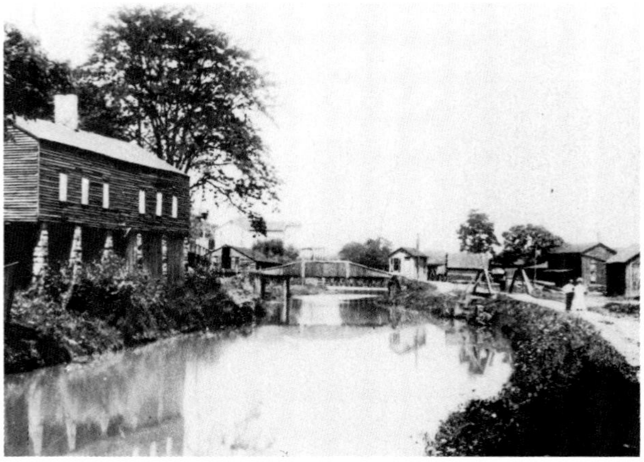

The Gleason Lumber Yard, on the canal at the north end of Roscoe. Mile: 136. (CSO)

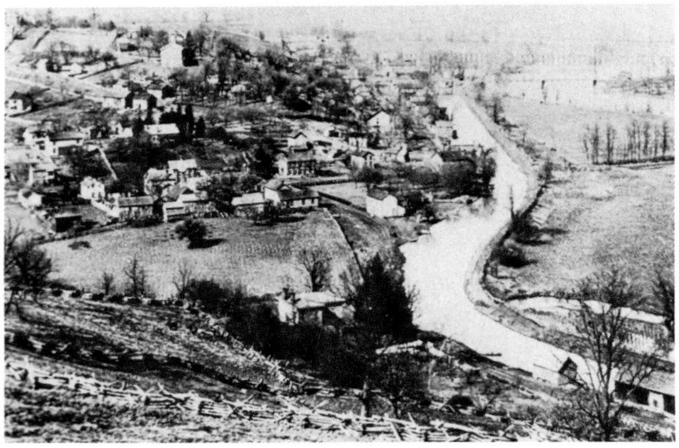

From Hoop's Hill it is apparent why citizens of Roscoe called it the Grand Canal. At the junction of the Ohio & Erie with the Walhonding Canal, Roscoe built its economy on the canal system. Mile: 136. (RVF)

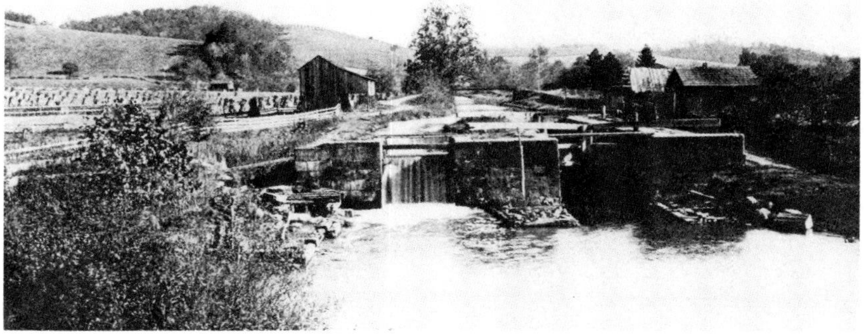

Wild Turkey Lock north of Coshocton is the first lock above today's Roscoe Village. Mile: 123. (CSO)

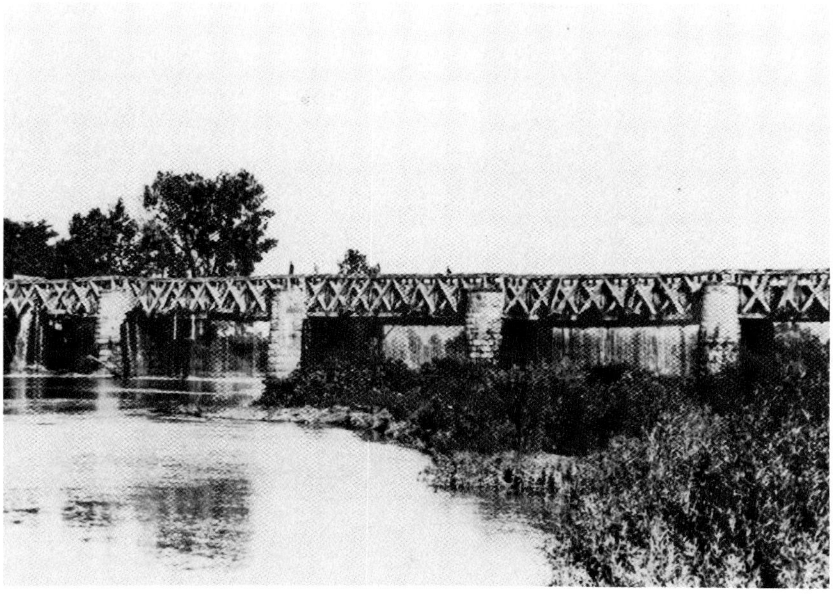

At the level of the Middle Basin, formerly called Mud Basin, the Ohio & Erie Canal crossed the Walhonding River on an aqueduct supported by four stone piers. Mile: 135. (CSO)

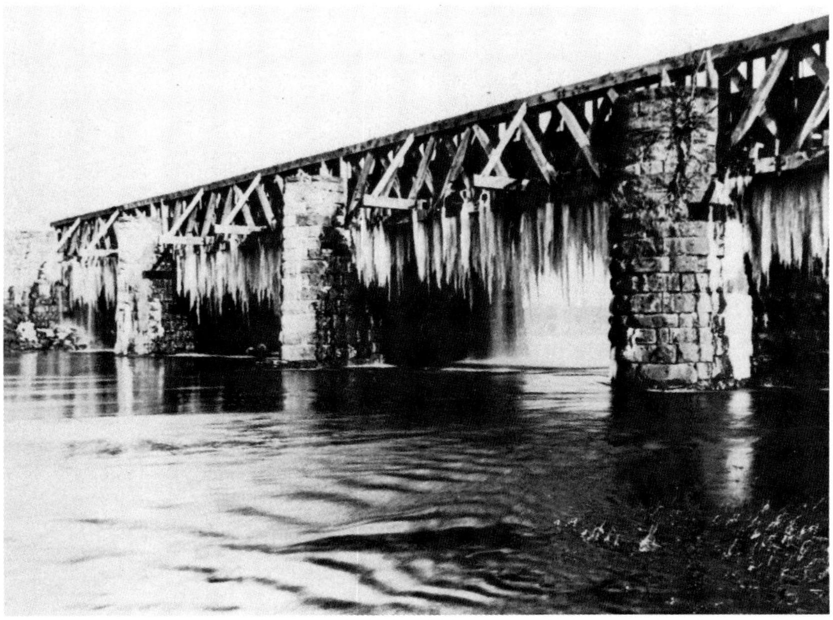

As winter approaches, aqueduct leaks become icicles. Ultimately they will form a wall of ice across the river. (RVF)

Below, The Ohio & Erie aqueduct entered two locks on the south bank of the Walhonding River. Here, the Double Locks, locks 26 and 27, are seen at the north end of the Lower Basin, where they lifted boats to the Middle Basin. The towpath runs along the shore at the right, and the railroad trestle in the foreground carried the Wally Flyer (RVF). *Top right,* The Double Locks dropped the canal level to that of the Lower Basin in the foreground. Using the stonework at the right of this lock for a foundation, Pearl Nye later built his Camp Charming on this site; 1880s. Mile: 135. *Middle right,* A closer view of the Double Locks when they were still operational. (CSO). *Bottom right,* At the south end of the Lower Basin, the Triple Locks dropped the Walhonding Canal into the basin to join the Ohio & Erie. The Triple Locks were the site of Roscoe's two mills, the Lily of the Valley and the Lone Star mills. The piling in the water at right is remains of the Wally Flyer railroad trestle. Mile: 135+. (From a color slide by Jack Gieck, 1971.)

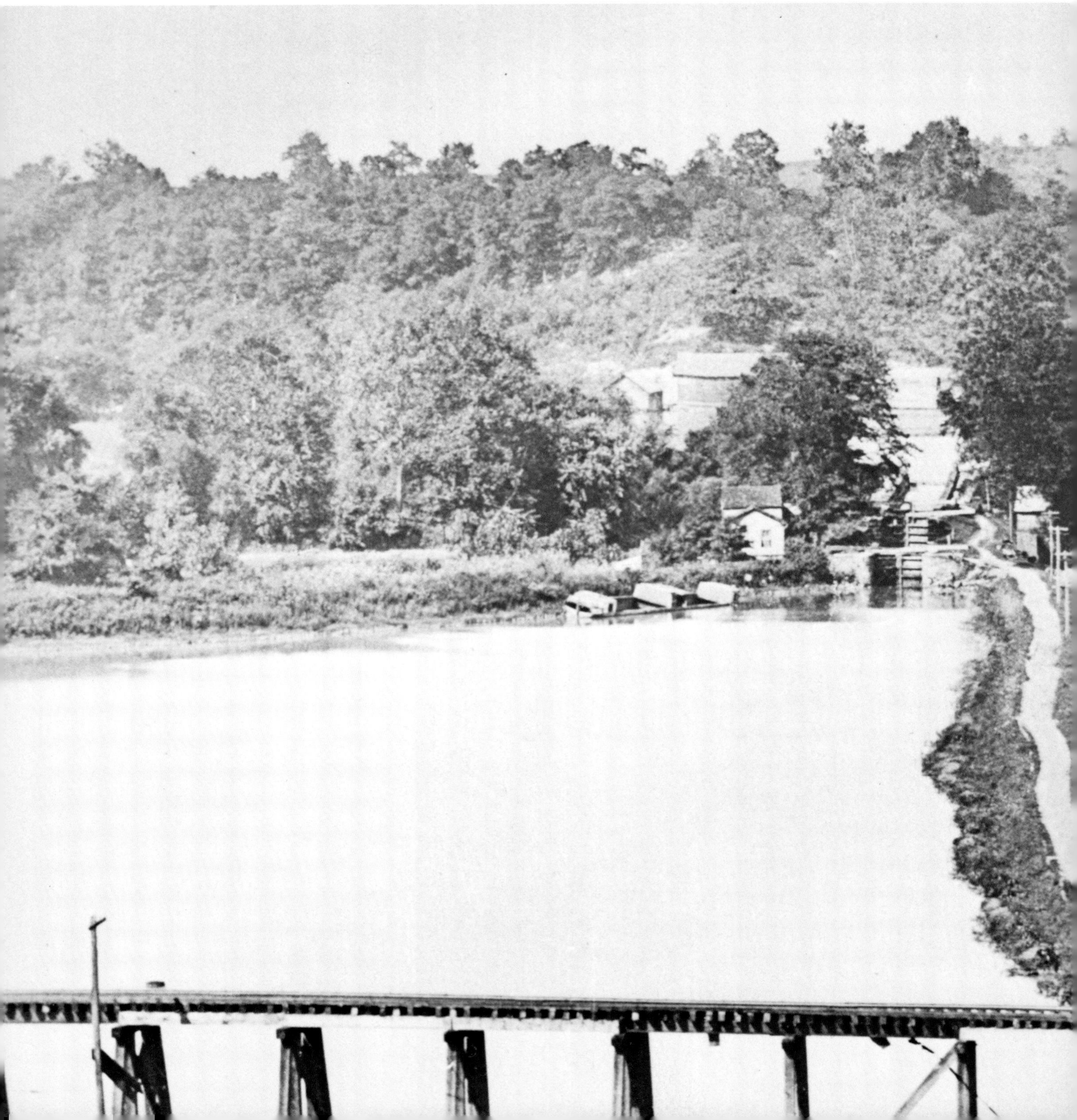

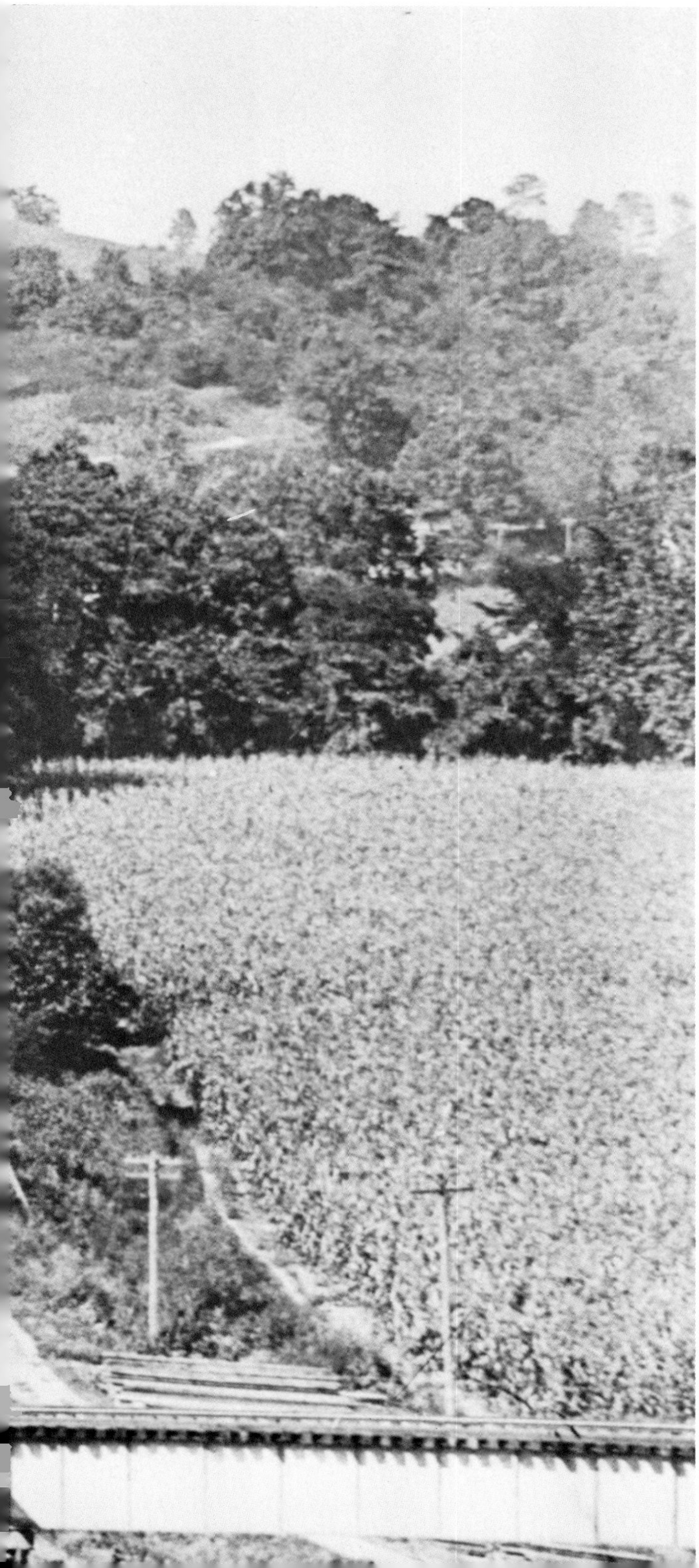
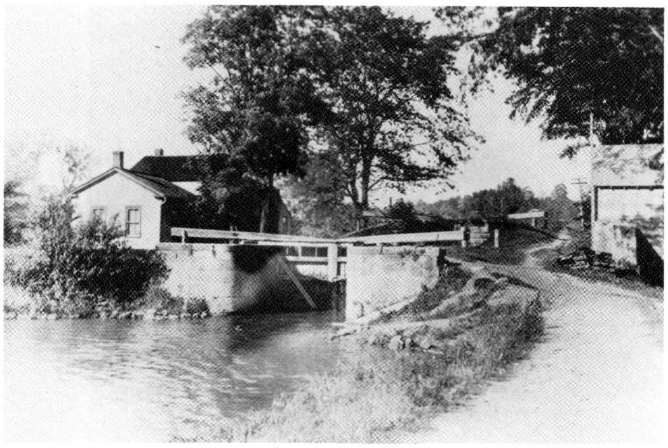
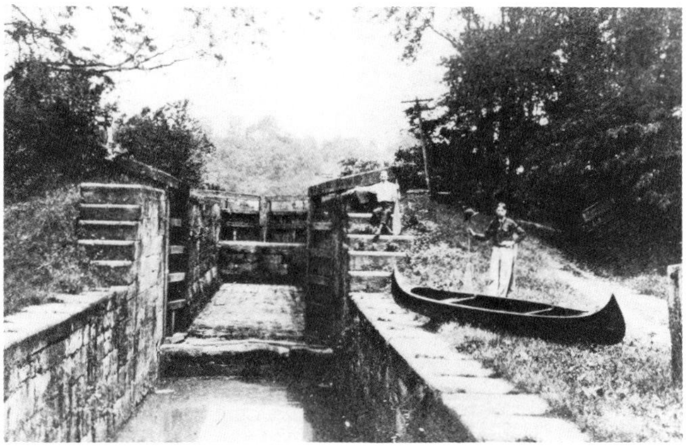
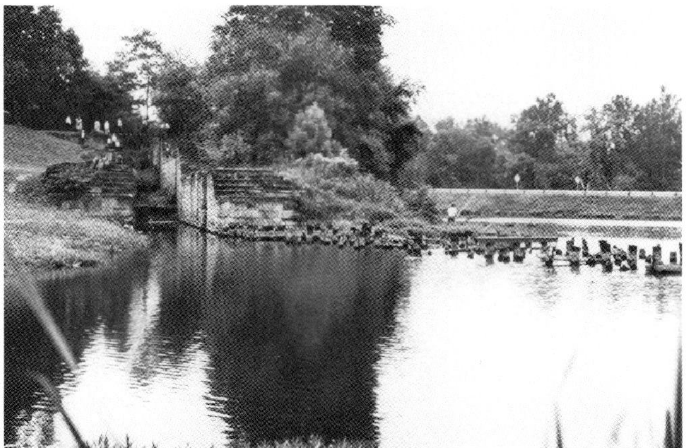

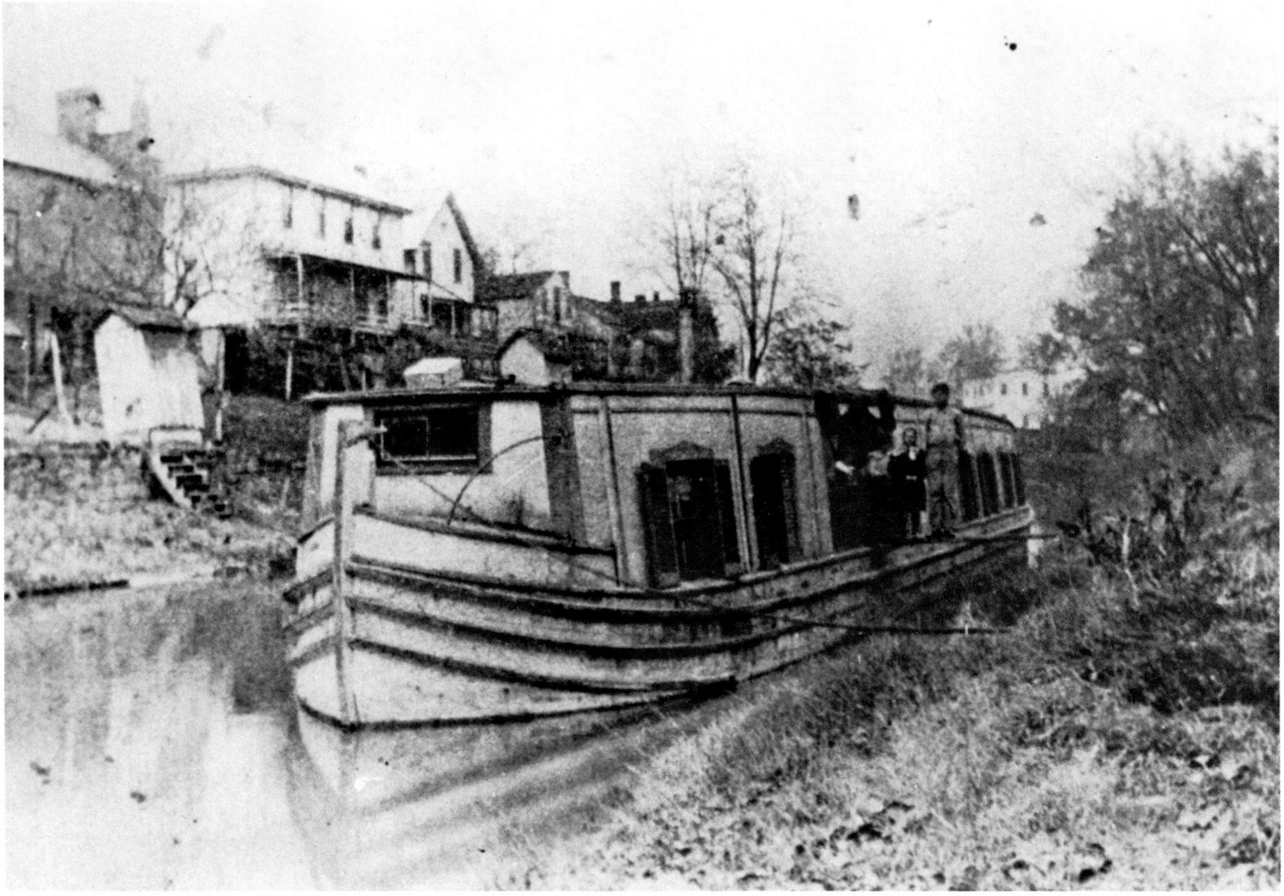

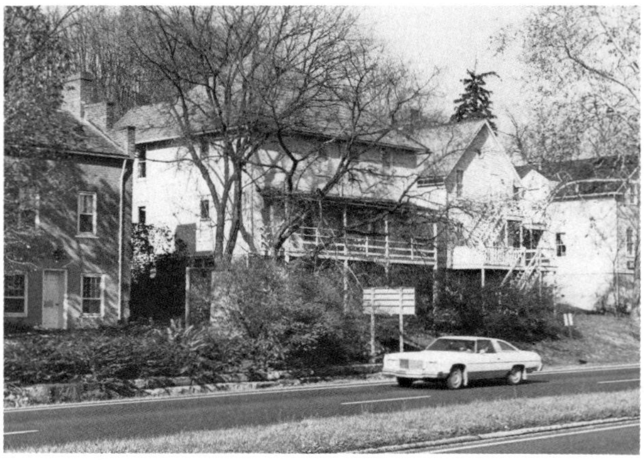

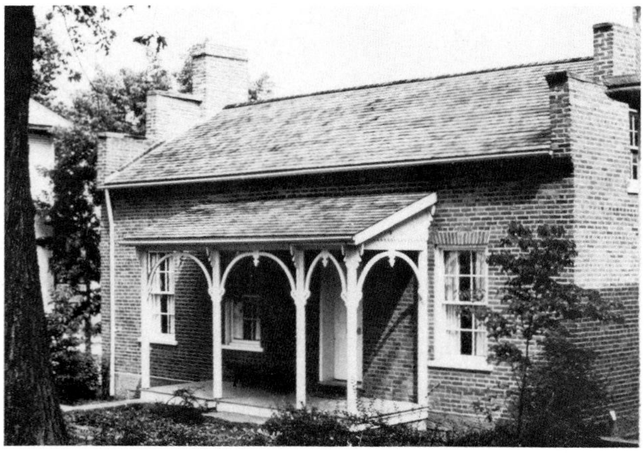

A latter-day passenger packet (top) at Roscoe in a vintage photograph which also shows the distinctive roof line of the tollkeeper's house at the upper left; also the tollkeeper's outhouse (CSO). In this almost identical scene (left) more than a century later, the same roof line on the restored tollkeeper's house may be identified at the upper left, as may three other buildings to the north. Today, State Highway 16 runs in the bed of the canal. Mile: 136. (Photo by Jack Gieck.) Another view (right) of the tollkeeper's house, next door to Roscoe's Old Warehouse Restaurant. Note the unusual roof treatment; 1971. (From a color slide by Jack Gieck.)

Photos on the facing page show a popular attraction just north of Roscoe Village above Roscoe's Upper Basin. The *Monticello II,* a canal boat replica, is operated by the Coshocton City and County Park District. Here, visitors may take a forty-minute ride on a one-mile restored section of the Ohio & Erie Canal. Mile: 135. (CSO/Jack Gieck)

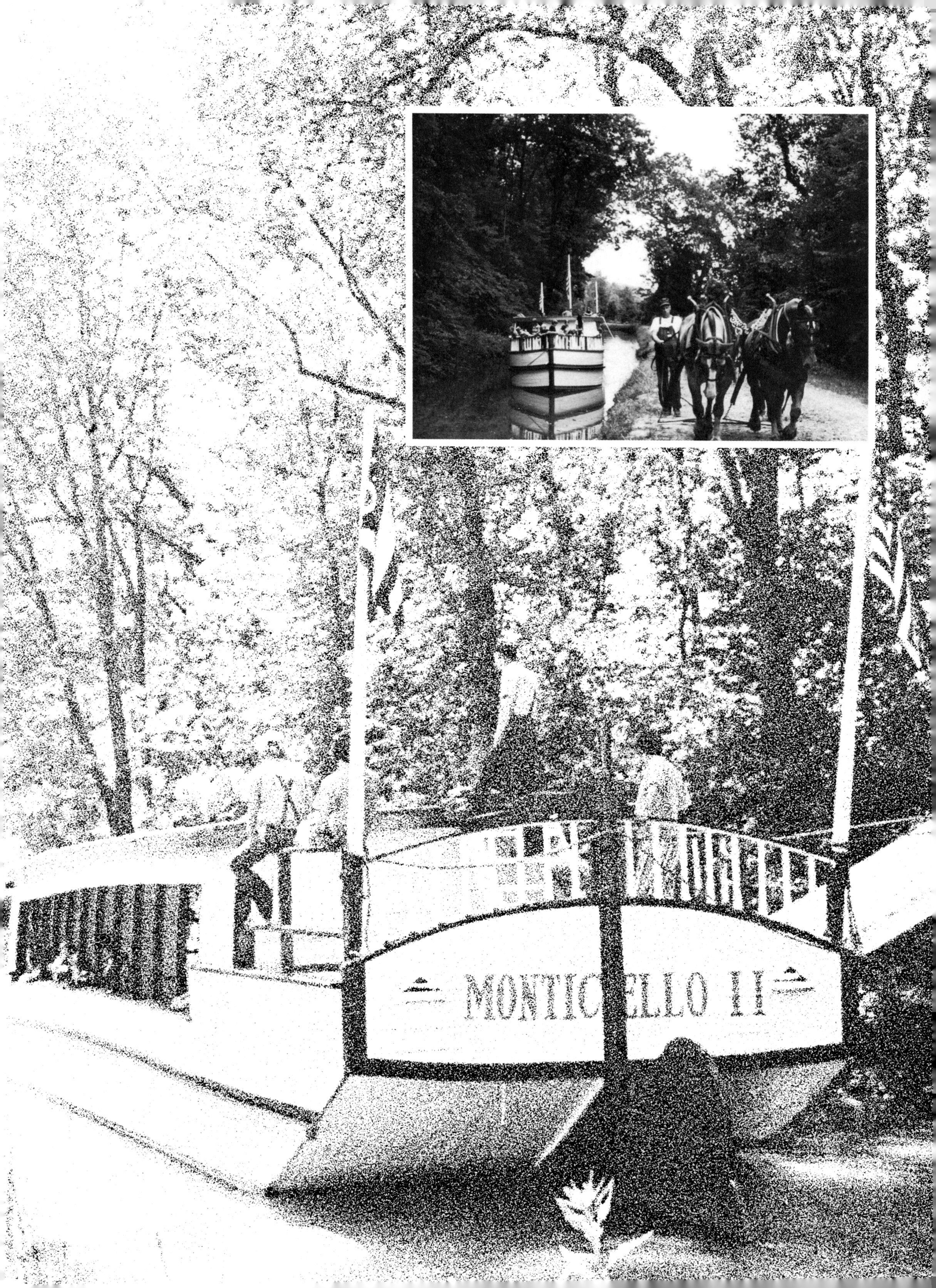
MONTICELLO II

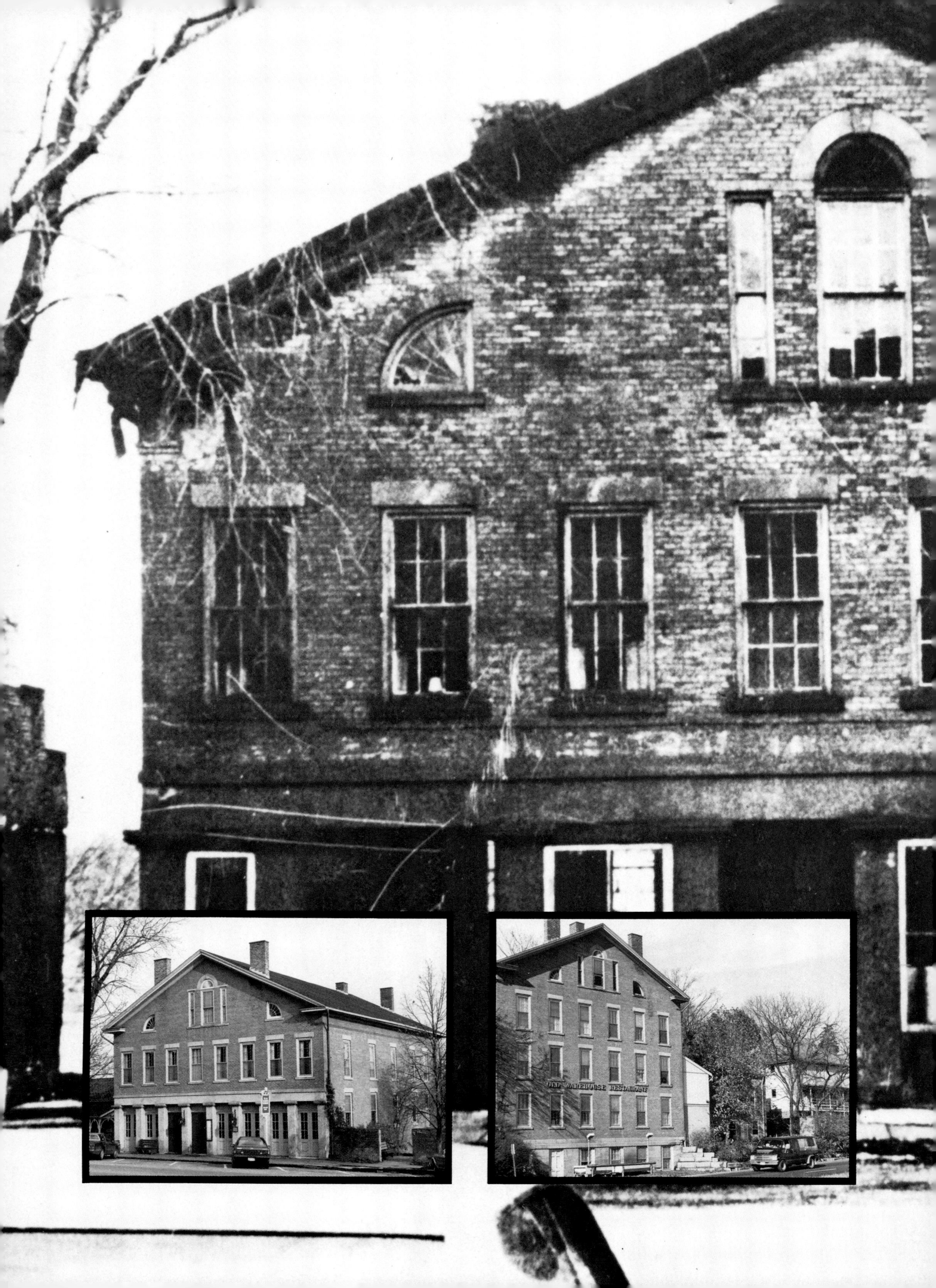
OLD WAREHOUSE RESTAURANT

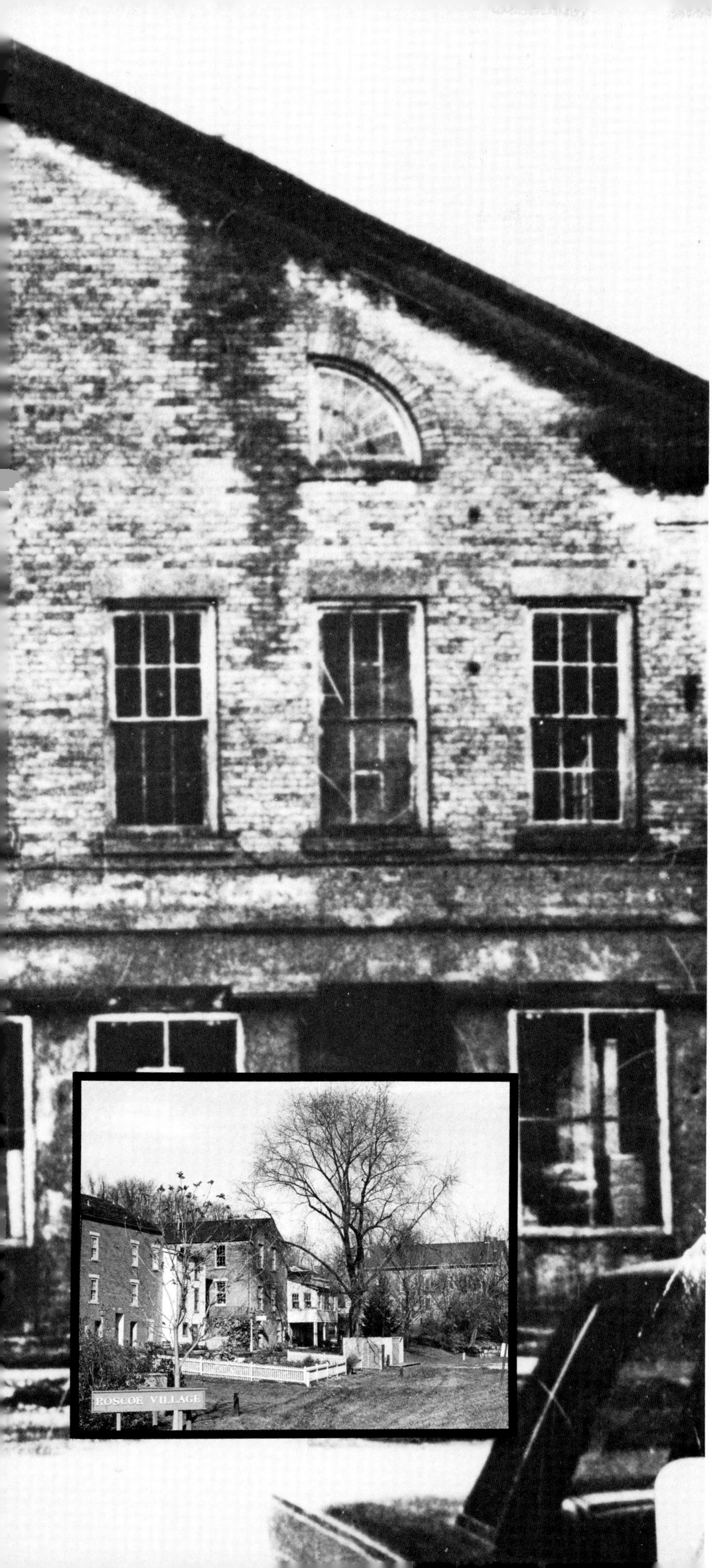

By the mid-twentieth century Roscoe had fallen into the decay typified by the condition of this old canal warehouse (RVF). Beginning in 1968 a restoration of the canal village was spearheaded by Edward and Frances Montgomery, who also created the Roscoe Village Foundation. After restoration, the structure became the Old Warehouse Restaurant, a principal attraction of the restored Roscoe Village along Whitewoman Street (a translation from the Delaware name for the adjacent Walhonding River). The small building at extreme left in the photo at bottom left is the tollkeeper's house. Now composed of dozens of restored buildings, the living museum is open daily year round. The rear of the canal warehouse at Roscoe was at canal dock level. The tollkeeper's house is obscured in the center photo by the tree behind the van traveling in the canal bed, now Route 16. At the south end of Roscoe Village, the depression in the right foreground of the photo at right may be identified as the former canal bed, running behind the restored canal buildings on the left bank. (RVF/Jack Gieck, 1986)

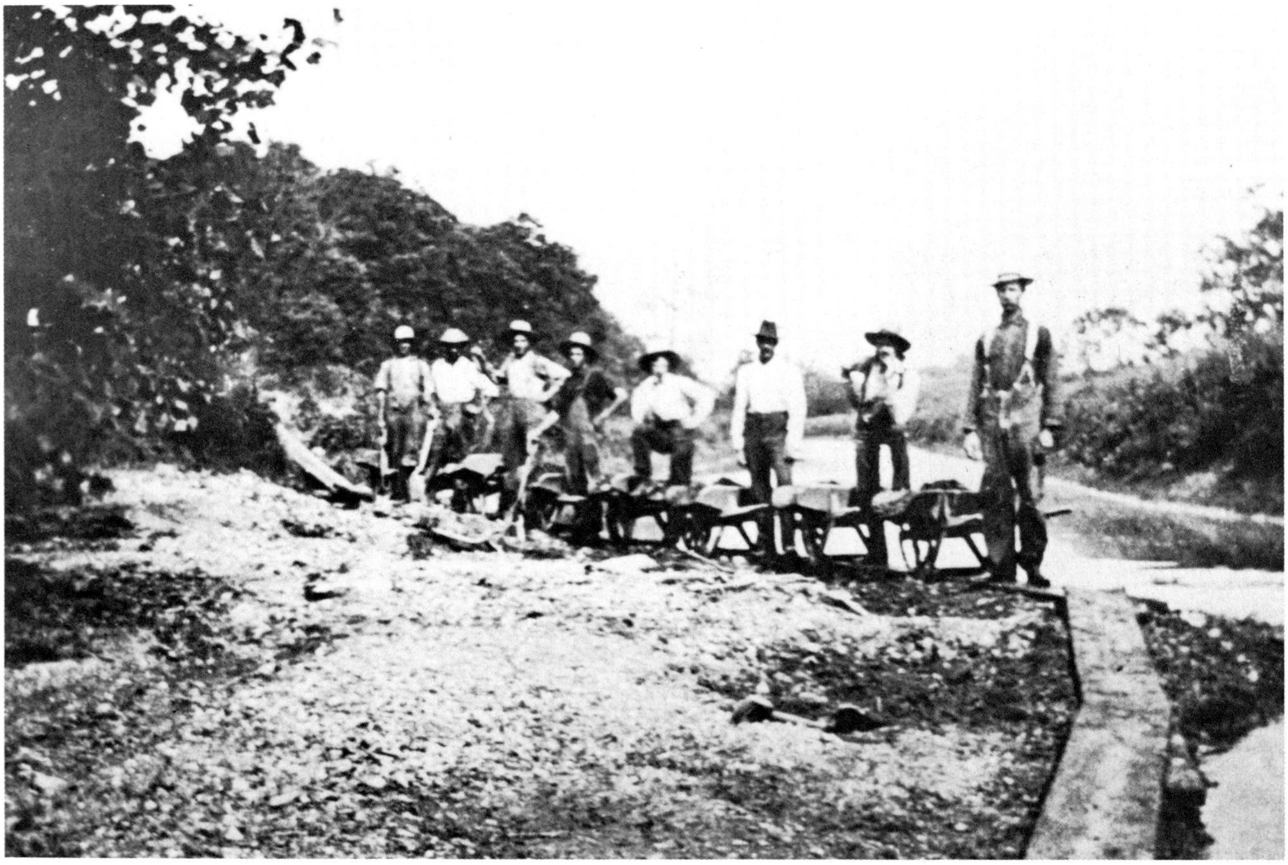

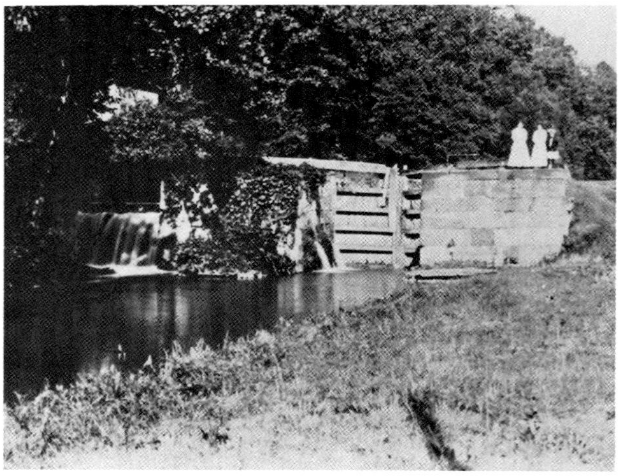

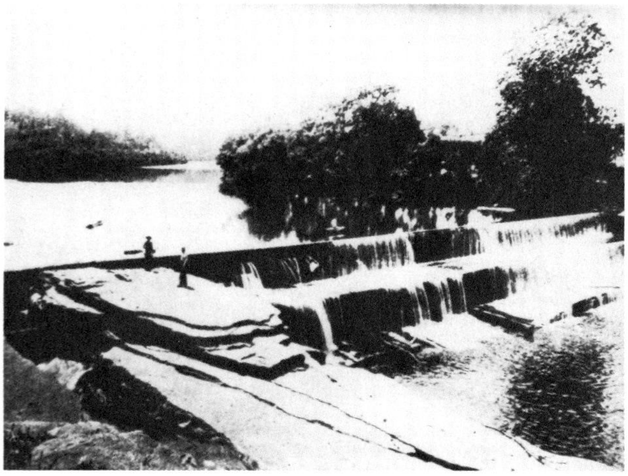

Top, Pearl Nye's caption for this scene in his collection was "State robbers at McGuire's Bar." Translated, he meant that it was a State Boat crew who had stopped to rest and maybe have a drink; "McGuire's Bar" was anywhere they engaged in such idling activities. The photograph is near Roscoe. (Canal Visitor Center, CVNRA, courtesy Cloea Thomas.)

Above left, Ladies pose on Lock 30 at Adams Mills—the last lock in the descent from the Portage Summit at Akron; from here the canal ascended to the Licking Summit, the Ohio & Erie's second summit near Newark. Just south of this lock was the junction with the Dresden Sidecut. Mile: 147 (CSO). *Right,* A vintage photo of the dam on the Licking River that provided a slackwater at Licking Narrows. Mile: 166. (OHS)

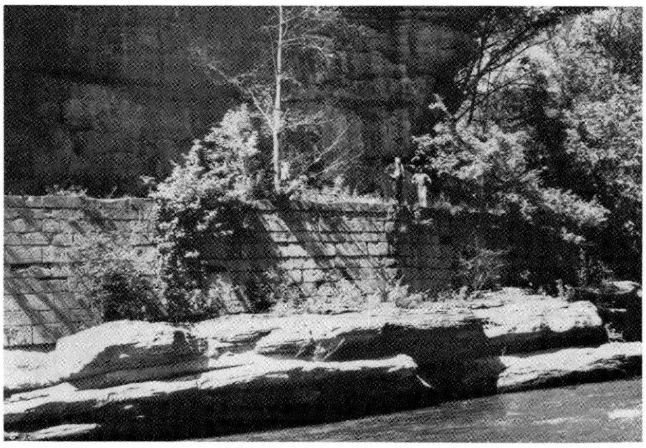

The built-up towpath on the face of Black Hand Rock at Licking Narrows permitted mule teams to tow the boats through the slackwater at Black Hand Gorge. In this photo, with the dam no longer in place, the river has dropped to its original level. Mile: 166. (Photo by F. W. Trevorrow, 1966.)

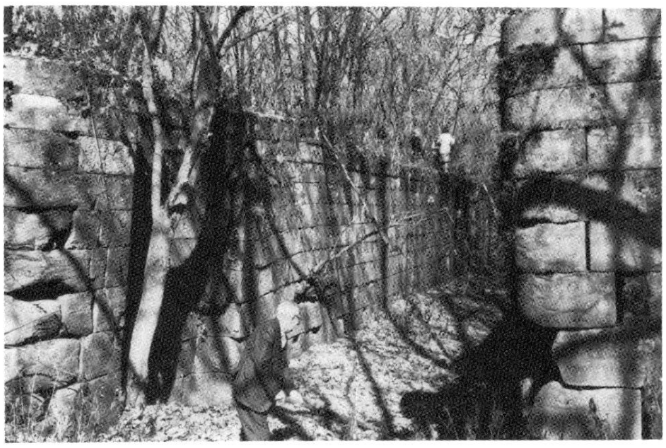

A guard lock and members of the Canal Society of Ohio at the lower end of the Black Hand Gorge slackwater; 1974. Mile: 166. (Photo by F. W. Trevorrow)

The canal in Newark in the 1890s. Suspended above the swing bridge is an electric street light—an innovation for its time. Mile: 176. (CSO)

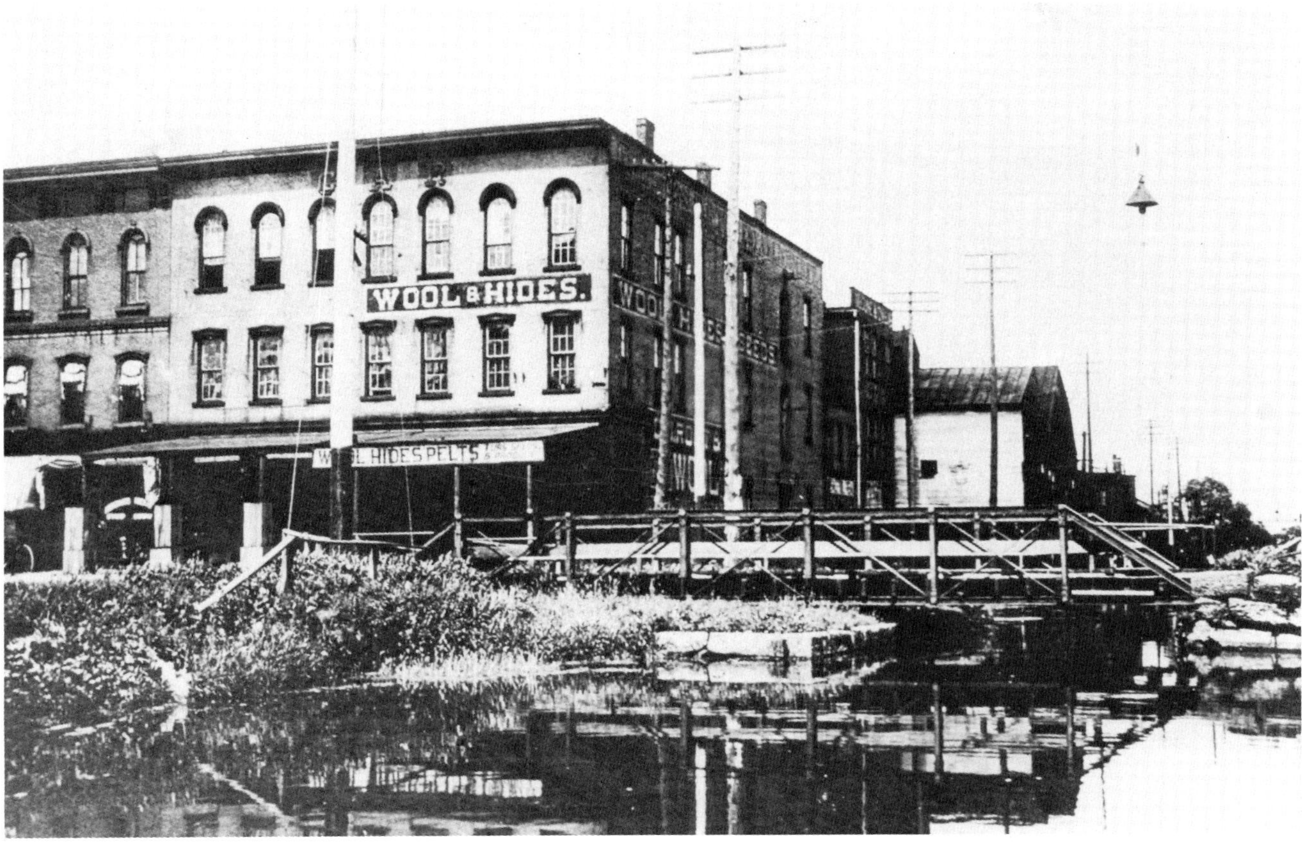

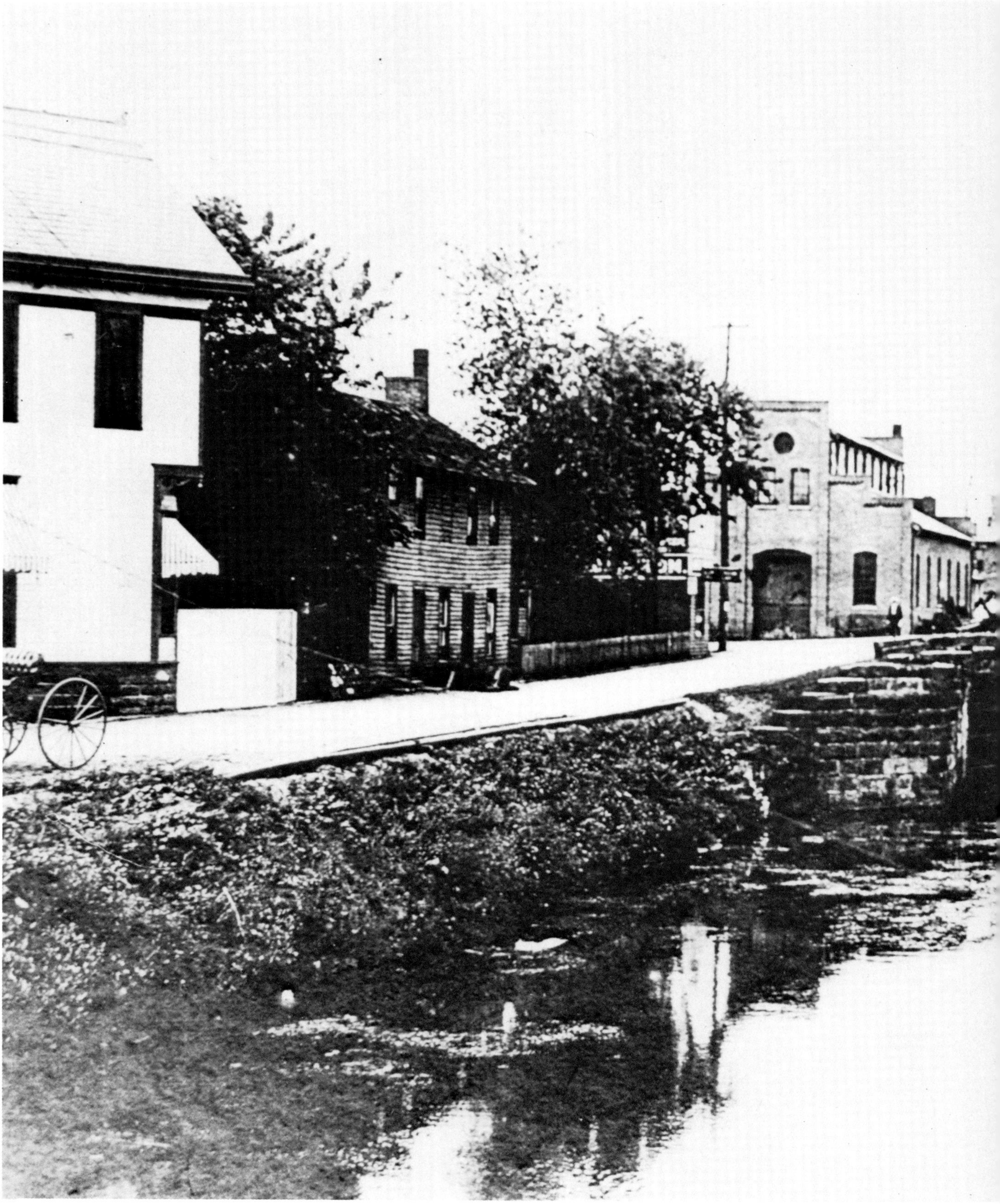

Locks in Newark; the tiny black object above lock gates is the town's new electric street light; c. 1890. (CSO)

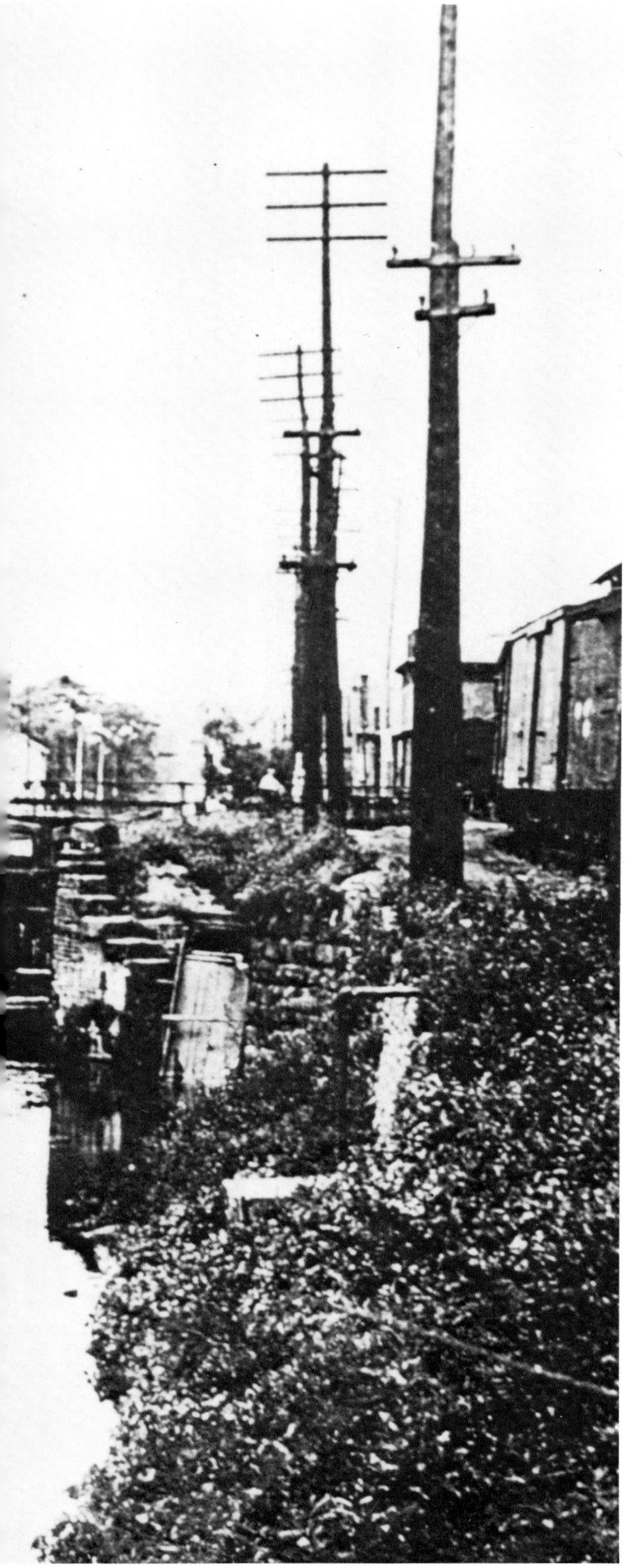

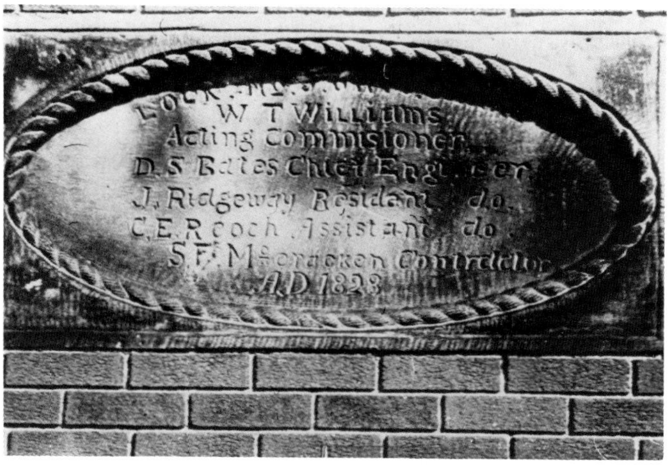

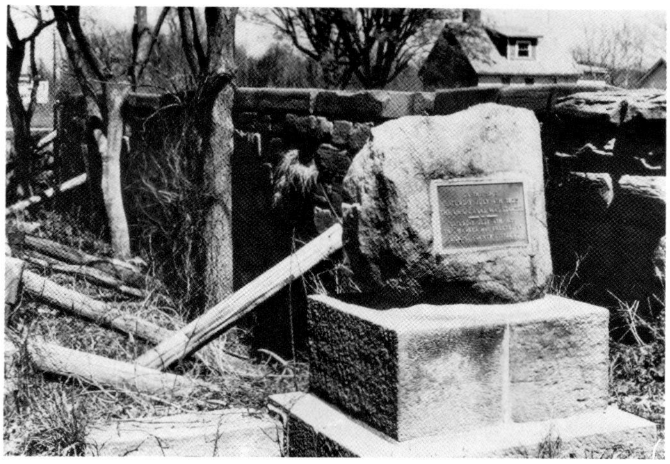

A lock plate dated 1829 (top), on Lock 9 at Newark, honors the canal commissioners and engineers responsible for local construction. Mile: 176. (CSO)

This monument at the lock on the site of the 1825 groundbreaking ceremonies (bottom) marks the beginning of construction on the Ohio & Erie Canal at Newark; 1978. Mile: 181. (From a color slide by Jack Gieck.)

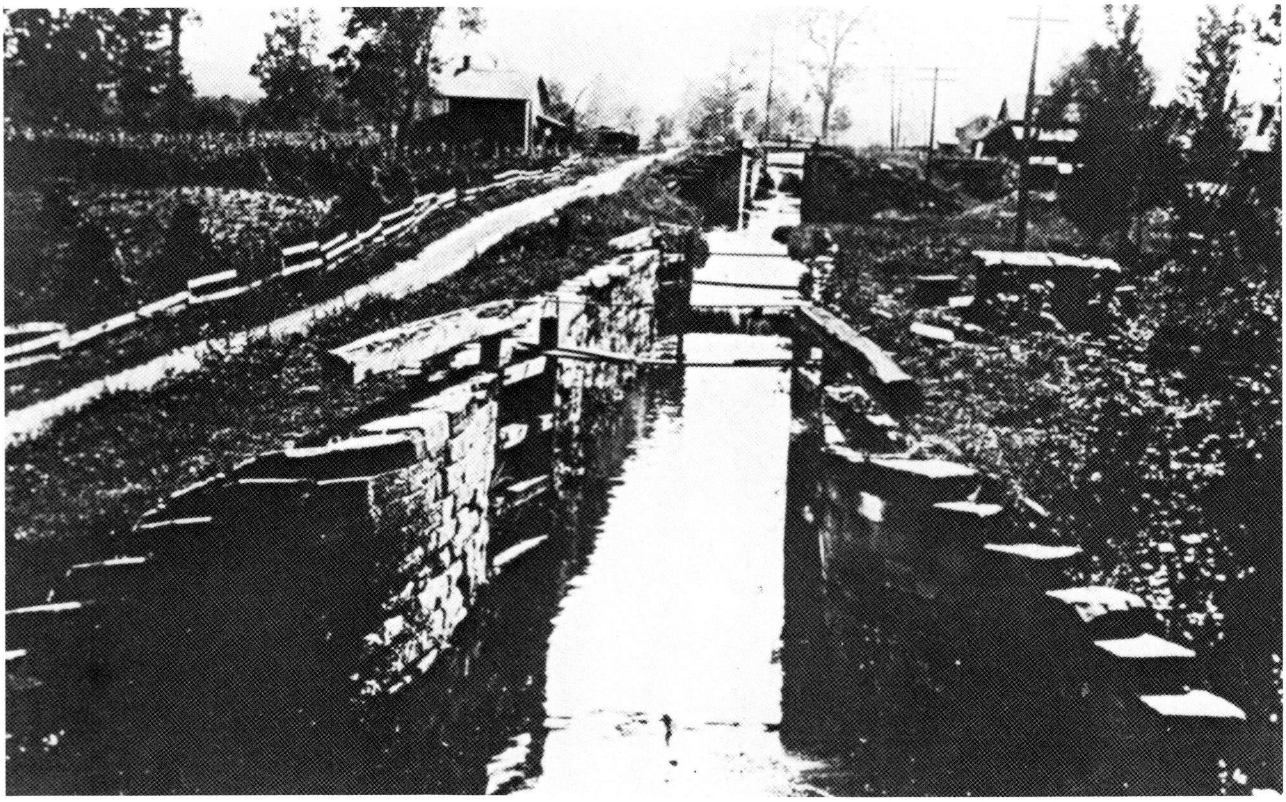

The vintage photo above shows the ascent to Lock One at Licking Summit, Newark (CSO). The remains of canal bed in the Newark area are shown in the photo at right. Mile: 181. (From a color slide by Jack Gieck.)

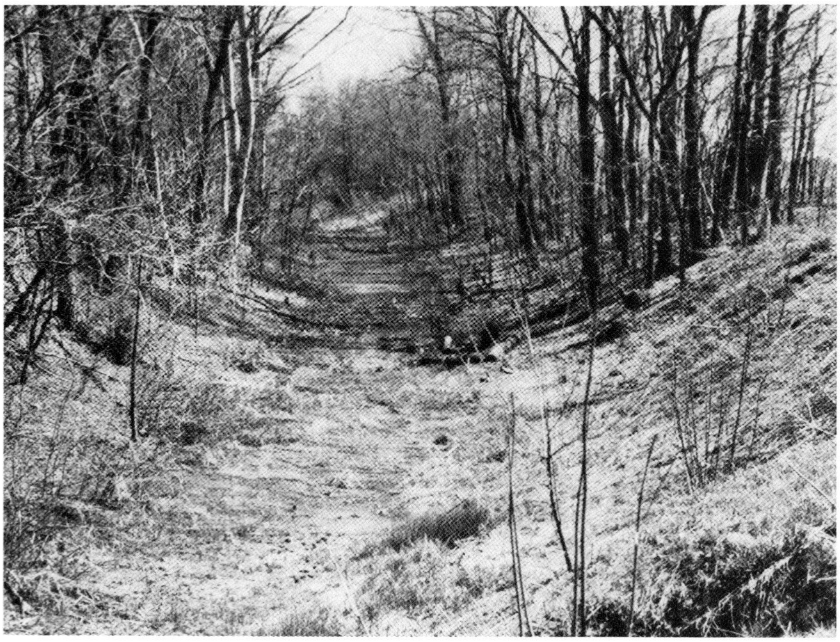

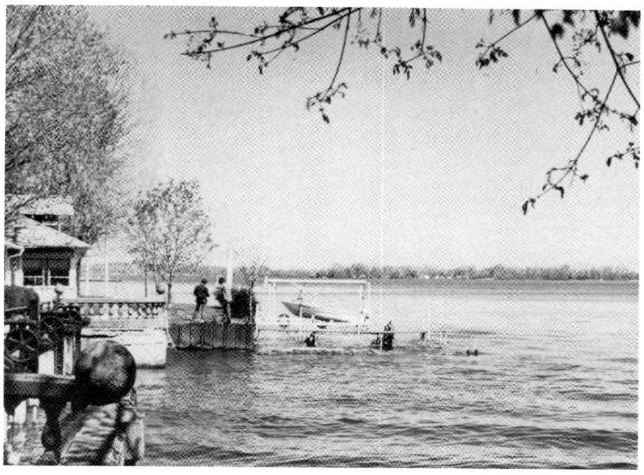

South of Newark the canal traversed Licking Reservoir, now called Buckeye Lake. Like the Portage Lakes, this reservoir has become a recreational area; the bulkhead which dams the lake may be seen at left; 1978. Mile: 189. (From a color slide by Jack Gieck.)

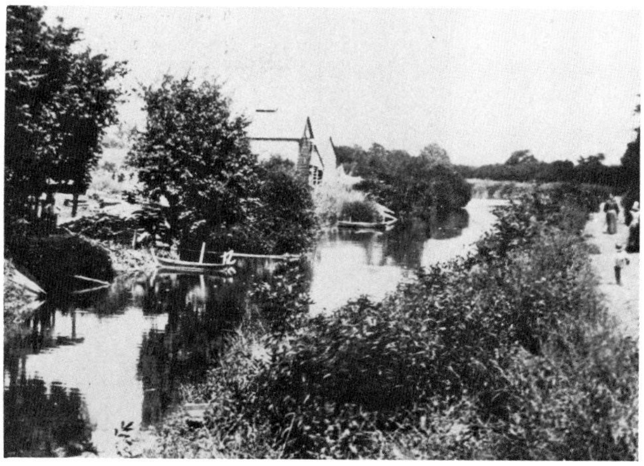

The canal on Licking Summit near Millersport, looking toward the Granville Feeder. The towpath at right is a thoroughfare for local pedestrian traffic. Mile: 191. (CSO)

The restored Pugh residence at Baltimore in 1974 (Photo by F. W. Trevorrow). Inset is the same scene in the 1860s, as reproduced in an engraving in Evert's 1875 *Fairfield County Atlas*. The dock and barn elevator are the property of canal pioneer Byron Pugh, who made his fortune shipping lumber and grain, and whose handsome Victorian mansion stands prominently in the background. The canal at Pugh Lock, Baltimore, is still watered in 1978. Pugh Lock was also known as South Lock, Deep Cut Lock, and Summit Lock. Mile: 196. (WRHS)

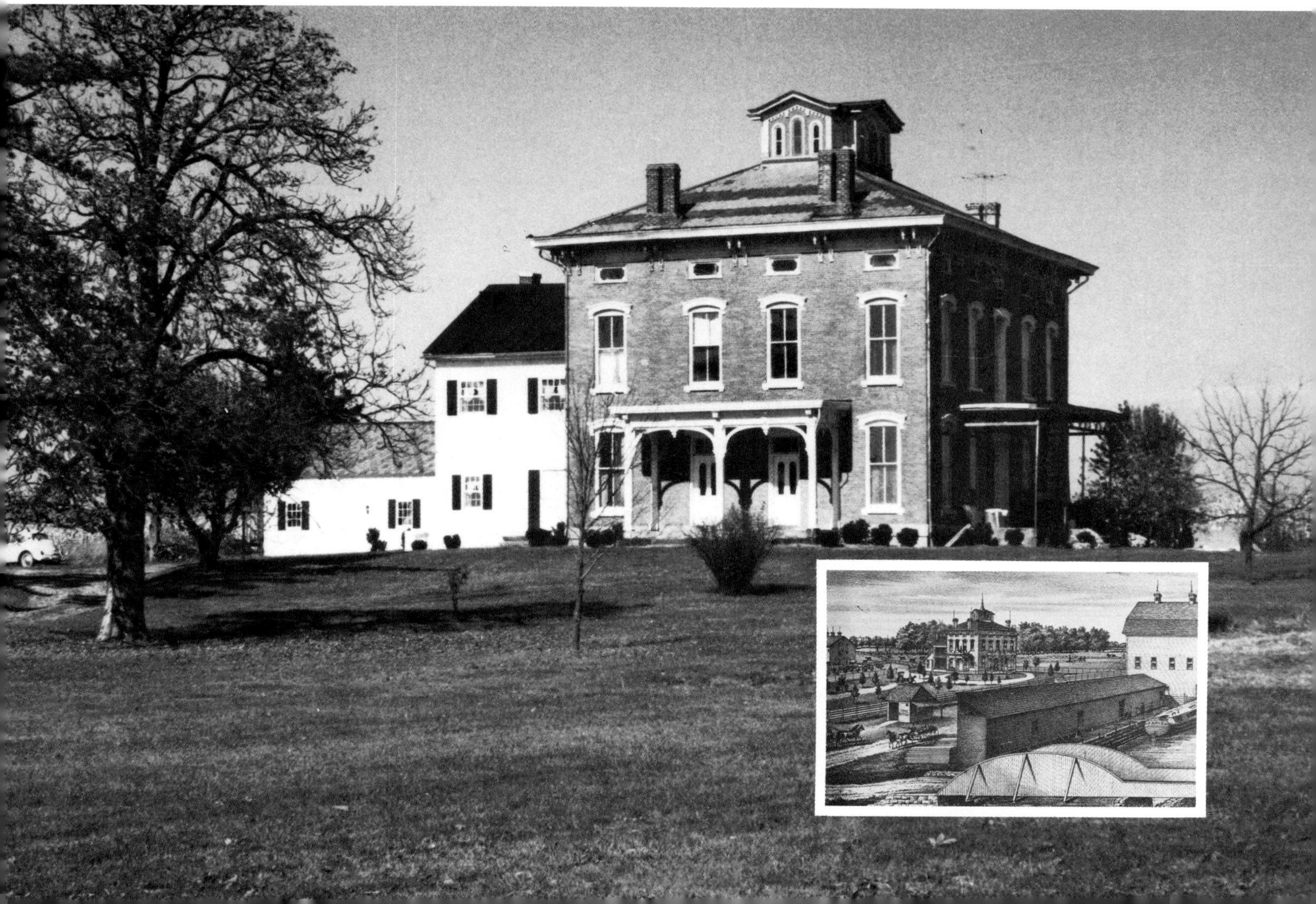

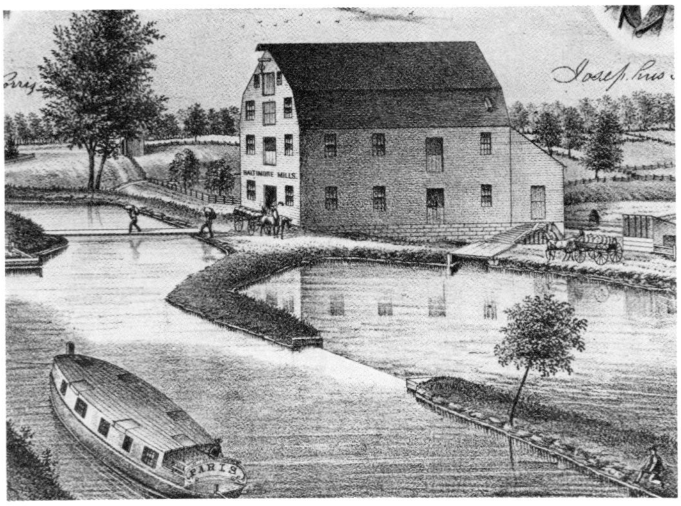

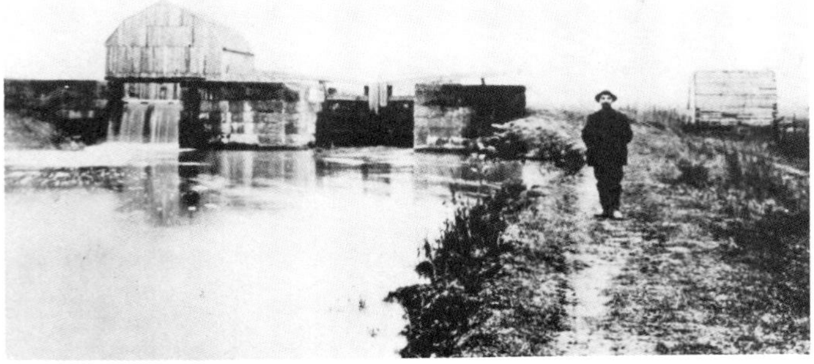

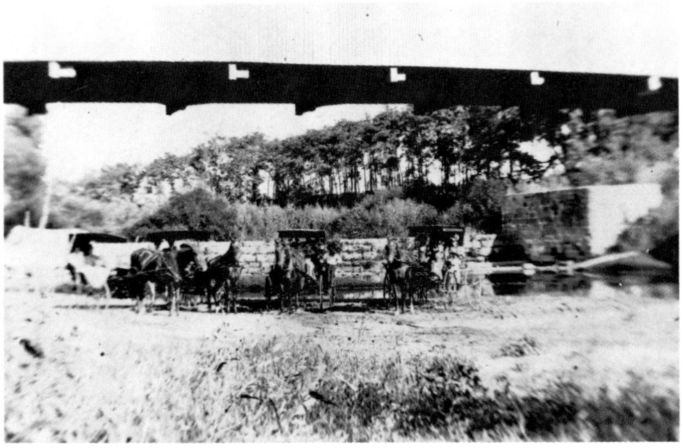

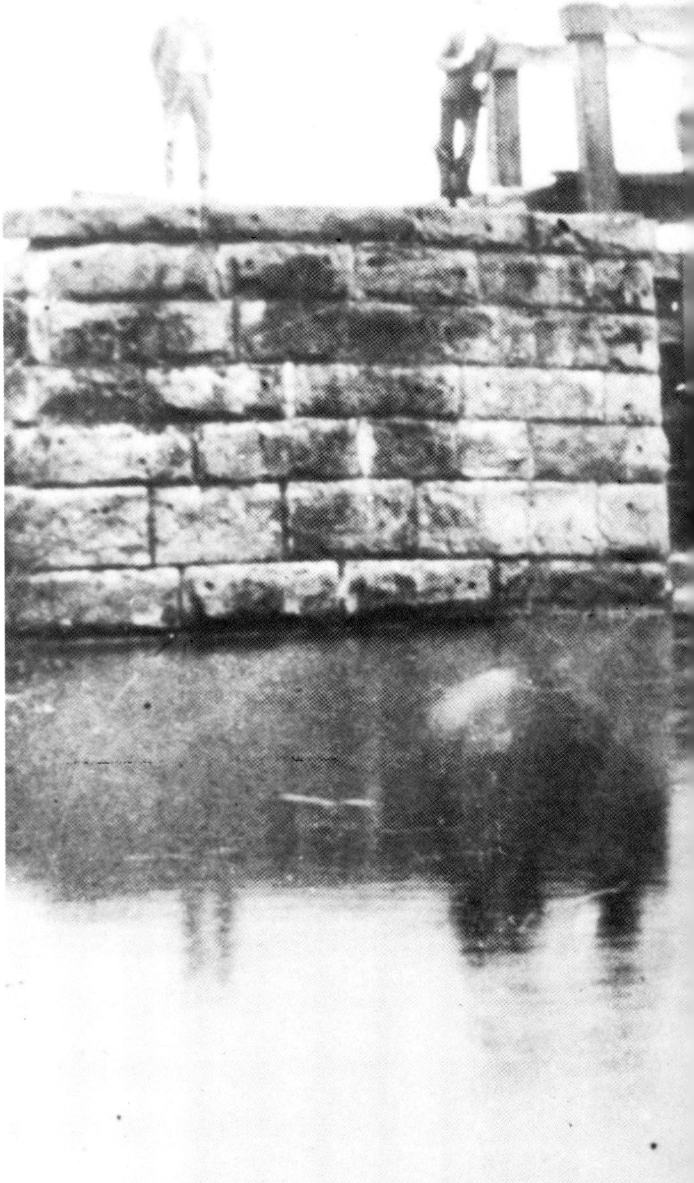

Top, Nearby the Pugh residence was Baltimore Mills, operated by Josephus Norris. Mile: 196. (Evert's 1875 *Fairfield County Atlas,* courtesy WRHS).

Center, Distillery Lock, Lockville, in 1880. Mile: 206. (Pockrandt Collection, SCHS)

Bottom, This photographer's lens was center sharp (only!) as he recorded this scene of couples posing in their buggies beneath a covered bridge at the State Dam on Walnut Creek south of Lockville. The dam was a part of the hydraulic system for Walnut Creek Feeder. Mile: 207. (OHS)

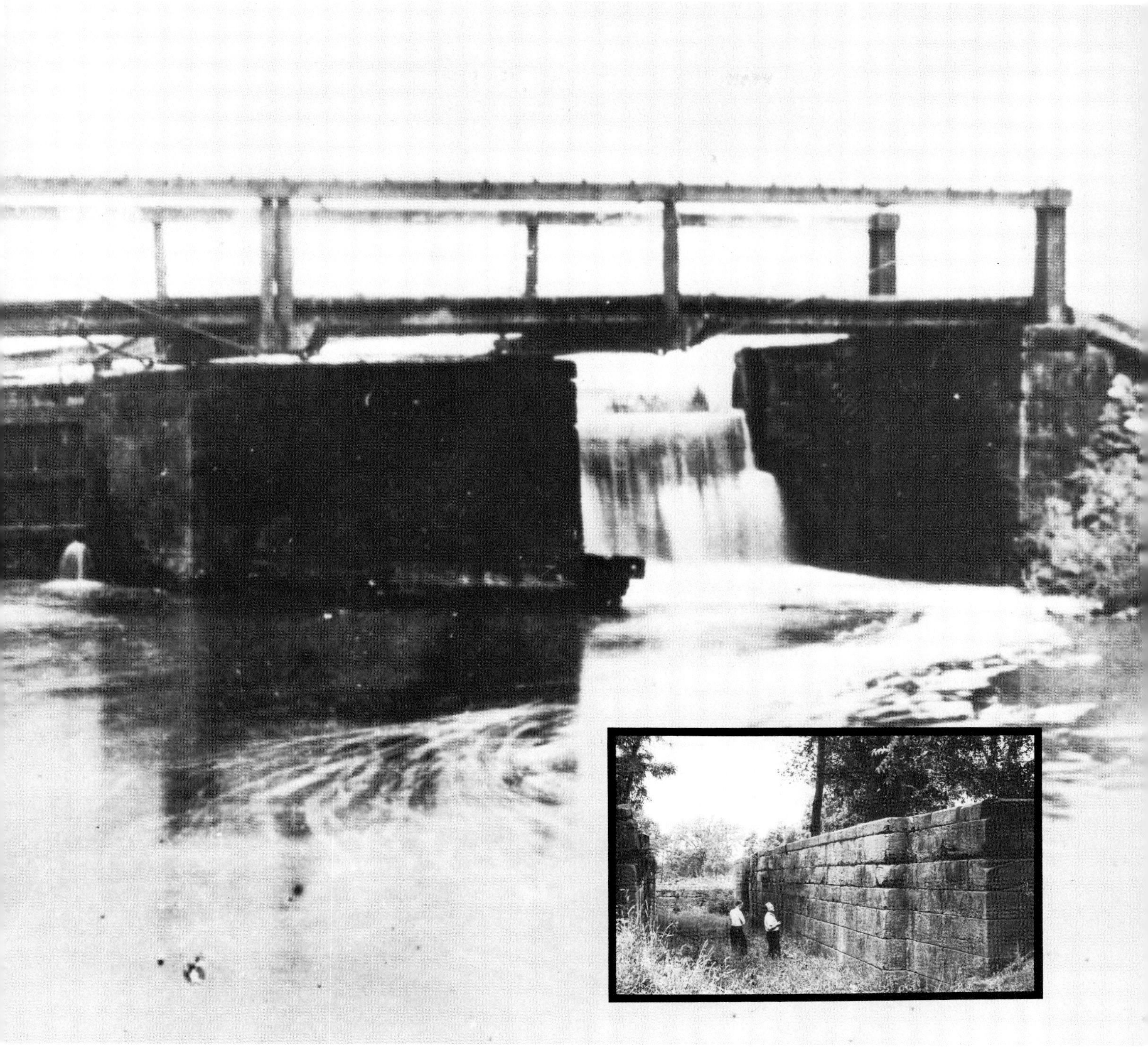

Grist Mill Lock at Upper Lockville, had a towpath change bridge across both the lock and the tumble (spillway), so that mules could cross over the canal as the towpath moved to the other side (OHS). The remains of Grist Mill Lock (inset) are explored by members of the Canal Society in 1966. Mile: 206. (Photo by F. W. Trevorrow.)

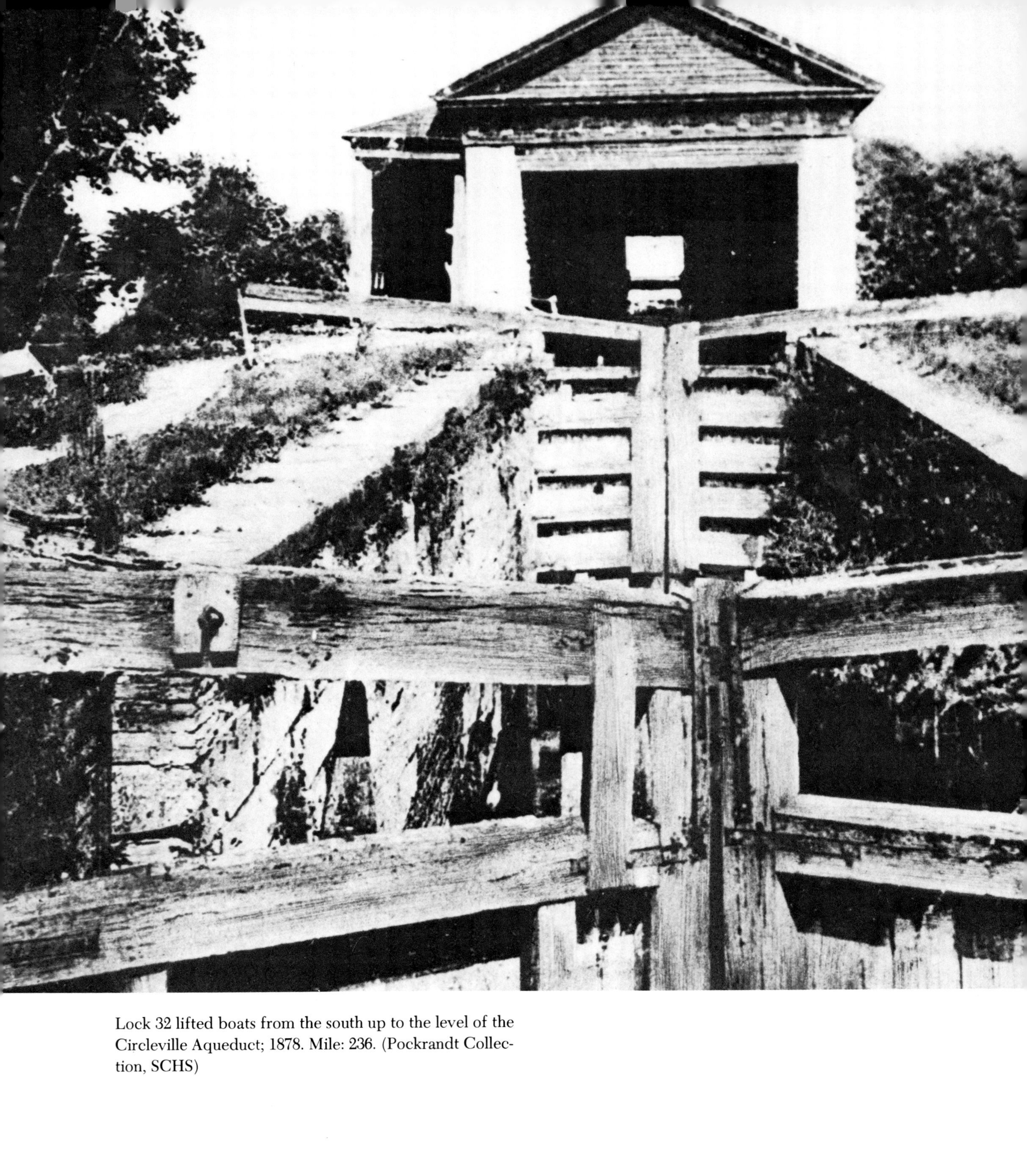

Lock 32 lifted boats from the south up to the level of the Circleville Aqueduct; 1878. Mile: 236. (Pockrandt Collection, SCHS)

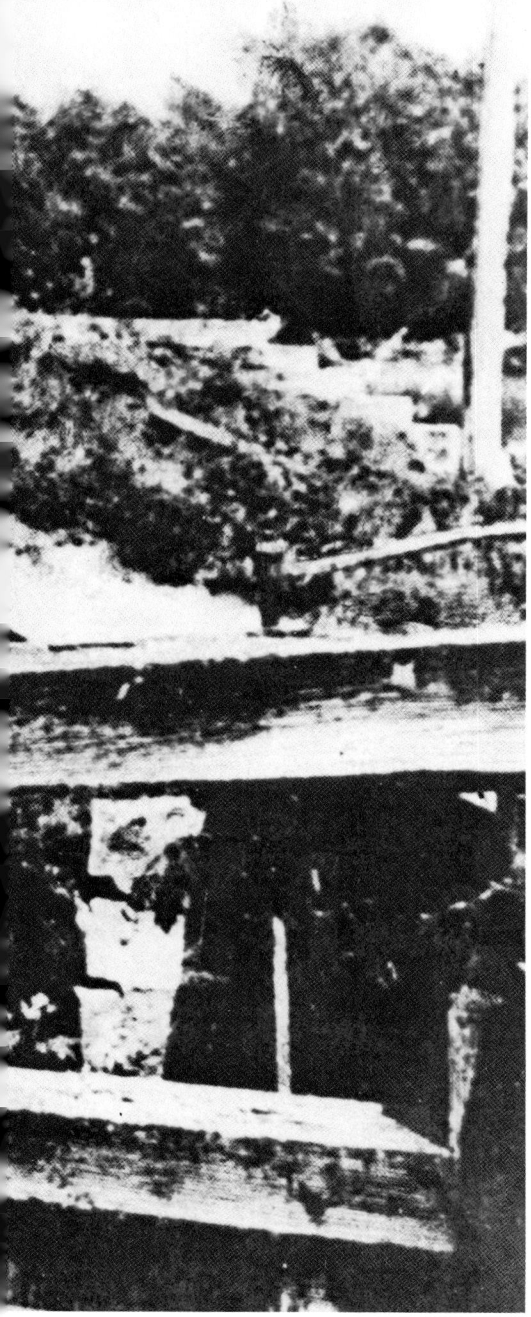

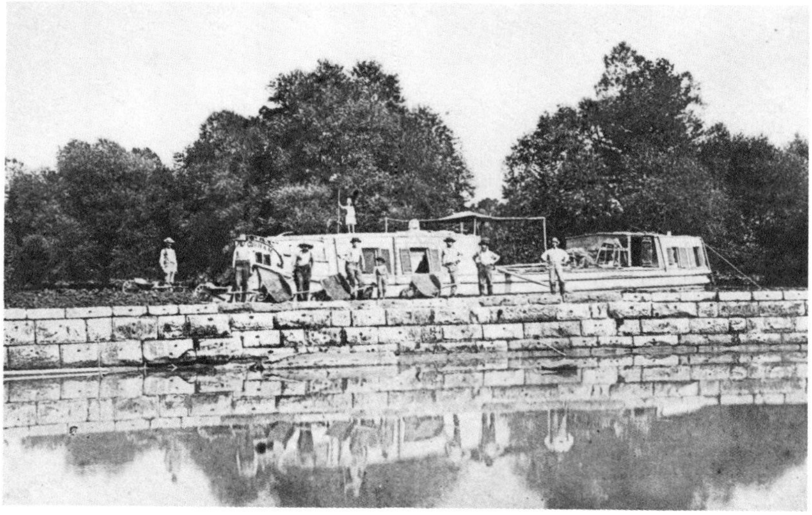

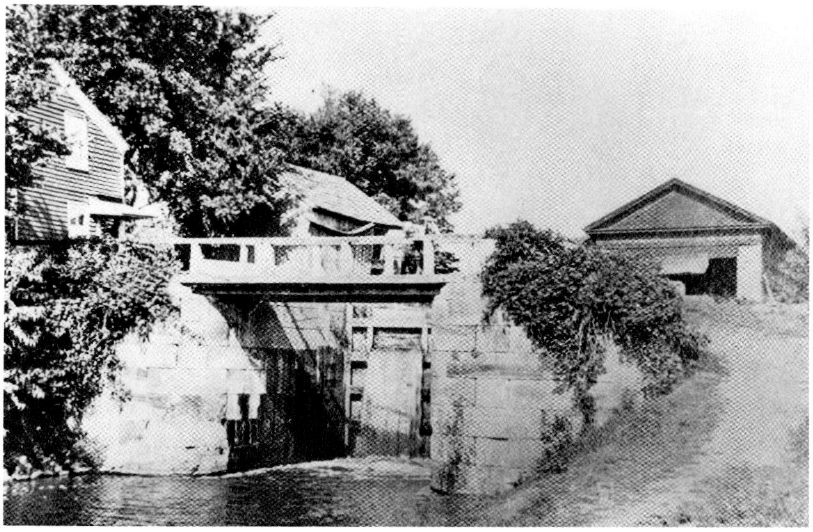

A state boat at Lockville (top). Children "help" the crew on the *Dick Gorham*. (Pockrandt Collection, SCHS)

The lockmaster's house (bottom) at south end of the Circleville Aqueduct can be seen projecting above the towpath at upper right. Mile: 236. (Canal Visitor Center, CVNRA, courtesy Cloea Thomas.)

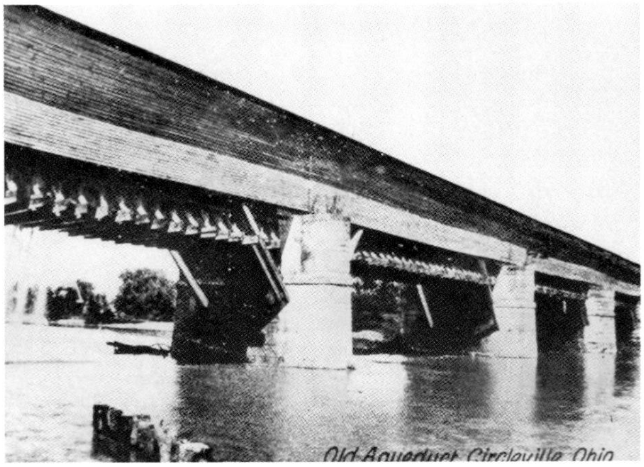

The 425-foot Circleville Aqueduct over the Scioto River, most expensive structure ($10,000) on the canal, was built by the Conrad Building & Bridge Company. After the canal was abandoned, the structure became a dance hall and later a gambling casino. The aqueduct burned and fell into the river in April 1915. Mile: 236. (From an old postcard.)

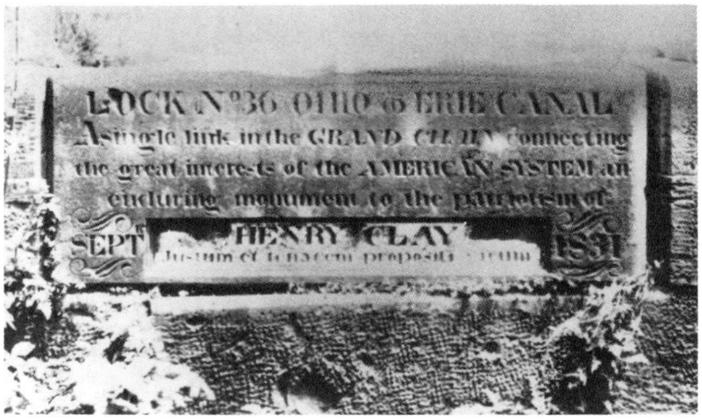

The lock plate on Lock 36 at Lower Marfield shows its 1831 completion date. Former Secretary of State Henry Clay of Kentucky ("I would rather be right than president") was nominated for president for the second time in 1832. Mile: 257. (CSO)

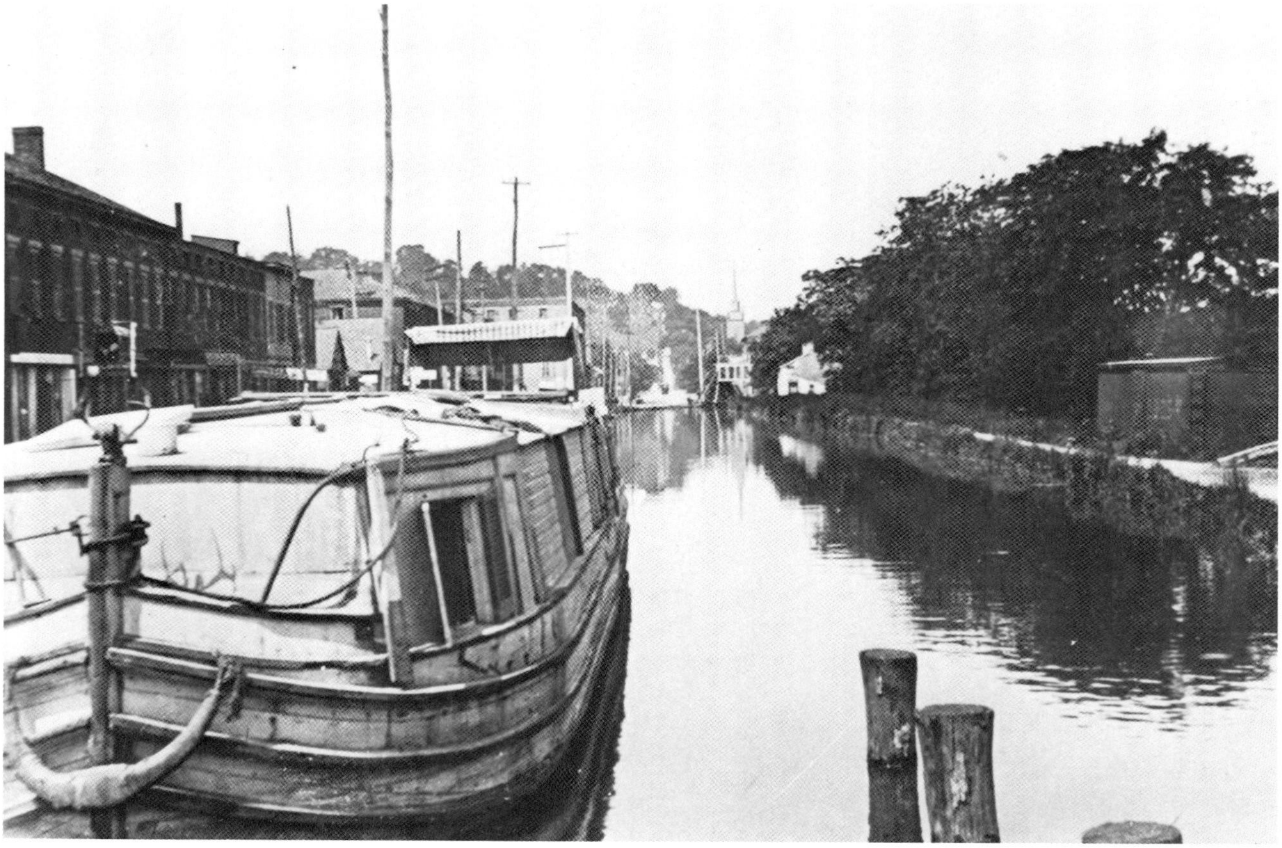

A view down Water Street looking west. The canal brought prosperity to Chillicothe as the city became a center of Ohio's pork-packing and milling industries. Mile: 258. (CSO)

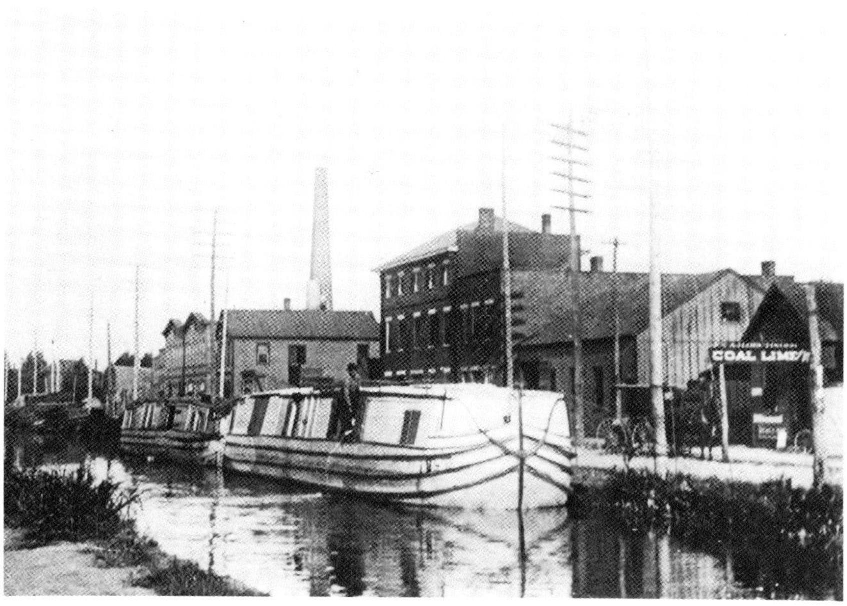

Canal boats along Water Street (looking east from Paint Street) in Chillicothe. Mile: 258. (CSO)

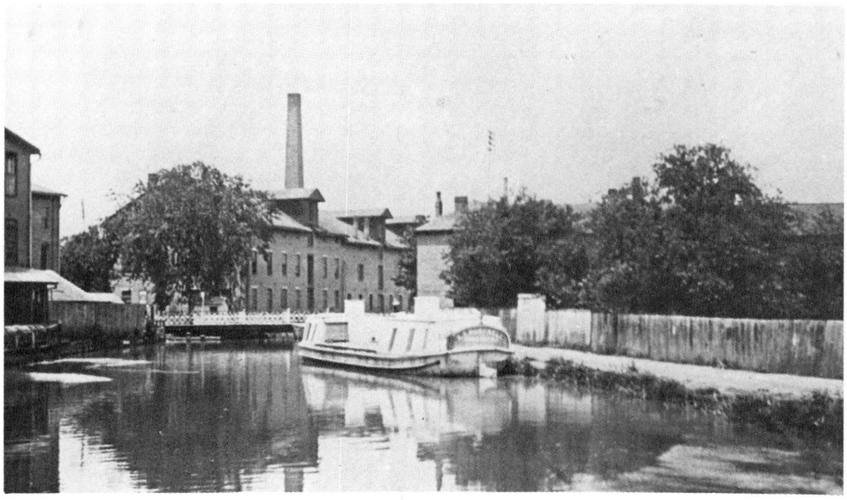

Looking north from 4th Street Lock in Chillicothe. Mile: 258. (CSO)

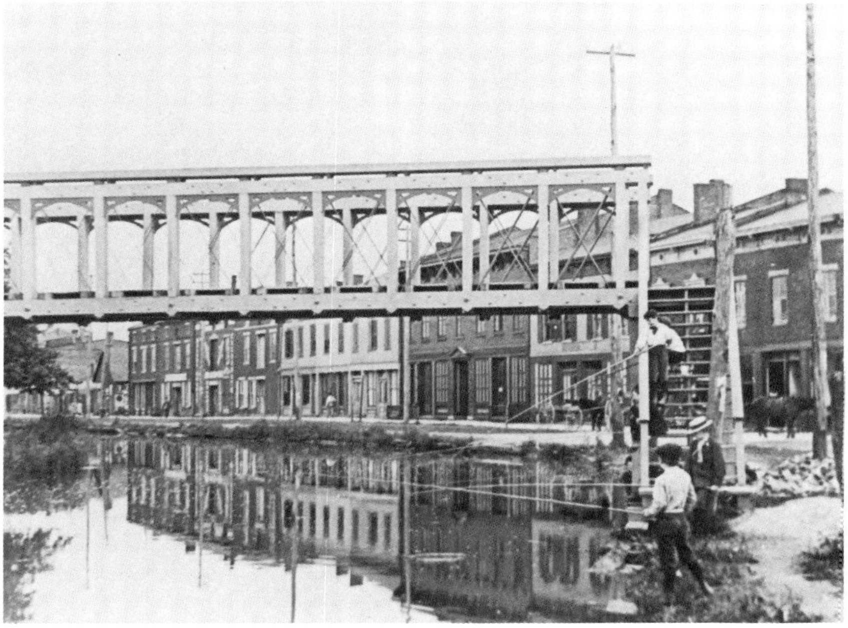

An iron footbridge over the canal in Chillicothe. Following a fire in 1852, all of the buildings along Water Street were moved back fifteen feet to provide a roadway parallel to the canal. Mile: 258. (CSO)

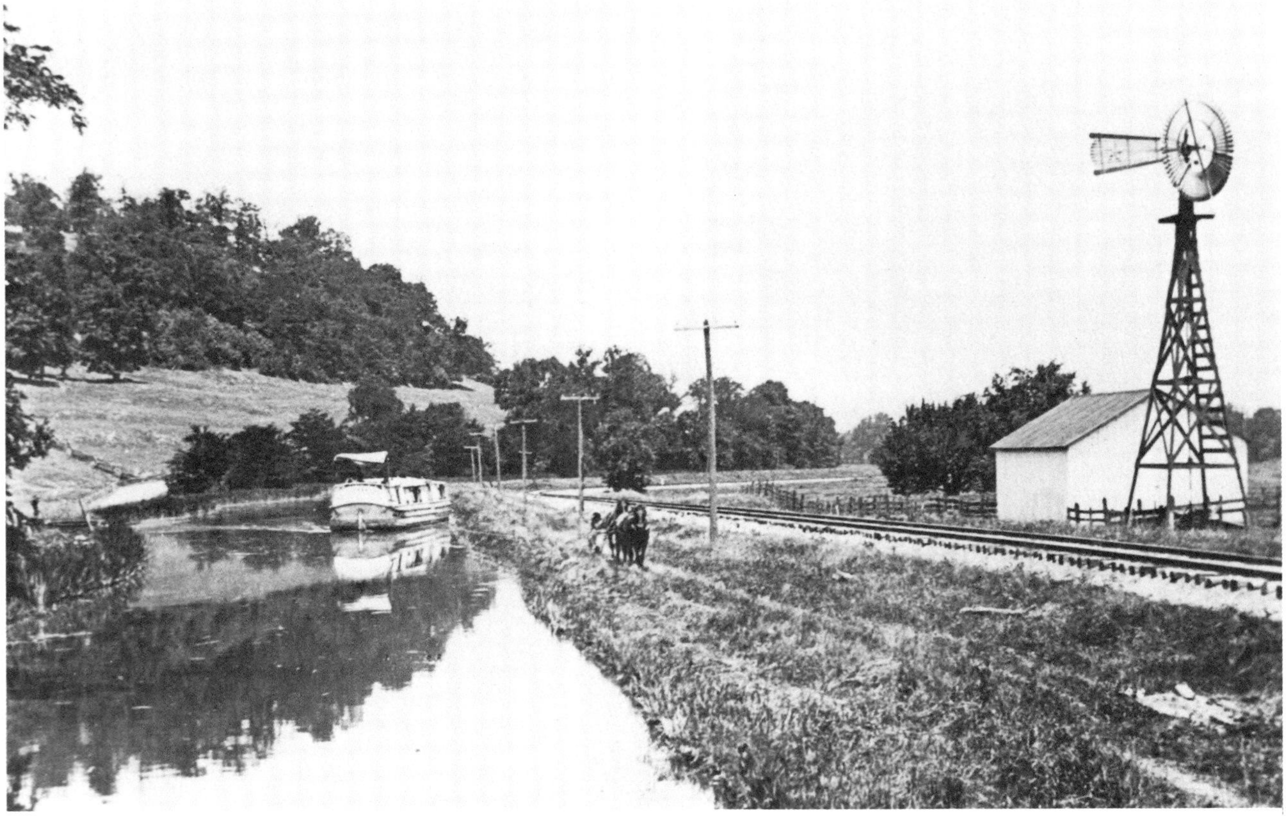

A freight boat (above) passes Kidnocker's farm south of Chillicothe, c. 1880s. Mile: 260 (CSO). A woodcut (below) portrays the State (feeder) Dam on the Scioto River south of Chillicothe, with Tomlinson's Three Locks on the canal at left. Mile: 264. (CSO)

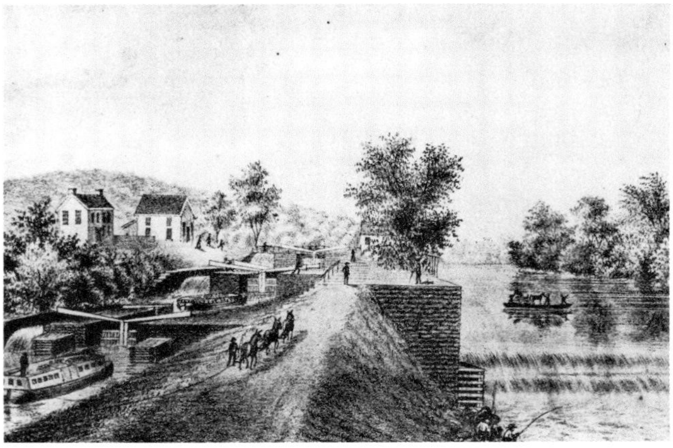

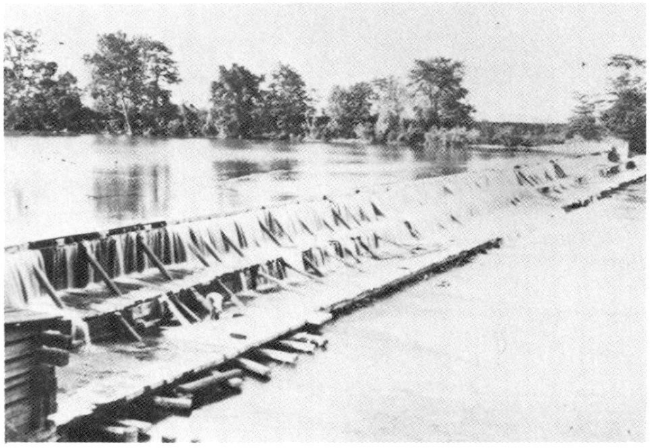

A vintage photograph of the State Dam on the Scioto. (CSO)

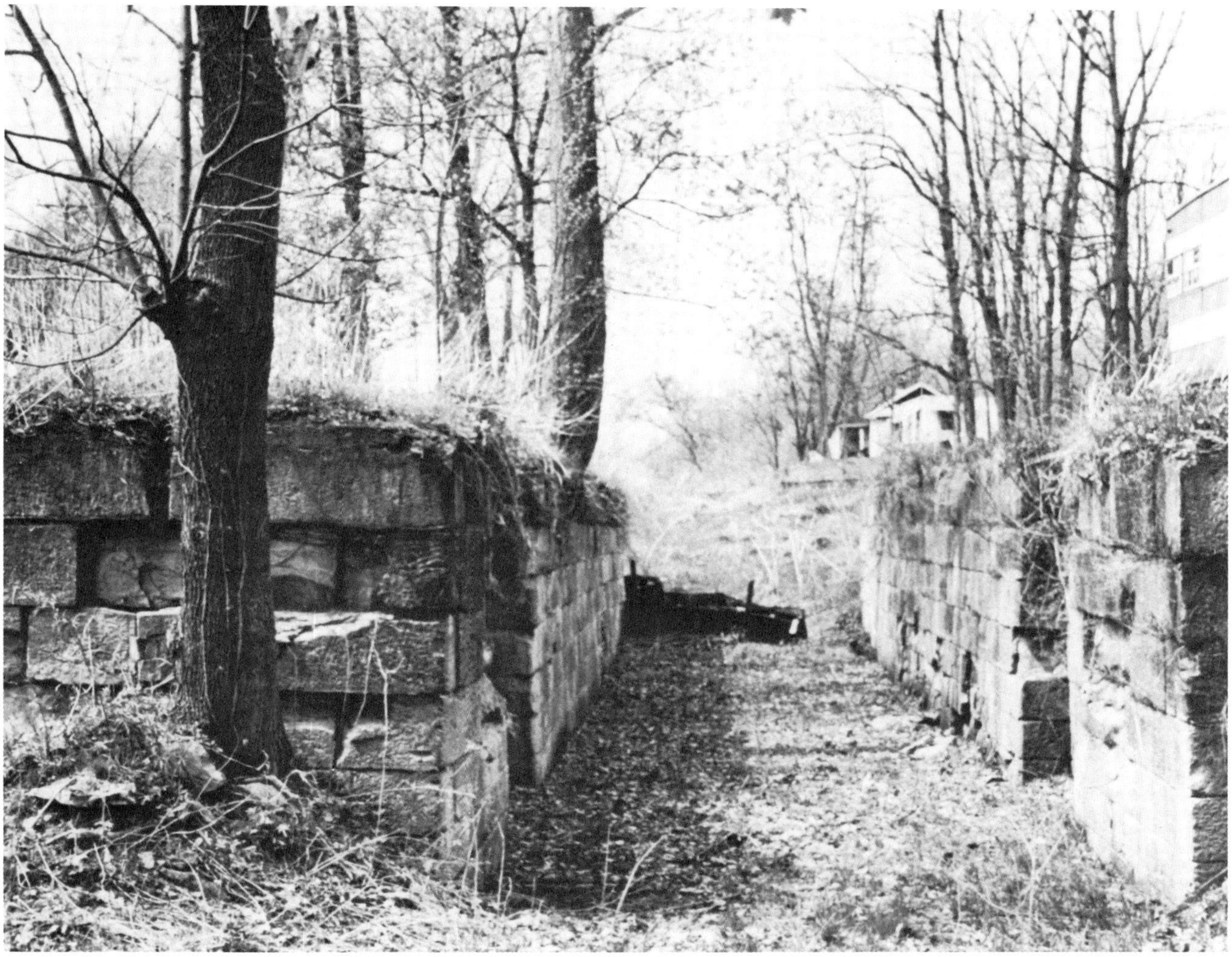

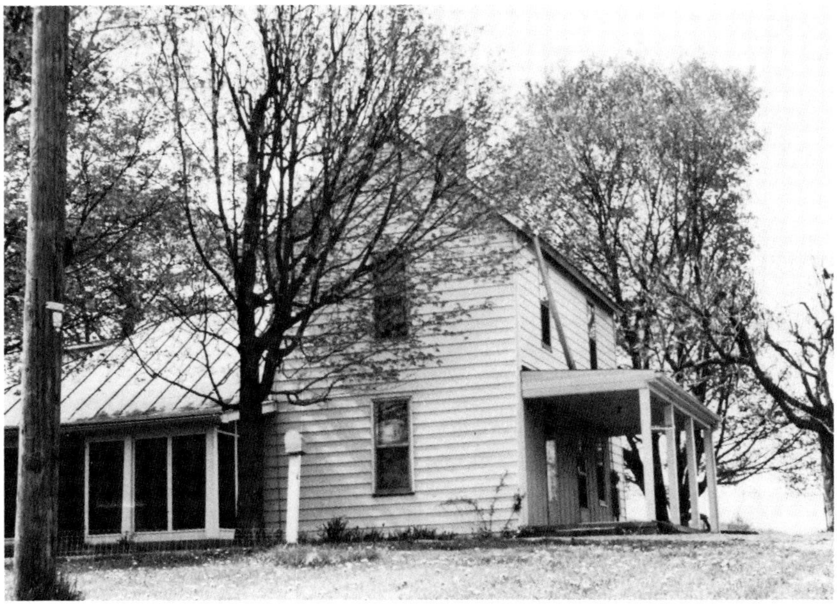

Above, A portion of Tomlinson's Three Locks as the structure looked in 1978. *Left*, The residence and store of Joseph Hirn and brother, who may also have been lockmasters for Tomlinson's Three Locks. From appearances, it is likely that the house was built from the same plans as the lockmaster's residence on the Miami & Erie Canal near the Mad River Feeder at Dayton; 1978. Mile: 264. (From color slides by Jack Gieck.)

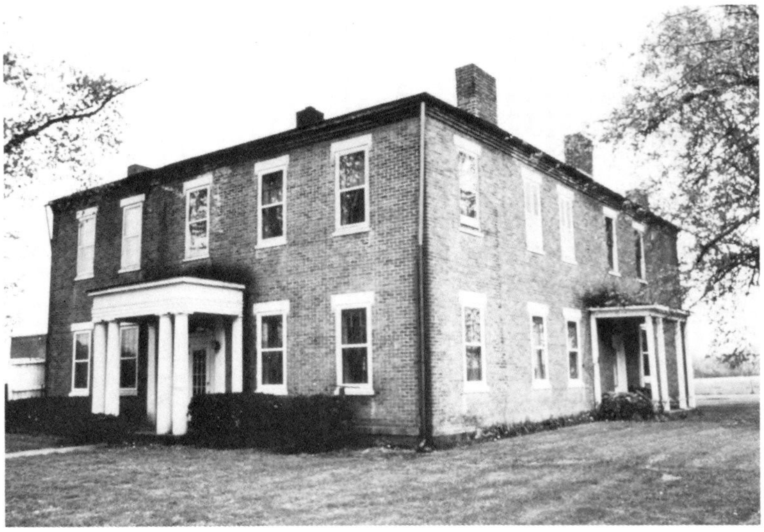

Above, The Higby homestead, north of Waverly, was residence of a canal entrepreneur who made his fortune shipping canal produce to New Orleans by *raft*. *Below*, This guard lock on the Higby property was built of stone only at the ends as support for the gates. Mile: 269. (From color slides by Jack Gieck, 1978.)

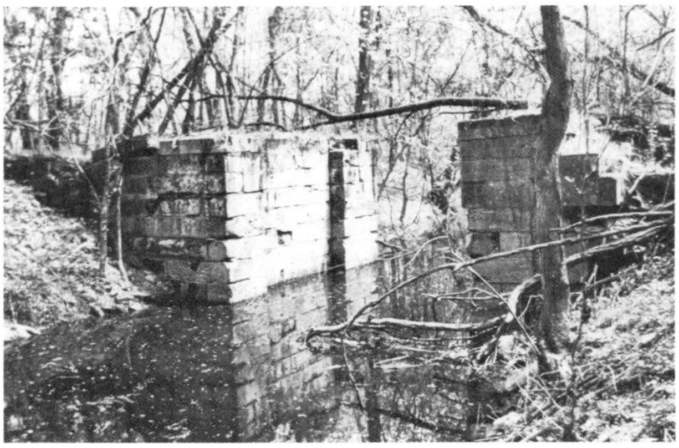

Below, This double culvert over Sunfish Creek north of Coopersville is one of the Ohio & Erie's few culverts. The axle assembly above the culvert may be an antique stump puller; 1970. Mile: 286. (CSO)

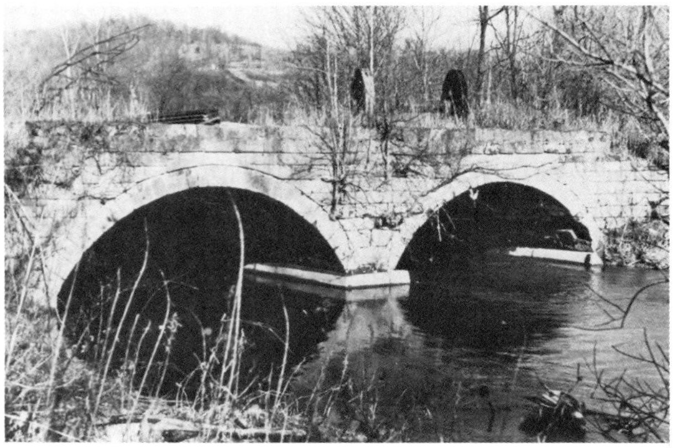

Below, John Hayes's boat, the *Friendly*, in Lower Pee Pee Lock south of Waverly, near Pee Pee Creek Aqueduct. Locks were invariably good places to fish, and the photographer caught no less than six fishermen plus a kibitzer in this early photo. Mile: 282. (Canal Visitor Center, CVNRA, courtesy Cloea Thomas.)

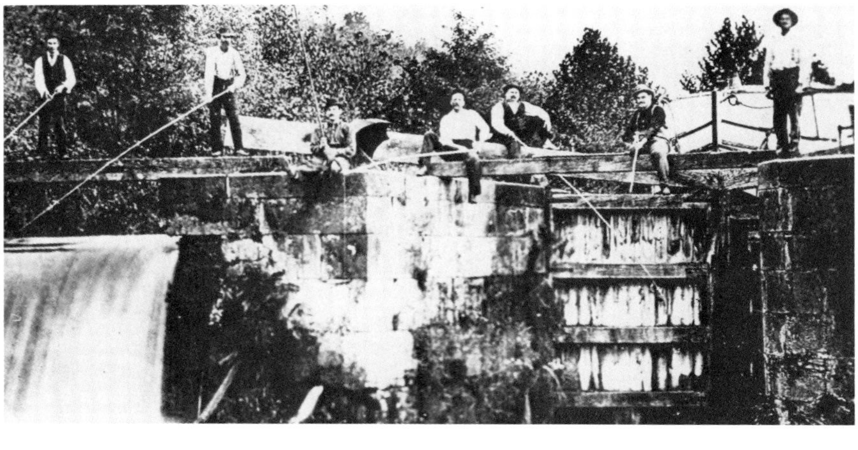

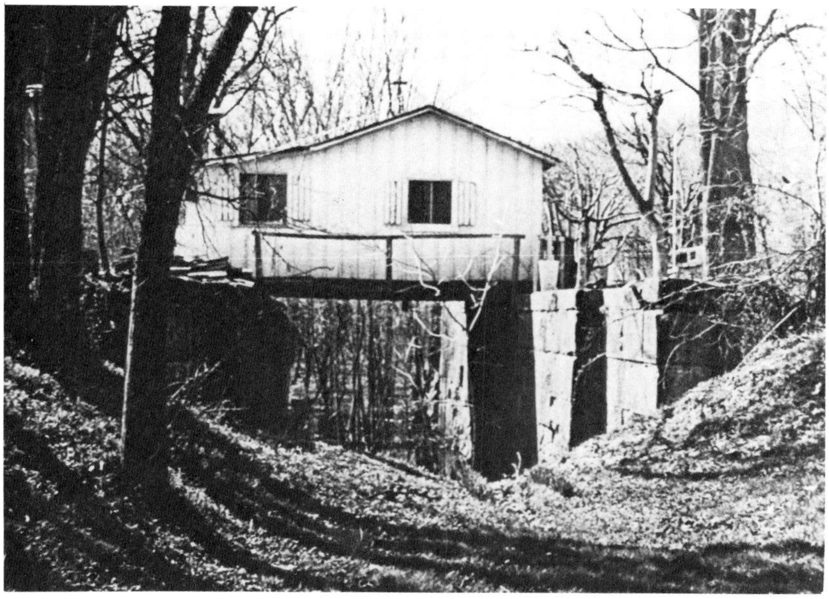

Reminiscent of Nye's Camp Charming, this cottage uses Lock 49 at Rushtown as its foundation. Mile: 301. (CSO)

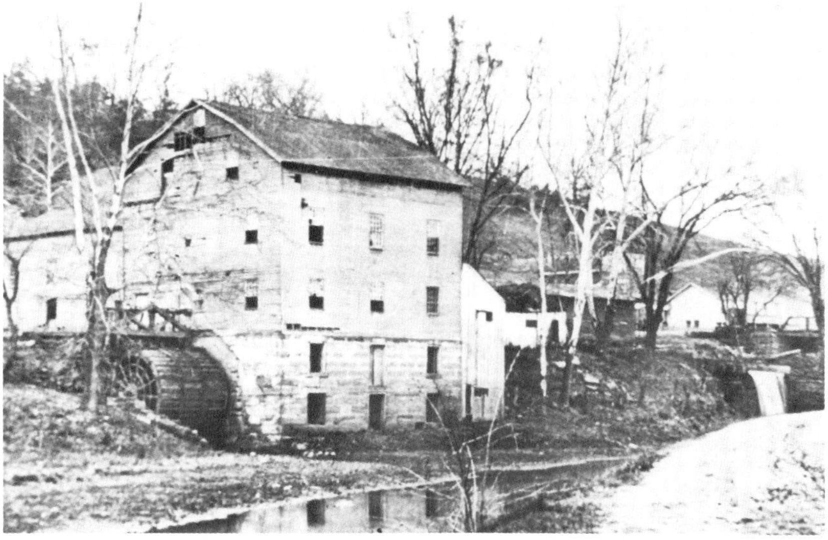

The five-story Union Mills north of Portsmouth, with its overshot waterwheel. Mile: 304. (Truman C. Throckmorton)

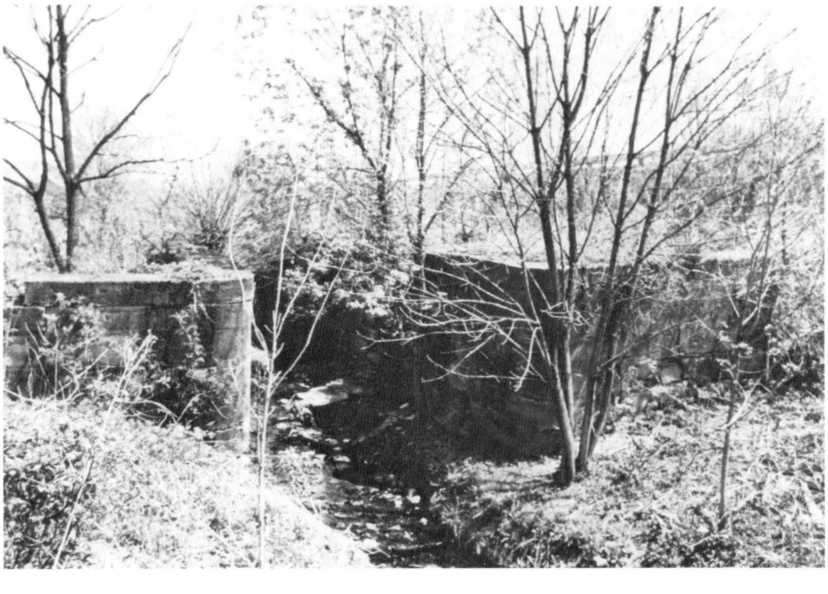

Lock 50 at Union Mills in 1978, last surviving lock near the southern end of the Ohio & Erie. Mile: 304. (From a color slide by Jack Gieck.)

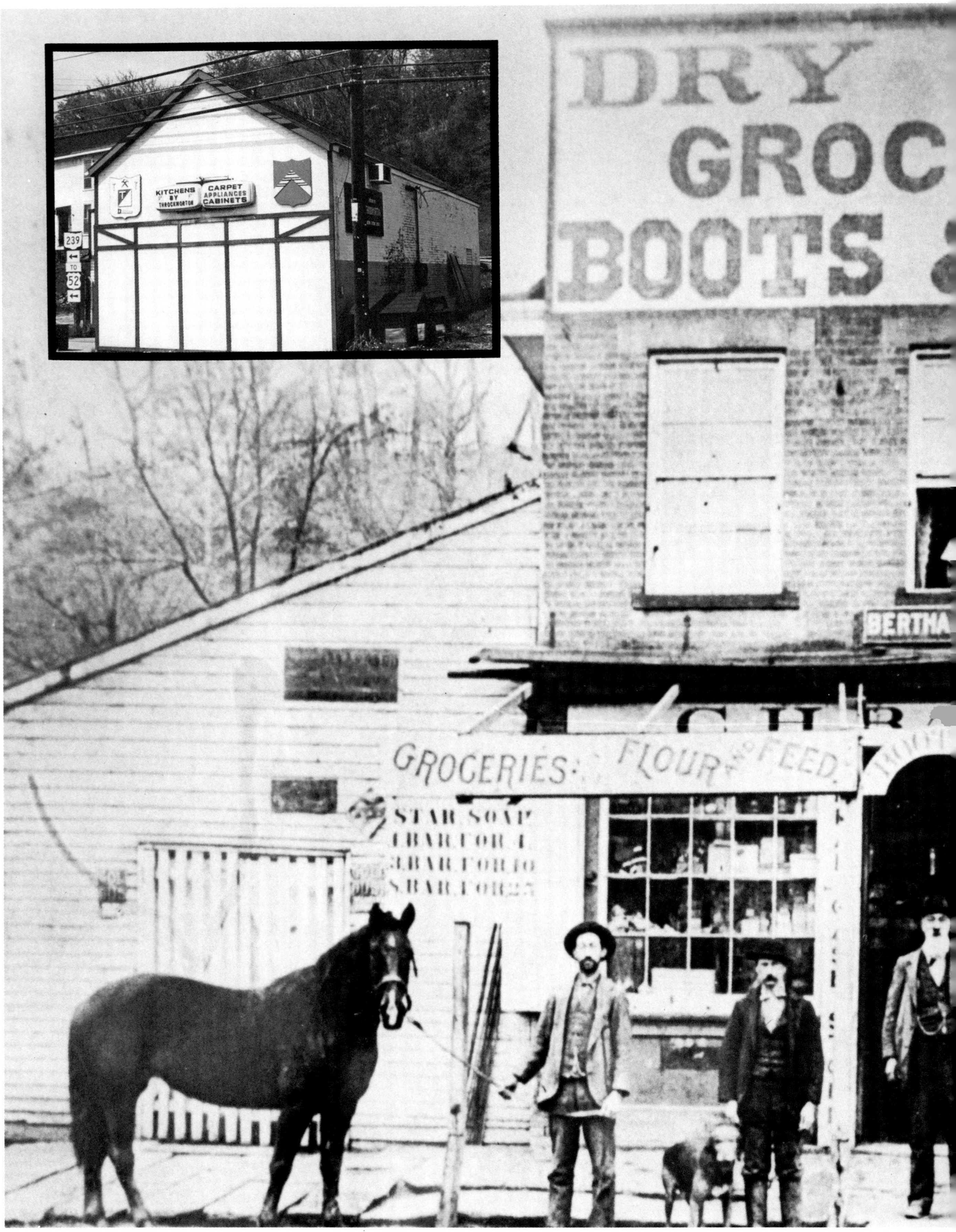
KITCHENS BY THROCKMORTON
CARPET APPLIANCES CABINETS
239
TO
52
DRY
GROC
BOOTS
BERTHA
GROCERIES
FLOUR AND FEED
STAR SOAP

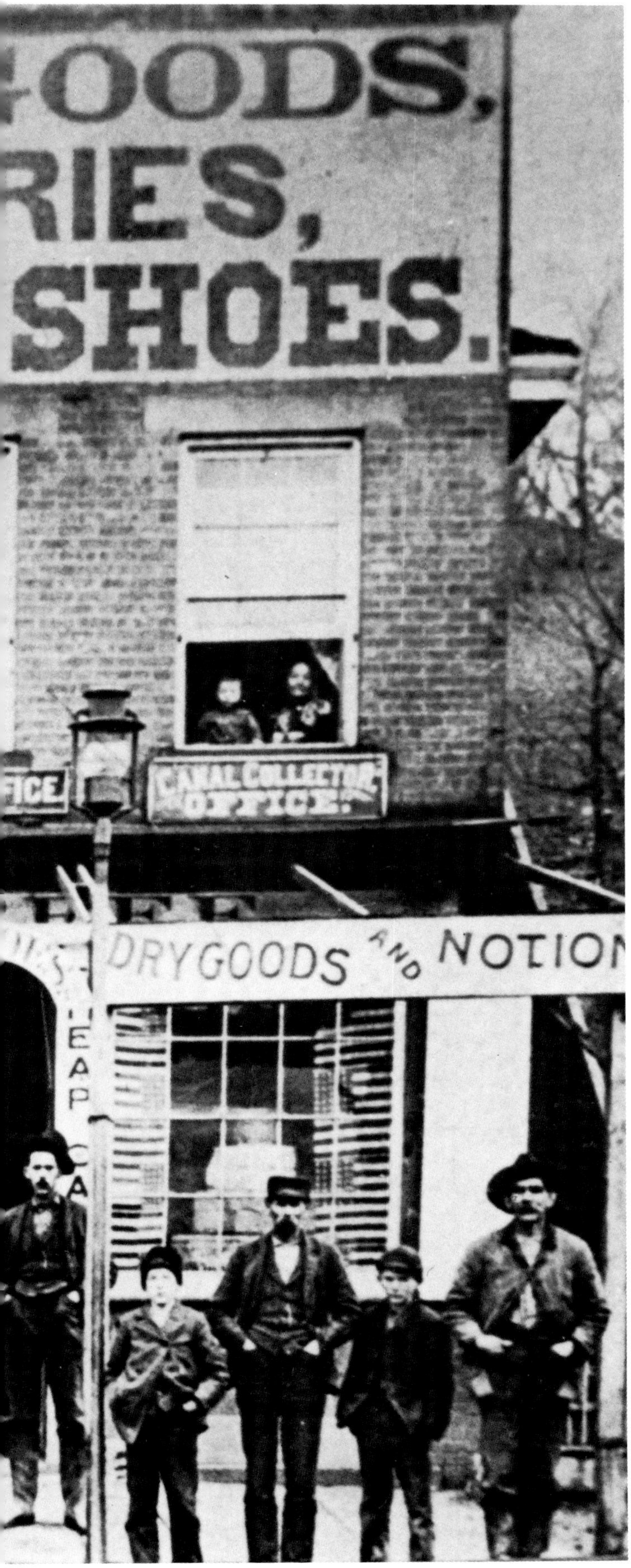

This tollkeeper's house at Union Mills Locks was also a general store. It is apparent that the entire community knew this photograph was going to be made (Truman C. Throckmorton). The tollkeeper's house (inset) as it was in 1978, after losing its upper story to a fire many years earlier. Mile: 304. (From a color slide by Jack Gieck.)

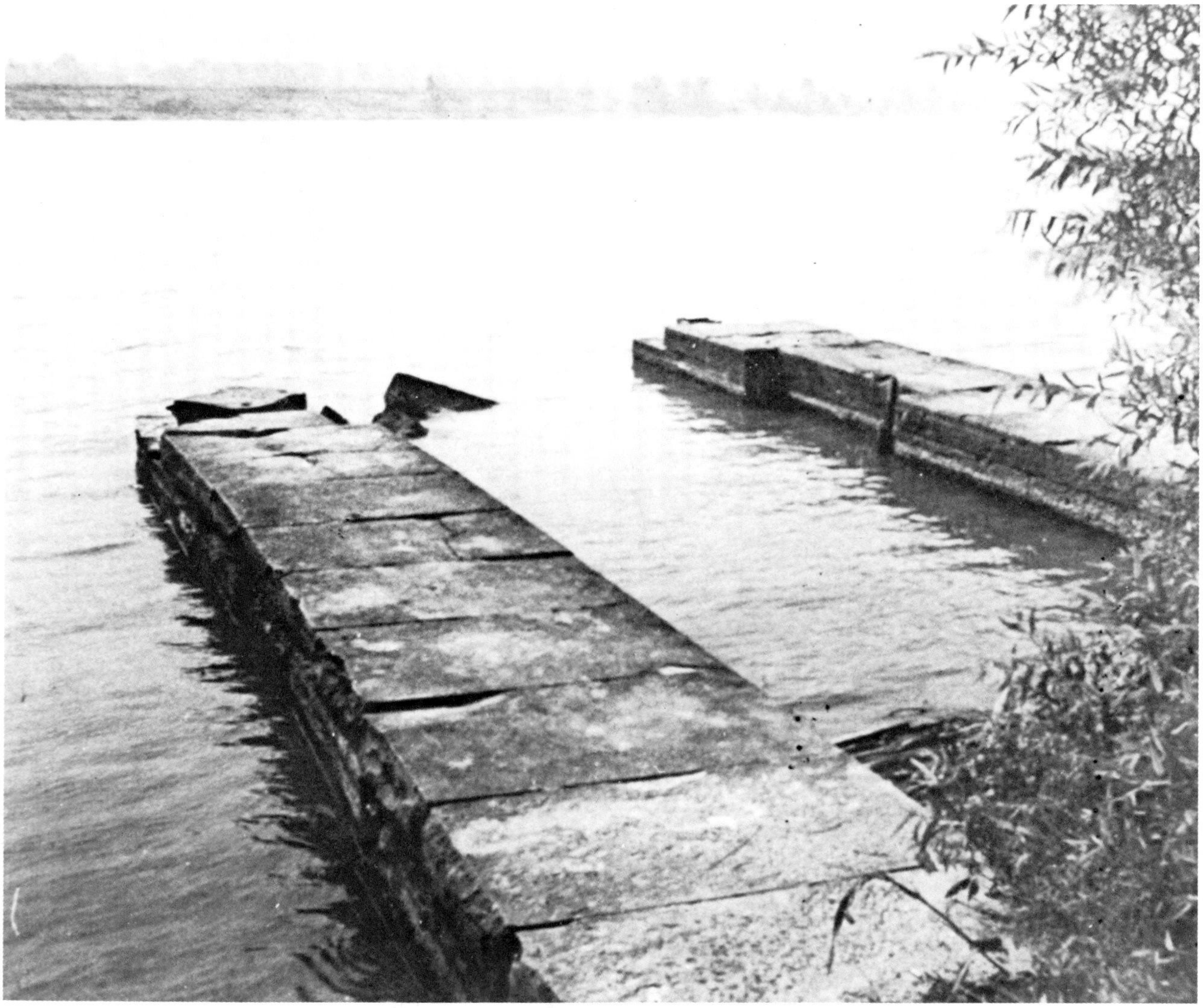

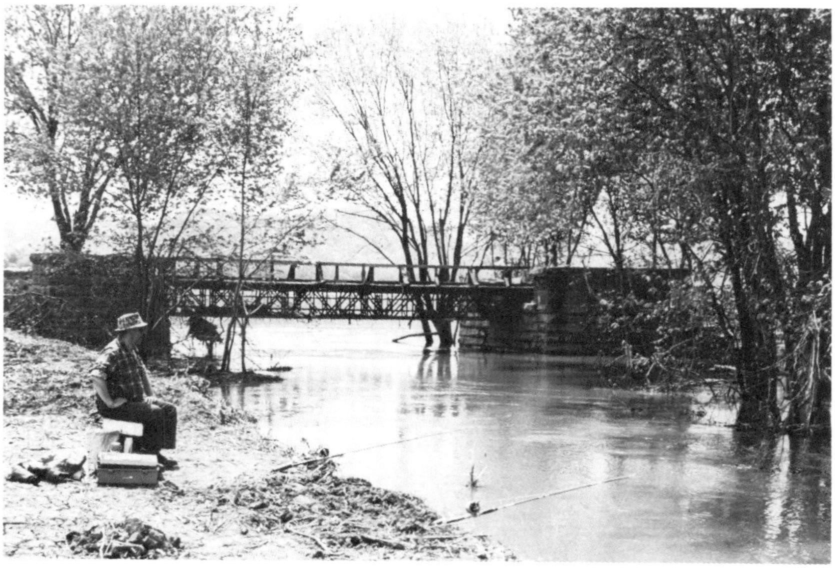

The remains of outlet lock (above) into the Ohio River, visible only when the river is at a low stage. (CSO)

The southern terminus of the Ohio & Erie, locking down into the Scioto River near its mouth, was abandoned in favor of a cut directly into the Ohio River. Planners had hoped that Ohio River boats could exchange cargoes in the Scioto mouth as they did in the Cuyahoga River at Cleveland—an objective which proved impossible because of seasonal Ohio River flooding and silting of the Scioto mouth. (From a color slide by Jack Gieck, 1978.)

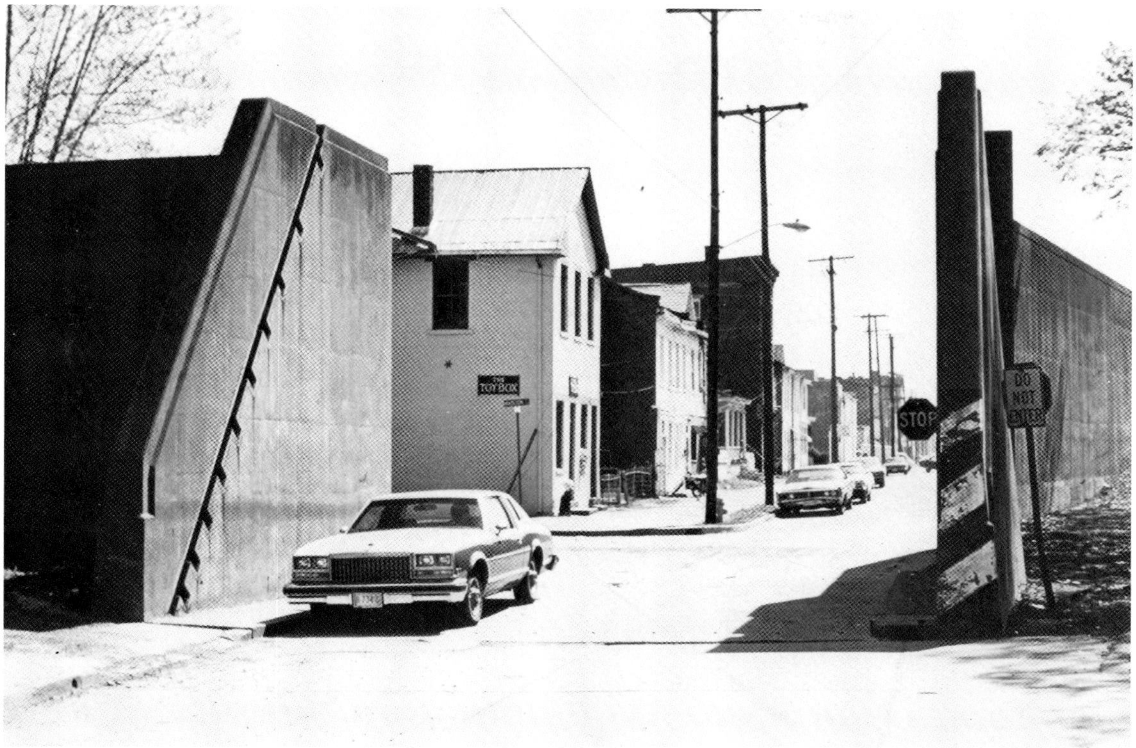

Flooding of the canal's southern terminus prevented Portsmouth from ever achieving the prominence and success of its northern counterpart, Cleveland. Even this massive 65-foot reinforced concrete floodwall (above) three miles in length, constructed by the WPA in 1934, failed to protect the city in 1937, when the river crested at seventy-one feet and inundated the city once again. (From a color slide by Jack Gieck, 1978.)

The city of Portsmouth (below), on the Ohio River—the southern terminus of the Ohio & Erie; 1978. (From a color slide by Jack Gieck.)

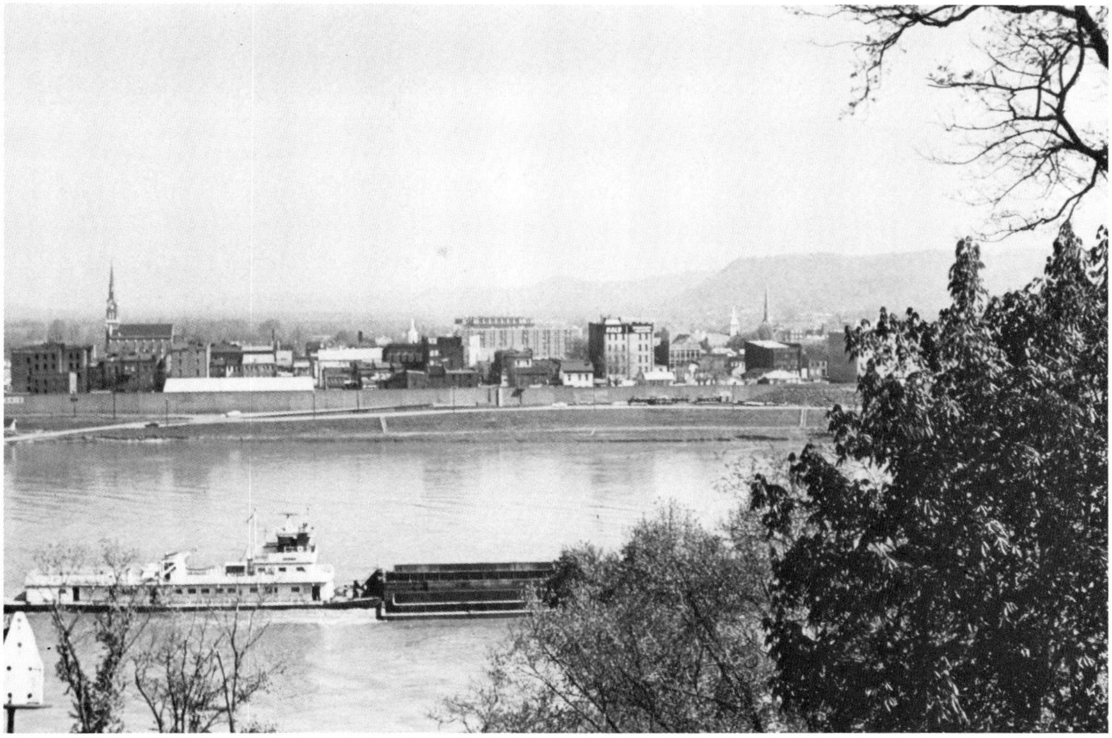

7 The Miami & Erie Canal

Unlike the Ohio & Erie, the Miami & Erie Canal was not initially conceived in a single, bold master plan providing a route from Lake Erie to the Ohio River. As explained by researcher Edith McNally:

"The Miami & Erie Canal was a composite of three separately authorized canals: the Miami Canal, authorized in February of 1825, the Miami Extension Canal, authorized in 1831, and the Wabash and Erie Canal, authorized in February of 1833 [after a political skirmish with the state of Indiana; see below].

"At Middletown, where ground had been broken in 1825 for the Miami, water was let into the canal on July 1, 1827, and by July 4 was deep enough to float the canal packet *Samuel Forrer* [named for the Miami Canal's famous chief engineer, later a canal commissioner]. It was not until early December, however, that the canal was navigable between the canal basin north of Cincinnati to Middletown, and the boats *Washington* and *Clinton* executed their official voyages carrying bands, politicians, some of the canal commissioners, engineers, contractors, and townspeople. A more auspicious celebration occurred on March 17, 1828, as the same boats and four others arrived at the basin in Cincinnati, where they were greeted by the salute of an artillery company, a bust of DeWitt Clinton, flags, and cheers."[1]

The *Liberty Hall and Cincinnati Gazette* reported on the event three days later:

"The weather was fine and many thousands of our citizens paraded along the banks to witness the interesting spectacle. Each boat was crowded with passengers who had proceeded to the basin in the morning for the purpose of being of the number who should make 'the first trip in canal boats to Cincinnati.'

"The Artillery Company, commanded by Captain Vinton, was stationed at the angle of the canal. At the same place was elevated a bust of DE WITT CLINTON [who had died a month before on February 11] and the national banner. The moment the *Washington* crossed the corporation line of the city, a national salute commenced by the artillery and continued till all the boats had passed to the place of landing, near Main Street. The citizens who lined the banks of the canal for nearly a mile in length then gave three cheers. . . .

"As an indication of a new direction of the business of the city, the Draymen, to the number of 30 or 40, with their horses and drays, on one of which was hoisted a flag, came marching up Main Street in single file to the landing of the boats. The sight was pleasing and contributed much to the interest of the occasion."[2]

Before his bust a solemn toast was publicly drunk to "the memory of DeWitt Clinton—the Erie Canal, an imperishable monument of his fame." There followed thirteen more formal toasts, each recorded for posterity by a stenographer, and many more which, it may be assumed, became progressively less formal. As with similar events replicated all over the state, the twin trademarks of the Cincinnati canal festivities were inevitably gunpowder and alcohol.

•

1. Edith McNally, "The Rise and Fall of Ohio's Canals, 1825–1913," *The Gamut* (Fall 1983).
2. *Liberty Hall and Cincinnati Gazette*, March 20, 1828.

As completed in 1828, the Miami Canal from Howell's Basin at the northern city limits of Cincinnati to Dayton was sixty-five miles in length. But it took another year to reach Cincinnati's Main Street and until September 1834 to finally link up with the Ohio River, descending along the present route of Cincinnati's Eggleston Avenue through ten locks—each with a lift of eleven feet (the final river lock had a nominal lift of thirteen feet, which varied with the stage of the river). Seemingly endless delays in completing these final locks were due, variously, to severe weather conditions, lack of stone, and outbreaks of cholera.

Unlike its Cleveland counterpart on the Cuyahoga River, however, the canal's direct egress into the river at Cincinnati was little used. Silting up of the locks nearest the river, together with seasonal Ohio River flooding, made it more practical to transship by wagon to the steamer docks on the riverfront from Lockport Basin, at the end of Eleventh, or Canal Street (now Central Parkway). The last two miles west of the basin were lined with warehouses, freight forwarding companies, and boat line offices.

To the north, after leaving the city limits of Cincinnati, the canal entered the valley created by Mill Creek, following it for some twenty miles while passing through the communities of Cumminsville, Carthage, Lockland, Crescentville, and Rialto. The canal then entered a series of natural ponds that were the source of Mill Creek, and connected its valley with that of the Great Miami River. Three aqueducts were required in this section, two carrying the canal over Mill Creek from the west to the east and back to the west bank north of Carthage, the other taking the canal across Beaver Run near Crescentville.

Entering the Great Miami Valley, the canal followed the east bank of the river through the towns of Port Union, Lesourdesville, Excello, Amanda, and Middletown. Wooden aqueducts crossed Gregory's Creek at Lesourdesville and Dick's Creek at Amanda. Middletown was the Miami Canal's junction with the Warren County Canal, an ill-fated branch canal providing Lebanon with access to the main trunk. Plagued with maintenance problems and sparse traffic, the Warren County Canal never earned its keep.

The lower division of the Miami Canal was forty-four miles long, and contained thirty-two locks to overcome a difference in elevation of 212 feet. Its upper division began at the feeder from the Great Miami River two miles north of Middletown, from which it followed the east bank of the river for twenty-two miles to Dayton, touching the towns of Franklin, Miamisburg, West Carrollton, and Alexandersville, terminating in a basin at Dayton.

Even though the southern terminus was not put through to the Ohio River for several years, the Miami Canal became operational in April 1830 between Cincinnati and Dayton, with 141 passages logged during the first month. By 1832, over a thousand passengers a week were being transported over the fifty-four-mile waterway.

Work began on the Miami Extension in 1833, after a seventeen-mile section from Dayton to a point south of Troy was authorized by the legislature. Additional segments to the north were parceled out in piecemeal enabling legislation until 1843, when the last thirty-three-mile stretch was let to contractors. Work on the entire system had been slowed by the panic of 1837, which resulted in irregular payments to the contractors. At last, in 1845, the entire 115-mile Miami Extension was opened to traffic.

The Miami Extension proceeded north from Dayton for nine miles on the east bank of the Great Miami River after crossing the Mad River on a wooden aqueduct. Here, the canal crossed over to the west bank of the Great Miami on the famous (in artists' renderings) Great Miami Aqueduct. Three miles above Piqua, the canal crossed Loramie Creek on another aqueduct and immediately ascended the dramatic staircase of five locks at Lockington, arriving at Loramie Summit, which was uninterrupted by locks for twenty-three miles—all the way to New Bremen, passing through Woodbourne, Newport, Berlin, and Minster along the way, and employing an exceptionally tall double culvert over Turtle Creek north of Lockington to maintain its elevation.

From Loramie Summit, the canal descended to Junction, so named because Indiana's Wabash & Erie joined the Miami Extension here. On the way, in fifty-seven miles, the canal went through the towns of St. Marys, Kossuth, Spencerville, Acadia, Delphos, Ottoville, Hamer, Mandale, St. Andrew's, and Charloe, requiring six aqueducts or culverts, and thirty-two locks to drop the canal a total of 217 feet.

A major hydraulic feeder entered the canal at St. Marys, eight miles from the northern end of the summit. Lake St. Marys, or Grand Lake Reservoir, was the largest artificial body of water in the country at the time. North of Kossuth was Deep Cut, a 6,600-foot excavation more than fifty feet deep that permitted the canal to cross the formidable St. Marys Moraine, a ridge of hard blue clay that separates the

St. Marys watershed from that of the Auglaize River. Deep Cut took a force of four to five hundred laborers nearly four years to complete.

The canal's link with the Wabash & Erie at Junction provided the Miami's needed access to Toledo and Lake Erie. The Wabash & Erie had been completed between Toledo and Fort Wayne two years earlier. In the state of Indiana, this canal served the towns of Fort Wayne, Wabash, Peru, Logansport, and Lafayette; by the time it was put through to its southern terminus at Evansville, the route stretched over 462 miles—375 miles in Indiana, the balance in Ohio. It was, in fact, the longest canal in the nation.

In its Ohio section the Wabash & Erie made extensive use of slackwaters along the Maumee River, with major dams on the river at Providence and Independence. The canal also had provided the Swan Creek (Toledo) and Maumee sidecuts between the canal and the channel of the Maumee River.

A major political problem had arisen between Indiana and Ohio during the planning stages of the Wabash & Erie as a result of an 1827 act of Congress, granting Indiana millions of acres of government land adjacent to the proposed canal route to help finance its construction—a typical congressional gratuity at the time to encourage canal building. But this action would have resulted in Indiana's owning and controlling some 250,000 acres of land within the state of Ohio, creating an odd state of affairs. The issue was ultimately settled by a compact between the two state legislatures in which Ohio agreed to finance and construct the portion of the canal within its own borders.

It was in 1845 that Ohio's Board of Public Works first suggested that "for the sake of convenience," the entire canal route from the Ohio River to Lake Erie be designated by the name, "Miami & Erie Canal." This ingenious political expedient, summarily appropriating and renaming eighty-seven miles of the Wabash & Erie, was ratified by the state legislature on March 14, 1849. Thereafter, Ohio could advertise that it had a second major canal, some 250 miles in length.

Near the mouth of the Maumee River, there had been enormous competition between local communities for the profitable status of being the northern terminus of the canal. The political compromise eventually reached was that both Maumee and Toledo would have sidecuts from the canal to the river, with the canal's Lock One, or River Lock, being located at Manhattan—a community just north of today's downtown Toledo, long since swallowed up by the city. As it turned out, however, the direct river access was not extensively used because of the difficulty of controlling canal boats in the currents of the broad Maumee River. As with Cincinnati, most merchandise was off-loaded and transshipped from the canal to lake steamer docks. In fact, the channel through Toledo from Swan Creek Sidecut to Manhattan was ultimately abandoned in 1864.

The first boat to make the entire run from Cincinnati to Toledo was the *Banner*, under the command of Captain Ben M. Ayres. The *Banner* was a passenger packet operated by the highly successful Doyle and Dickey Line, which managed fleets in both Ohio and Indiana. Under the impending threat of railroad competition, the line had become obsessed with speed. Canal historian, educator, and author George C. Crout, writes about the trip:

"The 'Banner' left Cincinnati on Tuesday, June 24, 1845, at 2:30 P.M. and passed through Middletown early the next morning. A group of leading political figures as well as canal officials, headed by Samuel Forrer, Chief Canal Commissioner, were on board.

"Ceremonies had been held as the 'Banner' departed from Cincinnati, then the largest city in the West, for Toledo on Lake Erie. The editor of the Toledo Blade, J. W. Scott, described the arrival on June 27 at the Lake Erie port as follows: 'The splendid packet boat, *Banner*, of Doyle and Dickey's Line . . . had the honor of being the first boat to pass through the Miami Canal from Cincinnati to Toledo. She was welcomed by such demonstrations as the short notice of her coming enabled us to make. The long delayed day of the consummation of the great work has arrived and it well becomes us of Western Ohio to rejoice over it.'

"A few days later, Captain Denver of the 'Rose' completed the first southward trip between Toledo and Cincinnati . . . the 'Alfred Ely' claimed to be the first to make the round trip. . . . The exact date on which the 'Banner' left on its return trip is not known, but a passenger who preferred to be remembered as 'W.H.S.' published a detailed account of that trip:

" 'The *Banner*, the best of Doyle and Dickey's fine line of packets has few superiors for commodiousness of cabins, richness and variety of fare prepared by the best and most obliging stewards, and the quiet behavior of the men employed. The line is well supplied with good horses, there being relays through the route and there are now 100 men engaged. Upon the whole, there can be no fault in the plan of operations.' The line used 300 horses with relays every 10 to 12 miles in order to keep up the speed."[3]

As with Ohio & Erie photographs, excellent collections of Miami & Erie pictures are found in a number of locations, but the richest finds are near the northern and southern termini. Since operations on Ohio's western canal began with completion of its southernmost Miami Canal section, the photographs in this chapter are arranged in reverse order (south to north) compared with those in the Ohio & Erie chapter; but estimated mileages are still measured from Lake Erie.

3. George C. Crout, from an unpublished manuscript.

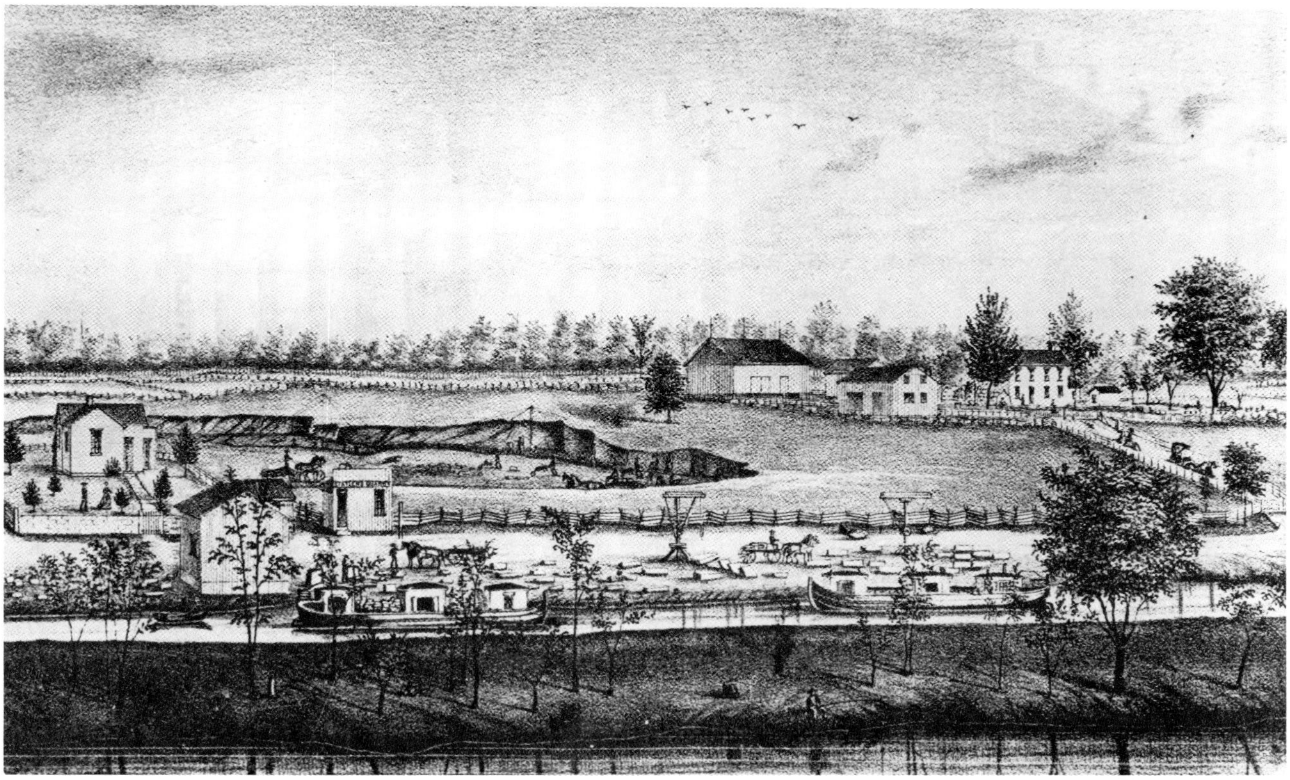

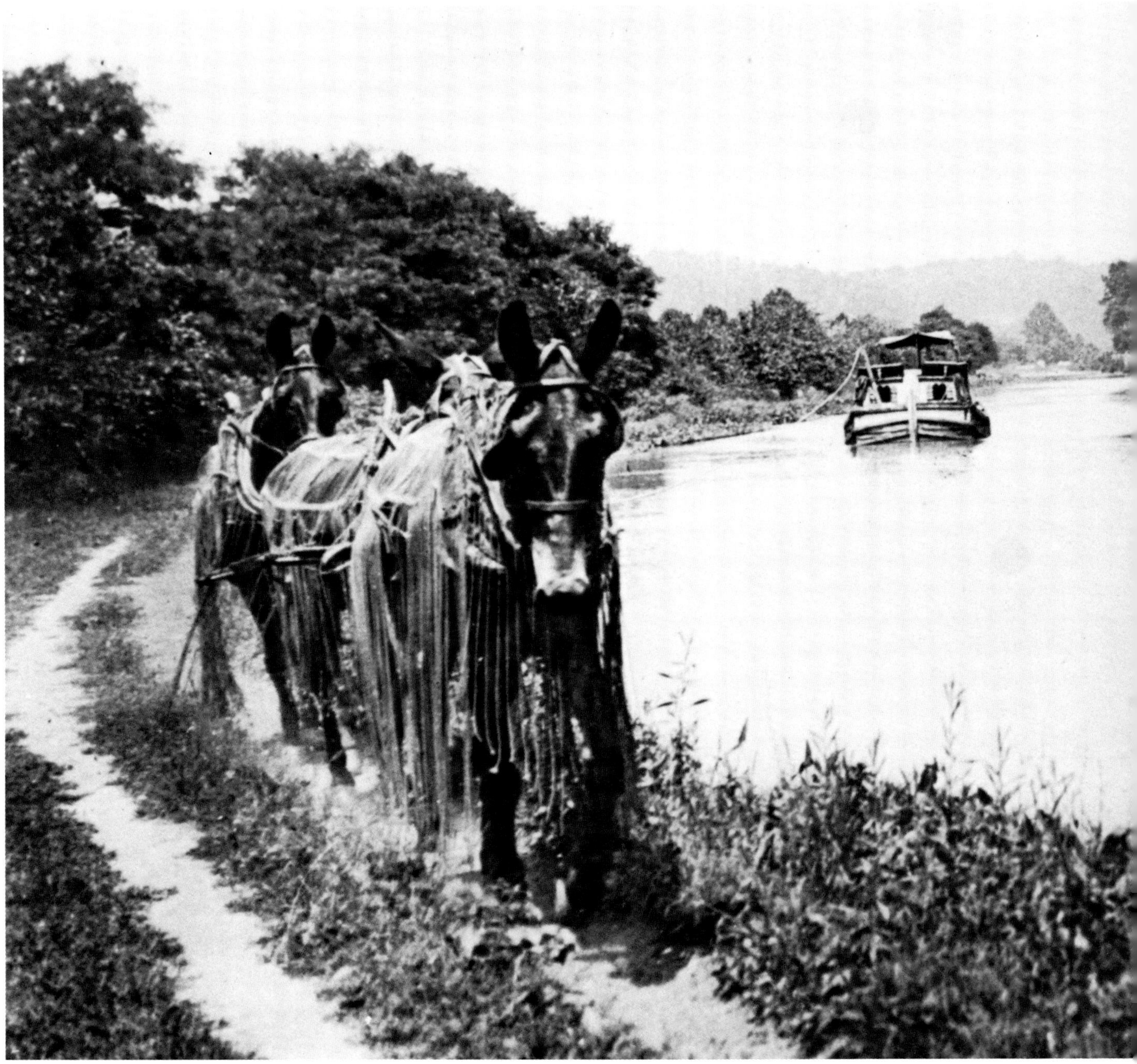

Below, This profile shows the changes in elevation of the Miami & Erie Canal between Lake Erie and the Ohio River, which required a total of 106 lift locks. Loramie Summit, which included the famous Deep Cut, occurred between New Bremen and Lockington.

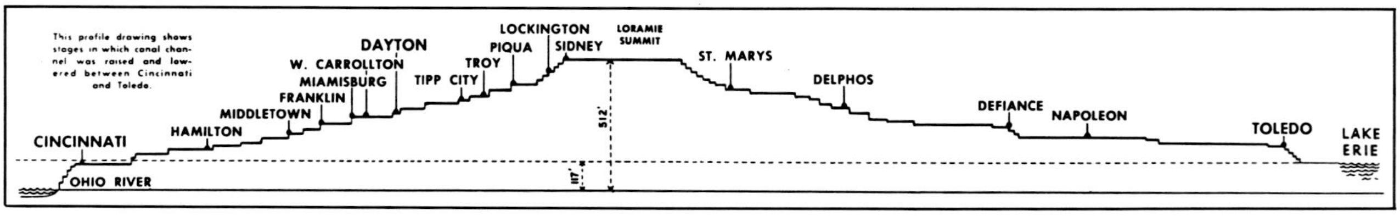

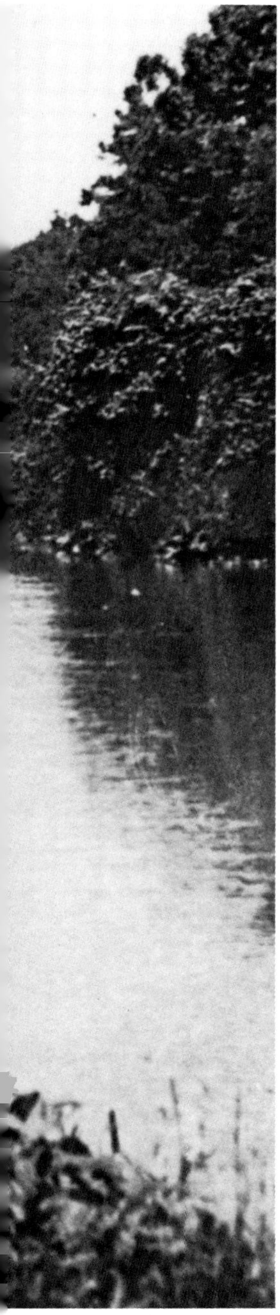

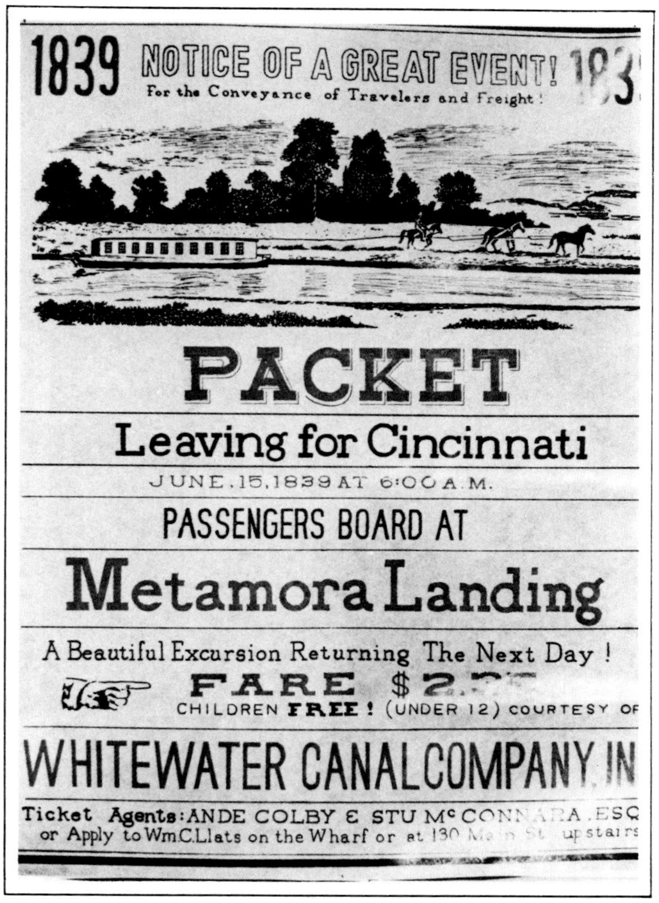

Some mules wore "fly nets," predecessors of woven fly sheets: a fringe of swaying strings weighted by knots chased insects while permitting ventilation for the animals' comfort. Bred for their superior work capacity and endurance, these hybrid creatures often had minds of their own about their working conditions. (CHS)

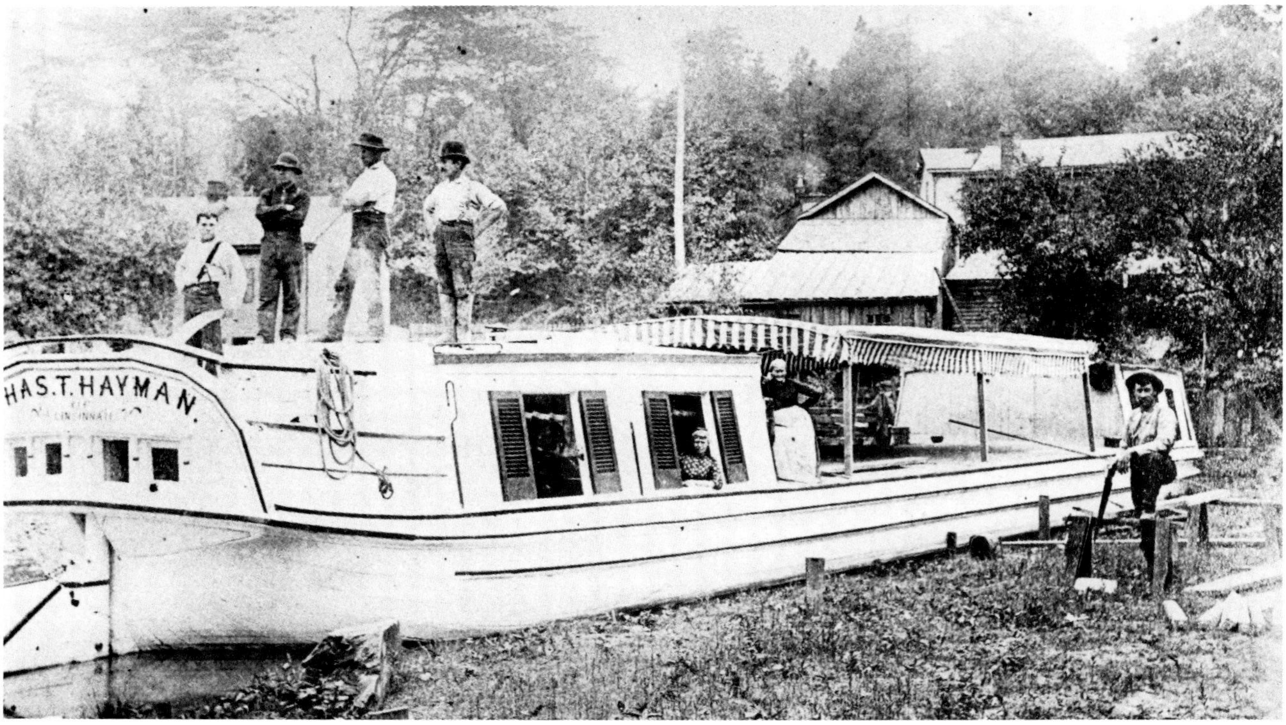

The *Chas. T. Hayman* of Cincinnati—a state boat that may have become a houseboat. (CHS)

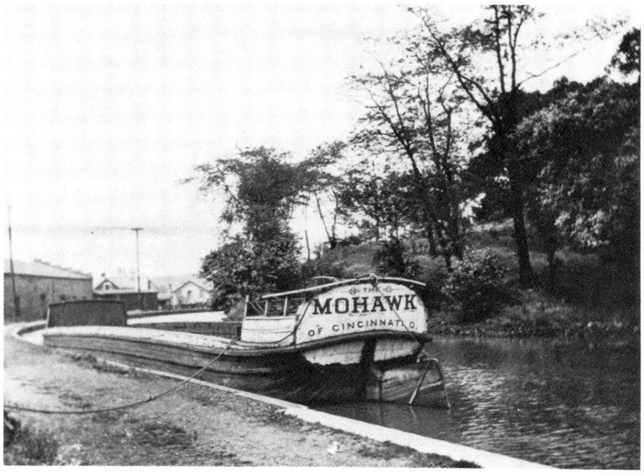

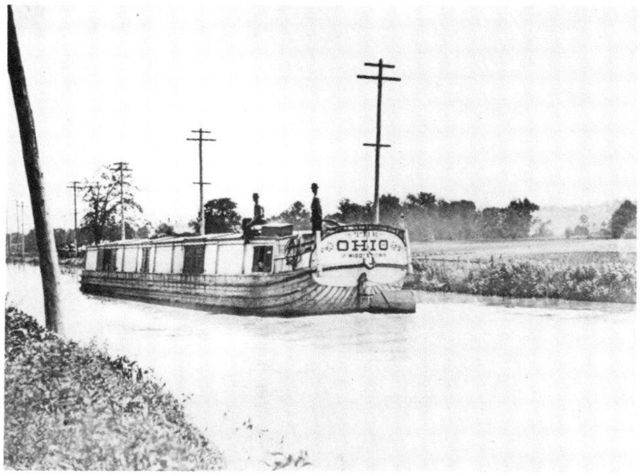

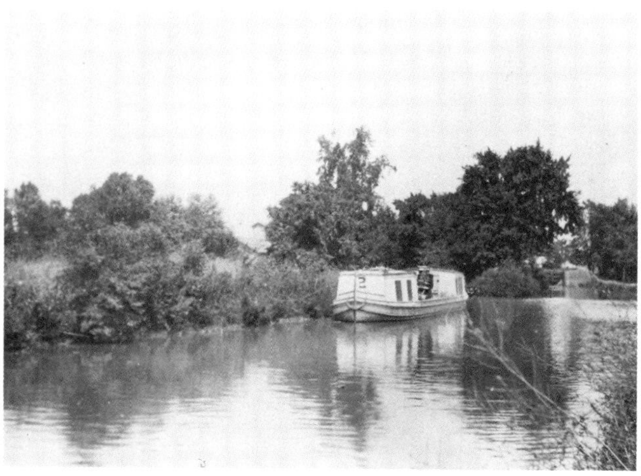

Clockwise from below left: A State Boat patrols the original Miami Canal section of the Miami & Erie Canal; the *Mohawk* of Cincinnati, a state boat; the *Ohio* of Middletown, a line boat in the 1880s. (CHS)

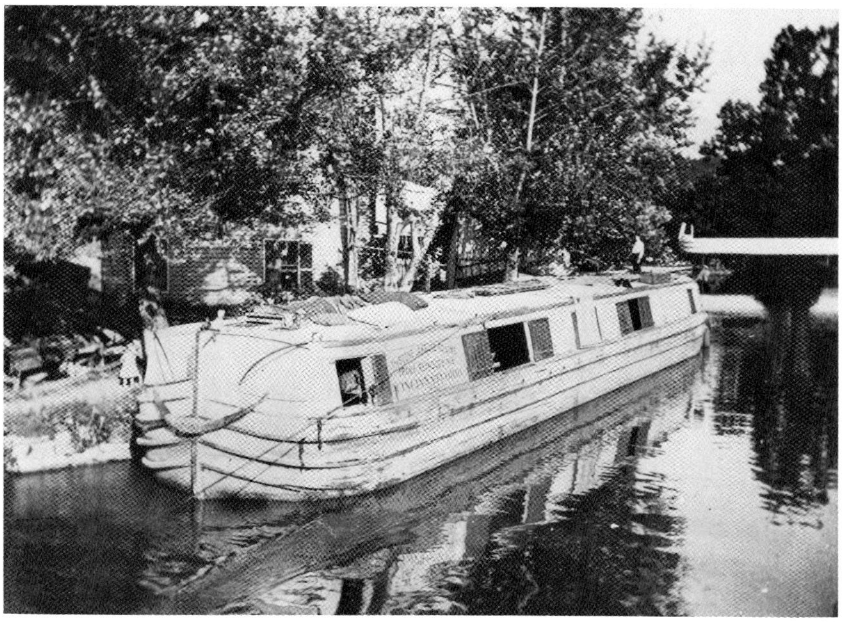

The Stone Lake Ice Company Line of Cincinnati, founded 1868, makes a delivery. (CHS)

Mule teams on a freight boat are switched north of Cincinnati. (CHS)

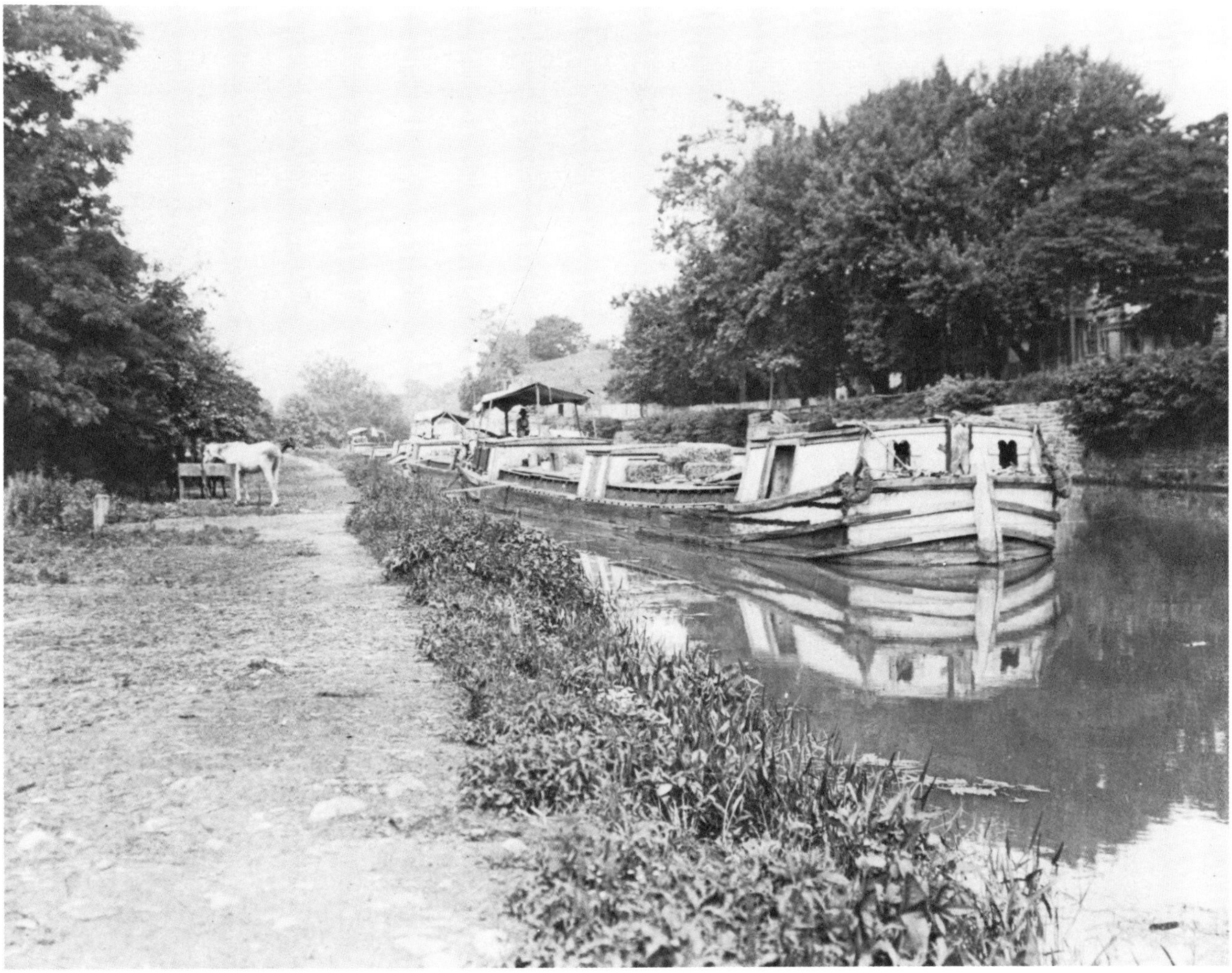

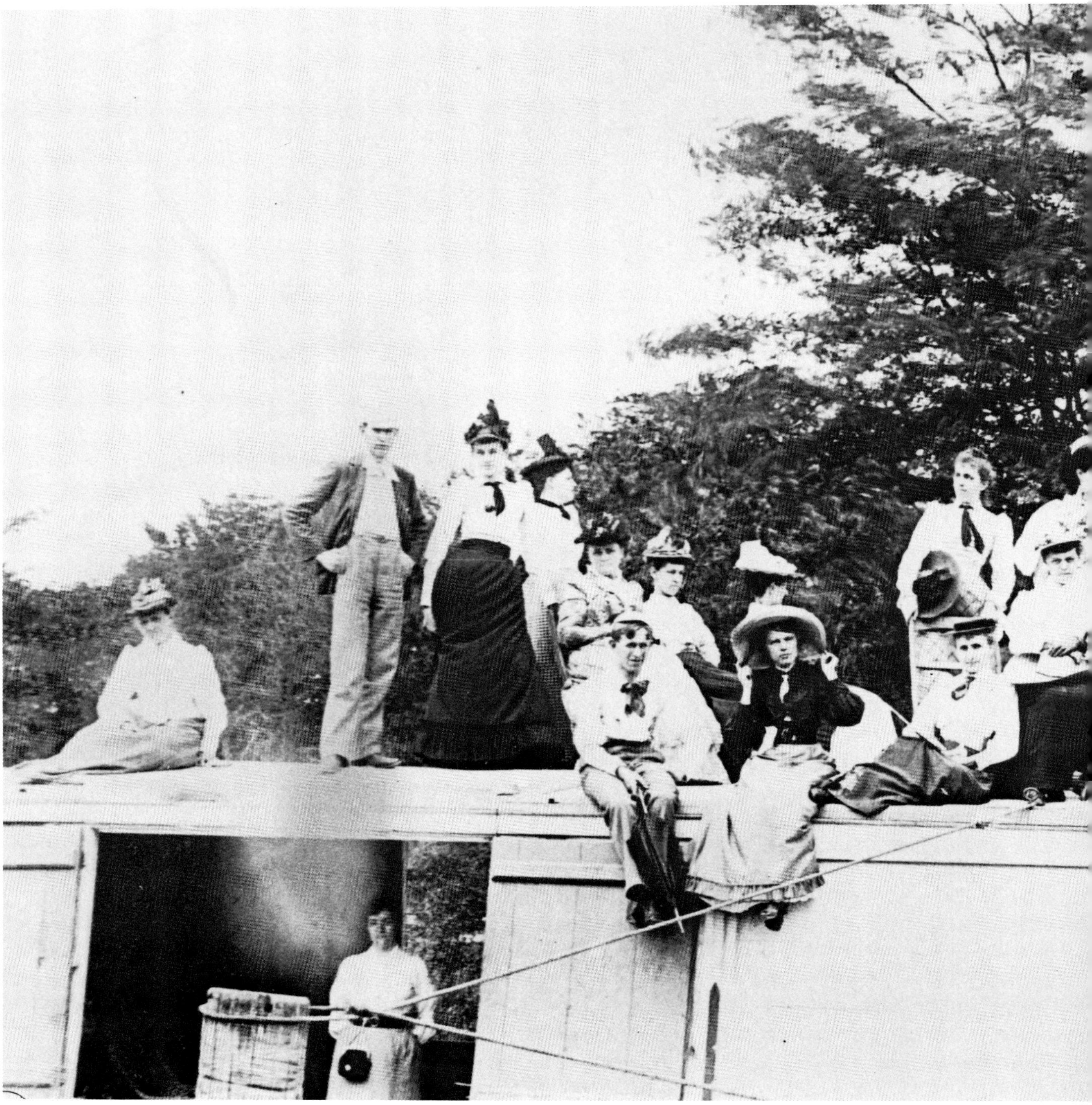

A teachers' outing (together with friends) celebrated the end of the school year in Troy in June 1892. (TRS)

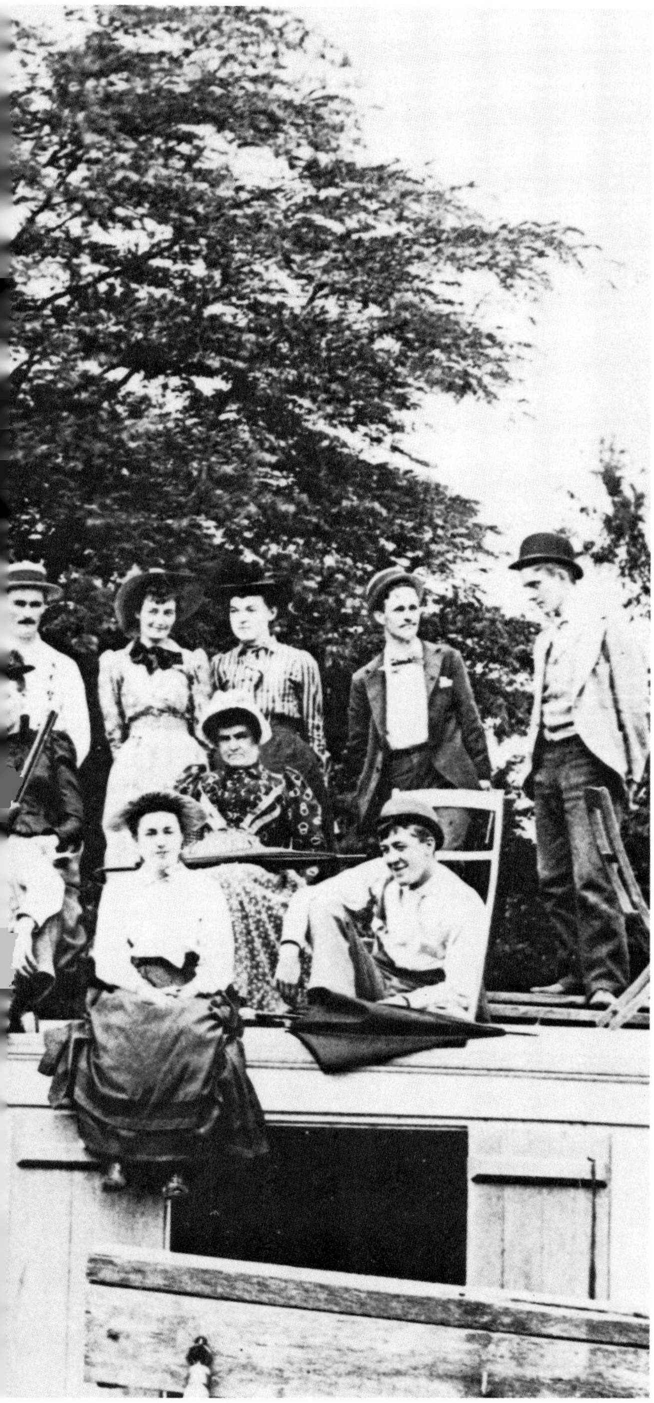

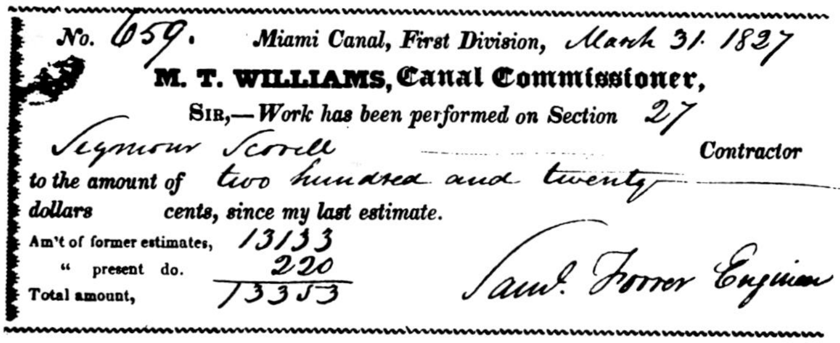

No. 659. Miami Canal, First Division, March 31. 1827

M. T. WILLIAMS, Canal Commissioner,

SIR,—Work has been performed on Section 27

Seymour Scovill Contractor

to the amount of two hundred and twenty

dollars cents, since my last estimate.

Am't of former estimates, 13133

" present do. 220

Total amount, 13353

Saml. Forrer Engineer

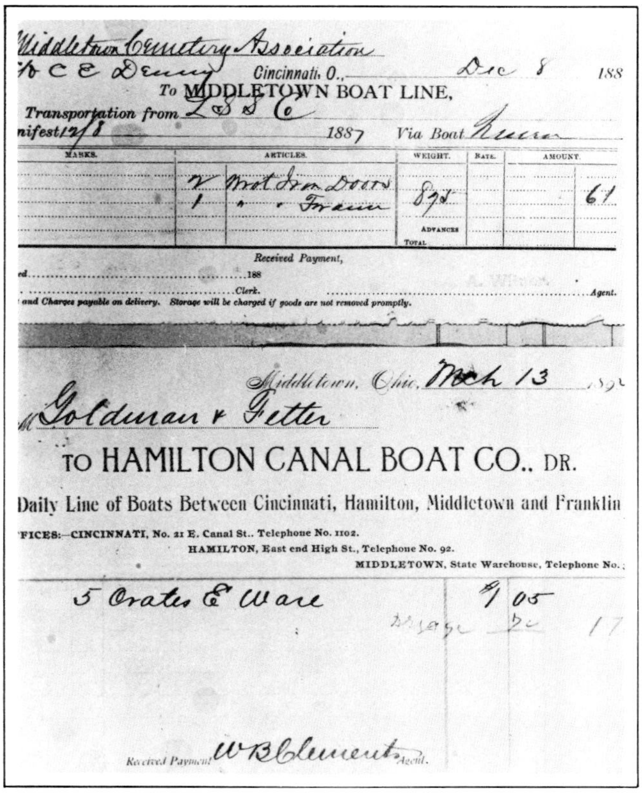

Middletown Cemetery Association

% C E Denny Cincinnati, O., Dec 8 188

To MIDDLETOWN BOAT LINE,

Transportation from L S & C

nifest 12/8 1887 Via Boat

MARKS.	ARTICLES.	WEIGHT.	RATE.	AMOUNT.
	2 Wrot Iron Doors	875		61
	1 " " Frame			
		ADVANCES		
		TOTAL		

Received Payment,

ed ... 188

... Clerk. ... Agent.

and Charges payable on delivery. Storage will be charged if goods are not removed promptly.

Middletown, Ohio, Mch 13 189

M Goldman & Fetter

TO HAMILTON CANAL BOAT CO., DR.

Daily Line of Boats Between Cincinnati, Hamilton, Middletown and Franklin

FICES:—CINCINNATI, No. 21 E. Canal St., Telephone No. 1102.
HAMILTON, East end High St., Telephone No. 92.
MIDDLETOWN, State Warehouse, Telephone No.

5 Crates E Ware 9 05 17

Received Payment W B Clements Agent.

An 1827 certificate (top) issued to contractor Seymour Scovill entitled him to a payment of $220 for work on Section 27 of the First Division of the Miami Canal. The certificate was signed by the famous Samuel Forrer, then resident engineer, later canal commissioner. Forrer served Ohio's canal system for more than 50 years—longer than any other person. (CHS)

Bills of lading (bottom) from 1887 and 1895. (George C. Crout)

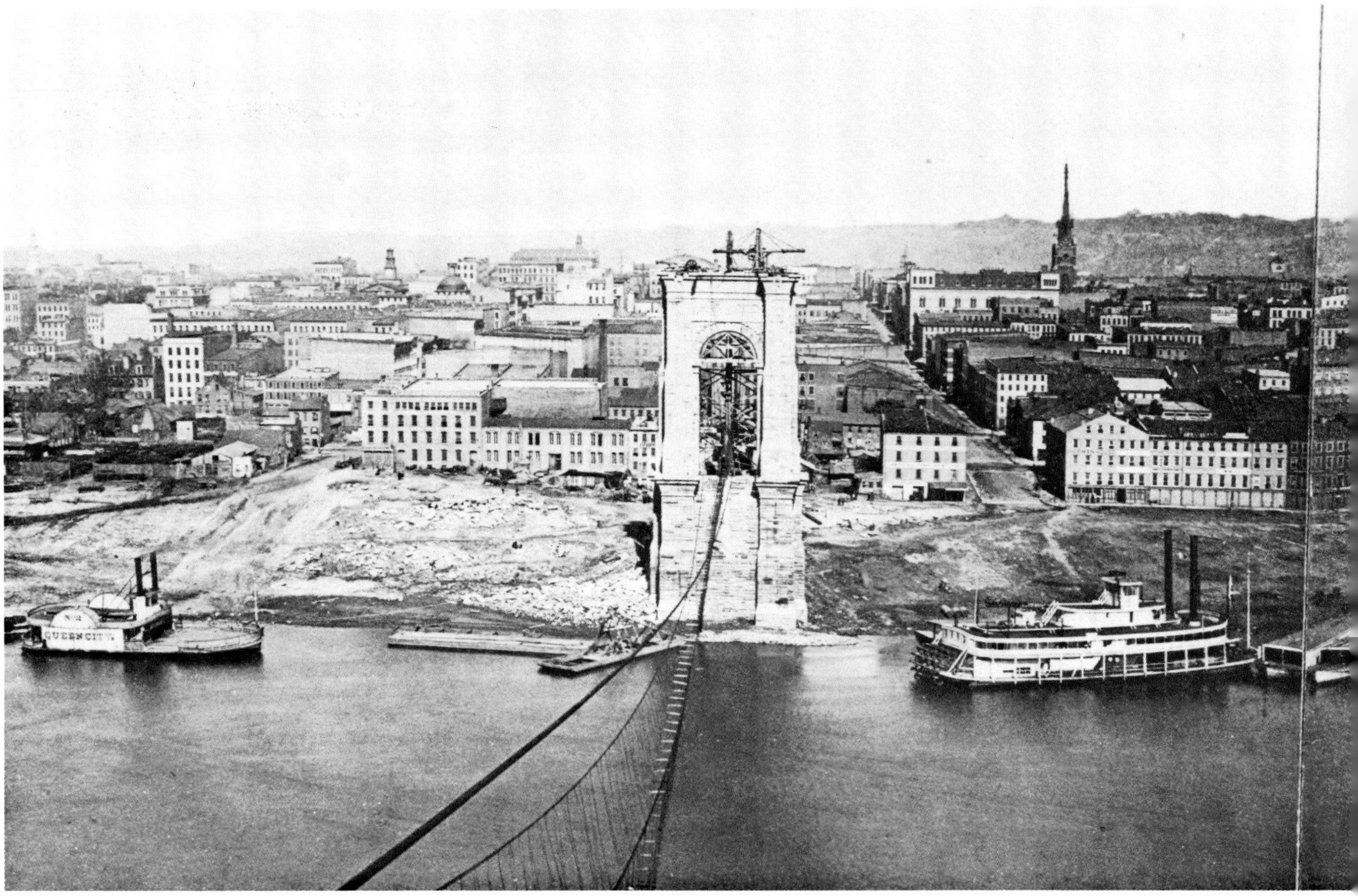

The Cincinnati waterfront in 1865. The canal locked down to the level of the Ohio River at extreme right; Mount Adams rises in the background. To the left is John Roebling's Great Suspension Bridge under construction. Connecting Cincinnati with Covington, Kentucky, it was the first bridge across the Ohio River. Mile: 250. (CHS)

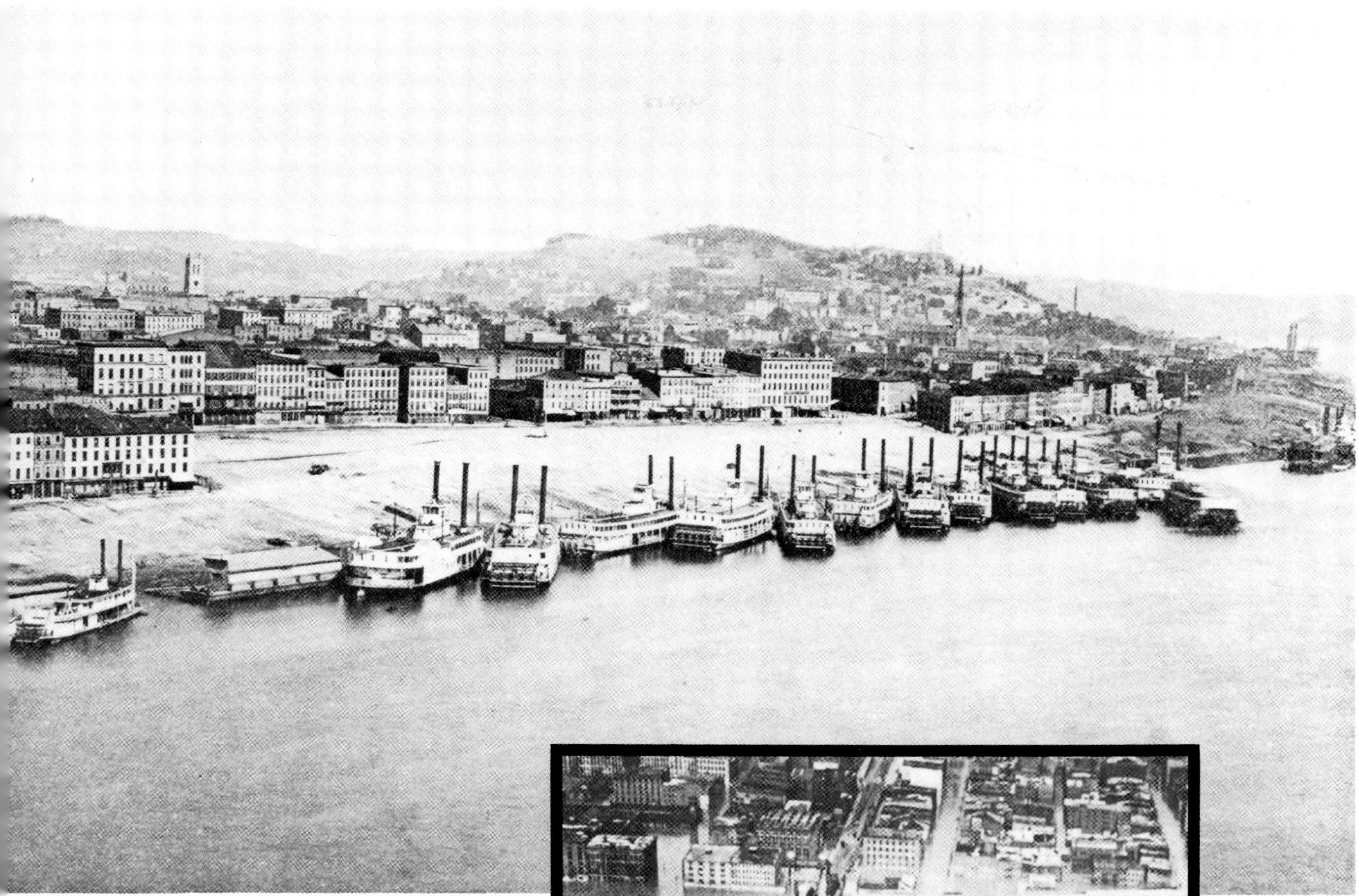

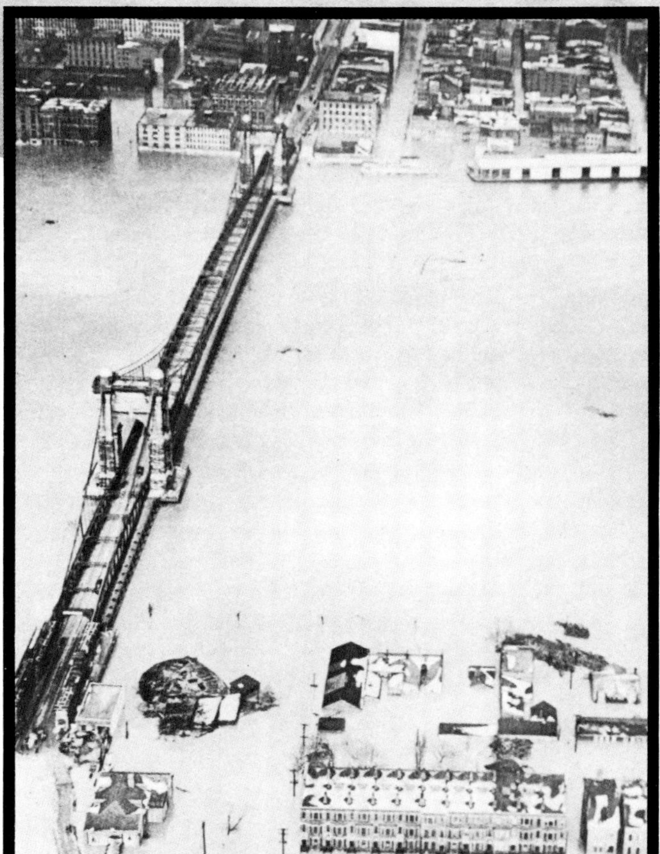

As was the case with the Ohio & Erie's southern terminus at Portsmouth, Ohio River flooding created major problems for the Miami & Erie Canal. This is Cincinnati's Great Suspension Bridge during the 1937 flood. (CHS)

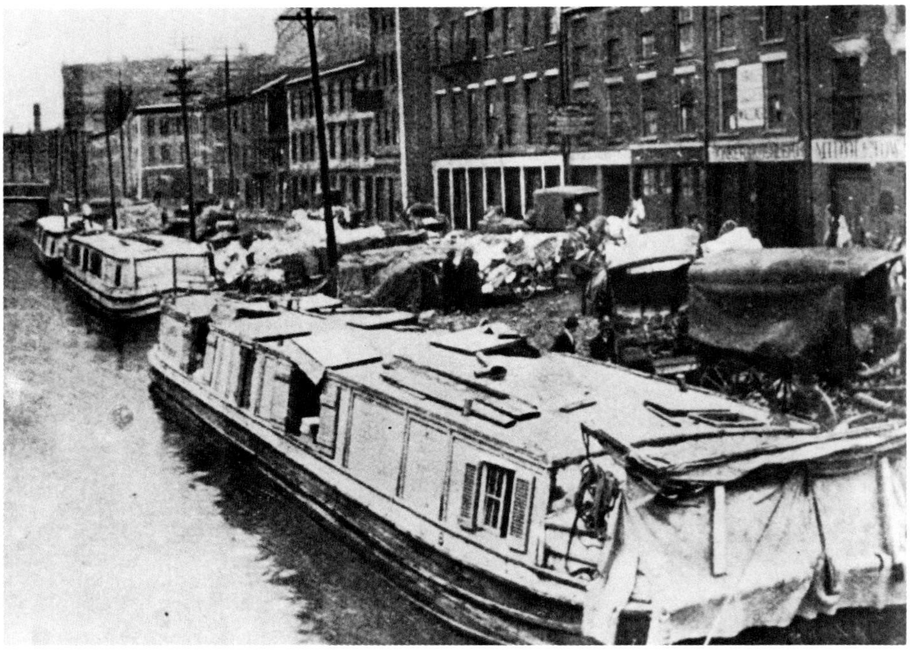

A scene along Cincinnati's Canal Street market district in 1900. Mile: 249. (CSO)

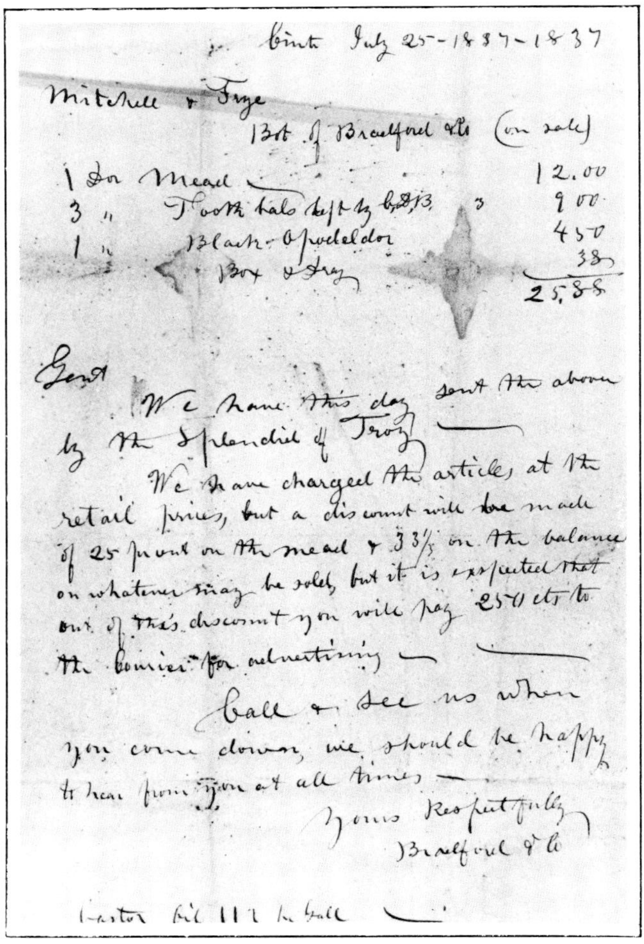

Cint. July 25 - 1837 - 1837

Mitchell & Frye

Bot of Bradford & Co (on sale)

1	Doz	Mead	12.00
3	"	Tooth Wash kept by B.D.B.	9.00
1	"	Black Opodeldoc	4.50
		Box & Dray	38
			25.88

Gent

We have this day sent the above by the Splendid of Troy —

We have charged the articles at the retail prices, but a discount will be made of 25 percent on the mead & 33⅓ on the balance on whatever may be sold, but it is expected that out of this discount you will pay 250 cts to the Courier for advertising —

Call & see us when you come down, we should be happy to hear from you at all times —

Yours Respectfully
Bradford & Co

Castor bil 112 lb ball —

A consignment bill of lading from 1837 offered the consignee a commission of 25 to 33 percent (less a discount for the shipper's advertising) for sale of the merchandise at its destination. (TRS)

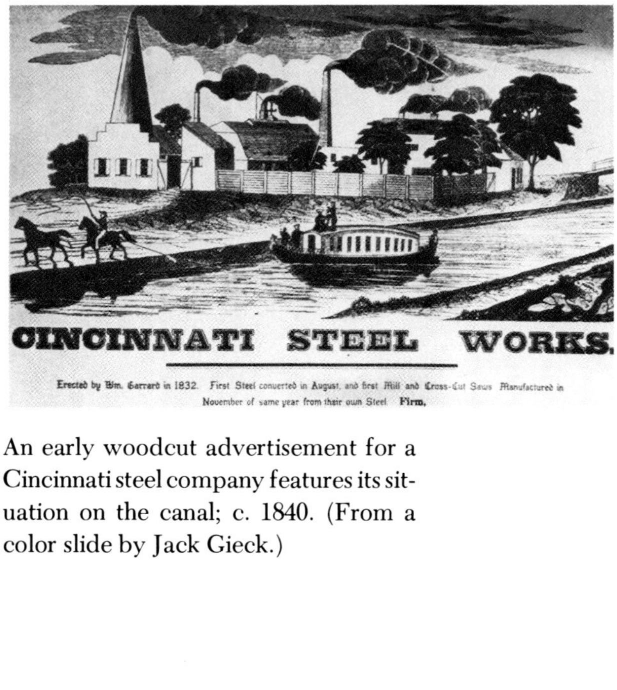

An early woodcut advertisement for a Cincinnati steel company features its situation on the canal; c. 1840. (From a color slide by Jack Gieck.)

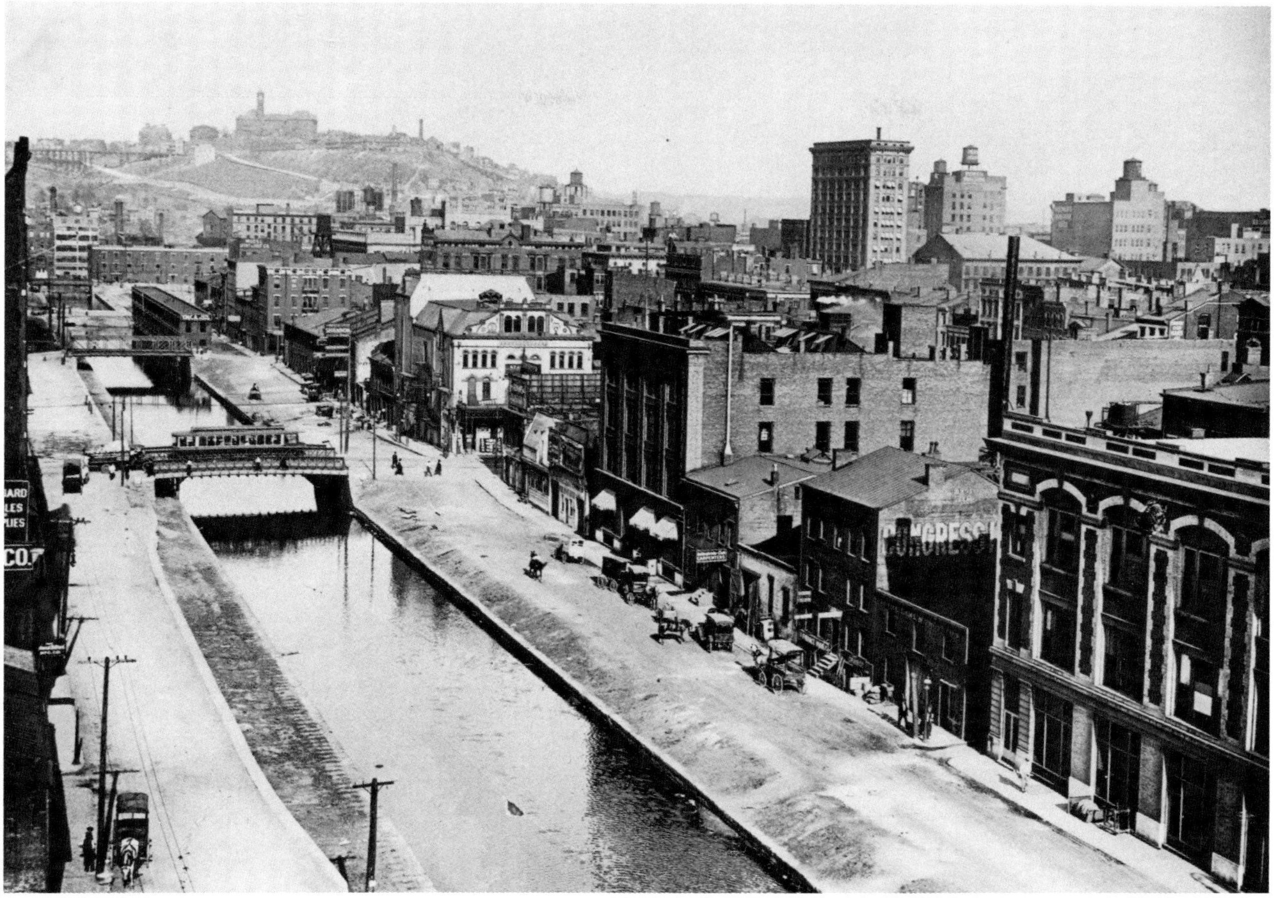

Looking east down Cincinnati's Canal Street (above) in 1916 (now Central Parkway), a street car crosses the Vine Street Bridge. The ornate building at its right is the Standard Theatre, which catered to the German population in this section of the city. This ethnic community regularly referred to crossing the canal here as going "over the Rhine," which became an accepted designation for this part of town. Mount Adams rises in the background. A romanticized early engraving (right) pictures the Rhine section. The architecture has become Victorian, and the canal boats in the foreground look a little like Venetian gondolas. Mile: 249. (CHS)

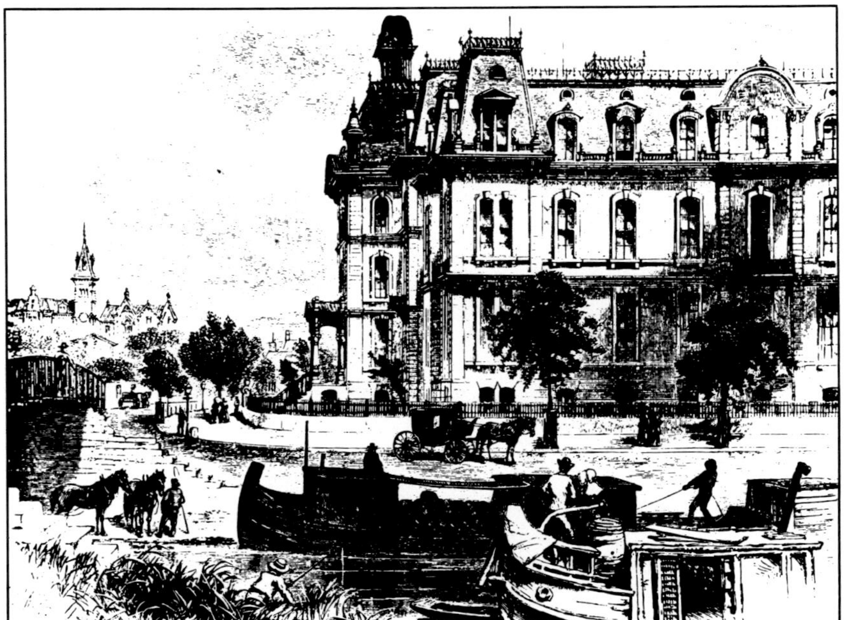

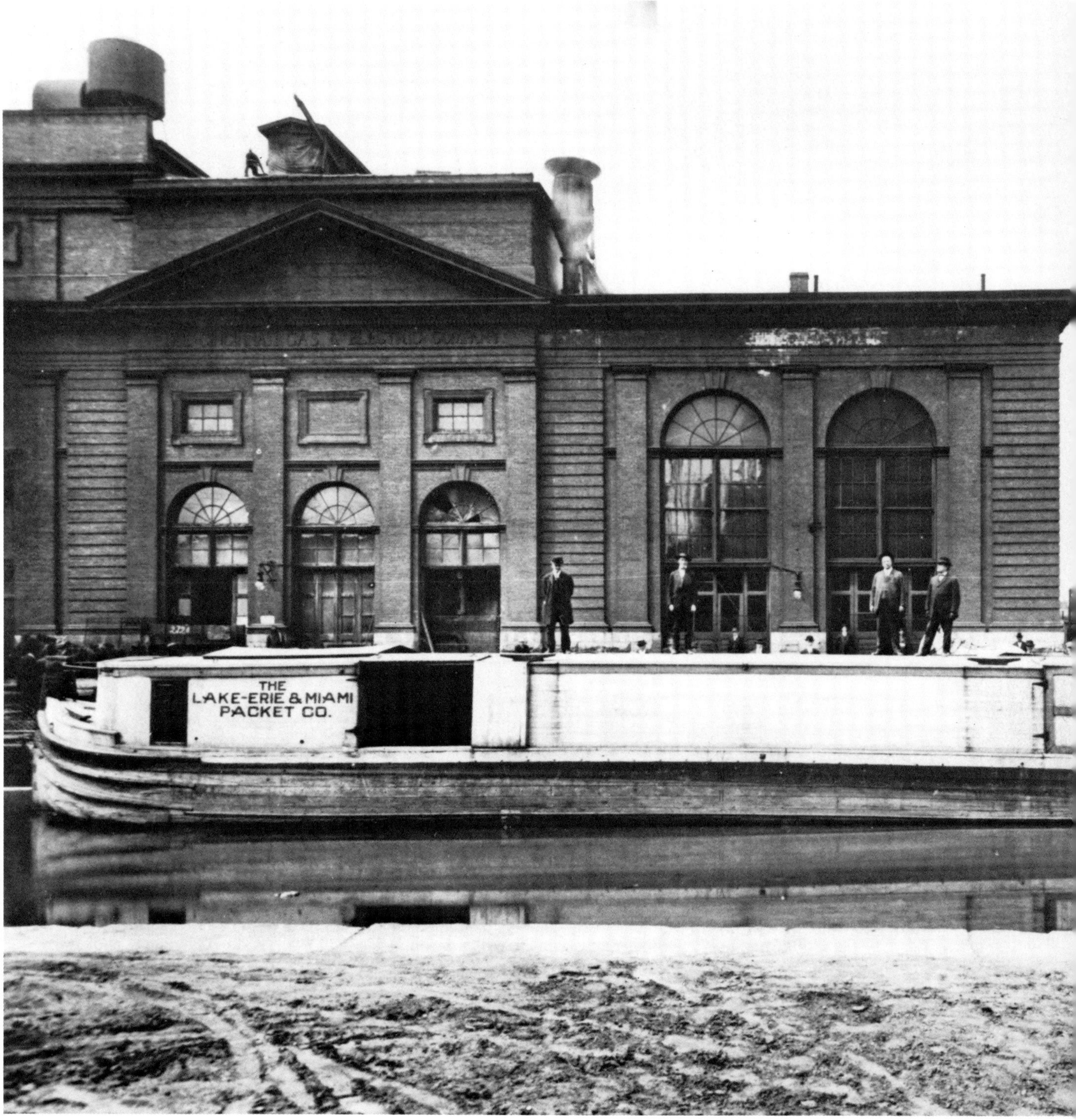

The *Lady Hamilton,* a line boat of the Lake Erie & Miami Packet Company at Cincinnati. Mile: 248. (CHS)

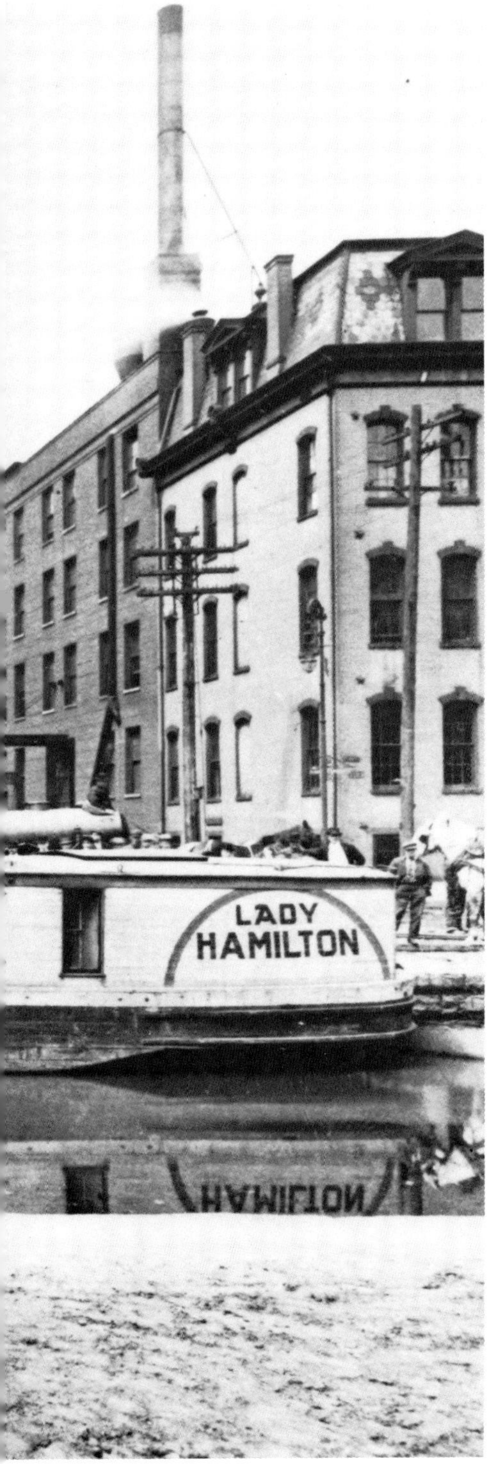

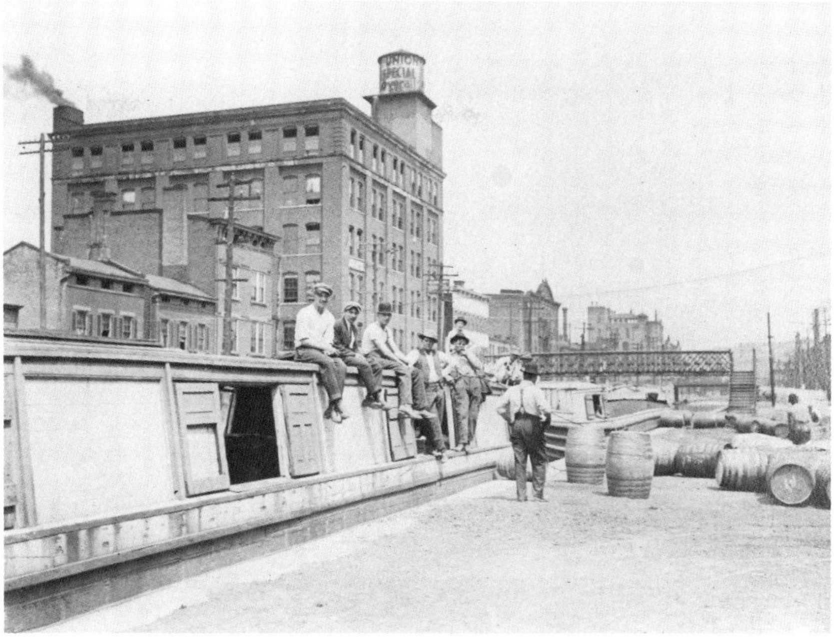

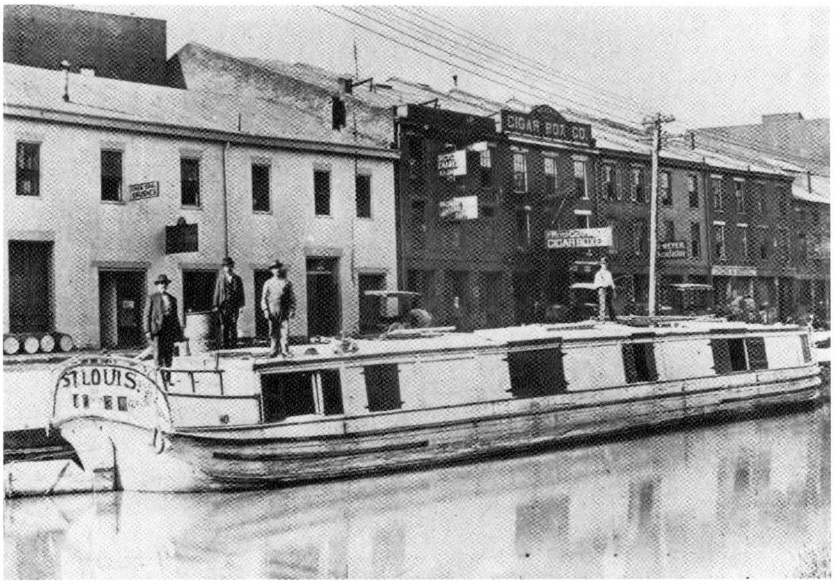

The *Carrollton* (top) and crew in Cincinnati. (CHS)

The *St. Louis* (bottom) docked next to a cigar box factory in Cincinnati. Mile: 249. (CHS)

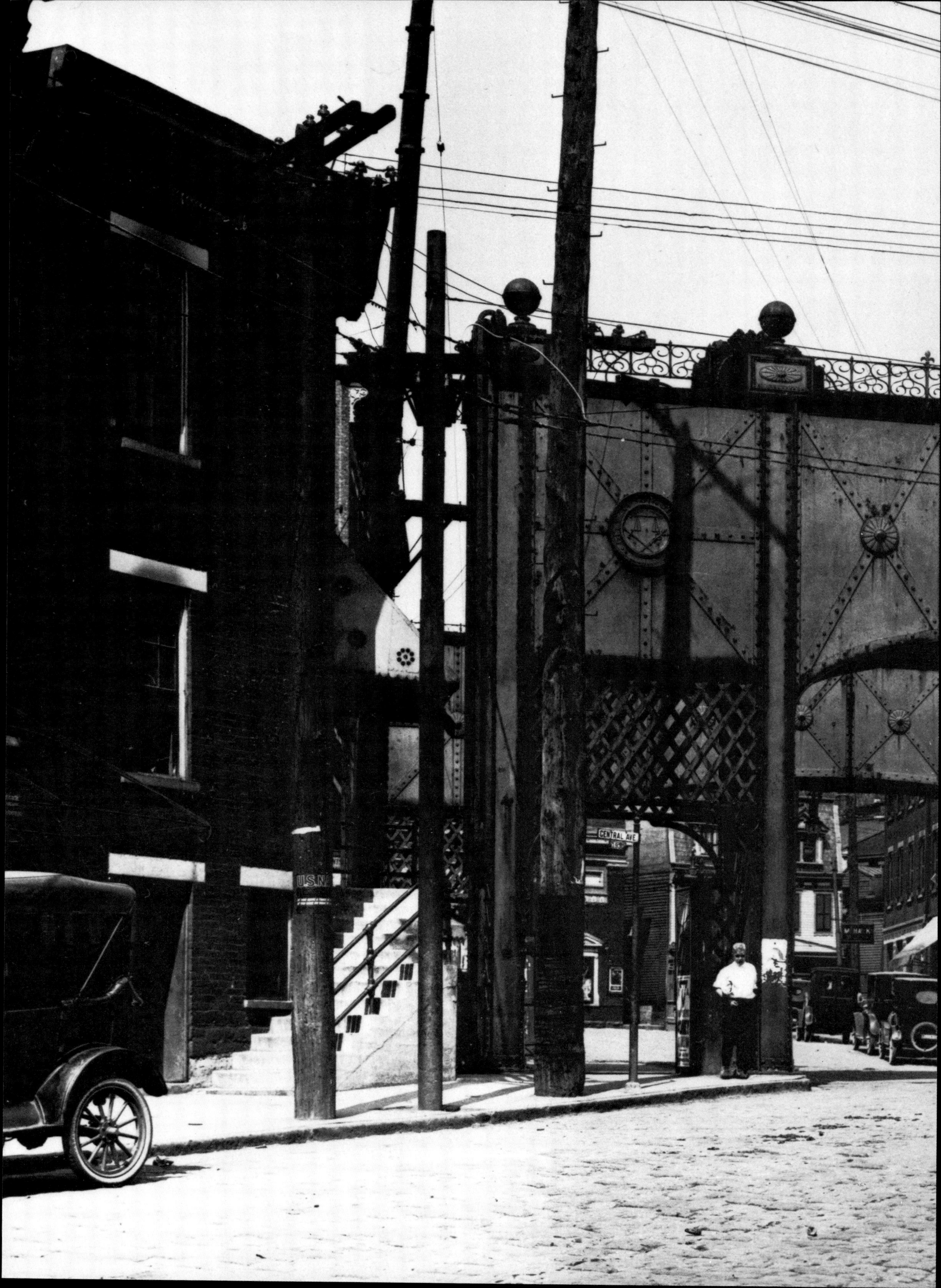
CENTRAL AVE

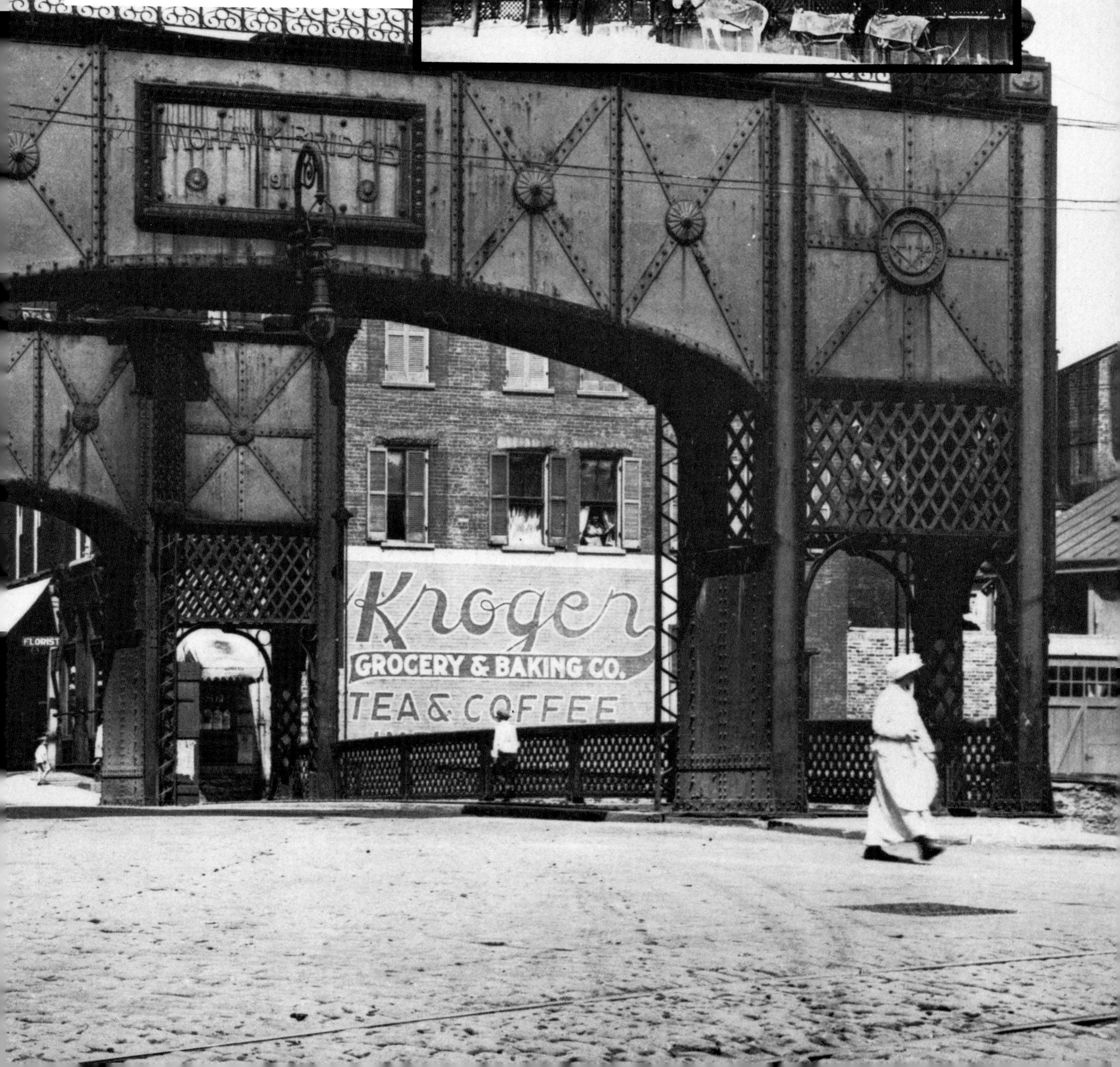

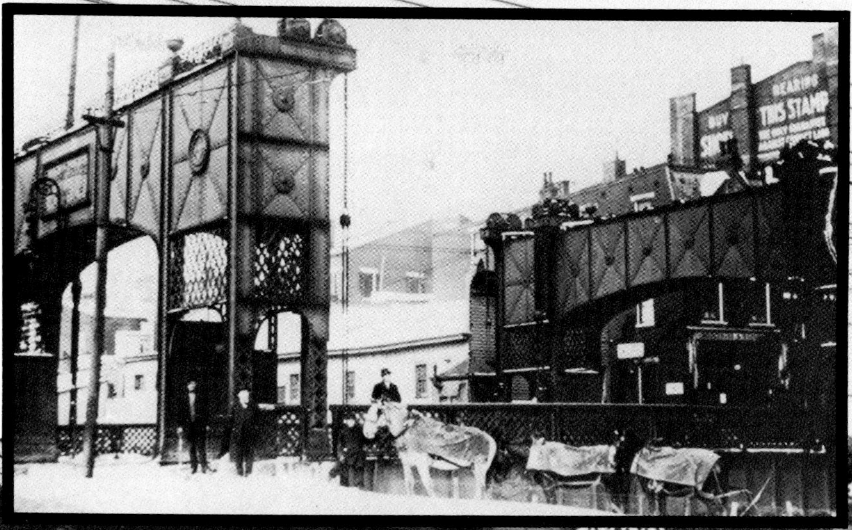

A steel lift bridge over the canal in Cincinnati. Its towers contained counterweights to permit the roadway to lift vertically so that boats could pass underneath; c. 1920 (OHS). *Inset* is an end view of one of the towers on the lift bridge showing its pulley and chain mechanism attached to the tower counterweight. With the bridge down, the mule team and driver have come up off the towpath to pose. Mile: 248. (CHS)

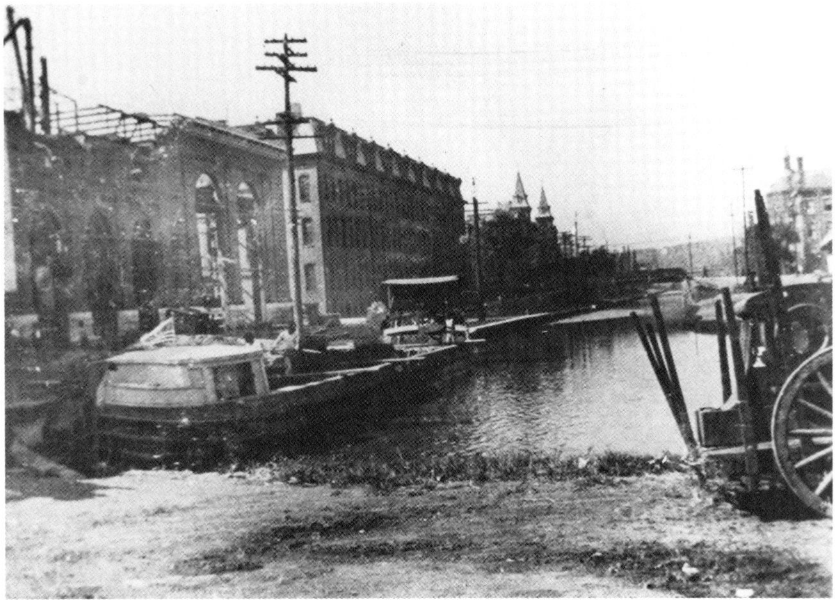

Left, Plum Street Bend looking north toward the Music Hall, where the canal abruptly turned north at the end of Canal Street; c. 1890. *Below,* an engraving shows Plum Street Bend (looking south); c. 1870. Mile: 248. (CSO)

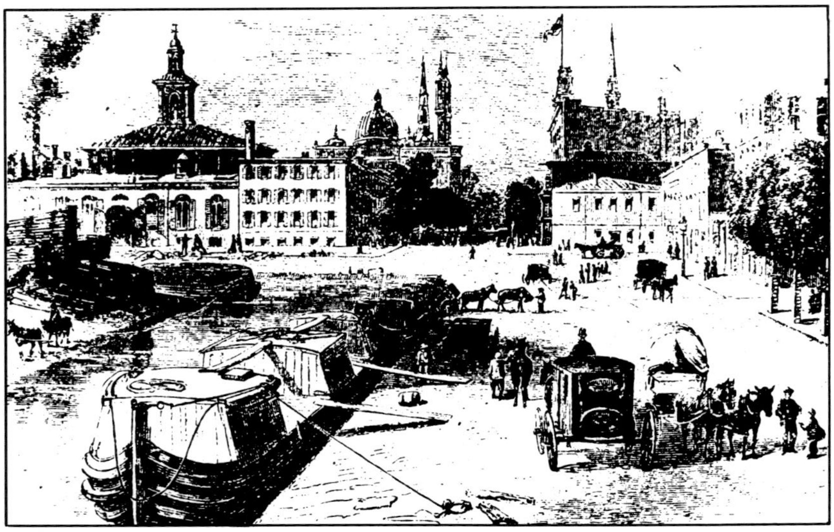

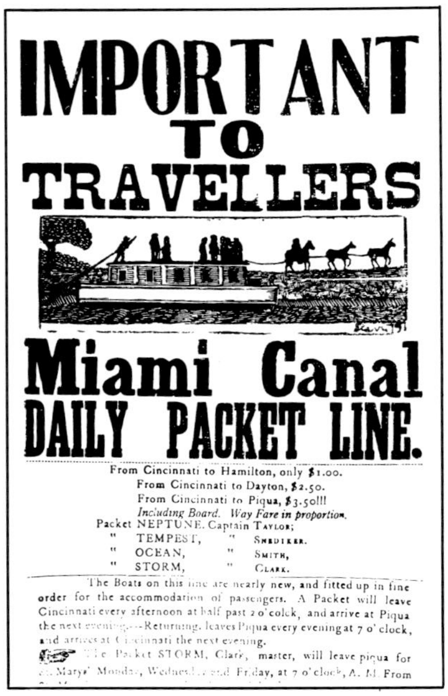

Right, "Venetian Canal in Machinery Hall," a painting by the famous Ohio artist, H. F. Farny, as he depicted the Cincinnati Centennial Celebration in *Harper's Weekly* (1888, vol. 32, p. 1647, courtesy F. W. Trevorrow). *Inset,* Boats are lined up just north of the Music Hall on the Plum Street section (looking north). During the Cincinnati Centennial Celebration in 1888 this section of the canal was covered over, becoming Machinery Hall. Mile: 248. (CHS)

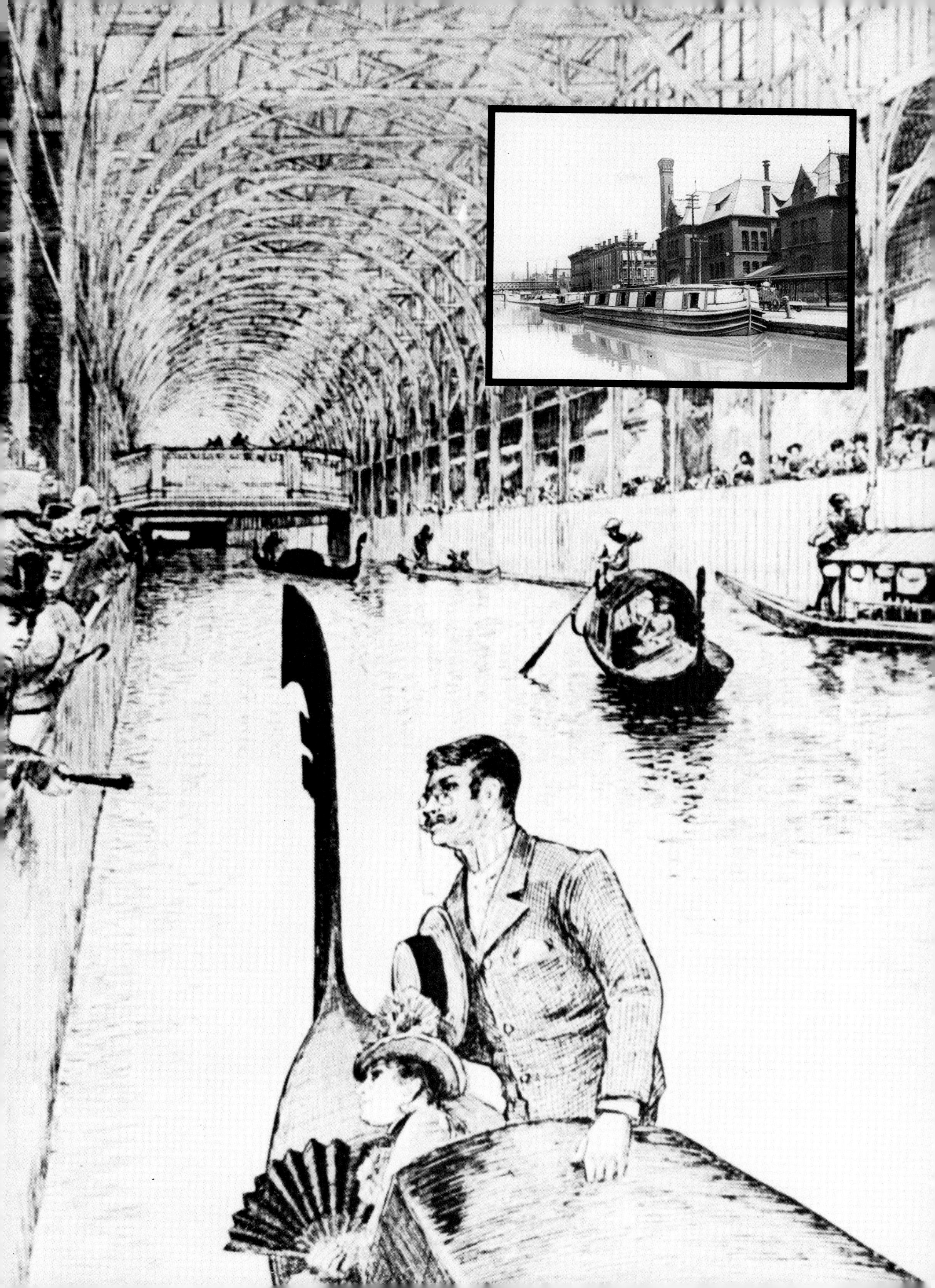

"A paid up dream and darkened tunnels are all that remain today of the Great Cincinnati Subway.

"The standard-gauge interurbans could not penetrate the downtown innards of the Queen City, and their passengers were forced to transfer to wide-gauge city cars at numerous outer terminals. The interurban companies, falling into hard times in 1916, were part of a pressure group that pushed a subway to ballot-box success.

"The bond issue raised six million dollars, and the Kaiser raised war clouds in Germany. Construction of the subway began in earnest in 1919, at the same time flivver production soared in Detroit. The subway started at Main and Central Parkway, and the bore continued east to Plum where it swung north. The line burrowed into the open to follow the old Miami & Erie Canal bed, and then after paralleling Colerain Avenue turned east through Winton Place to Norwood. Stations were completed, and the line was graded ready for track when money ran out in 1927.

"There were buses on the highways that needed no subway as rubber tires were not affected by change in track gauge. Buses worried not about track gauges or single or double trolleys. The interurbans were dead or mortally ill, with the exception of the Cincinnati, Hamilton & Dayton. The Cincinnati Street Railway wanted no part of it and CH&D officials were cool. Tall grass grew at the trackless high level stations, and green mold covered underground walls.

"Engineering consultants proposed spending more than ten million dollars to complete it. Further dreams included reboring twin railroad tunnels that were never used, and lie abandoned deep beneath Walnut Hills.

"The plans were shelved as the depression dealt despair in the Queen City. The Cincinnati & Lake Erie Railroad, the revitalized interurban running north, ordered 20 streamlined cars in 1931 with a door design to fit subway platforms. No one came forward to complete the tube, and a few years later C&LE revamped its cars to more conventional design and relied on its own buses to bring passengers to the Cumminsville railhead. All other interurbans were now junked. . . .

"When the C&LE halted in 1939, there was no hope for the subway. Modern expressways butchered the path, and the Central Parkway winds along the roadbed. . . ." (Harry Christiansen, *Ohio Trolley Trails*, vol. 1, 1971, WRHS.)

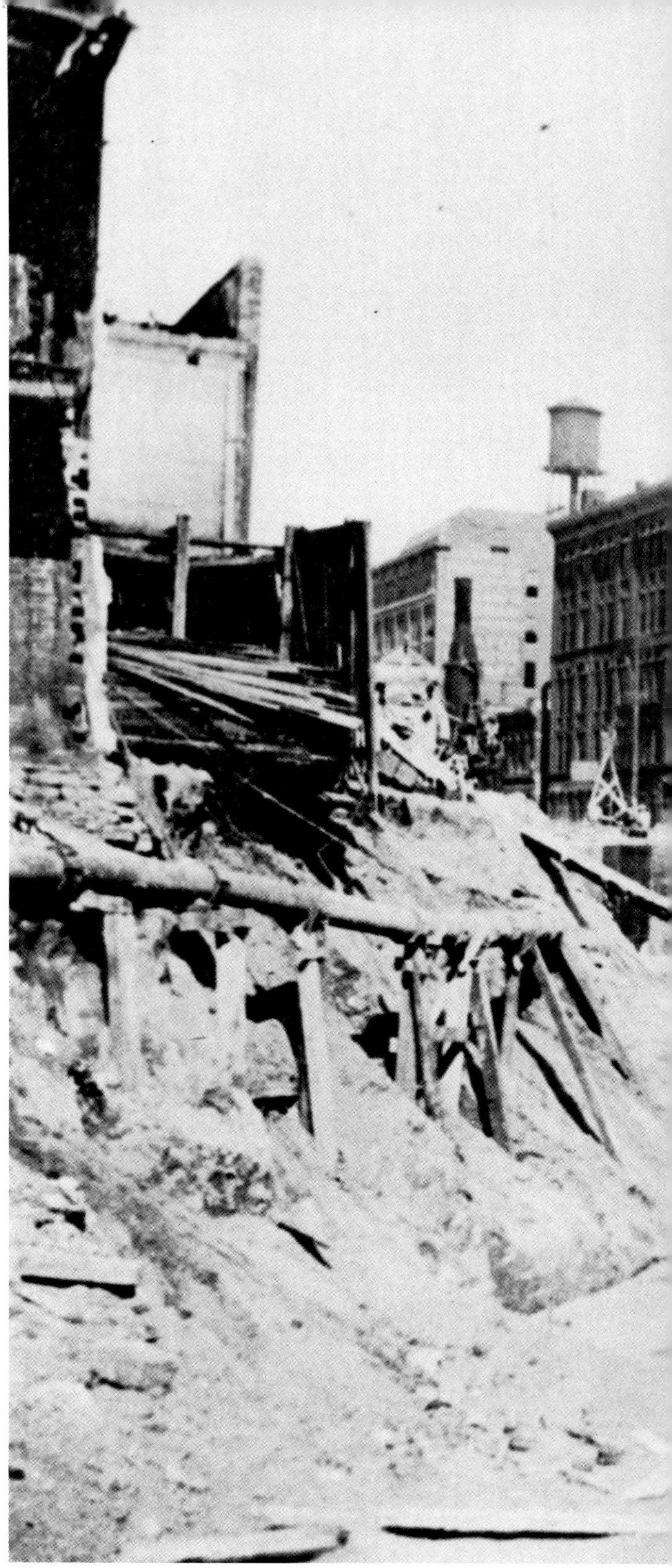

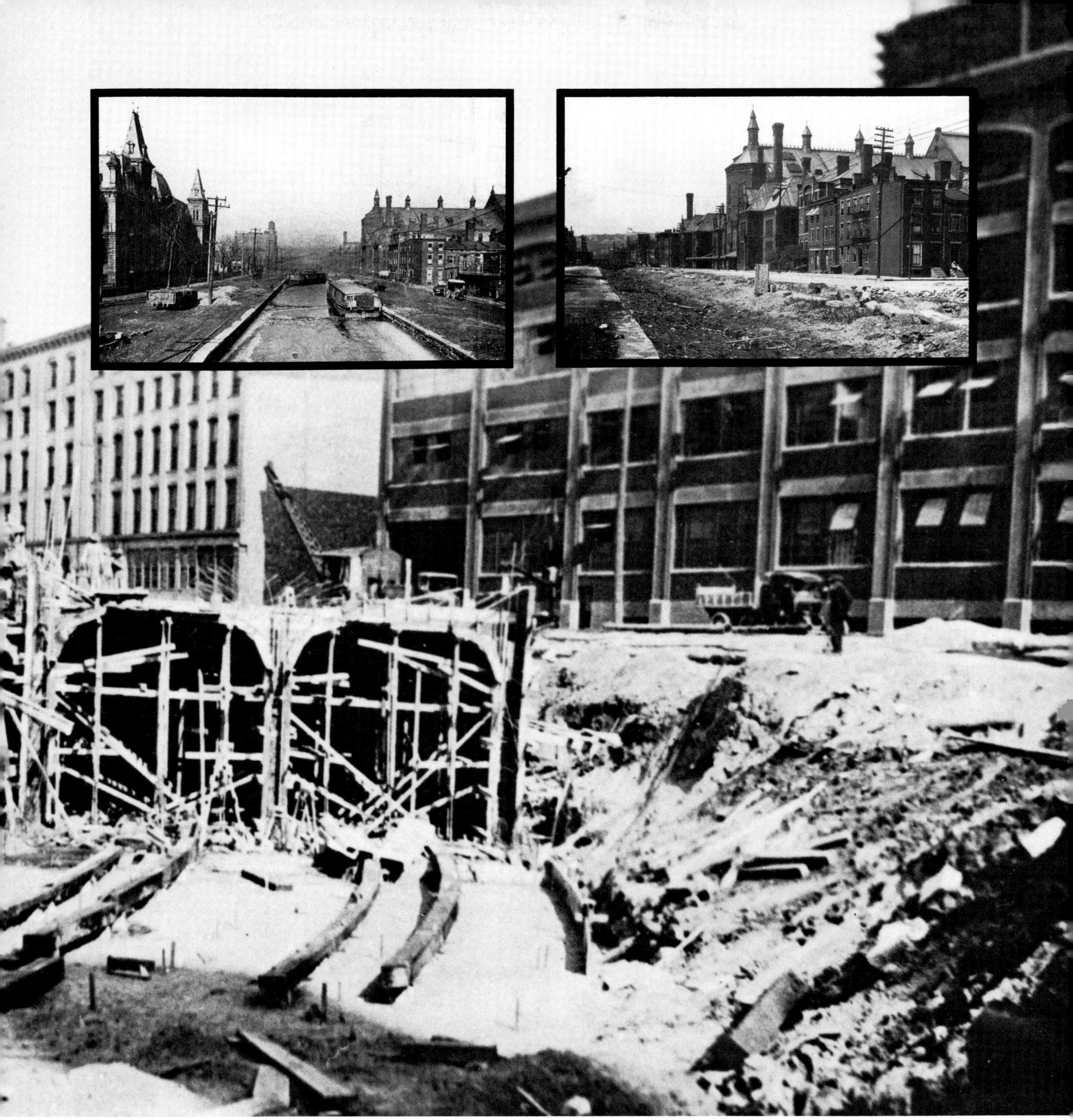

Plum Street Bend in 1921, when an unsuccessful attempt was made to build a subway in the canal bed. Mile: 248. (CHS)

Inset are photos of the area occupied by the Celebration's Machinery Hall which became the subject of this 1920-vintage propaganda piece (left), and a derelict canal (right) in front of Cincinnati's Music Hall, before it was covered over and paved to become Central Parkway; c. 1920s. (OHS)

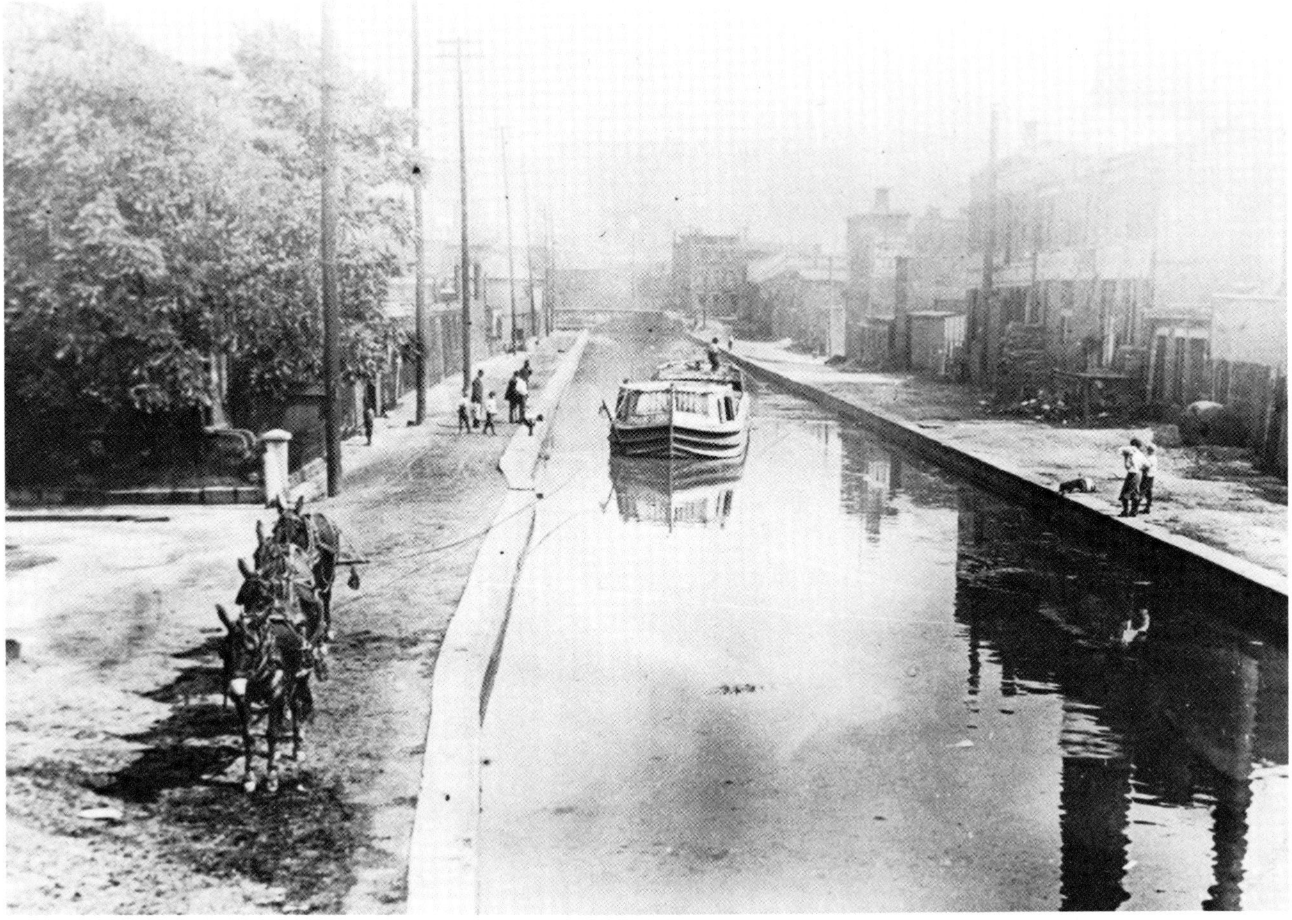

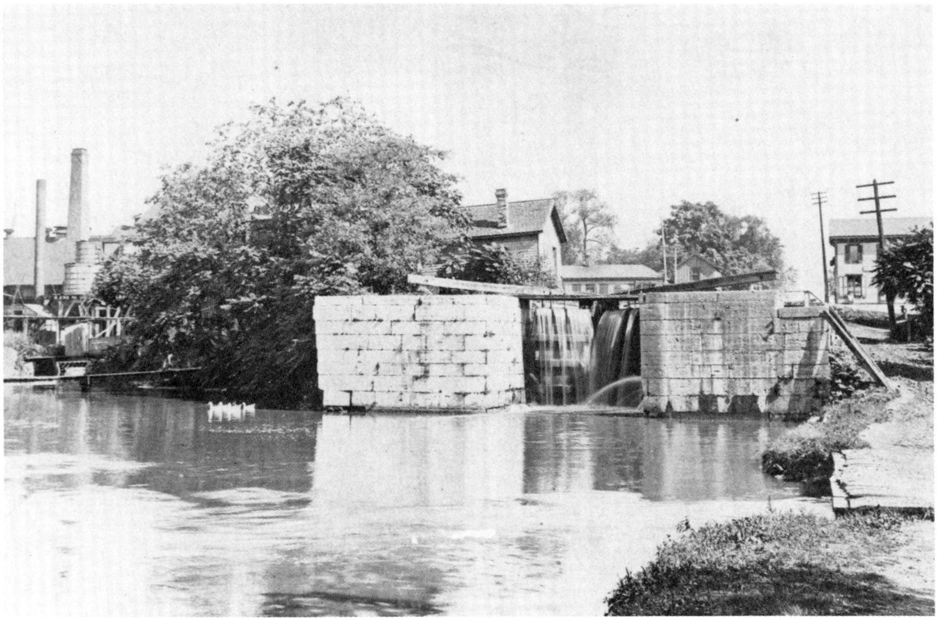

Above, A rare four-mule team in tight single file draws a boat out of Cincinnati's northern suburbs. Mile: 244–47. (CHS)

Left, Lock 41, Collector's Lock, was the center of the three locks amid Lockland industry and had a small basin at its lower level. Mile: 238. (George C. Crout)

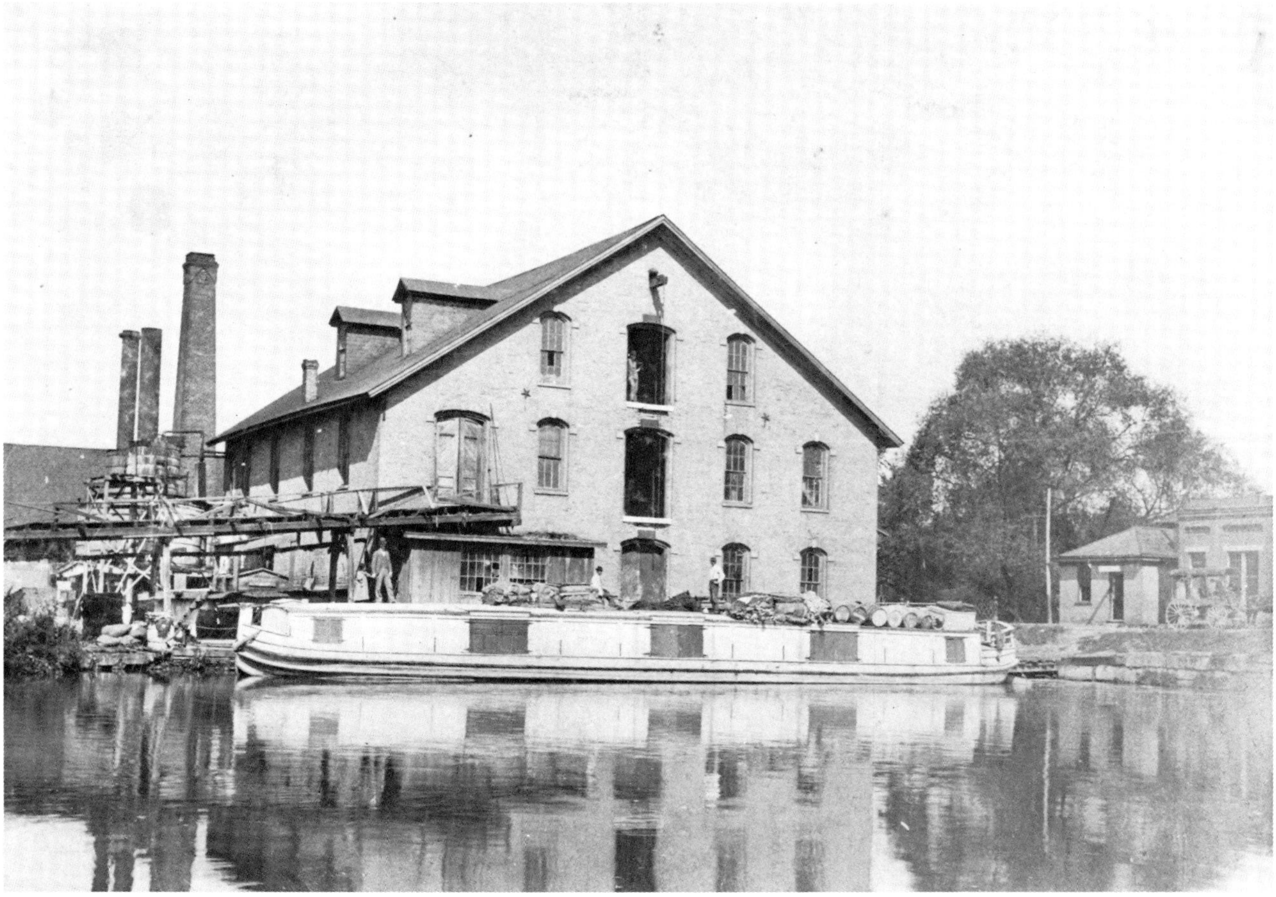

A line boat moored at the Fox Paper Company Dock. The paper company was organized in 1836. Mile: 237. (George C. Crout)

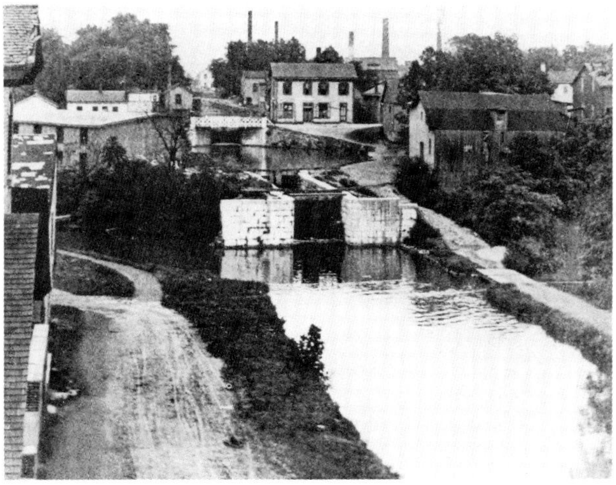

A staircase of three locks at Lockland; the absence of utility poles suggests that photo was made prior to 1880. Mile: 239. (CSO)

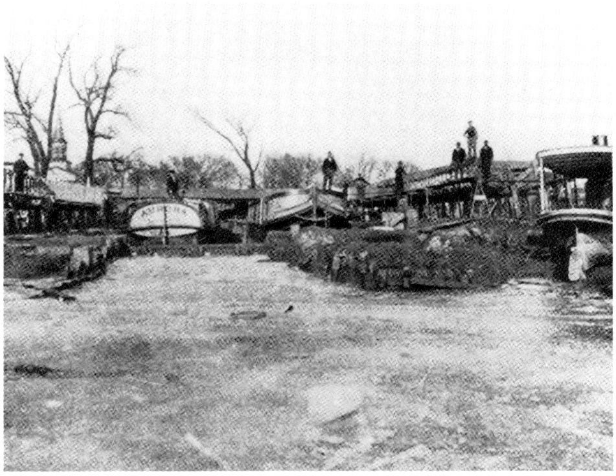

Lockland had a major boat-building facility. No less than five canal boats are in drydock undergoing construction or repair in this picture. (CSO)

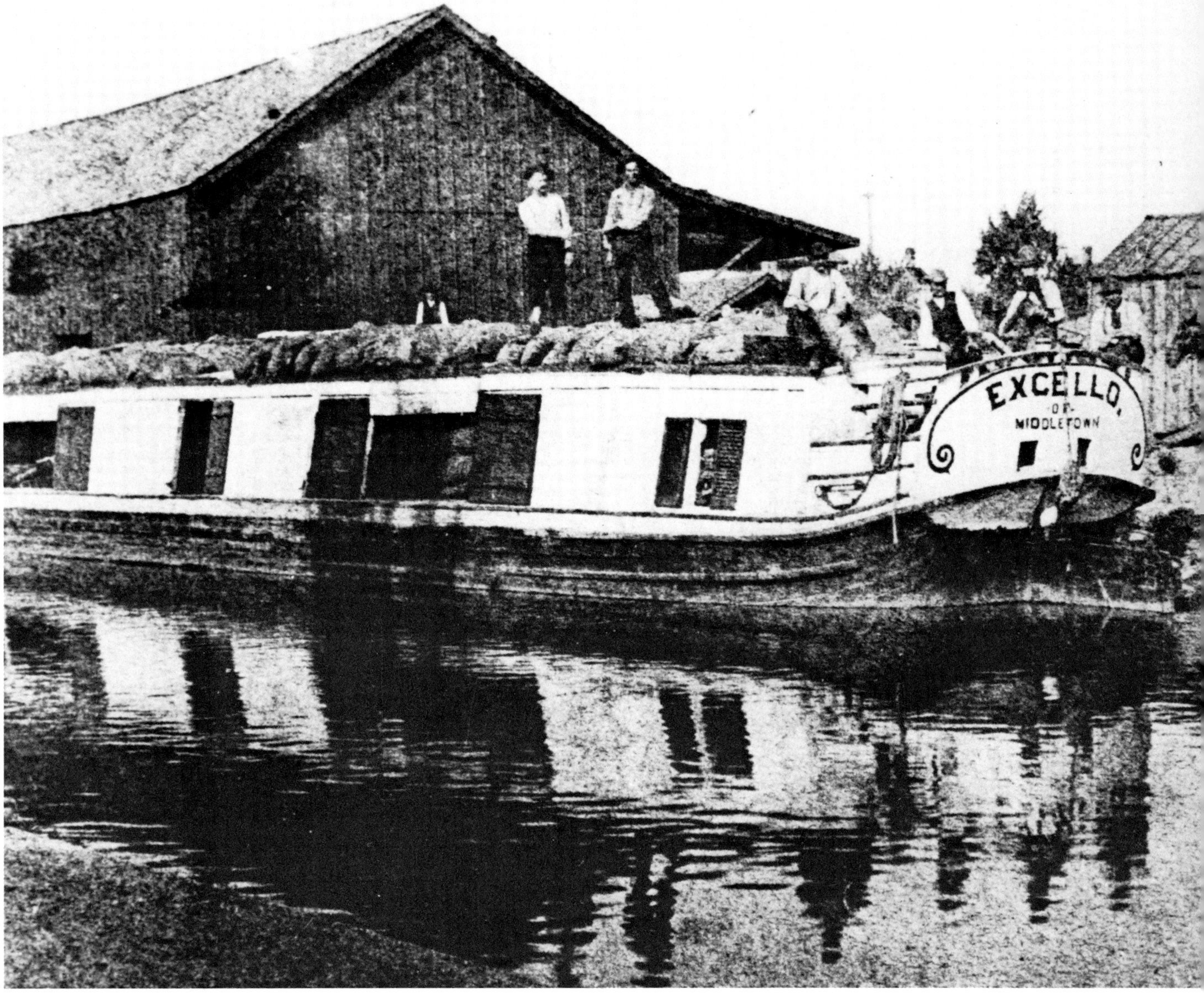

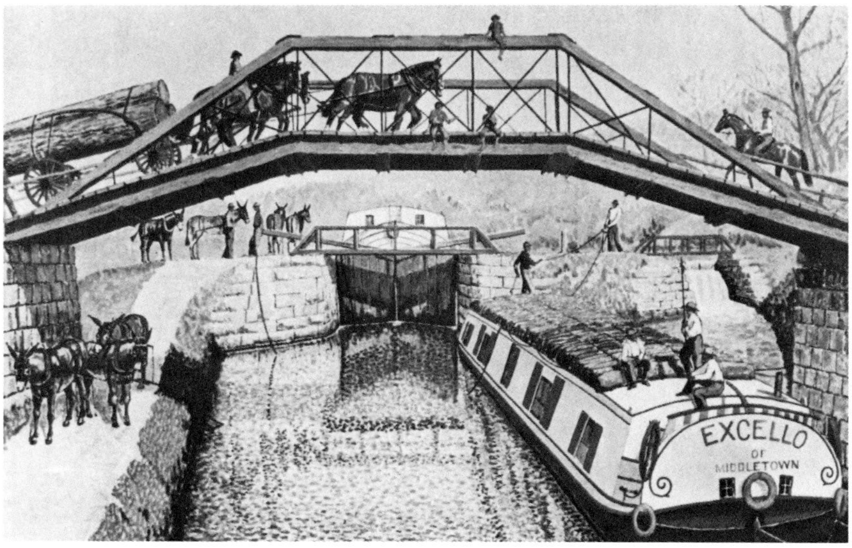

Above is a photo of *Excello* of Middletown, pictured in the Herbert W. Fall painting at left (George C. Crout). Fall documents the typical construction of a loading bridge (at Excello), which had a large hole in the center so that wagons could drop their cargoes directly into canal boats underneath. (Dust cover of *Butler County*, by George C. Crout, 1983.)

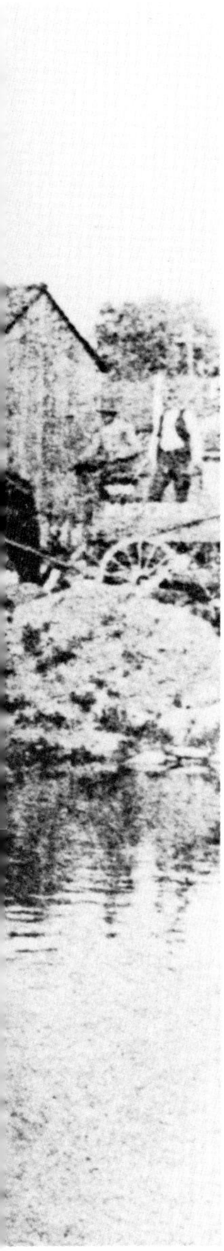

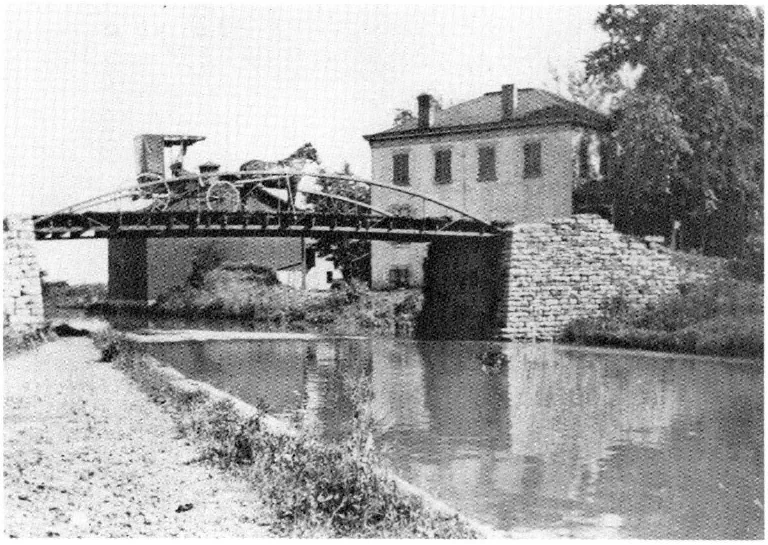

At Port Union, a buggy driver and his horse pause on bridge over canal to avoid blurring picture; possibly a loading bridge. Mile: 229. (CHS)

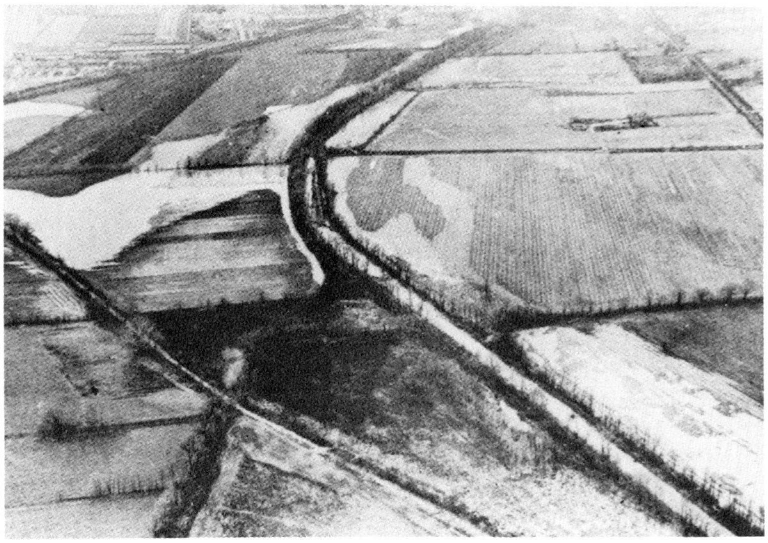

A rare early aerial view of the Miami & Erie Canal as it approaches Hamilton. Mile: 220. (Herbert F. Verity)

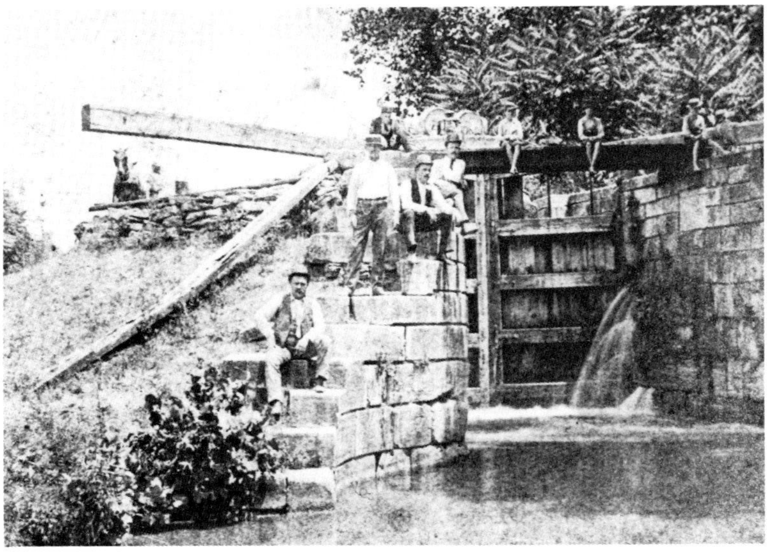

Constructed in 1826, Lock 34 (Lock 3 according to the original numbering system) at Excello, adjacent to Harding-Jones Paper Company, was the first lock completed on the Miami Canal. A boat is waiting to be locked down; a mule observing the photographer can be seen under the sweep handle at upper left. Mile: 212. (George C. Crout)

*"**An October weekend** in 1846 was to be long remembered by Middletown residents who watched as three canal boats stopped at Port Middletown. On these boats were the last remnants of the once-proud Miami Indian Nation. They finally had to leave their reservation along the Mississinewa River. The Miamis had not wanted to leave . . . and with so few Indians involved, even at the time many wondered why it was all necessary.

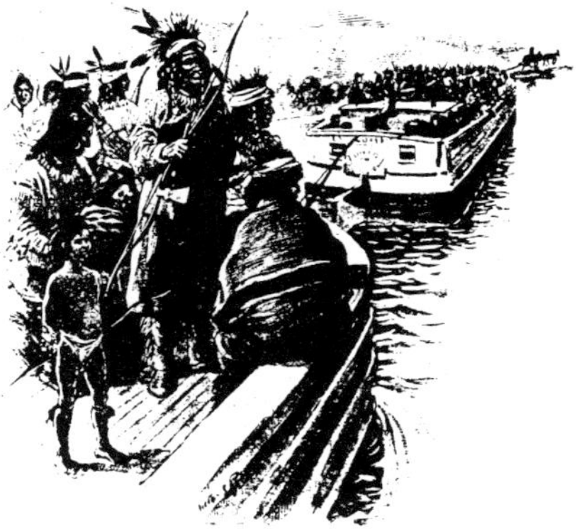

"The soldiers had to herd the Miamis to catch the canal boats at Peru and Fort Wayne. A few hid out and had to be captured by the military. Many picked up handfuls of earth from the graves of their ancestors, placing it in small containers for the trip.

"One eyewitness, Squire Fulwiler of Peru, Indiana, described the parting scene: 'Everybody was out to see the start as if it was a circus. I had a good many friends among the Indians, and moved about the somber assembly to say goodbye. It is true the Indian is stoic, but I saw tears in the eyes of many of the women and girls.'

"A fast packet boat went through Middletown on Friday, October 9. On board was Chief Lafontaine (last general chief of the Miamis), other Indian officials, and government agents. They were moving ahead of the main group, making final arrangements at Cincinnati for passage on a steamboat for St. Louis. . . . Then, on Sunday, October 11, three canal boats with the Miami Indians arrived at Port Middletown, stopping here for a rest. That evening the Miamis were fed from the kitchen of the U.S. Hotel, some staying there overnight and the others camping out in the streets and on the boats.

"Early the next morning, on the 12th, the Indians boarded their boats and the tow lines were fastened. . . . William Dean Howells, a nine-year-old youth, was later to recall in his book, *Stories of Ohio* [1897]: 'I, myself, when a boy living in Hamilton, saw the last of the Ohio Indians passing through the town on the three canal boats which carried the small remnant . . . out of the land that was to know them no more forever. It was quite time . . . for all I know they may have been tame farmers and mechanics, but in their moccasins and blankets, with their bows and arrows, they looked like wild hunters; and Ohio was no longer a good hunting ground.'

"By evening they had arrived at Cincinnati, and here they disembarked from their canal boats and, according to the newspaper, 'marched down Main Street to the river, where they embarked on the steamer, *Colorado*.'

"The Miami Indians reached St. Louis on October 20, 1846, boarded another steamship, the *Clermont II*, and went up the Missouri River to Kansas Landing, now Kansas City. From here they set out for the Osage Reservation, where the Indian agent certified the arrival of 323 Miamis.

"Historian Otto Winger, in his book, *The Last of the Miamis*, wrote: 'They arrived in the late fall, used as they were to hardships; yet without adequate home and food they could not withstand the Kansas winds and blizzards. But even so, homesickness was their great affliction. It is said from one third to one half of them died during the first year. Many of them started back east, begging their way as they went. Some begged white people to take care of their children and then wandered off, to be seen no more.' " (George C. Crout)

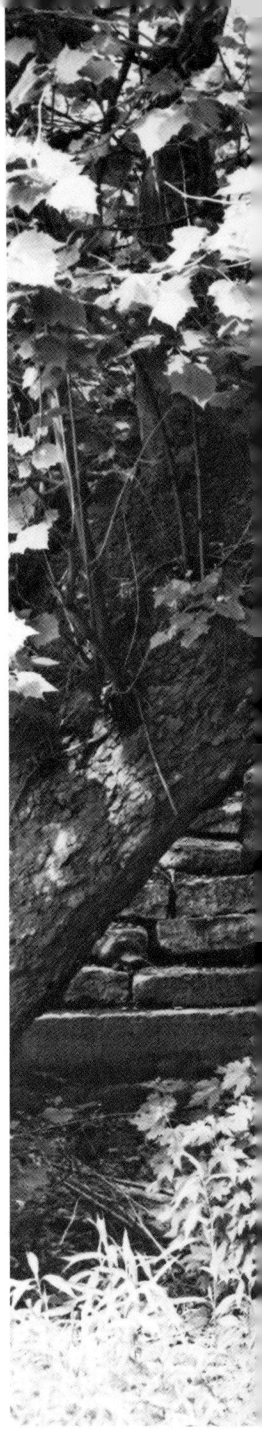

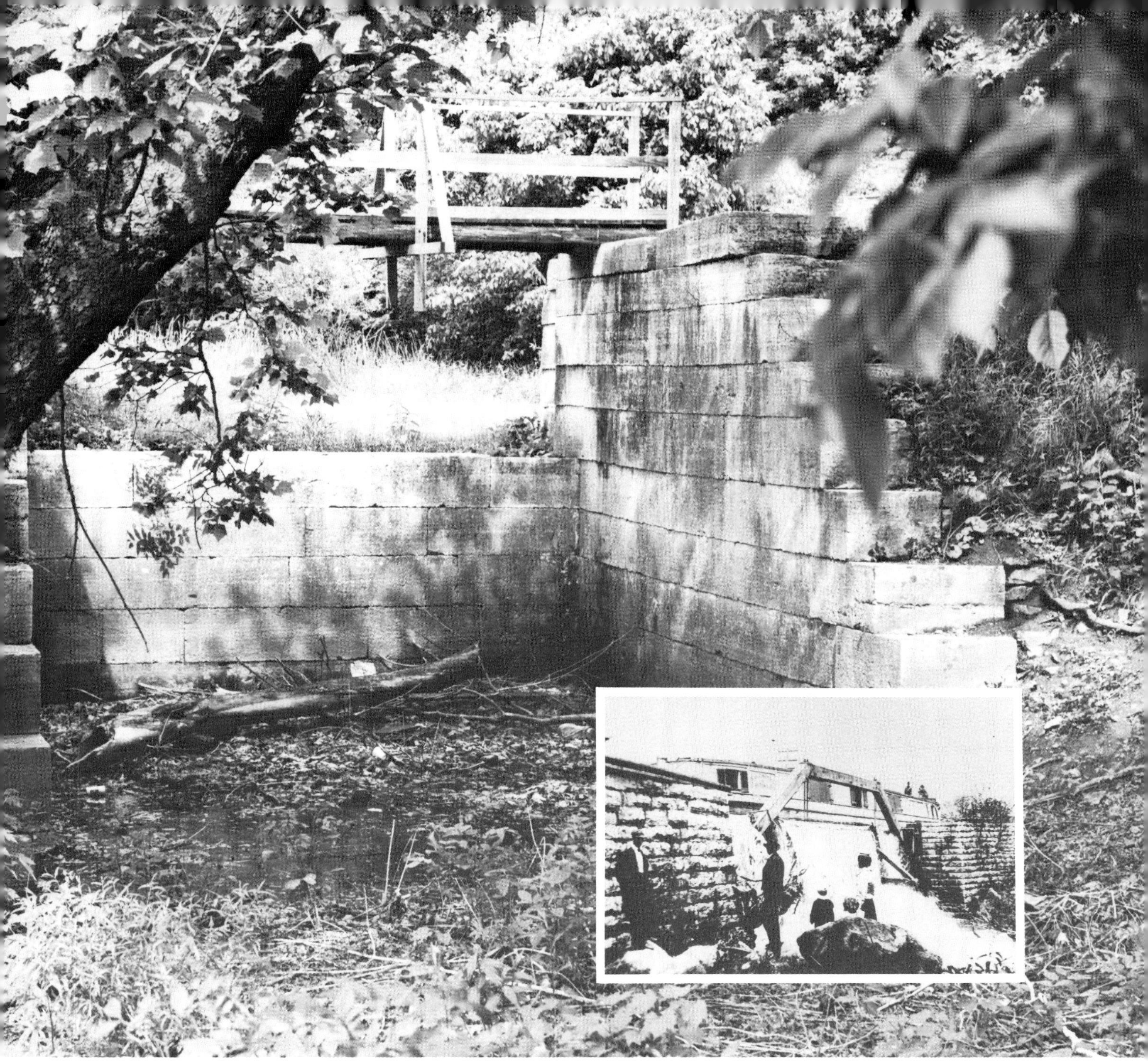

With its original stone construction still intact, the spillway at Excello Lock is part of a commemorative park dedicated in 1970 by the Middletown Historical Society in cooperation with the Miami Conservancy and the Butler County Park Board. Mile: 212. (Photo by Jack Gieck.)

Inset, A boat crosses the aqueduct over Dick's Creek at Excello. Mile: 211. (George C. Crout)

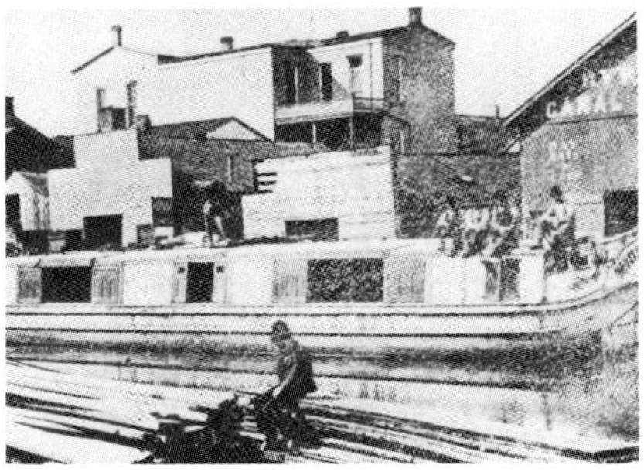

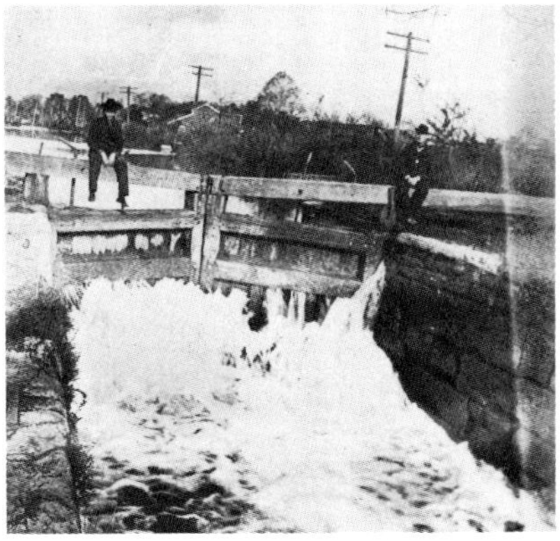

Far left, Port Middletown in 1880. The texture and shape of the photograph suggest it may be a Calotype, which is produced from a paper negative process. Mile: 208. (George C. Crout)

Left, Lock One at Middletown. Posing on the sweeps are Norman Nichols and Andrew Barnicle. Mile: 208. (George C. Crout)

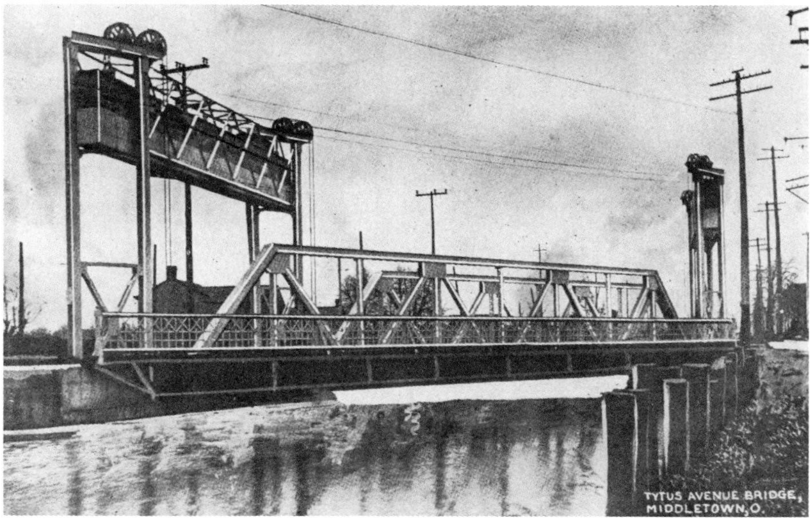

Above, Middletown also had a lift bridge, over Tytus Avenue. Massive counterweights span the road just below the pulleys. (From an old postcard, courtesy George C. Crout.) *Below,* Boys skinny-dipping in the Canal at Middletown. (George C. Crout)

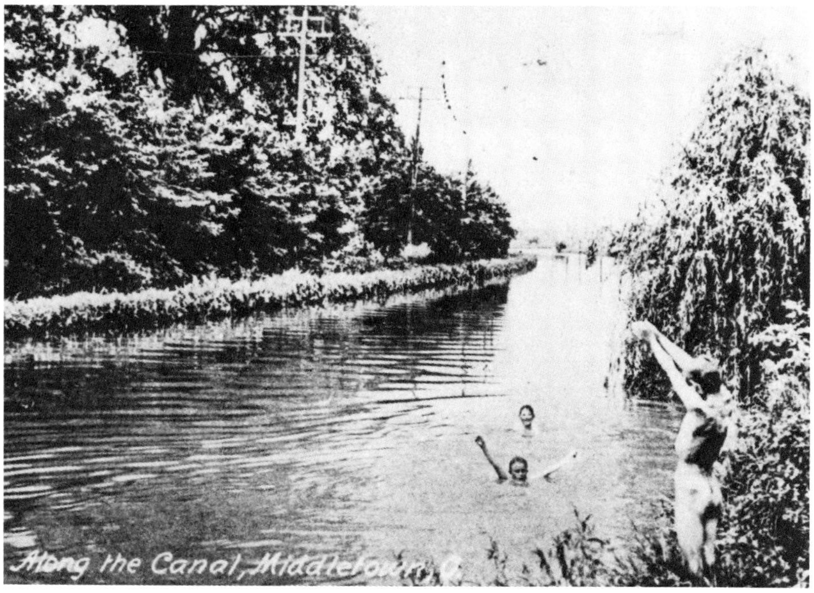

Right, The Middletown Hydraulic Canal was constructed in 1852 as a source of waterpower. Paralleling the Miami & Erie, the two canals were accessed by boats to serve local industry. As the *Middletown Emblem* observed on July 16, 1852, the Hydraulic Canal created "a power at Middletown equal to one hundred fifty run of four and a half feet millstones, or capable of driving sixty thousand spindles. There is, perhaps, no Hydraulic work in the State which will be more safe or permanent than this. . . . The canal varies in width from fifty feet top-water-line and six feet in depth to a width of one hundred feet and ten feet depth. . . ." The Hydraulic Canal, still in operation, is used by Sorg Paper Company's mill at Middletown; 1978. Mile: 208 (from a color slide by Jack Gieck). *Inset,* A reproduction of artist Herbert W. Fall's circa 1875 scene depicting boats of the Middletown Hydraulic Canal at the Sorg Paper Company. (George C. Crout)

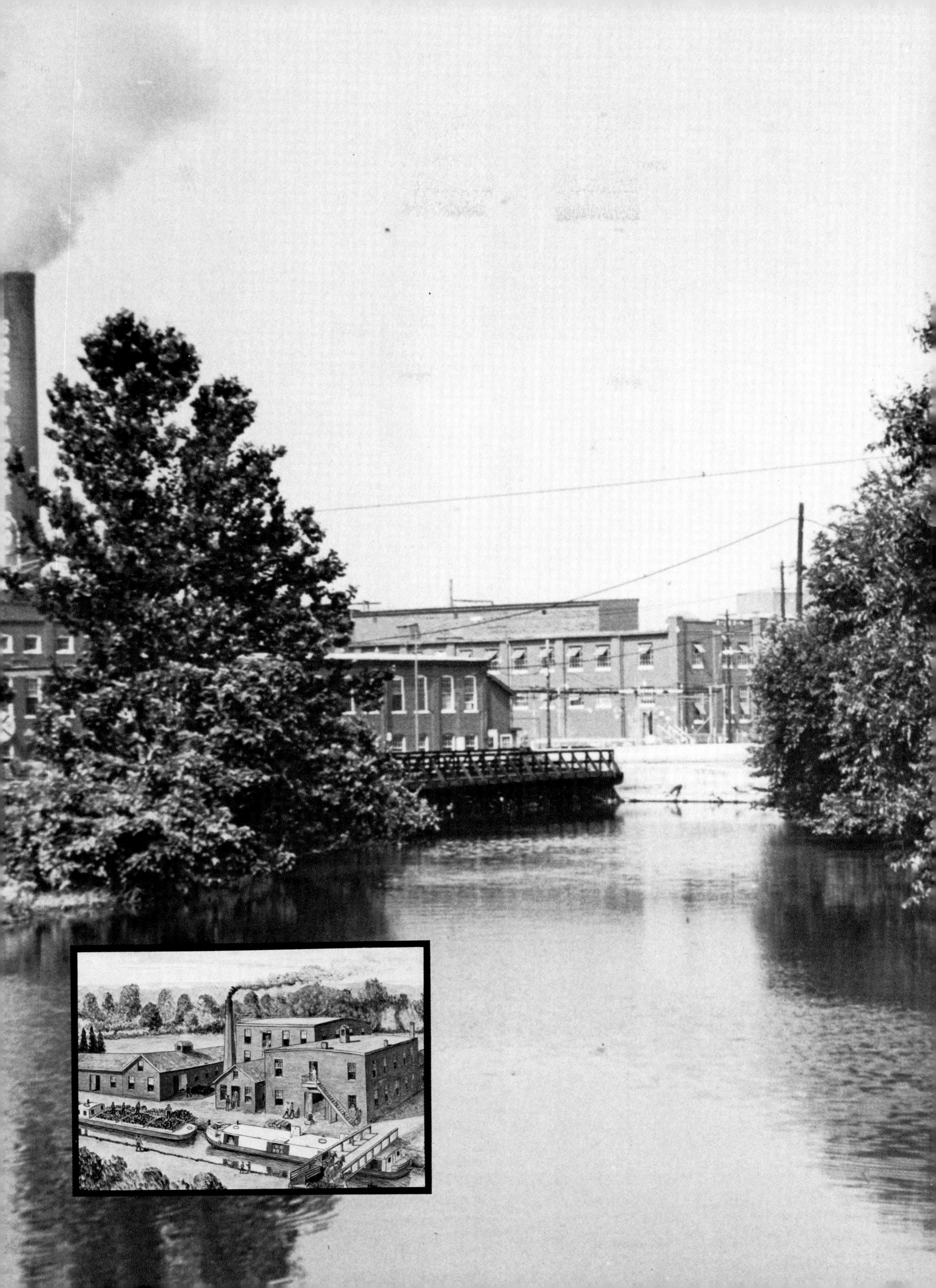

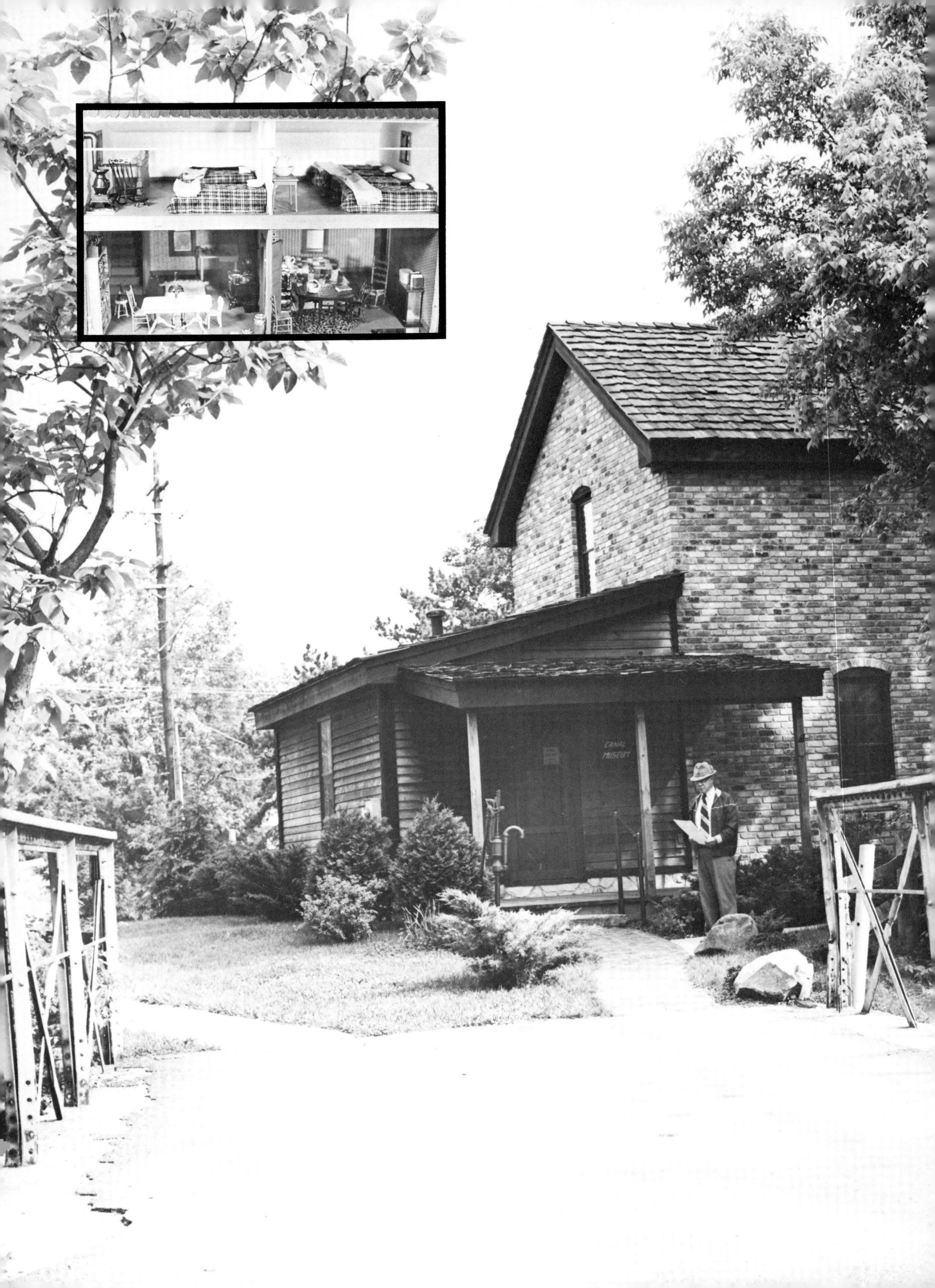
CANAL
MUSEUM

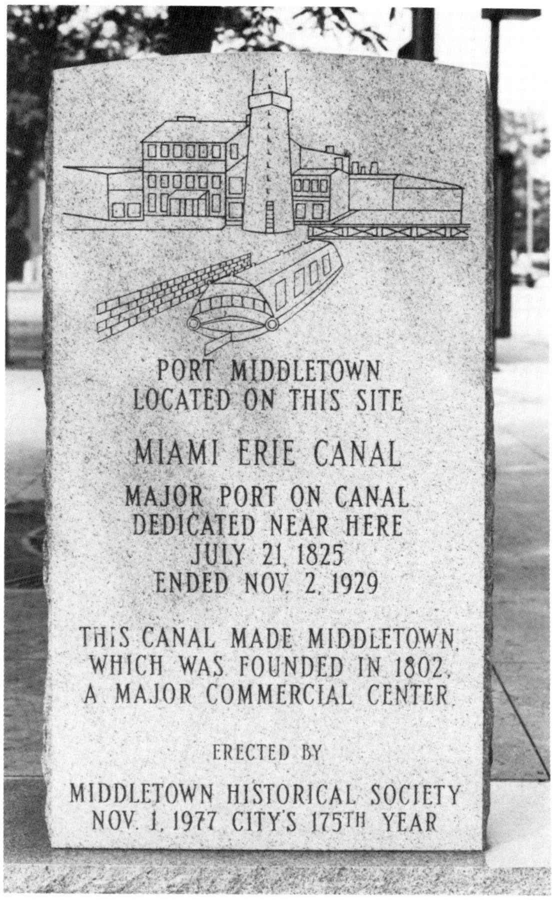

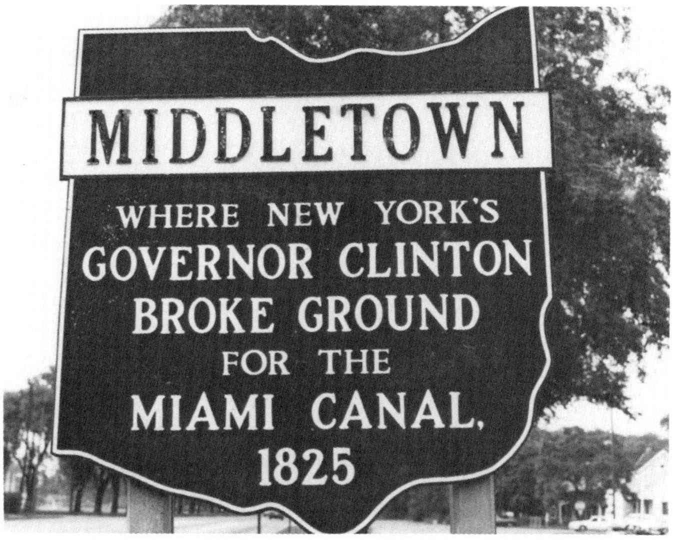

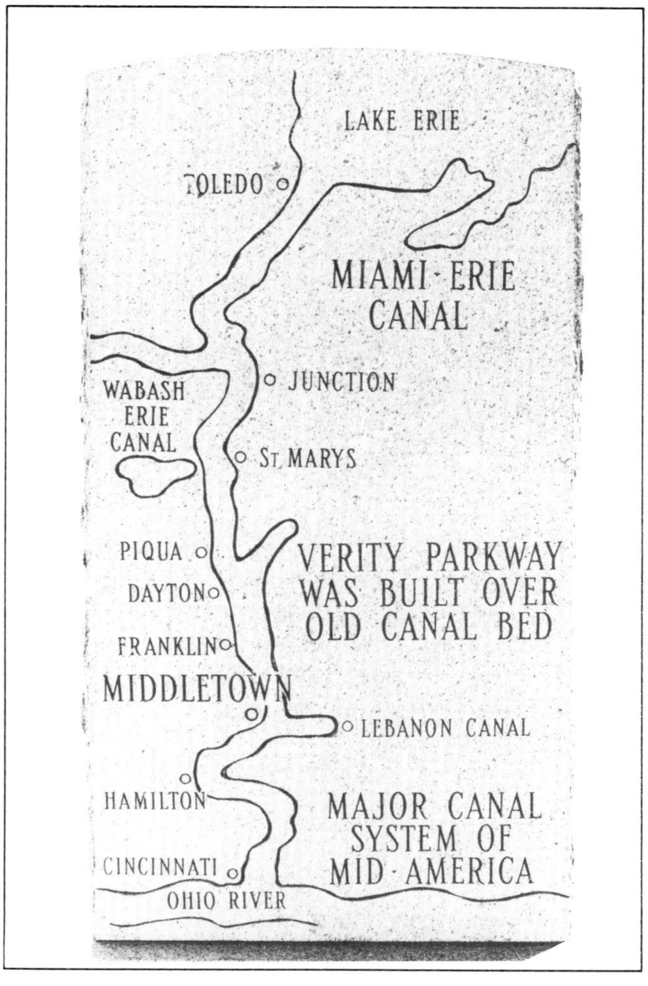

Left, On land between the Hydraulic Canal (beneath foreground bridge) and the former Miami & Erie channel (which ran behind the building), the Middletown Historical Society has constructed a Canal Museum as a replica of the locktender's house which stood at Lock 33, Amanda Lock. The building is constructed of materials recovered from many historic structures on the canal. Curator George Crout studies a broadside artifact for display. (Photo by Jack Gieck.) *Inset,* Originally created by a 1928 Middletown High School class, this Canal Museum display is a model of a locktender's house. Because of their meager pay ($10 a month), many locktenders maintained dormitory beds for rent to canal passengers and crewmen. (Photo by Jack Gieck, courtesy the Middletown Historical Society.)

The Middletown Historical Society has studded the city with mementos to Middletown's canal heritage. (1986 photos by Jack Gieck.)

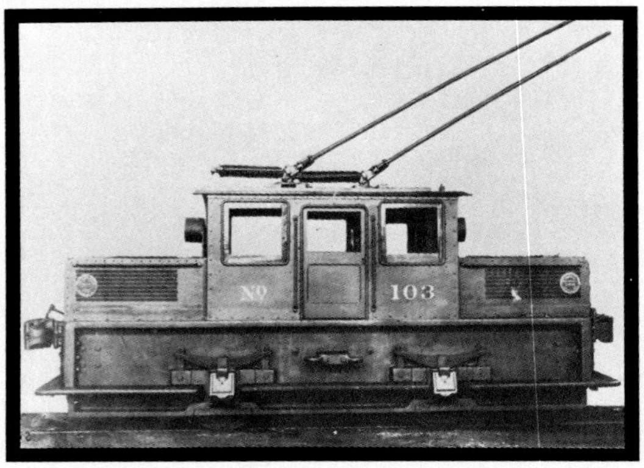

In an effort to modernize freight hauling technology at the turn of the century, Baldwin Locomotive was commissioned by the Miami & Erie Transportation Company to build a number of "Electric Mules," (inset) to tow canal boats between Cincinnati and Dayton. Although track was laid as far as Miamisburg by 1903 and reached Dayton in 1905, only the portion of the system between Hamilton and Middletown was ever operational. Here, it was soon learned that the clay banks could not tolerate the wake created by the speeds at which the electric locomotives operated, and after delivering only one Electric Mule, the balance of Baldwin's order was canceled. (Smithsonian Institution/CSO)

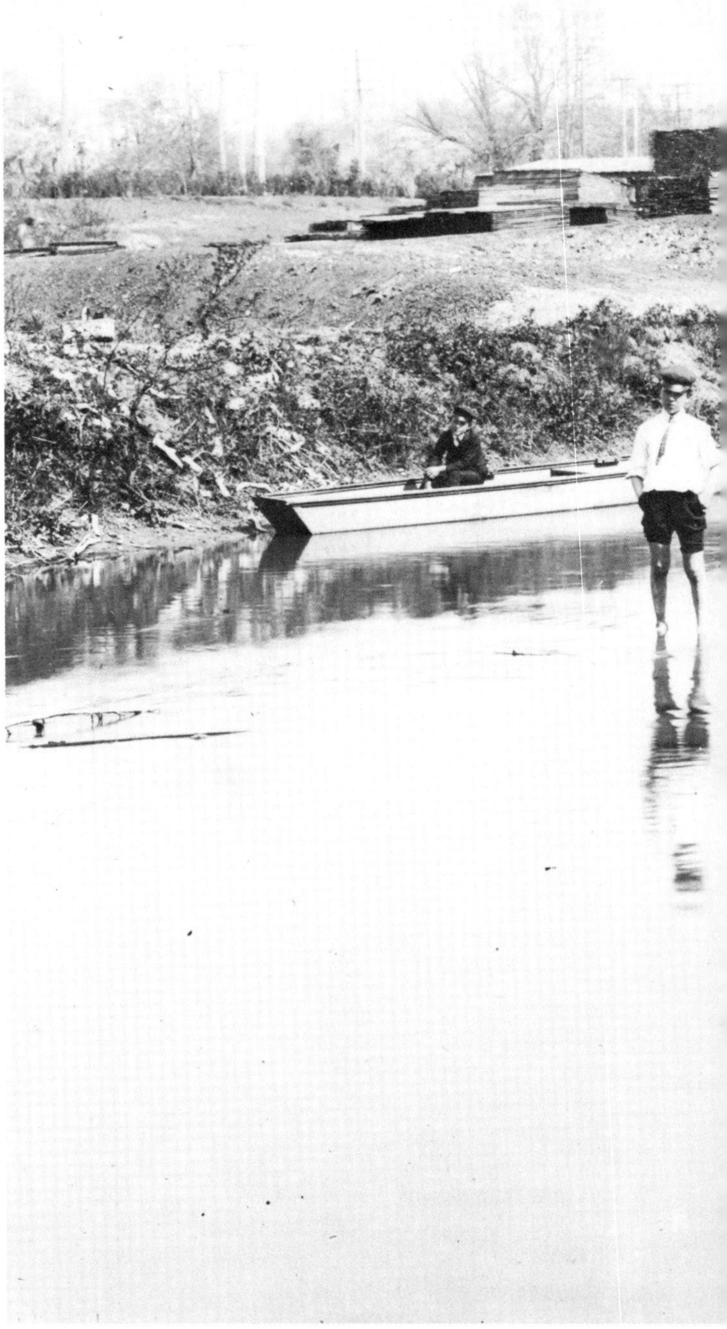

Electric Mule tracks parallel the canal south of Dayton in 1904. Although support poles for the trolley wire have been erected, it will be noted that the conductor has not been installed. In violation of the contractors' specifications, the spaces between the ties have not been graded to provide a smooth towpath, creating an intolerable trip hazard for mules still operating on the route. Mile: 185. (Photo originally made on 8 × 10 negative by B. G. Archembeault/courtesy RVF.)

Below, The oldest operating inn in Ohio, the Golden Lamb Inn at Lebanon, was the site of a historic celebration after ground-breaking ceremonies on July 28, 1825, at nearby Middletown. Lebanon and Middletown were connected by the Warren County Canal which was unsuccessful because the town was unable to generate adequate payloads; 1978. (From a color slide by Jack Gieck.)

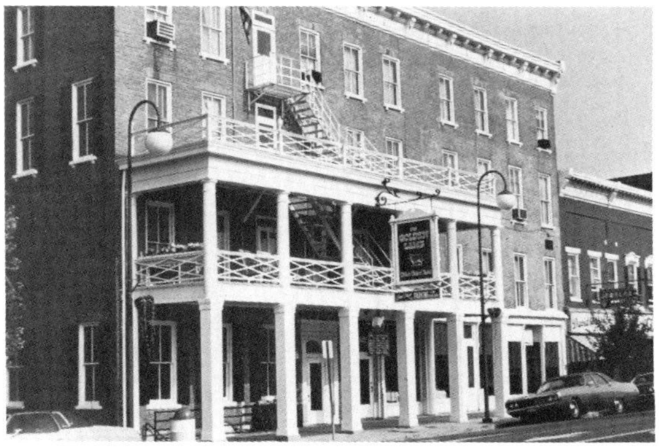

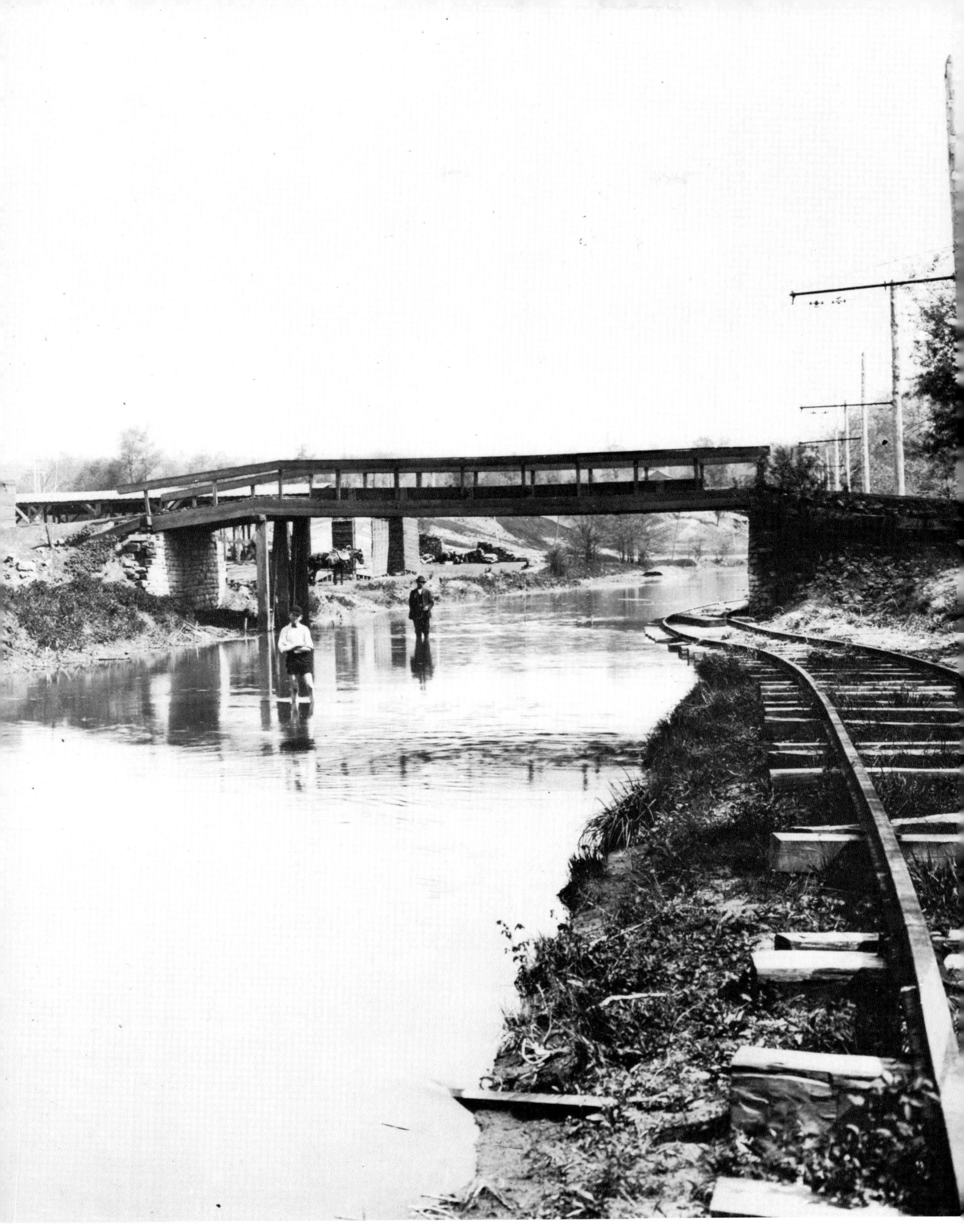

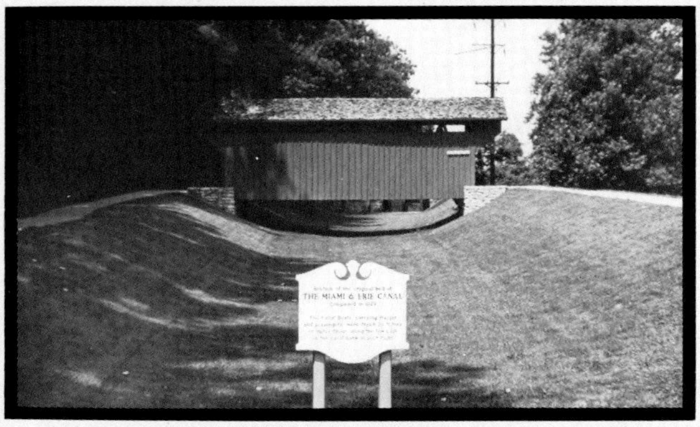
THE MIAMI & ERIE CANAL

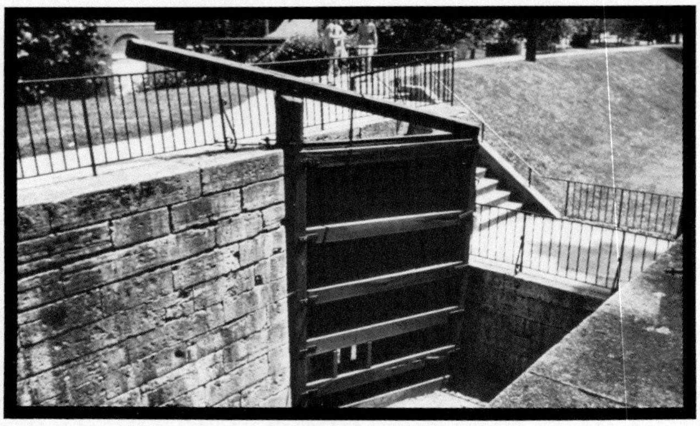

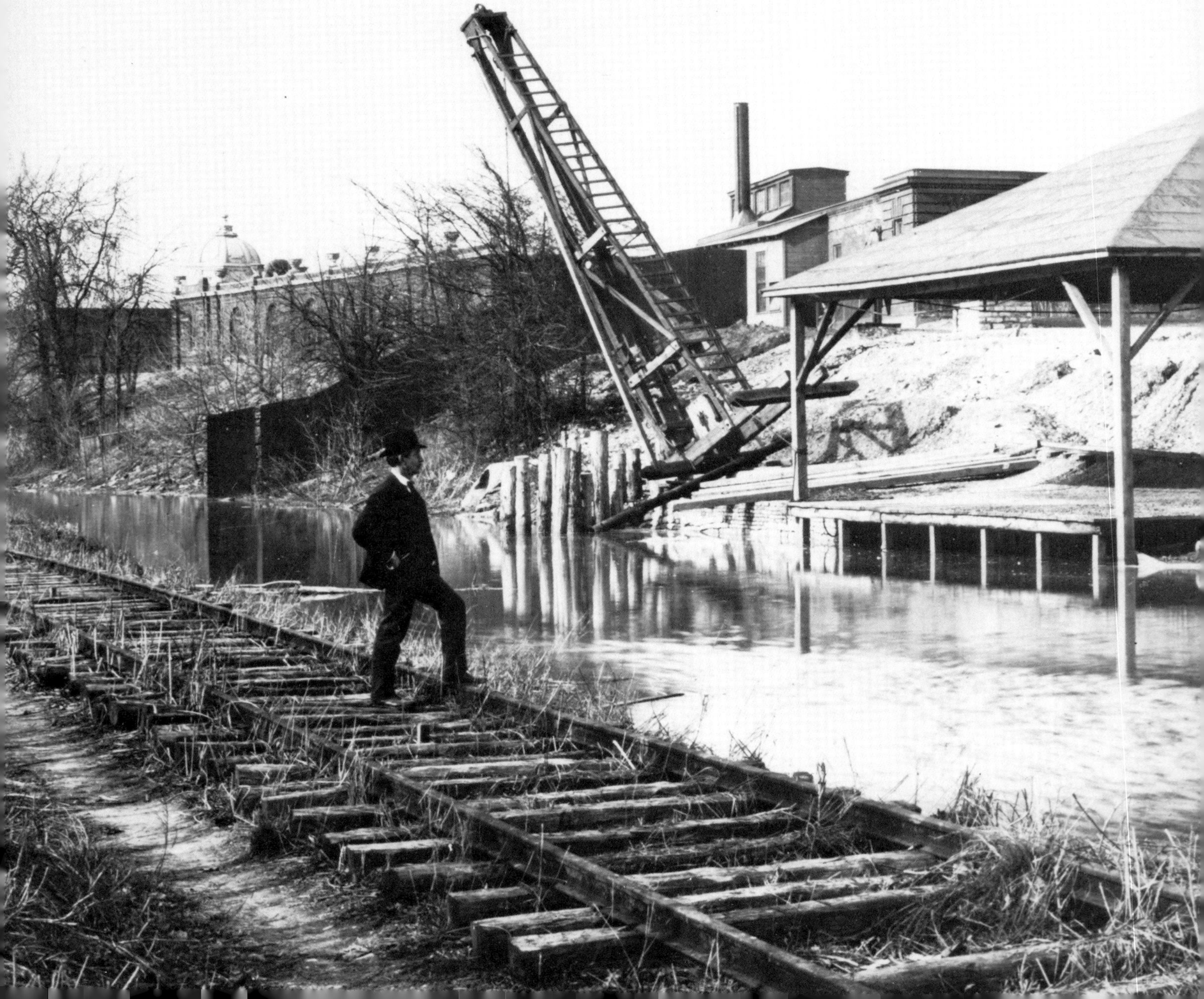

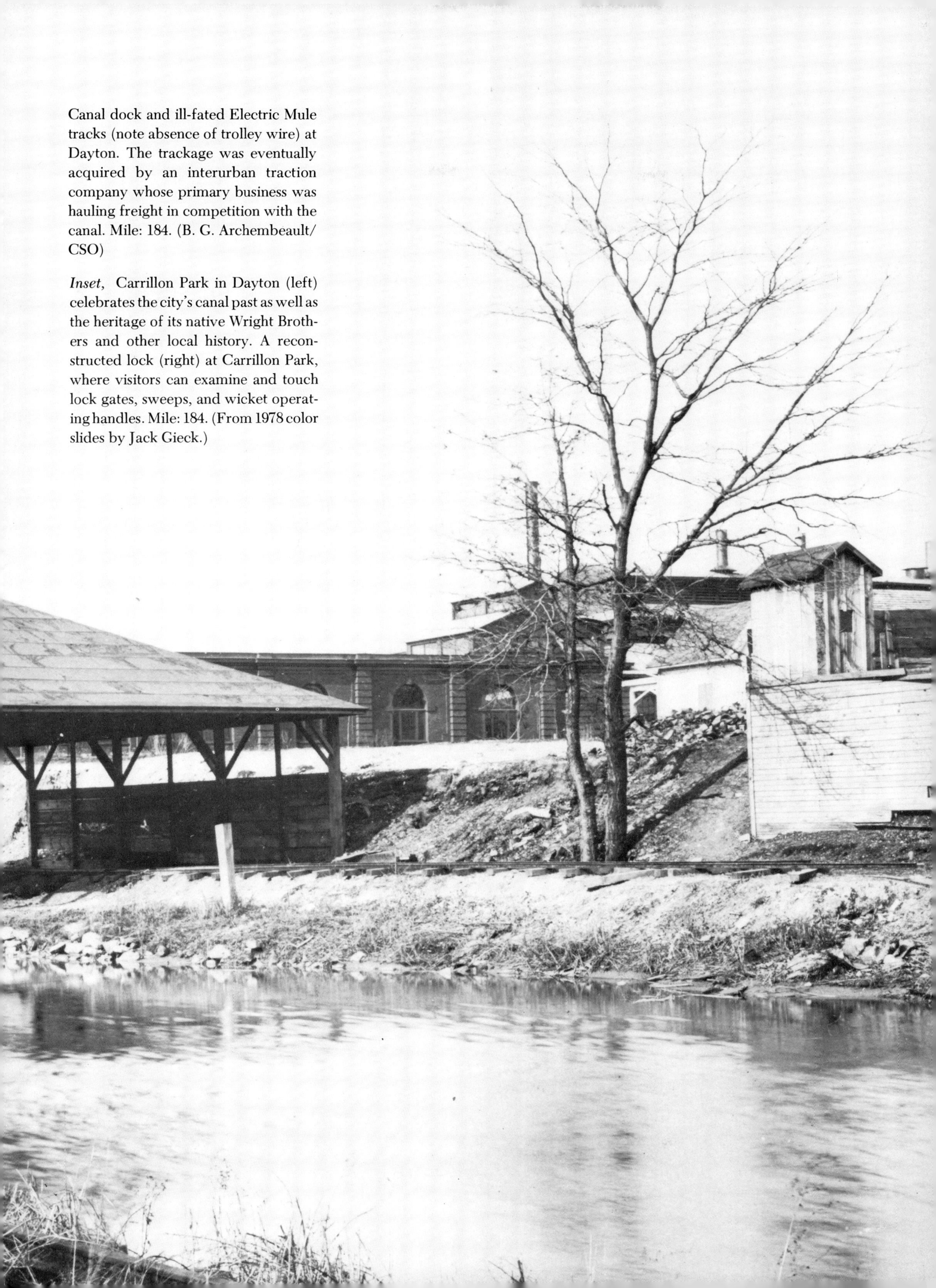

Canal dock and ill-fated Electric Mule tracks (note absence of trolley wire) at Dayton. The trackage was eventually acquired by an interurban traction company whose primary business was hauling freight in competition with the canal. Mile: 184. (B. G. Archembeault/ CSO)

Inset, Carrillon Park in Dayton (left) celebrates the city's canal past as well as the heritage of its native Wright Brothers and other local history. A reconstructed lock (right) at Carrillon Park, where visitors can examine and touch lock gates, sweeps, and wicket operating handles. Mile: 184. (From 1978 color slides by Jack Gieck.)

State boats moored behind Dayton Fairgrounds; 1904. Mile: 181. (RVF)

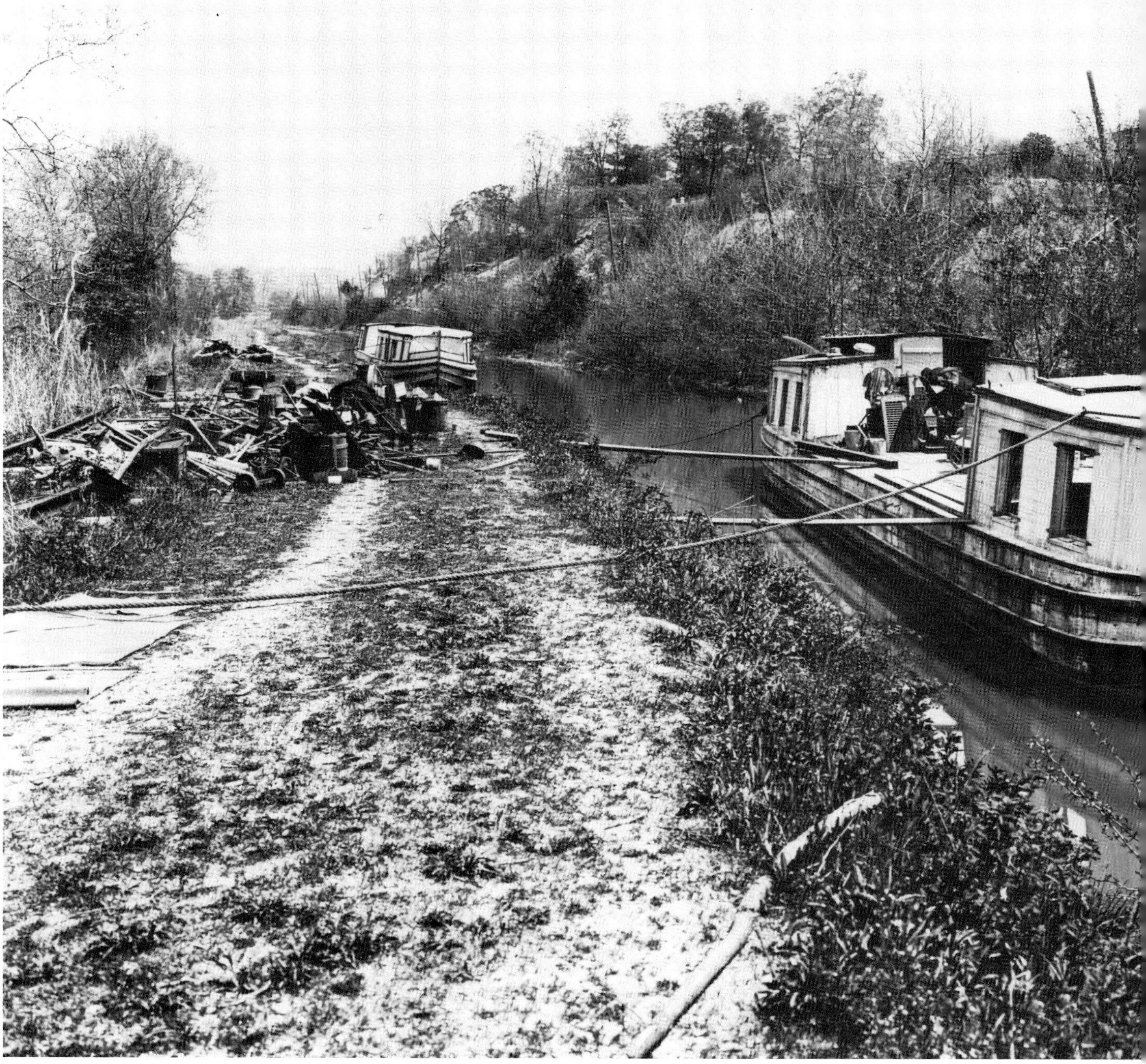

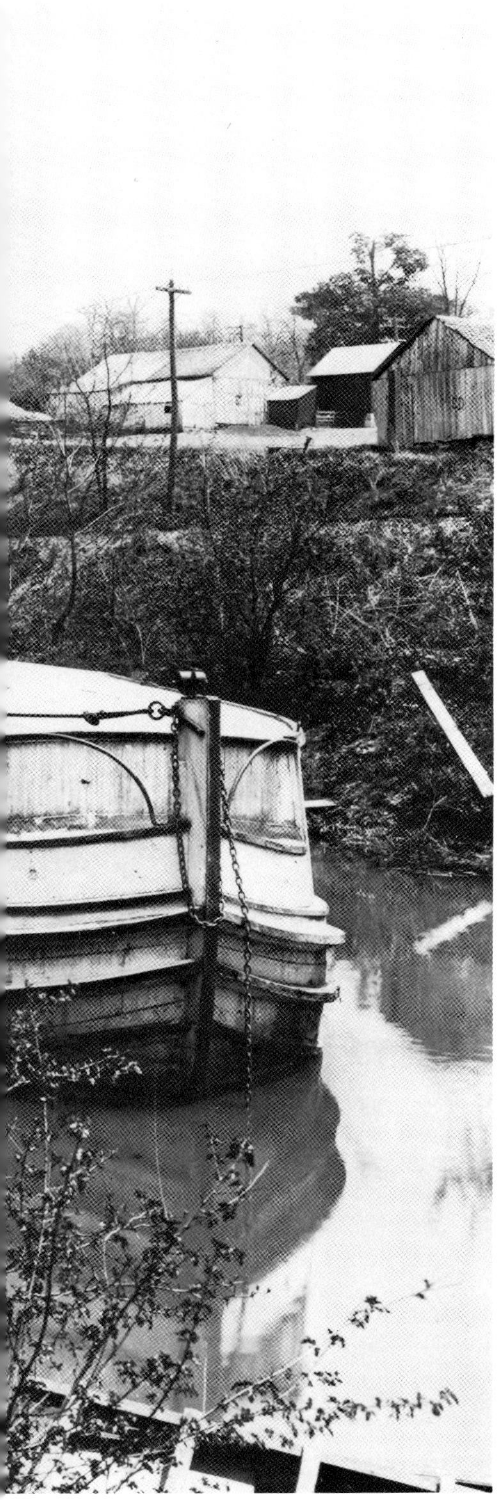

Taylor Street canal bridge; c. 1904. Mile: 181. (RVF)

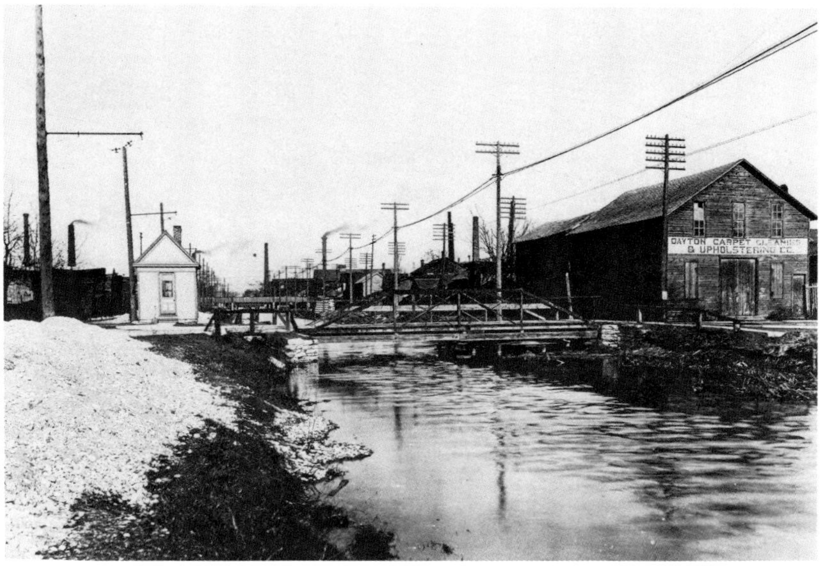

Canal boat *Dayton* at Apple Street; c. 1904. Mile: 181. (RVF)

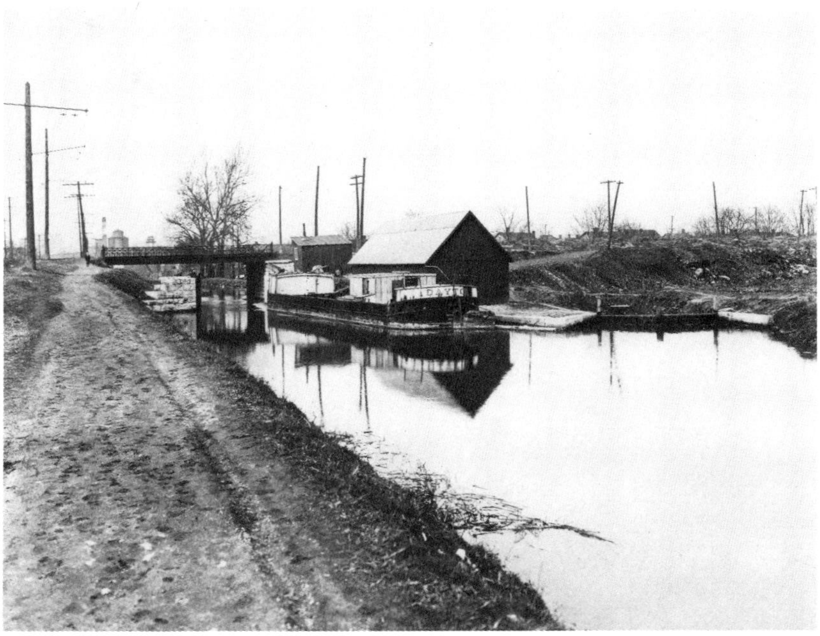

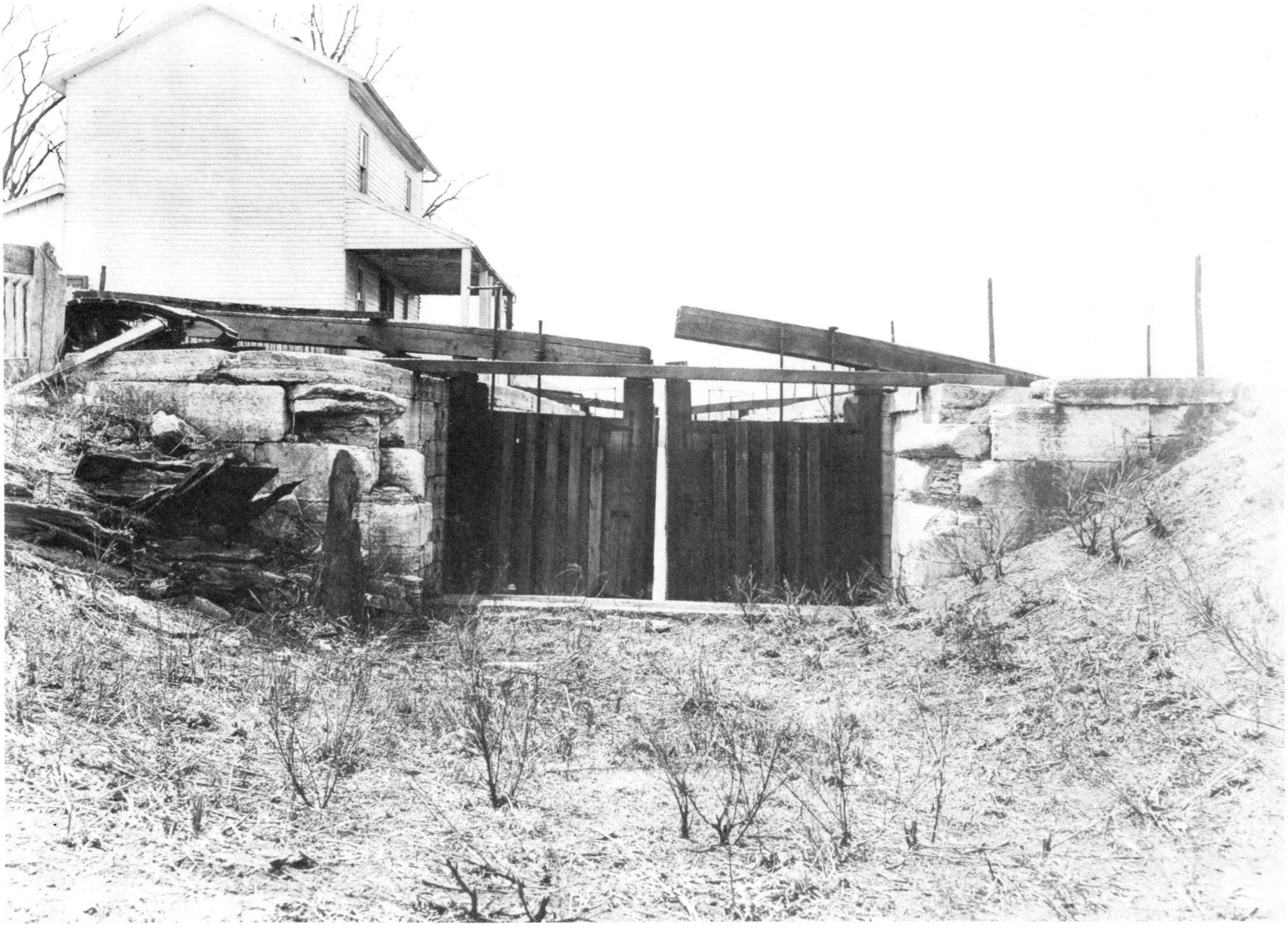

Above, Lock 20 at south end of Mad River Aqueduct near Dayton. Mad River feeder enters from opposite side of canal at center; Barney & Smith Works are to the right (OHS). *Right,* The north side of Lock 20 gates at Mad River Aqueduct after demise of the canal. Mile: 178. (CSO)

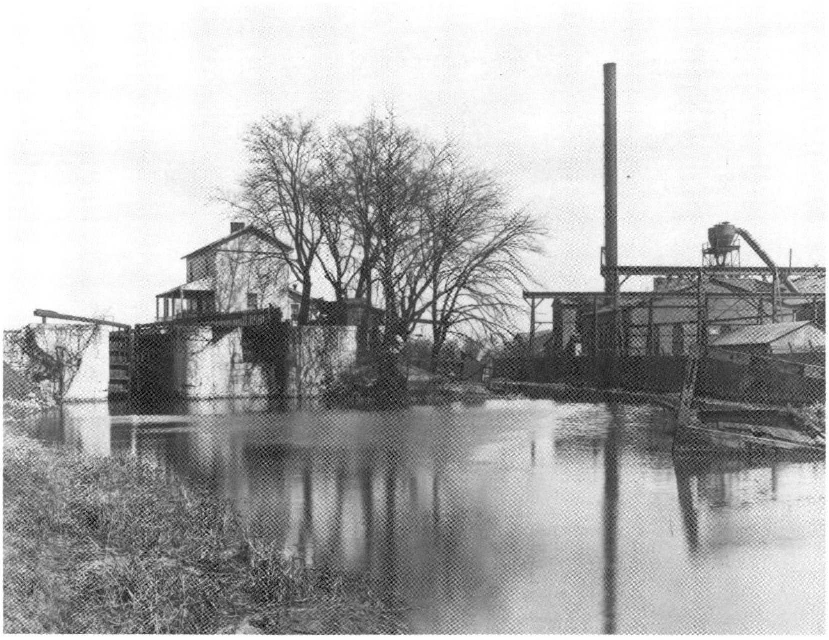

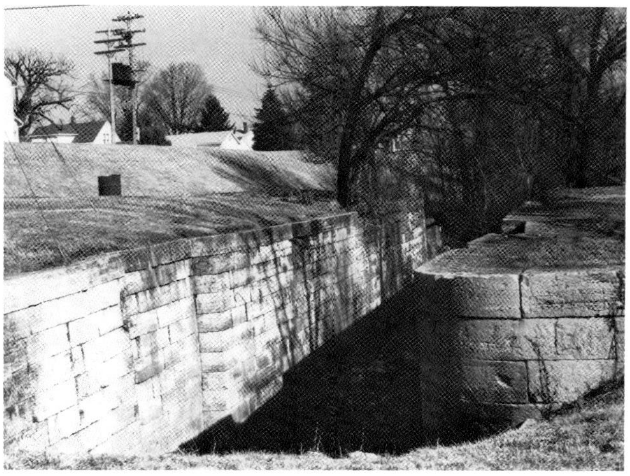

The lock at Tipp City exhibits rounded corners typical of Miami & Erie lock architecture; 1969. (CSO)

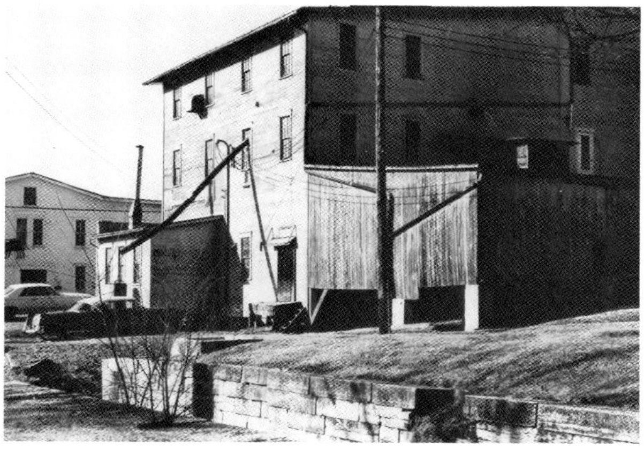

A mill adjacent to Lock 15, foreground, at Tipp City; 1969. Mile: 169. The aqueduct over Great Miami River was north of here. (CSO)

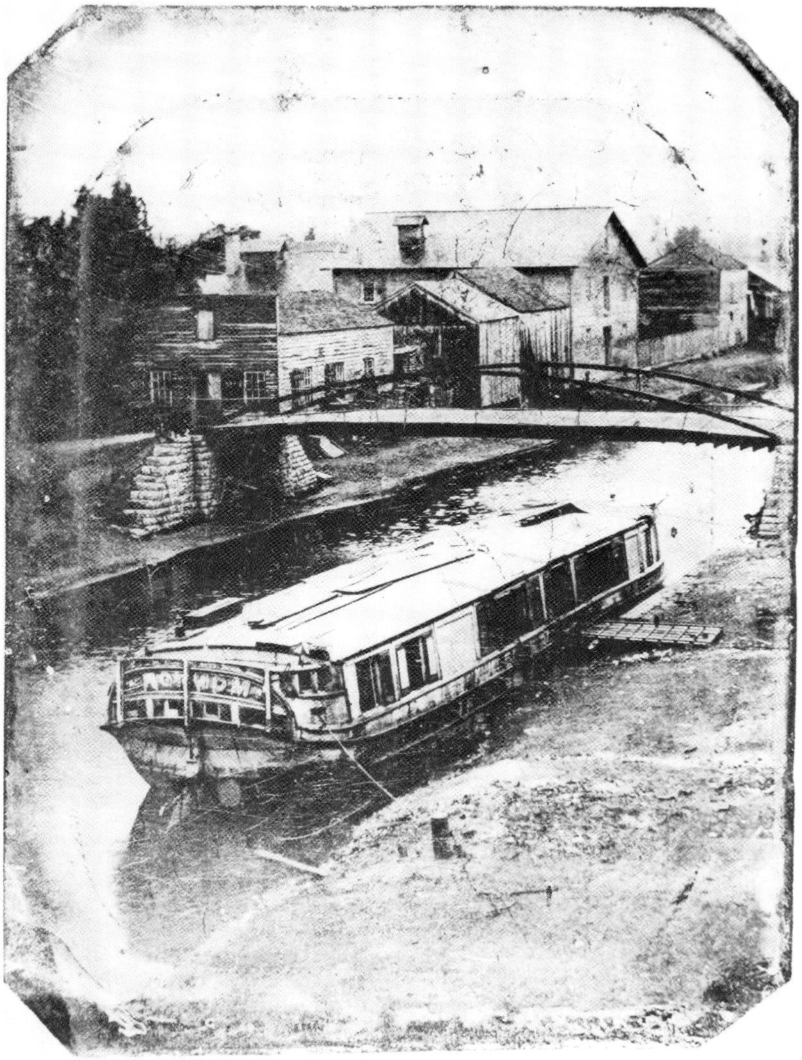

This photo of an original tintype pictures a line boat at Troy; c. 1860. A cheap descendent of the daguerreotype, the tintype produced a direct positive on a metal plate, which reversed the image (note boat name on the boat's stern). It was invented in 1856 by Peter Neff, Jr., who called his process melainotype, and also by Victor M. Griswold, who chose the name ferrotype. Both processes came to be called tintype many years later, probably a pejorative coined by serious glass-plate photographers. The original had a low-contrast, "muddy" appearance; the contrast of the print has been enhanced. Mile: 163. (Photo by Jack Gieck, courtesy THS.)

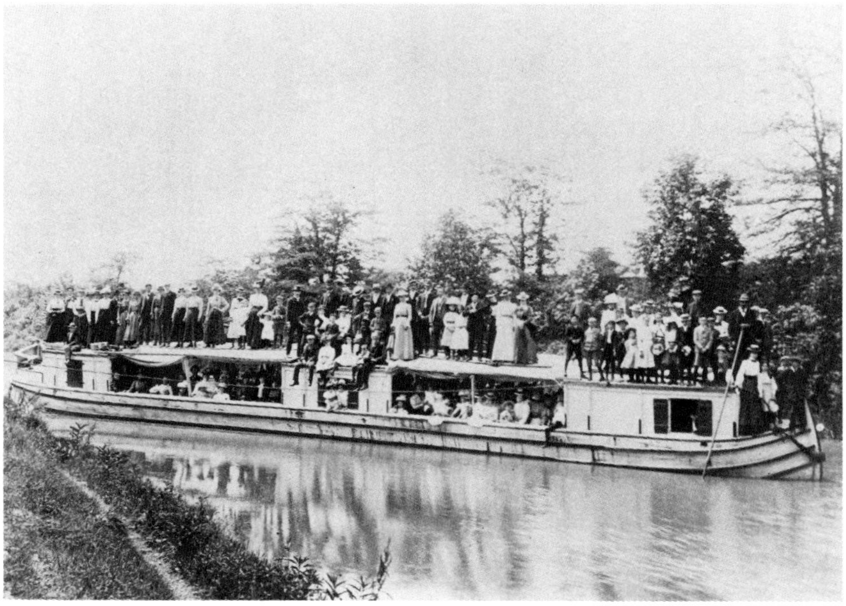

Only a few miles apart, the towns of Troy (right) and Piqua (far right) appear so similar on these postcards that they could have presented a problem to visiting canalers, especially since vintage maps indicate that saloons outnumbered all other business establishments in the towns. (Flesh Public Library, Piqua)

Visible from State Highway 25A are these remains (below right) of a twin arch stone culvert near Eldean on the Miami & Erie between Troy and Piqua. The structure has been placed on the National Register of Historic Places. Mile: 161. (CSO)

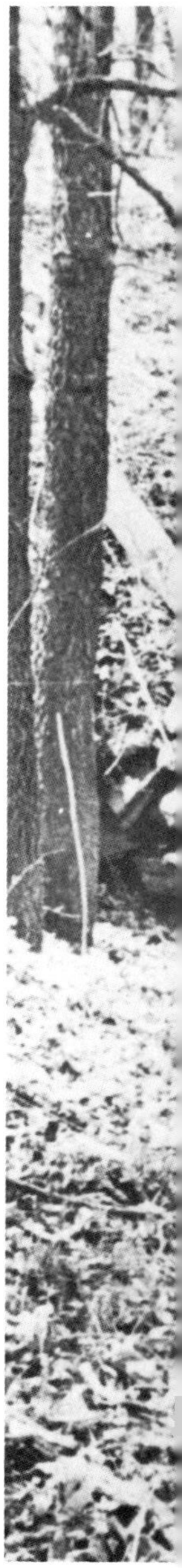

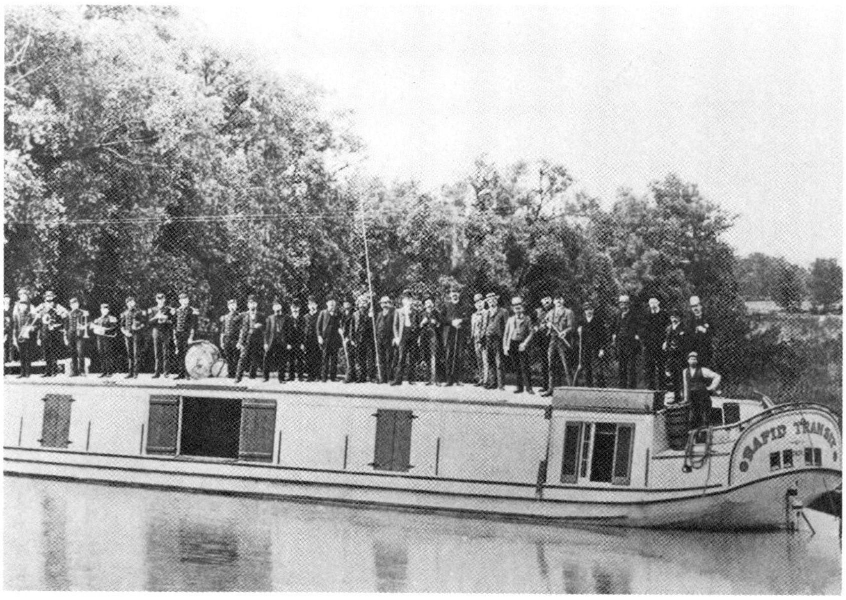

Hobart Brothers company picnic (top) at Troy includes enough employees' family members to more than load the boat to capacity. The company's canal tradition has continued for more than a century: with its headquarters astride the canal, Hobart Brothers' welding equipment factory today includes Lock 12 in its basement, as a part of the building's foundation. A tunnel linking two manufacturing buildings passes through the Troy Hydraulic Tail Race; c. 1882. Mile: 163. (THS)

Rapid Transit (bottom) of Troy entertains a distinguished party with entertainment that includes a uniformed brass band. Mile: 163. (CSO)

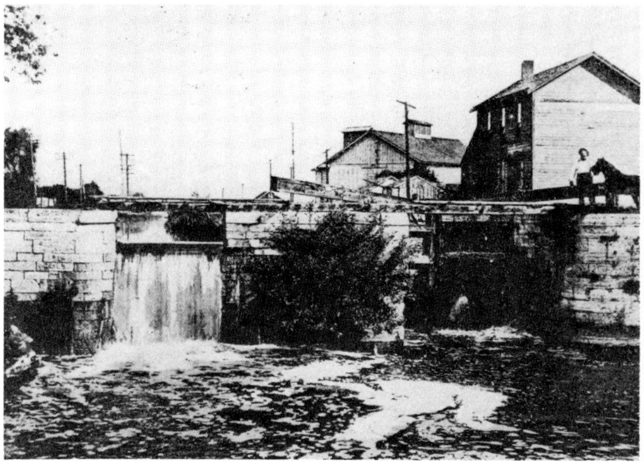

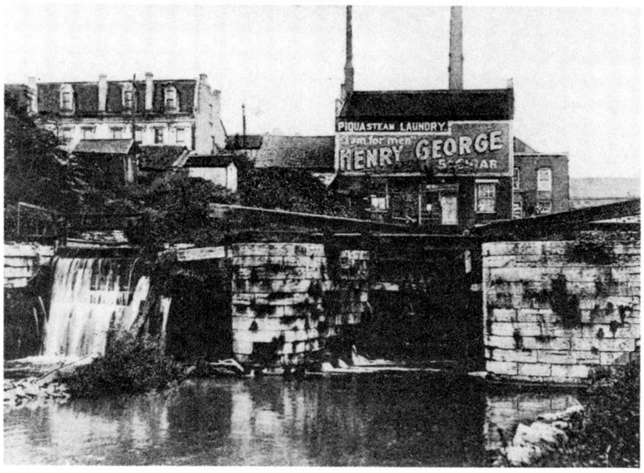
PIQUASTEAM LAUNDRY
HENRY GEORGE

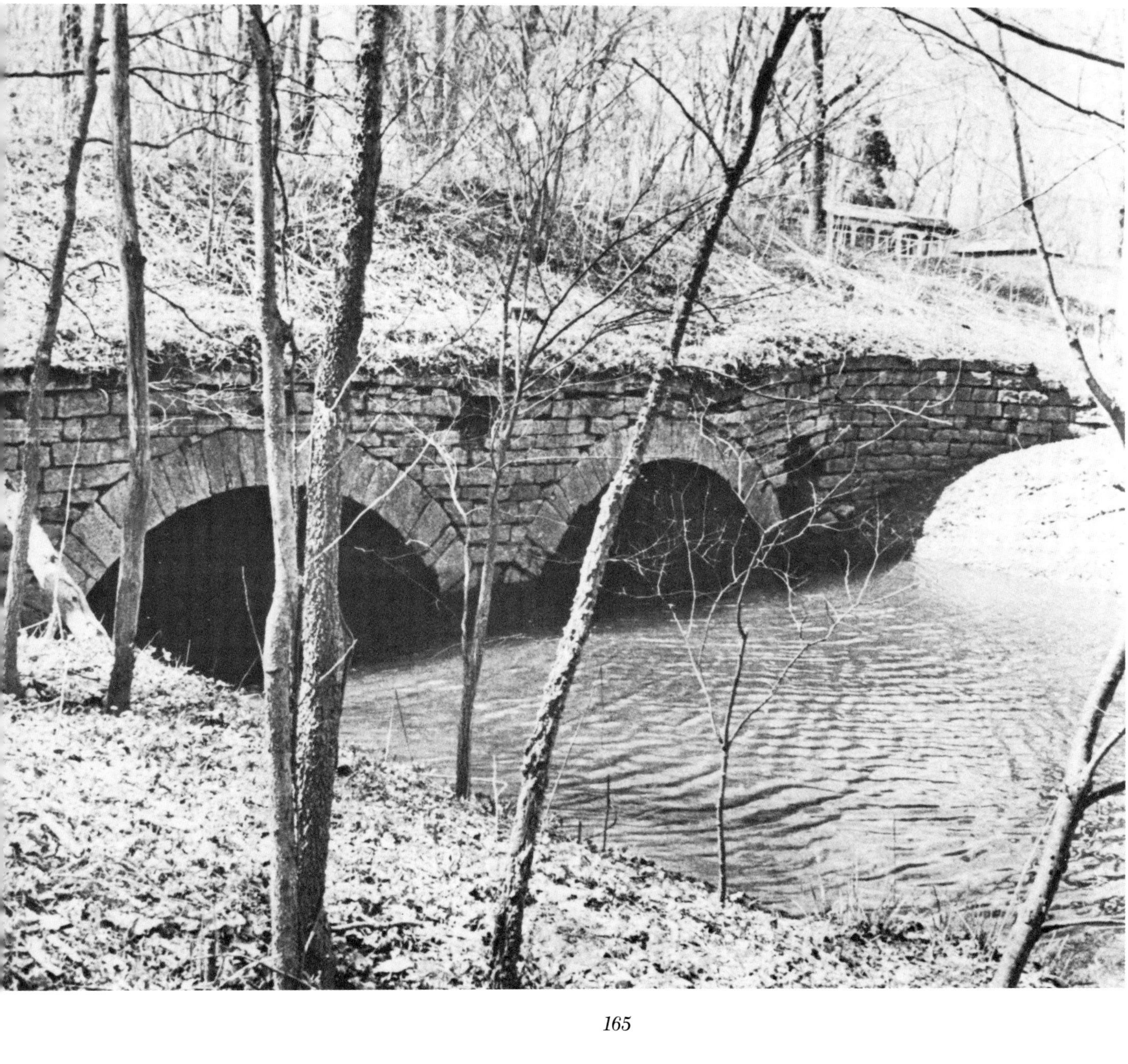

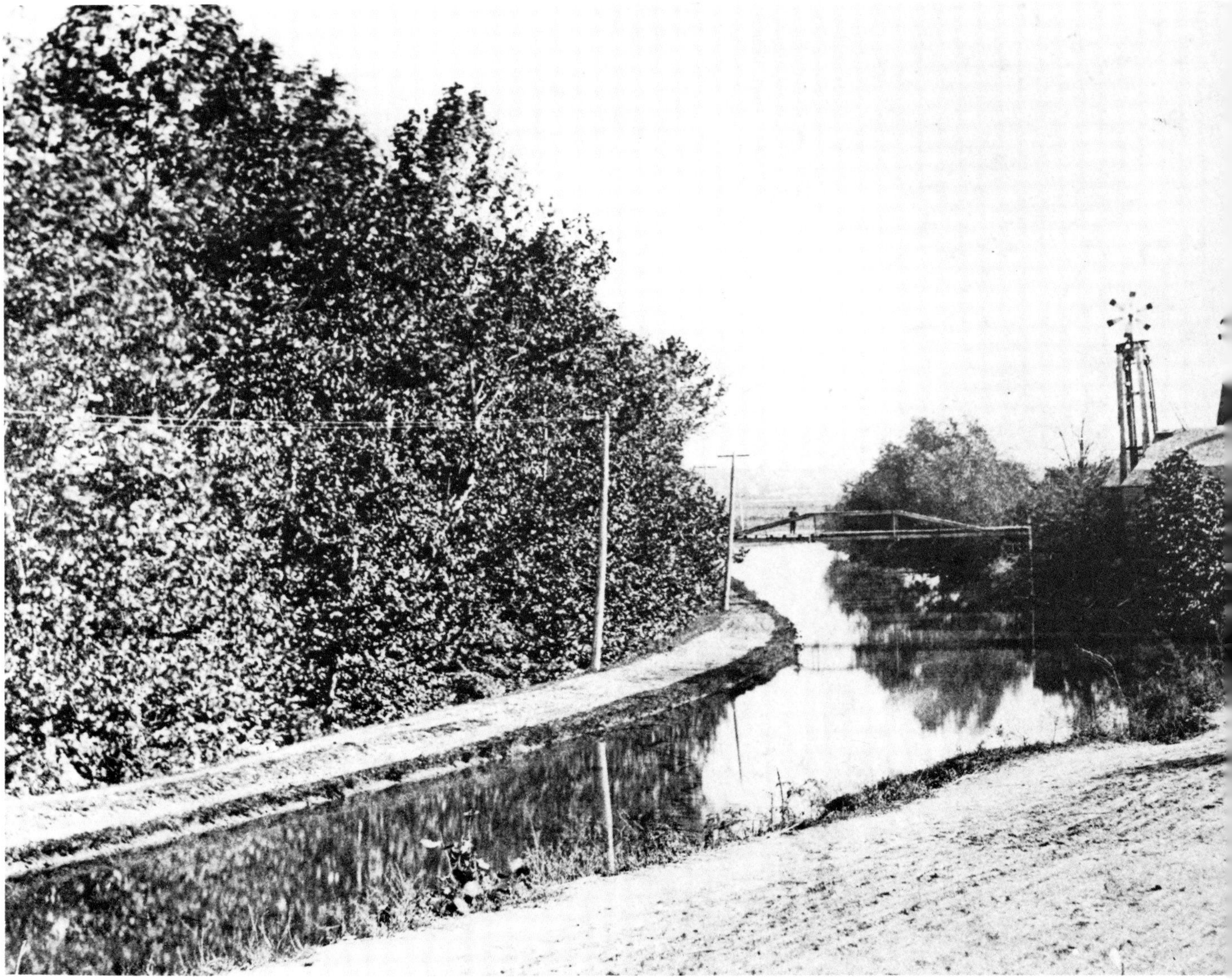

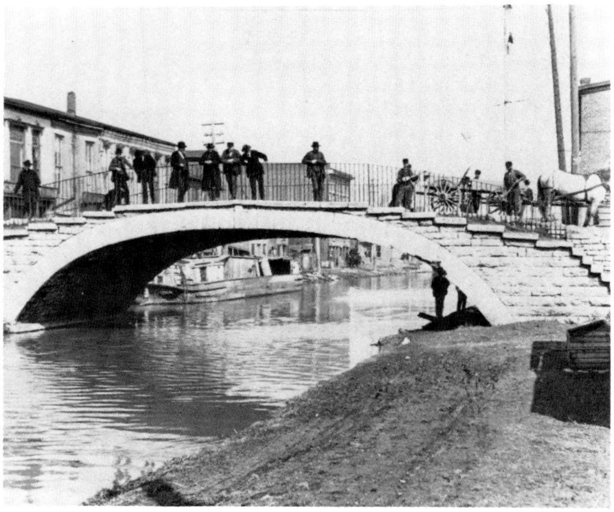

Above, Wiley Tannery on the canal at Piqua. Mile: 155. (Flesh Public Library, Piqua)

Left, A unique flat-arch stone bridge over the canal in downtown Piqua. A boat is moored just beyond the bridge next to the businesses fronting the canal. Mile: 155. (CSO)

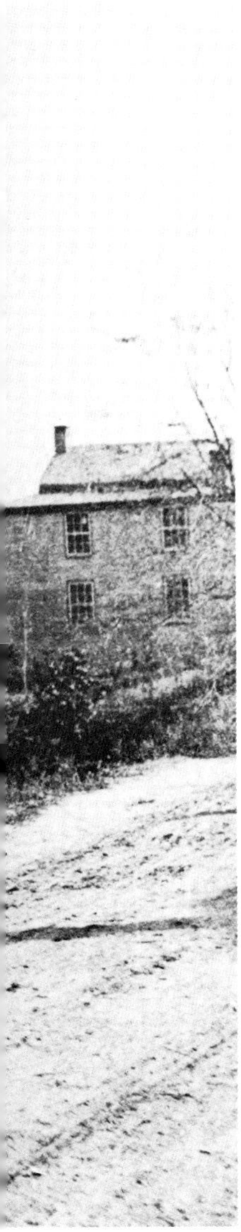

STORAGE.

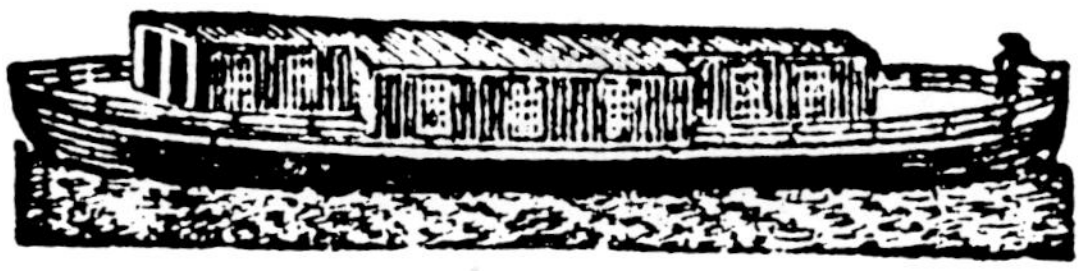

MIAMI CANAL,

Piqua, Ohio:

N. GREENHAM,

Respectfully informs country merchants and all those at a distance, that he will

RECEIVE & FORWARD,

All goods consigned to his care, from Piqua to and from Cincinnati and elsewhere, his warehouse being on the corner of the Basin, more convenient for landing, and unloading free from wharfage to all owners of Boats.

Due care and despatch will bo used, and prompt attention paid to all goods which may be placed under his care.

Piqua, July 5, 1837.

This vintage map shows the Miami & Erie canal paralleling the Great Miami River at Piqua. (Flesh Public Library, Piqua)

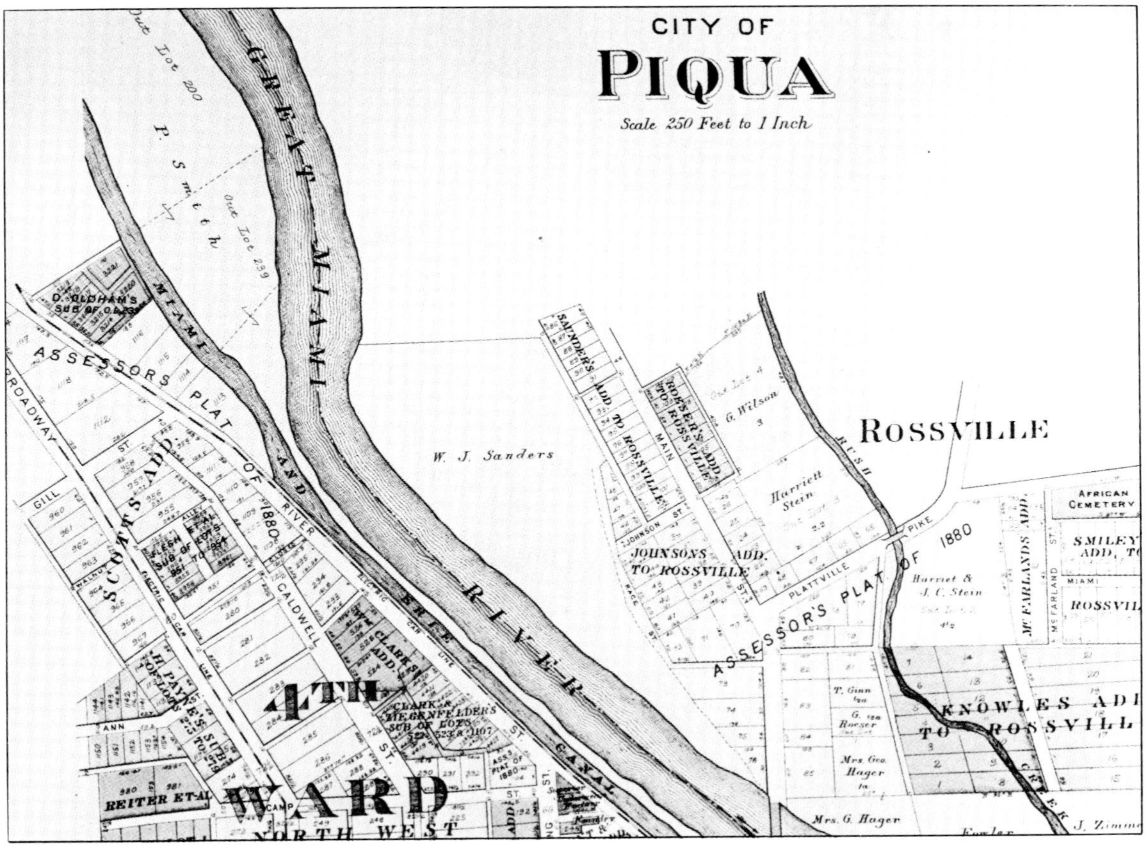

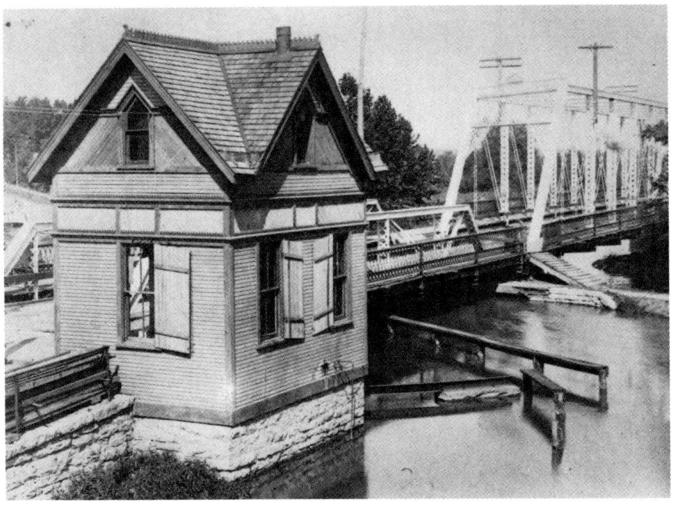

Piqua's steam-operated swing bridge over the canal on Main Street; c. 1910. The towpath is on far side of the canal; a ramp for the mules (slatted for traction) permitted the animals to walk up to the level of the roadway and down the other side. The small building contained the steam engine that operated the mechanism. (Fr. Dale Peterka)

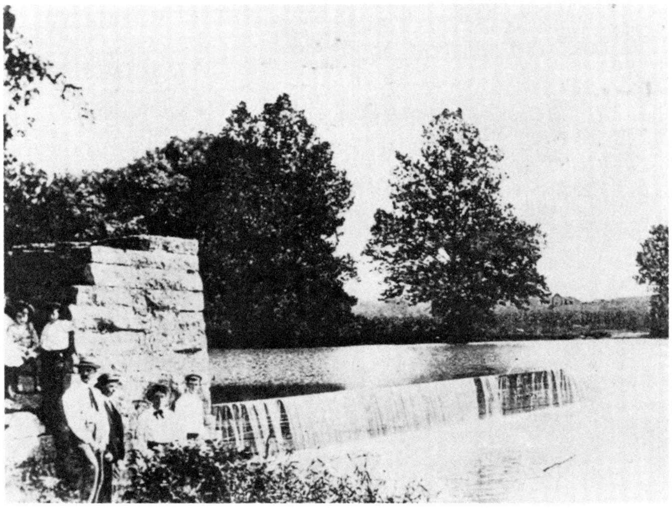

State Dam on the Great Miami River near Piqua, a popular picnic spot for more than a century; c. 1900. (CSO)

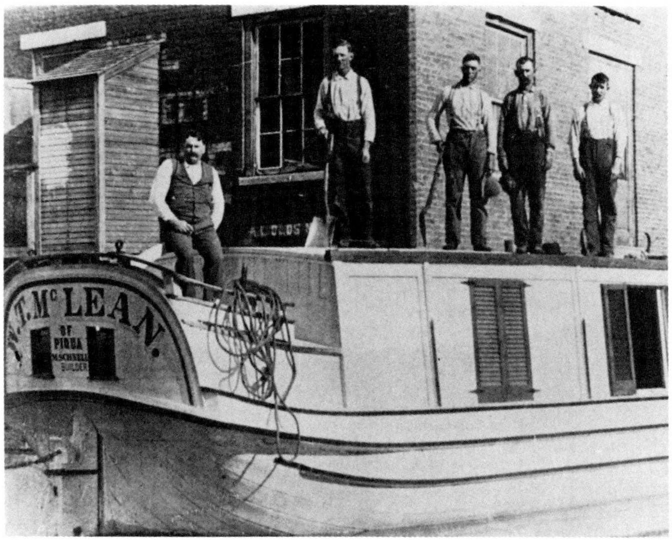

The *W. T. McLean* of Piqua, and crew. Proudly built by M. Schnell, who signed his work, the craft appears to be a state boat. Mile: 155. (Flesh Public Library, Piqua)

The Orr Linseed Oil Company (inset), a major Piqua industry, on the canal; 1897. The Orr Linseed Oil Company and canal after 1900 explosion. (Flesh Public Library, Piqua)

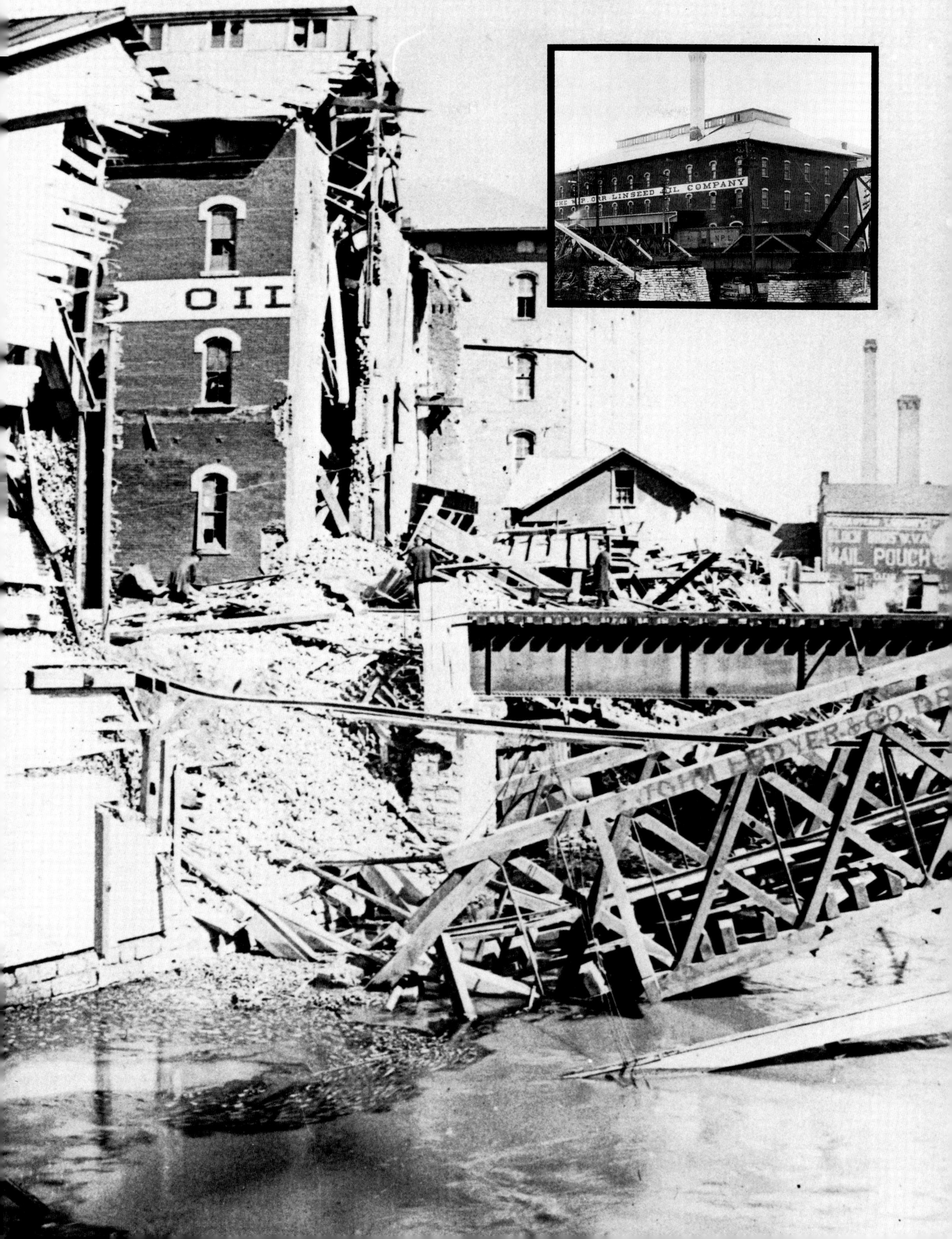
OIL
MAIL POUCH
COMPANY

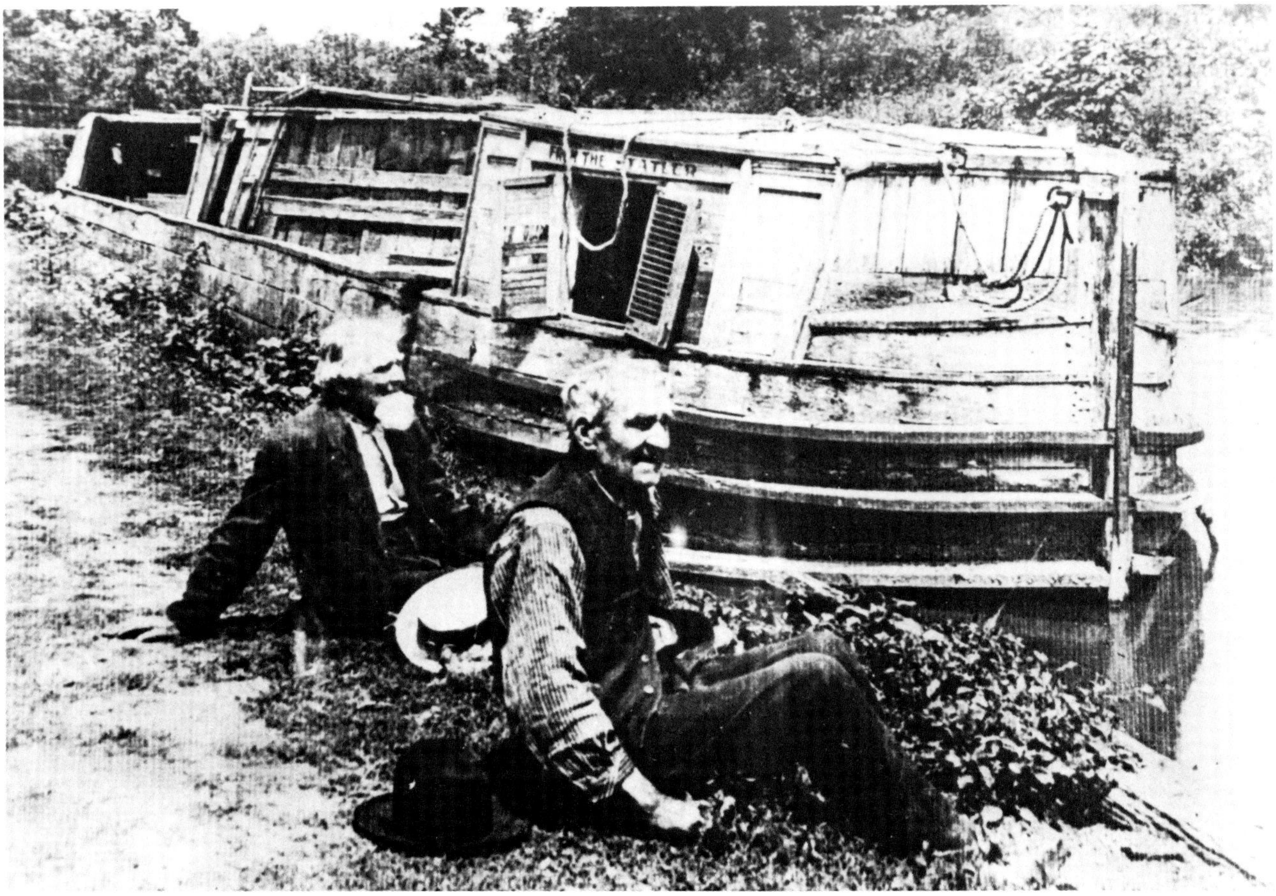

The last boat to pass through Piqua, the *DeCamp Statler*, a stone hauler serving the Statler Quarry in Piqua and the Snipp Gravel Pit in Newport. Canal boatmen are Adam Connover (left), who began his career as a mule driver in 1837 and had become captain of the *Arrow* by 1855; his companion is variously identified as either Eugene Piliod, of Newport, or William Combs, Sr., father of the boat's last owner; photo is dated 1912. Mile: 155. (CSO/research by Leonard U. Hall.)

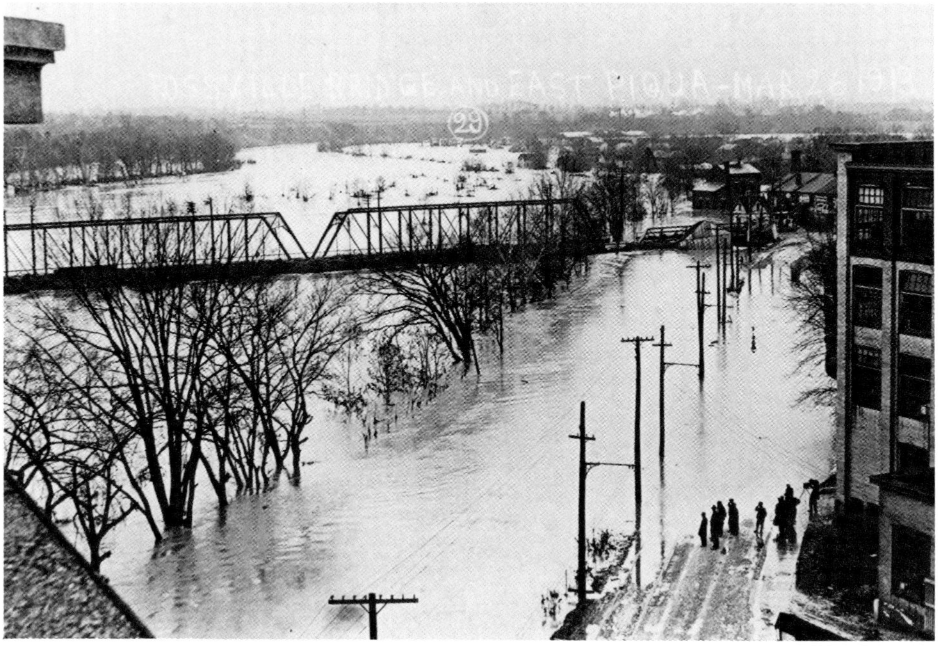

Piqua during the infamous 1913 flood (looking south). The diagonal row of trees marks the towpath that normally divided the Great Miami River, at left, from the canal channel at right. (Flesh Public Library, Piqua)

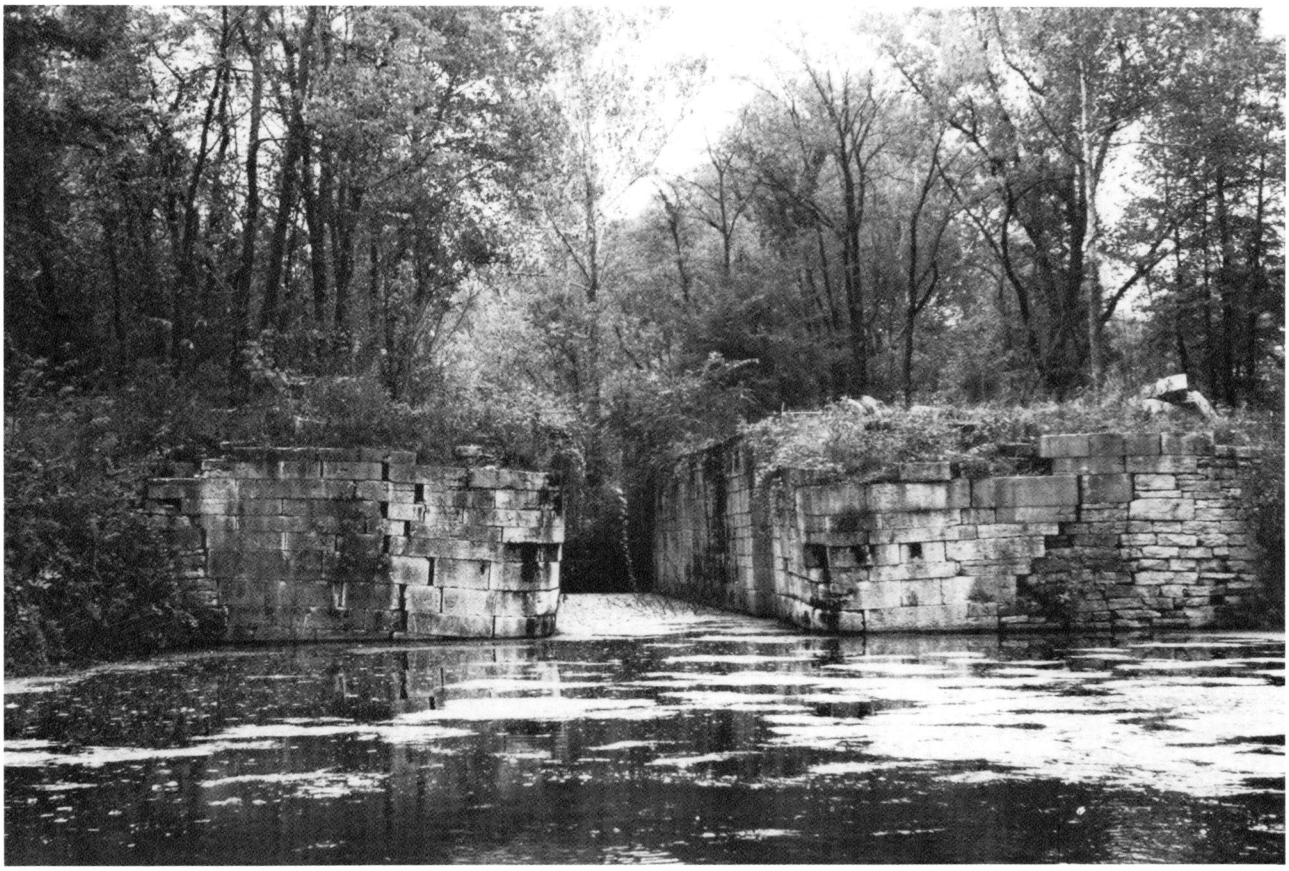

Lock 8, in the Piqua Historical Area, as seen from the *General Harrison*, which turns around at this point. Lock 8 as well as nearby Lock 7, Landman's Mill Lock, exhibit the typical rounded corners of Miami & Erie lock architecture; 1965. Mile: 152. (Photo by F. W. Trevorrow.)

The *General Harrison* (below right), an authentically built canal boat in the Piqua Historical Area, offers rides to visitors. Mules are on the towpath on the far shore; 1972. (Photo by Terry Woods.)

In the Piqua Historical Area, the Ohio Historical Society has restored this 1815 home (below left) of John Johnston, a federal Indian agent who became a canal commissioner and a well-to-do farm merchant who shipped his salt pork and apples on the canal which ran through the edge of his property here. (From a color slide by Jack Gieck, 1978.)

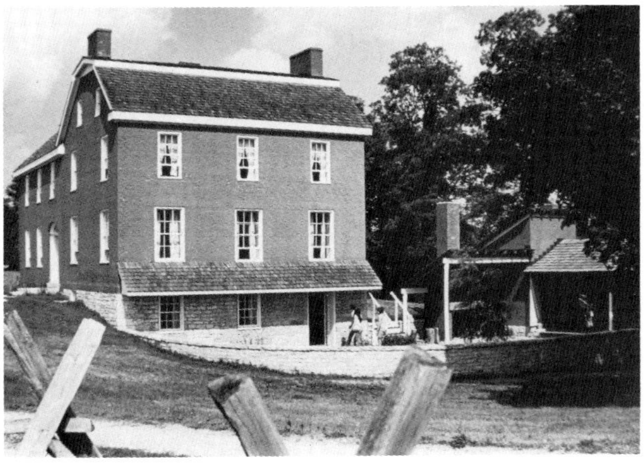

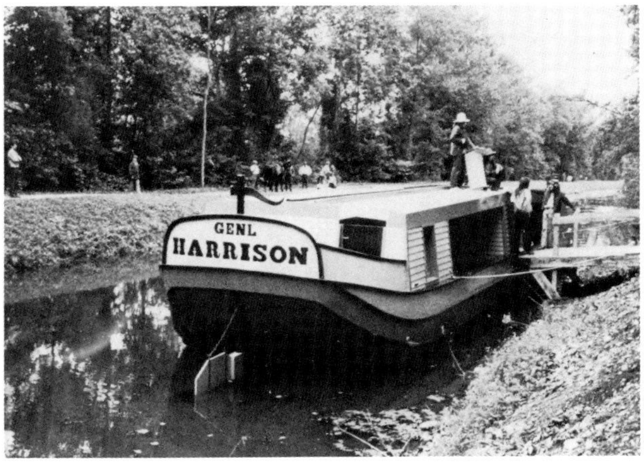

Locks 1, 2, and 3 in operation at Lockington about 1896, with a freight boat at left in the basin between Locks 2 and 3. During the heyday of the canal, Lockington boasted no less than 19 industries. (From *Art Work of Miami County*, W. H. Parish Publishing Co., Chicago, 1897, courtesy OHS)

Top, The north abutment of Loramie Creek Aqueduct, north of Piqua, at the foot of Lockington's staircase of locks; the south abutment is at right; 1978 (from a color slide by Jack Gieck). *Bottom,* In addition to being the time for annual canal repairs, spring also appears to be the silly season at Loramie Creek Aqueduct. Mile: 151. (Flesh Public Library, Piqua)

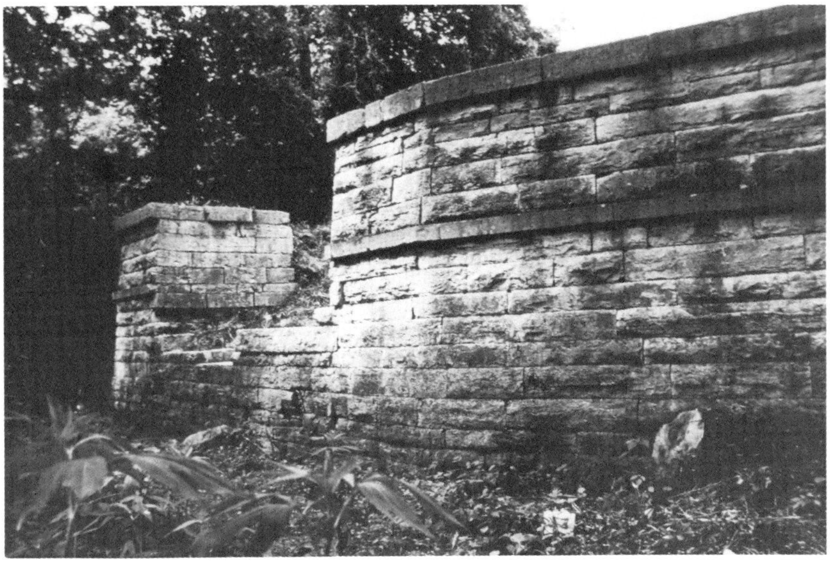

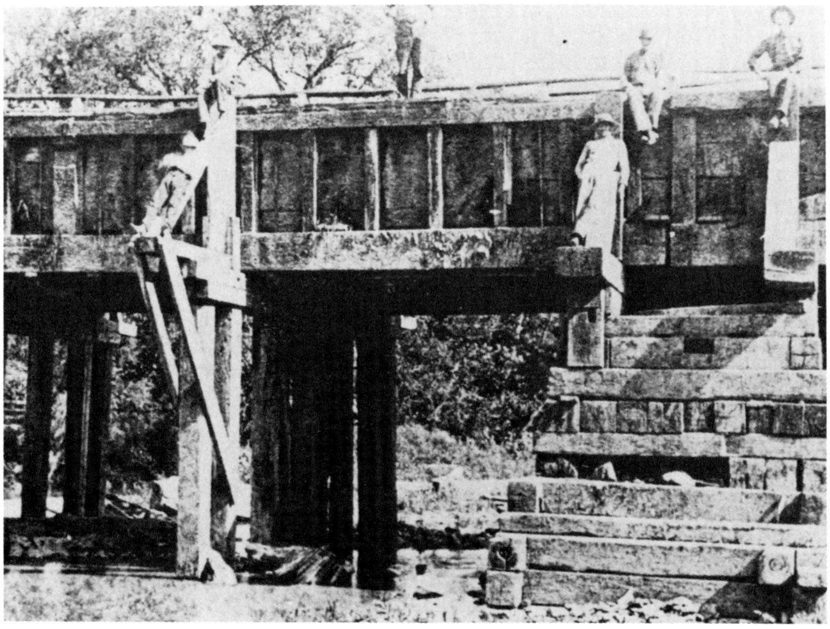

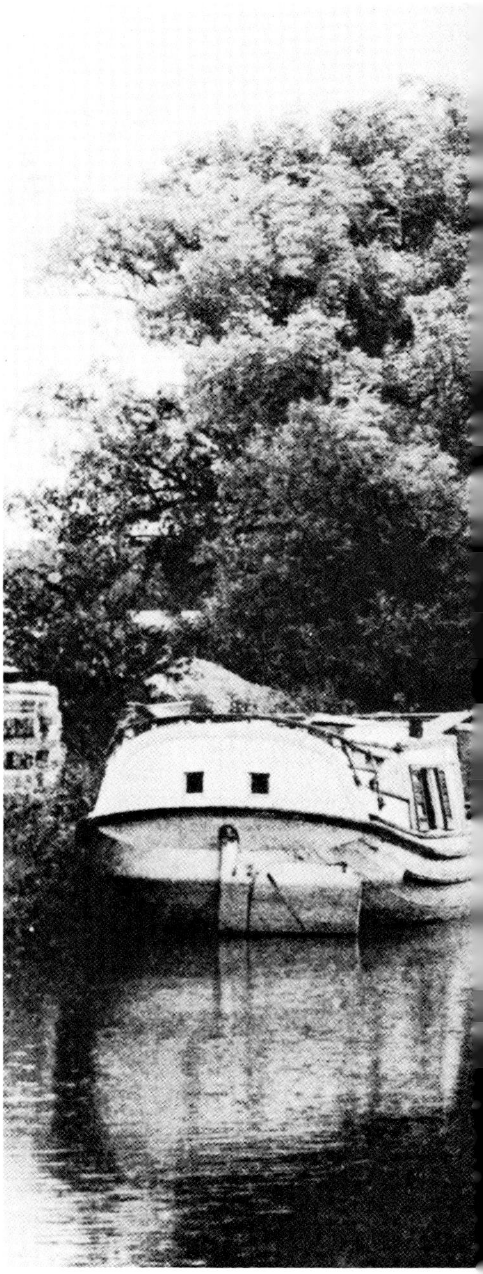

A view of Lockington's upper locks after the canal had been abandoned. The large two-story house at left was the lockmaster's residence. He obviously was responsible for handling more traffic than any other lockmaster on the canal; his responsibilities also extended to supervision of the drydock at the turn-around basin between Locks 1 and 2. The brick building peeking through the trees at the upper right was a tavern and brothel that accommodated crewmen while their boat was waiting to negotiate the locks. Mile: 151. (OHS/Lockington research by Terry D. Wright.)

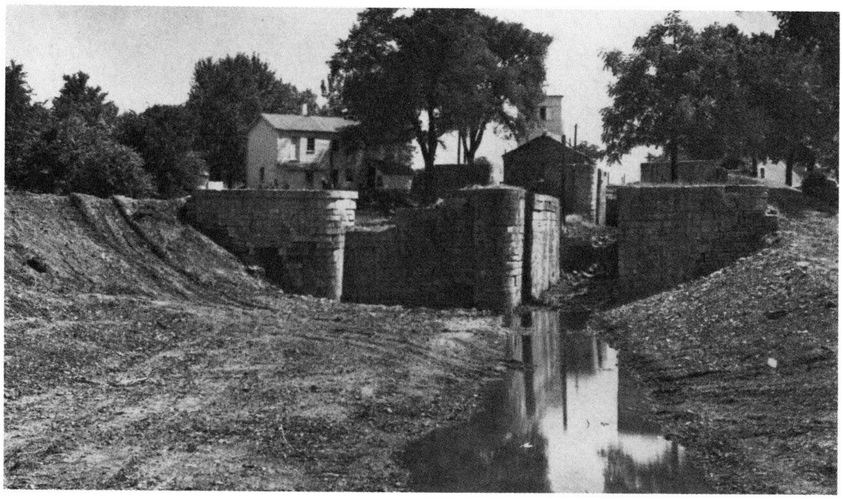

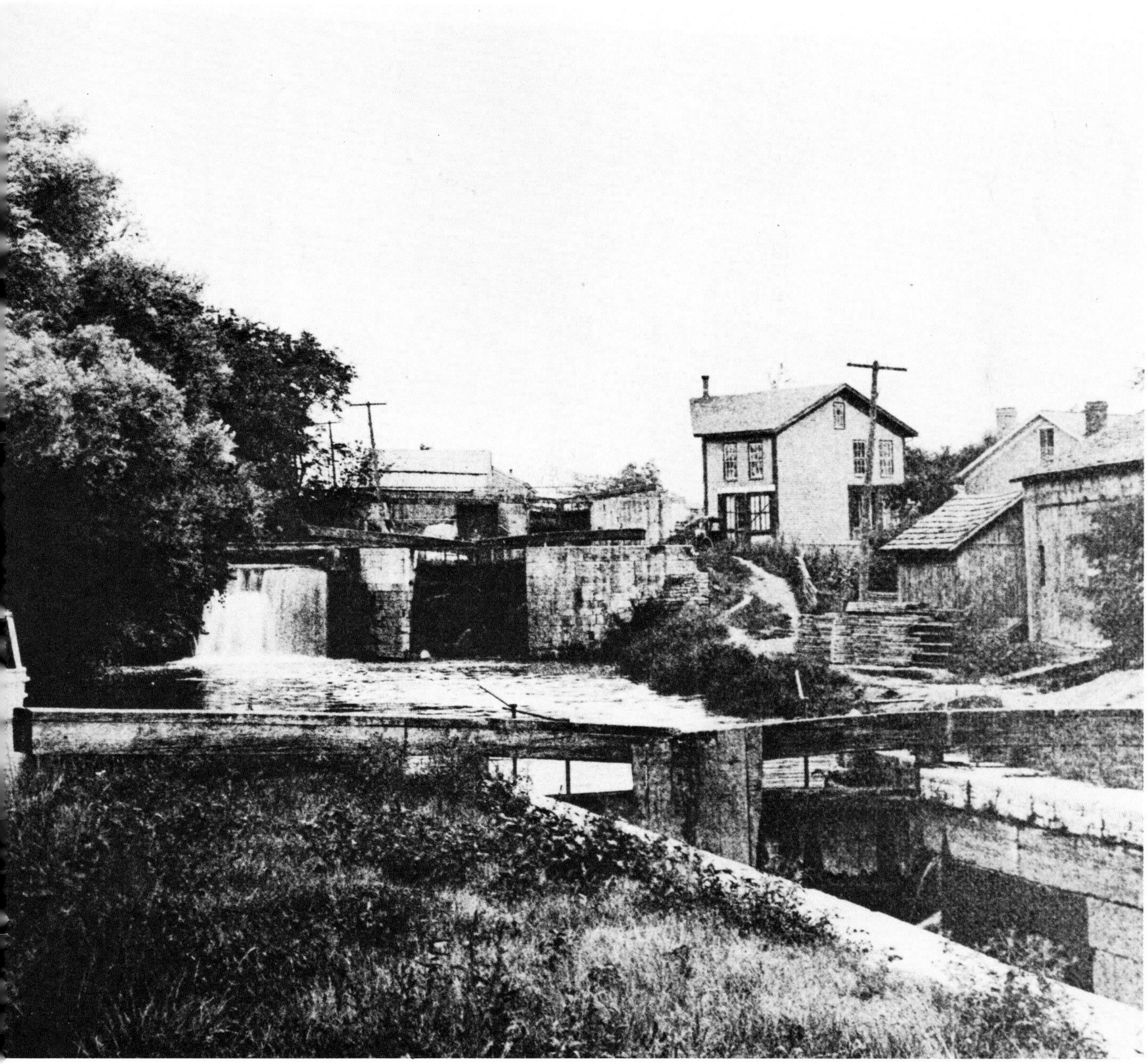

At the end of Loramie Summit, the Miami & Erie Canal began its descent of 512 feet to the Ohio River. Here at Lockington, five locks brought its level down to Loramie Creek Aqueduct (OHC). Inset is the same scene from the top of the Lockington staircase in 1978. Mile: 151. (From a color slide by Jack Gieck.)

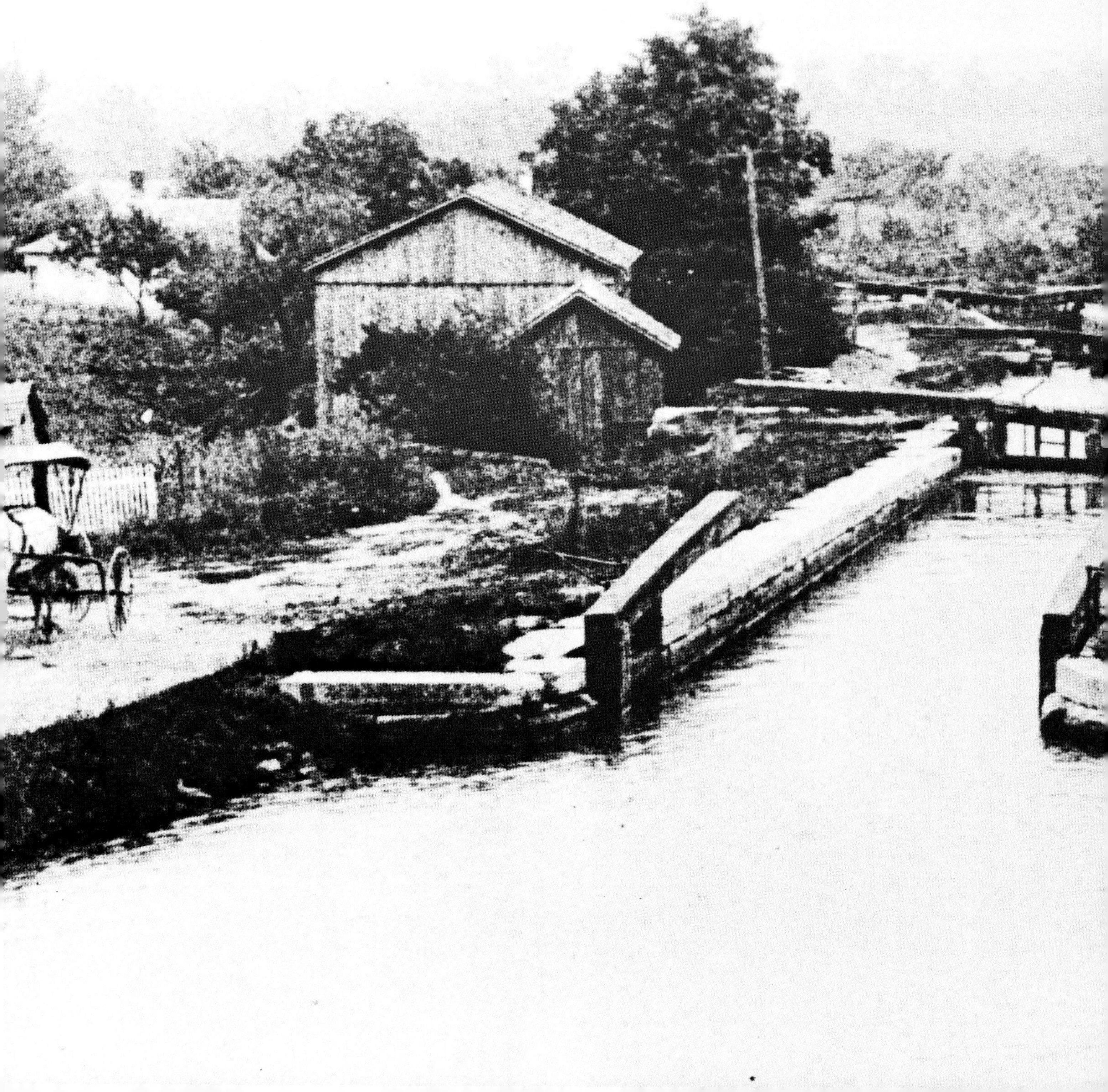

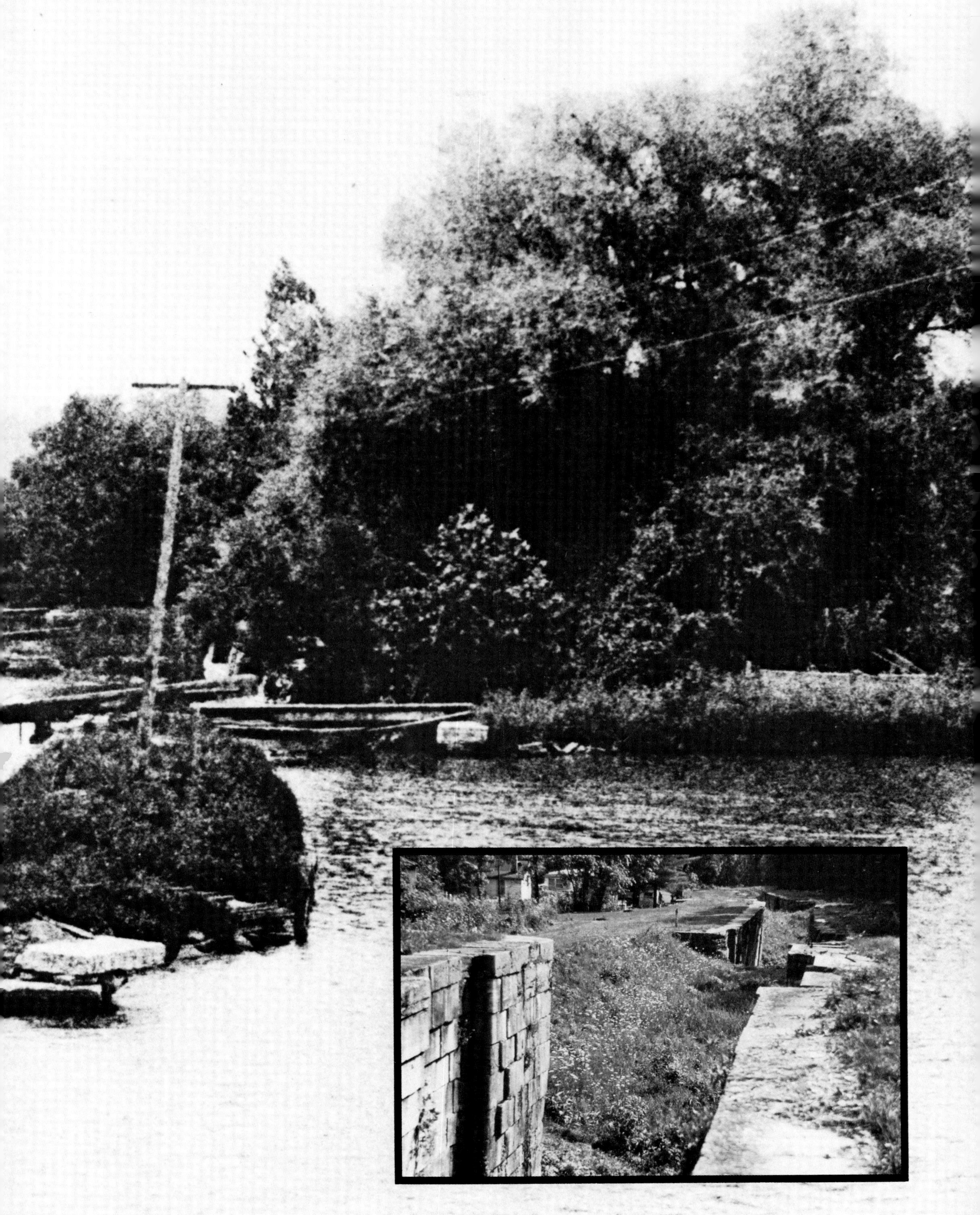

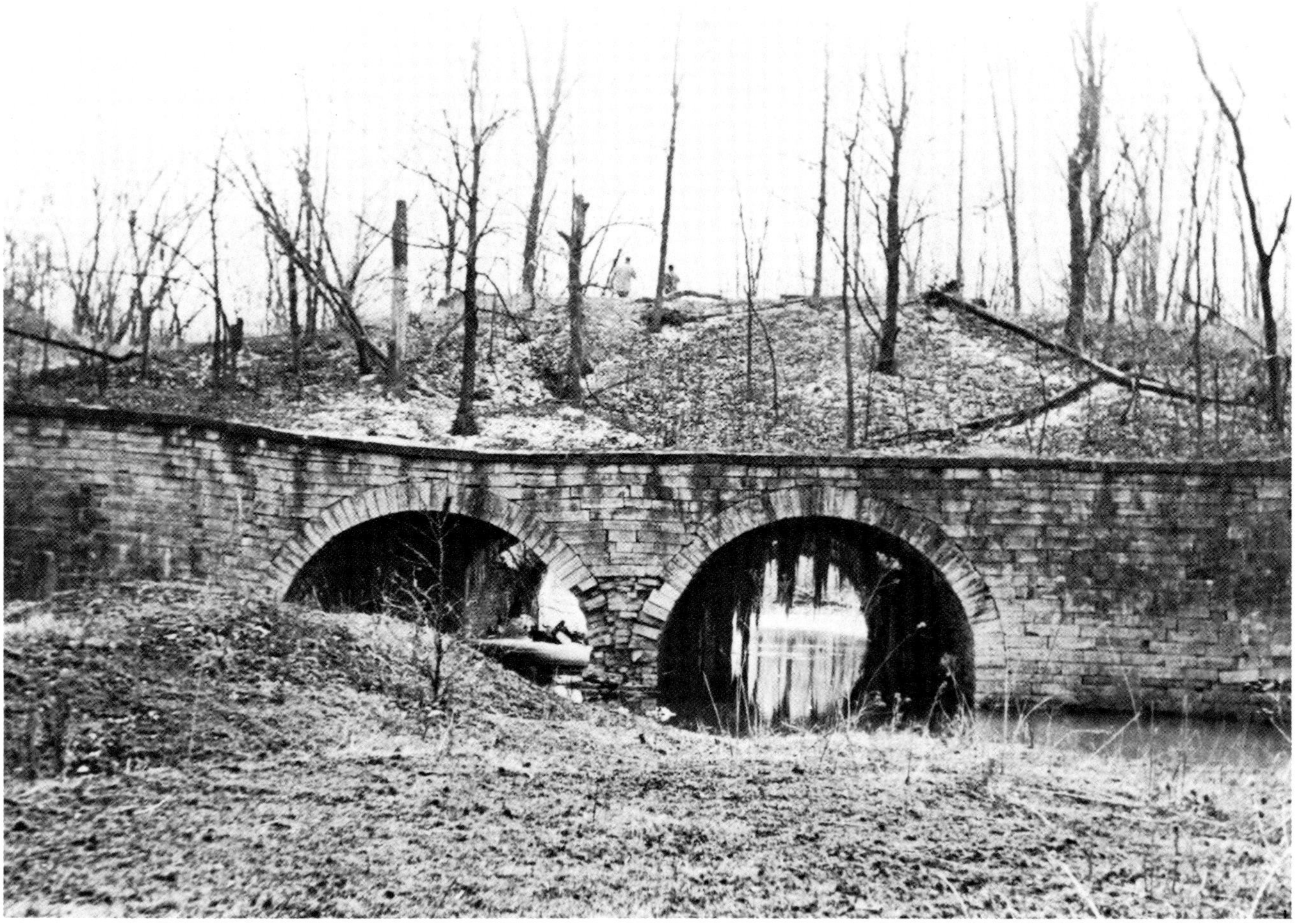

Turtle Creek Culvert helped maintain the elevation of Loramie Summit for 23 miles above Lockington. The structure was demolished in August 1982 to avoid damming and flooding Turtle Creek in the event of its collapse. Mile: 144. (CSO)

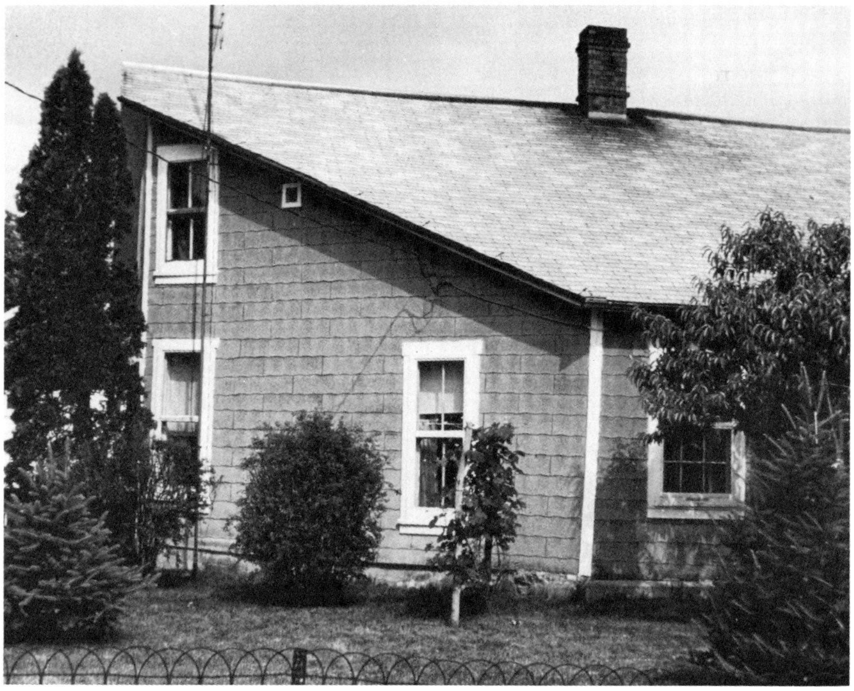

The odd shape of this residence adjacent to the towpath at the top of the locks at Lockington resulted from the owner's refusal to move his house when the canal went through. Instead, the house was rebuilt with the corner cut off. (From a color slide by Jack Gieck.)

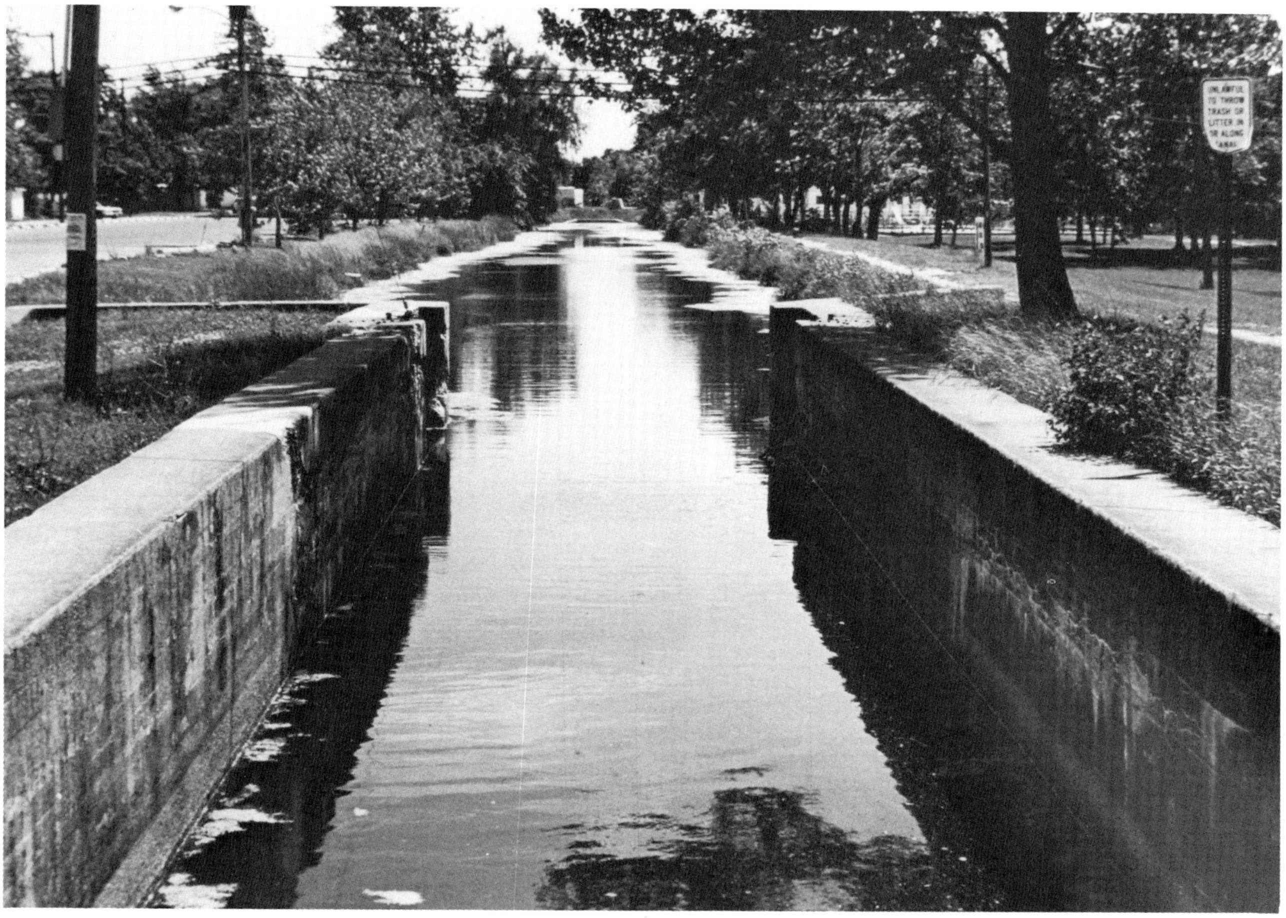

The north end of Loramie Summit, looking south, at the foot of Main Street in New Bremen. The canal is preserved in a park setting elsewhere in the town; 1978. Mile: 128. (From a color slide by Jack Gieck.)

This artist's rendition of the canal adjacent to Main Street in Berlin reflects the joy of the town's prosperity resulting from the canal. The mule, ridden by a driver who is dressed like a jockey, seems to be pulling the boat at a trot! Mile: 134. (*History of Shelby County, Ohio*, R. Sutton Co., Philadelphia, 1883, courtesy WRHS.)

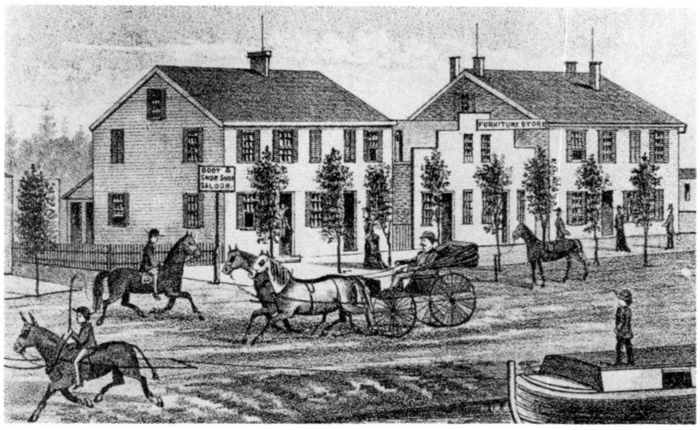

Loramie Reservoir, now Lake Loramie, with bulkhead at left, is one of three canal reservoirs serving the summit level of the Miami & Erie Canal. The others were Lewistown Reservoir, now Indian Lake, and Grand Lake St. Marys; 1976. (Photo by F. W. Trevorrow.)

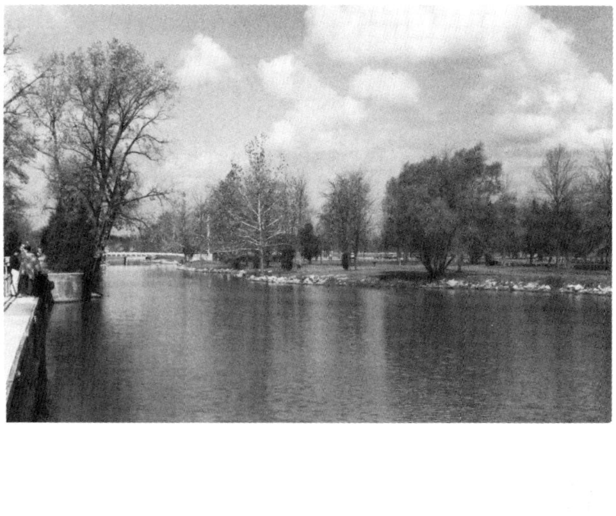

St. Marys Feeder (right) from Grand Lake St. Marys Reservoir, enters the Miami & Erie Canal at Mile 121; 1976. (Photo by F. W. Trevorrow.)

The aqueduct over the St. Marys River (far right) has been replaced by steel pipes to carry water into the city of St. Marys for industrial use; 1978. Mile: 120. (From a color slide by Jack Gieck.)

Two boats on the Miami & Erie (below right) near St. Marys at the Bloomer Brothers Box Factory and Western Strawboard works. The line boat *Gallant* is at right. Mile: 120. (Edith McNally)

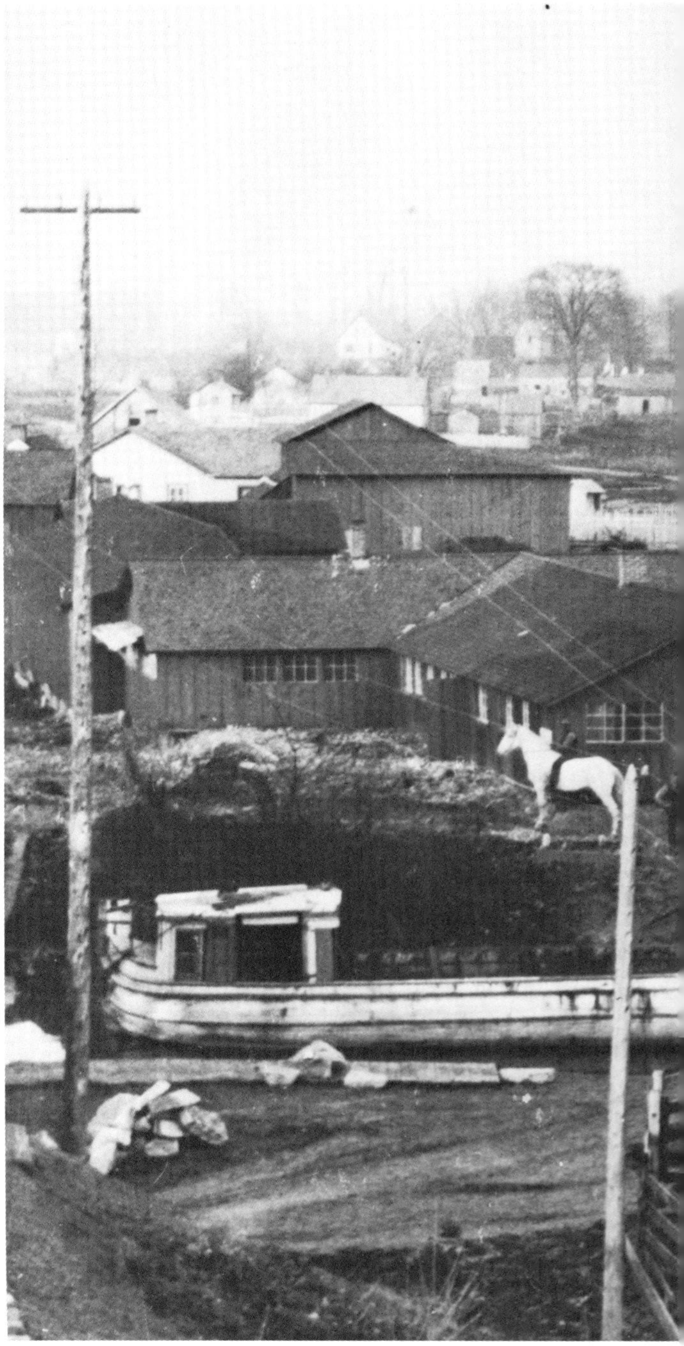

Top, Lock 8, Stone Lock, north of Loramie Summit. The canal is currently watered in this section, north through St. Marys as far as Delphos; 1978. *Bottom*, Near Stone Lock, "Bloody Bridge" is marked by a small monument next to the towpath: "Here, Bill Jones killed Jack Billings over the love of Minnie Warren." Mile: 123. (From color slides by Jack Gieck.)

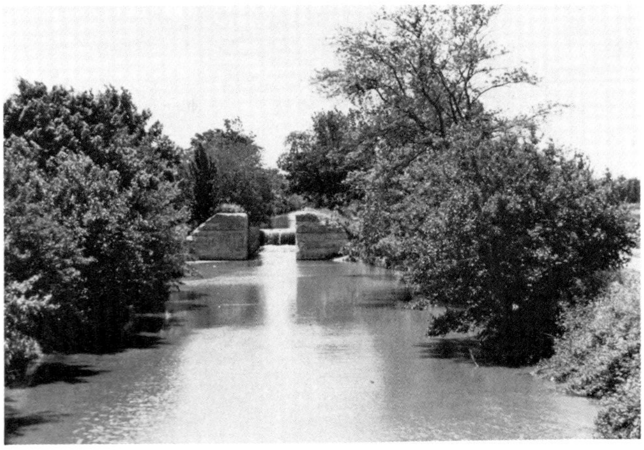

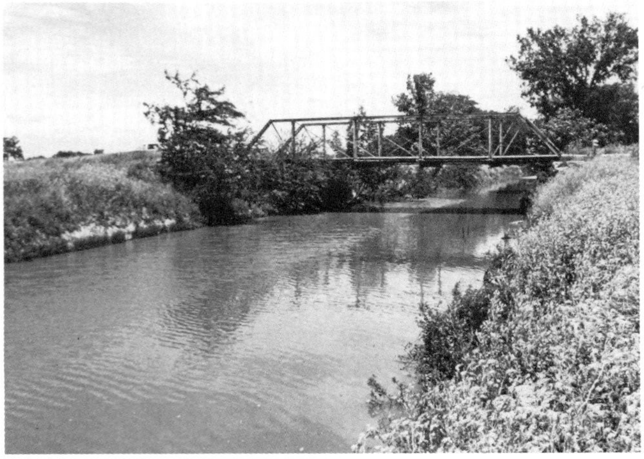

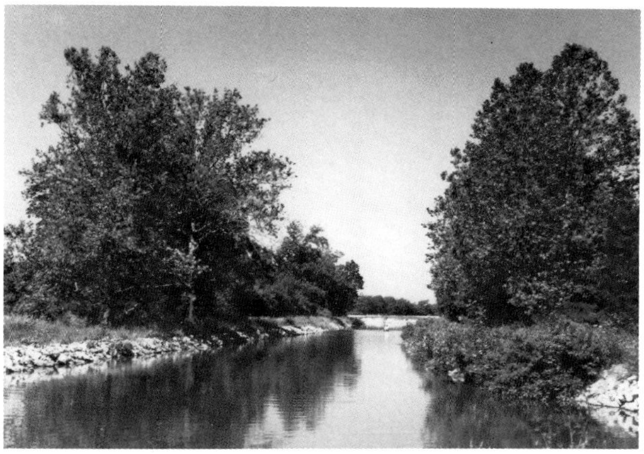 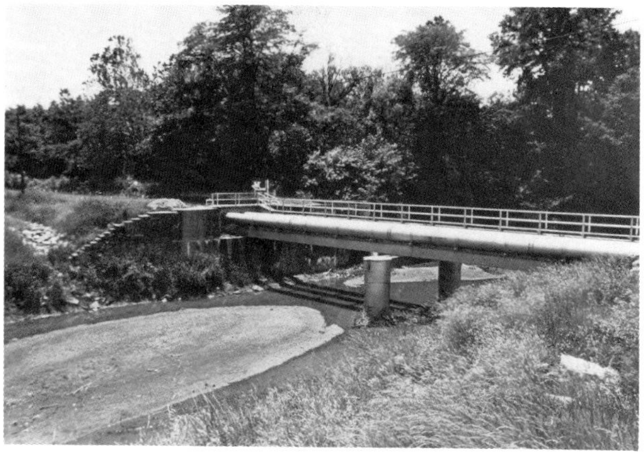

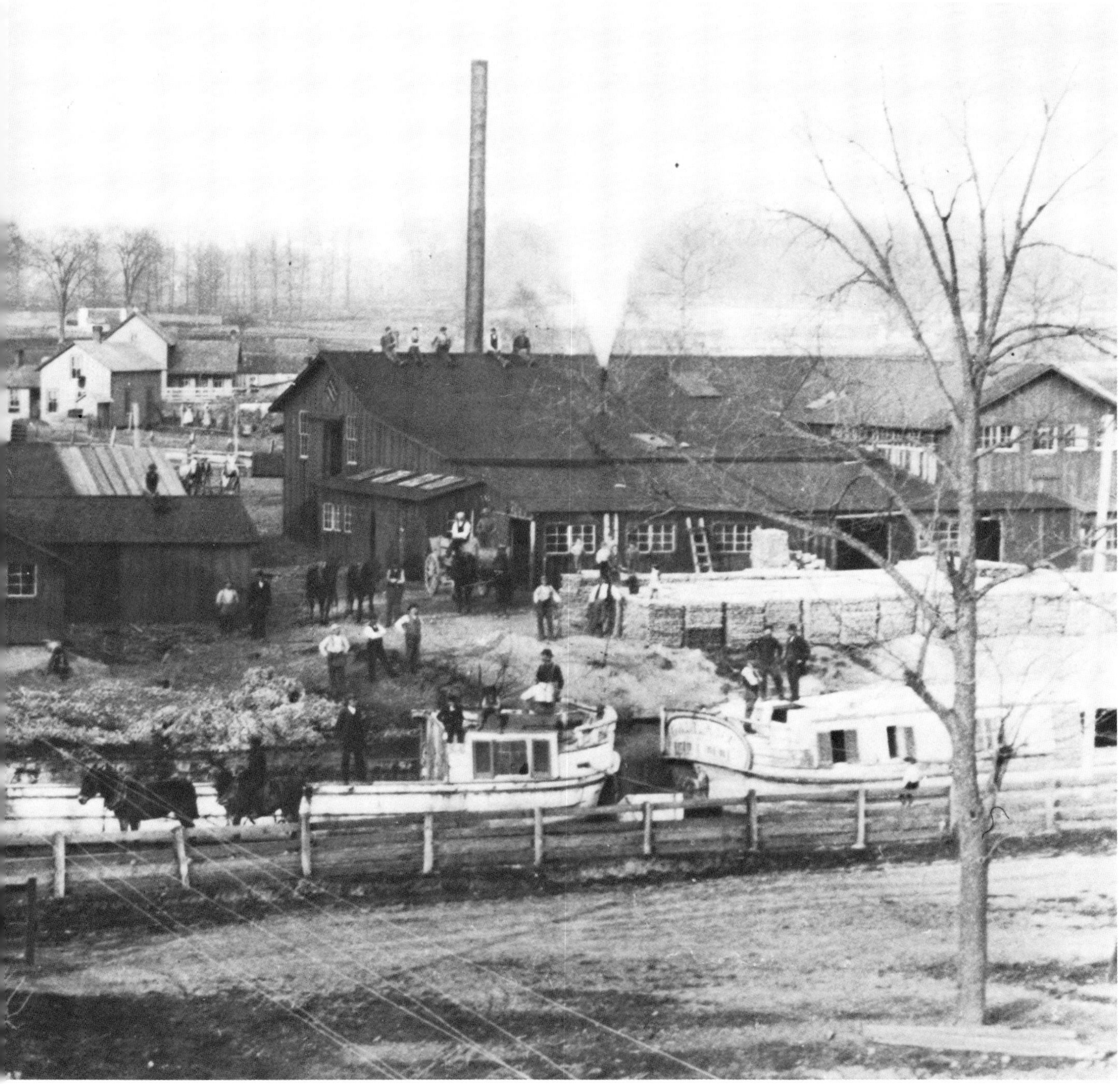

South of Spencerville is Deep Cut—a daring engineering achievement that eliminated the necessity of a chain of locks. Deep Cut is a continuous ditch from five to thirty-two feet deep, blasted through the blue-clay ridge that separates the Auglaize River valley from the St. Marys watershed. (CSO)

Inset, H. W. Meyers in Spencerville; 1906. (CSO)

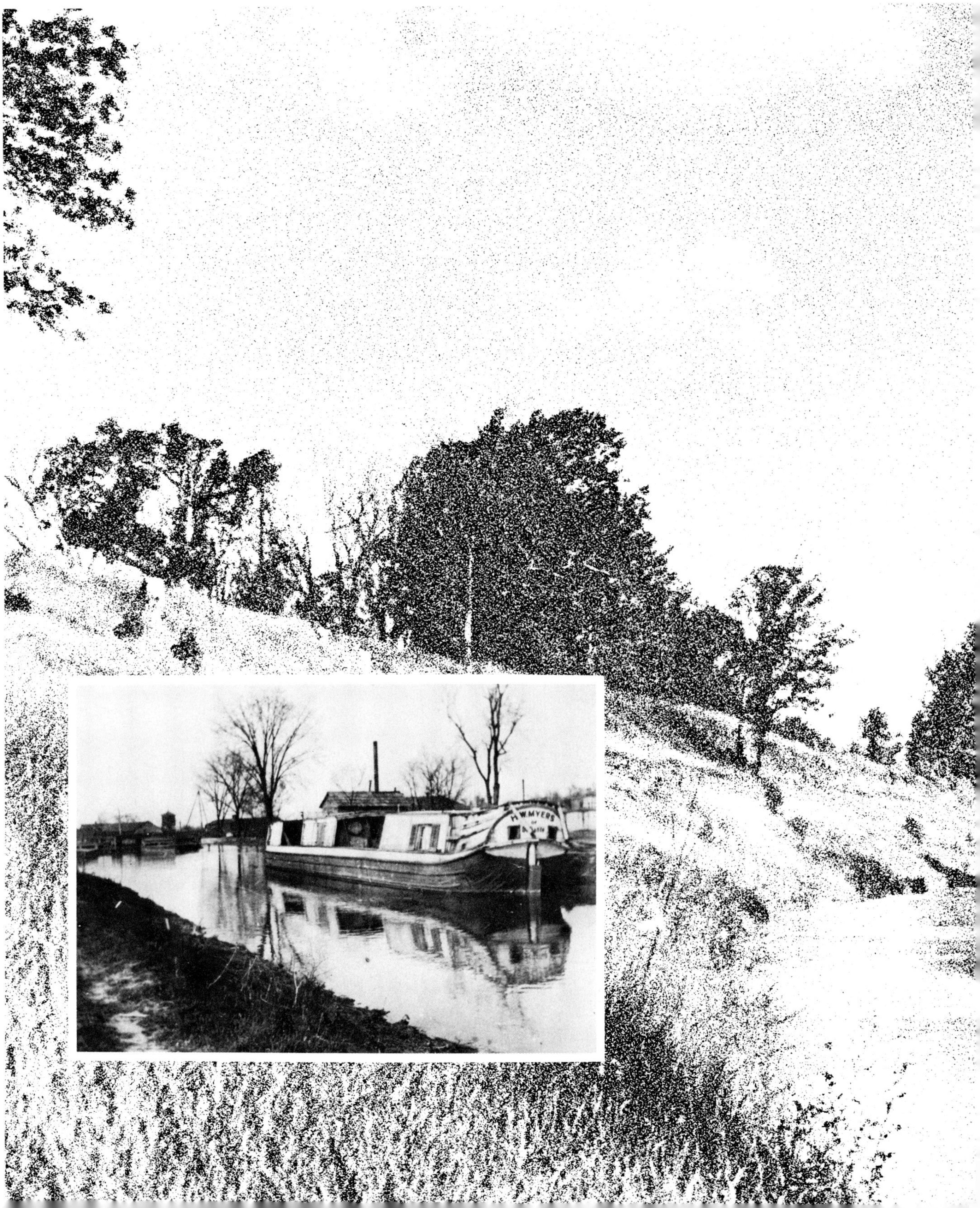

“***A ship wreck occurred*** on the ‘raging canal’ last Wednesday morning, resulting in the sinking of one vessel and the complete blockade of the harbor. Dave Norbeck had tied up one of his boats near the old mill race, and during the night one end got loose, and the boat swung across the canal, lodging on each bank; an upcoming boat drew the water out of the middle level to get through the lock, thus leaving the old boat supporting itself by the stem and stern on the banks, and the weight caused it to crack in the middle, soon filling up with waer, especially after being pushed into the channel so the upcoming boat could pass. The State Boat and dredge luckily were here and worked all day to get the old boat up and out of the way.” *(Spencerville Journal*, August 3, 1888)

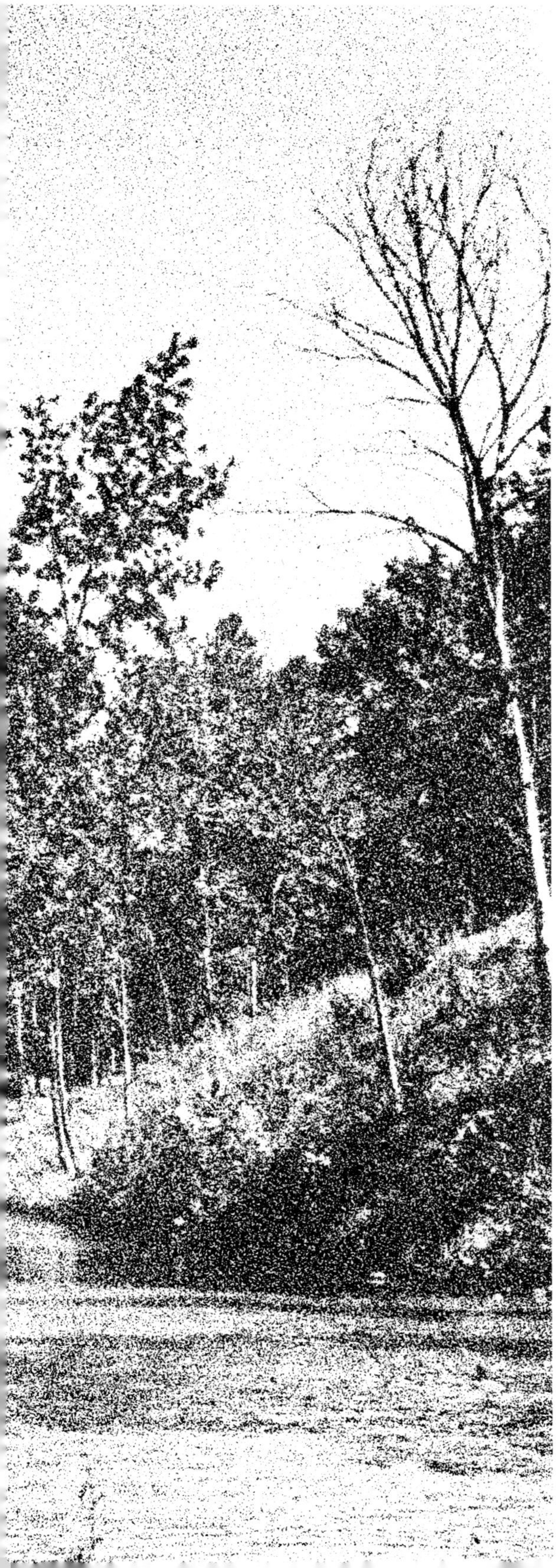

Lock 14, Saw Mill Lock, with its eroding concrete facing, is explored by a Canal Society tour group; 1975. Mile: 115. (Photo by F. W. Trevorrow.)

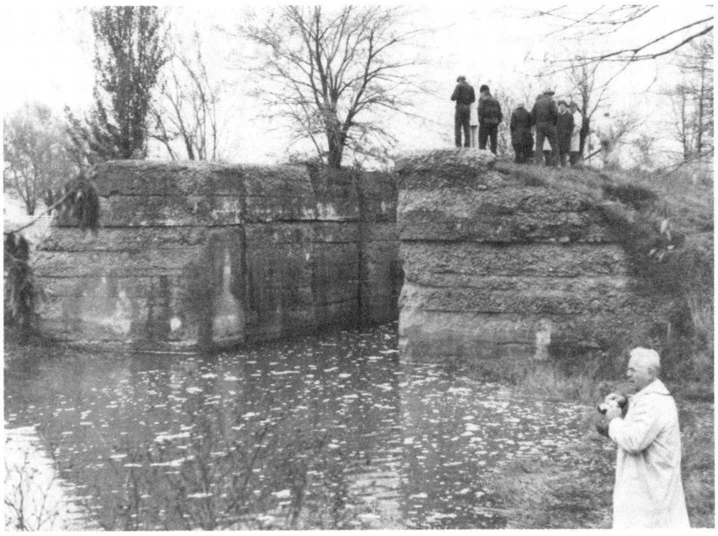

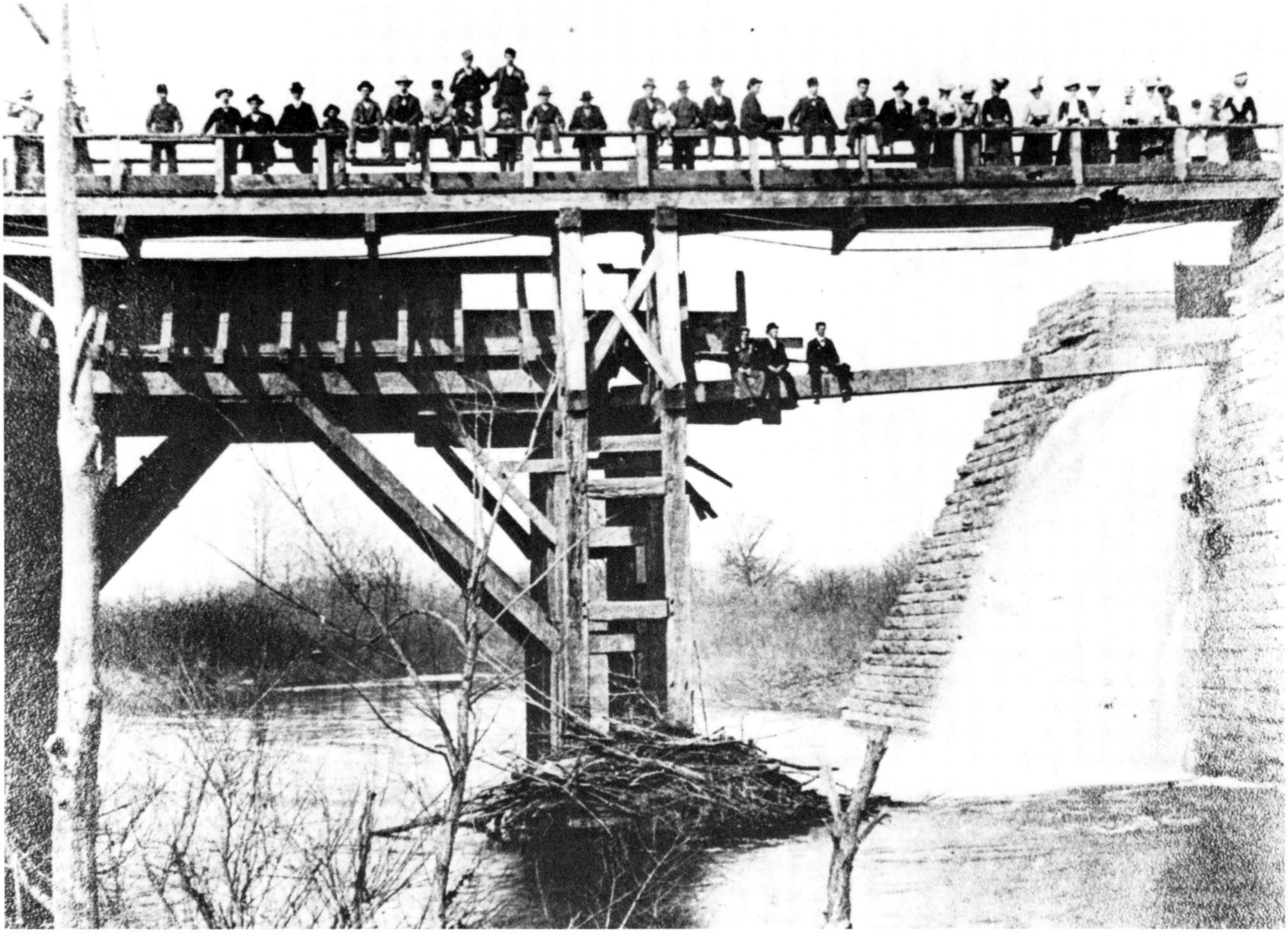

Spring flood damage brings out a crowd of sightseers at the Little Auglaize Aqueduct (above), where the entire canal stream cascades out of the abutment. In another view (below), the Little Auglaize Aqueduct has developed a substantial leak. Mile: 80 (Toledo/Lucas County Public Library)

The Delphos Chamber of Commerce preserves the town's canal heritage with this historical marker (below) in the canal bed; 1978. Mile: 97. (From a color slide by Jack Gieck.)

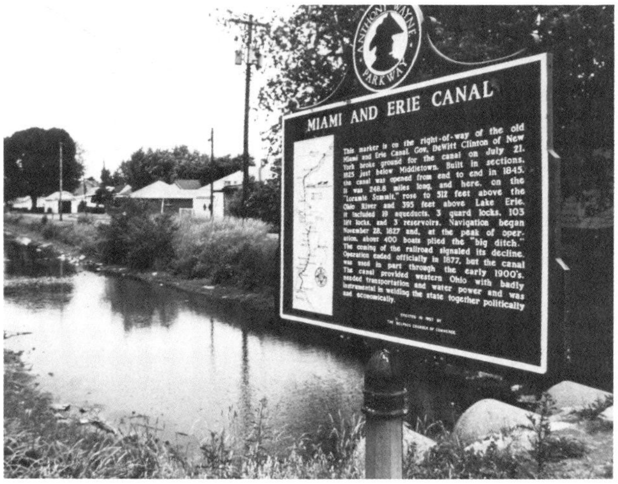

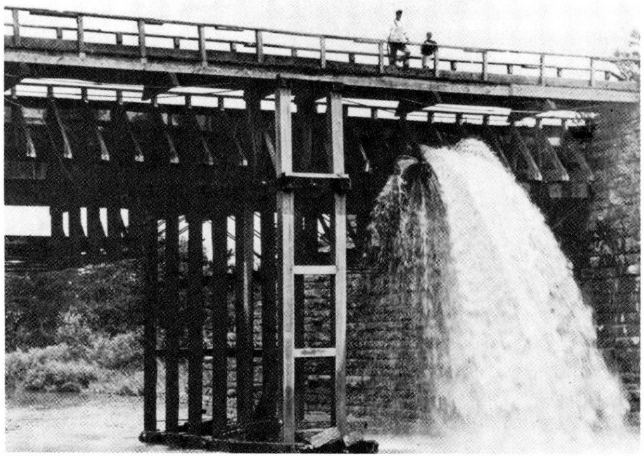

A seldom-seen limestone culvert (right) in dense woods near Junction. Mile: 70. (From a color slide by Jack Gieck.)

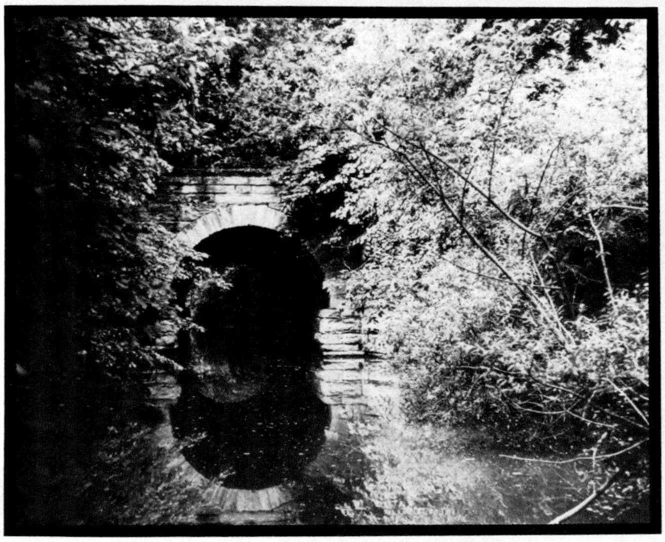

The stone 70-mile marker (below) at Junction, where the Wabash & Erie (an Indiana canal that connected Fort Wayne and other Indiana cities and towns with Lake Erie at Toledo) joined the Miami Extension to eventually become the northern section of the Miami & Erie Canal. (Vanderlip Collection, Toledo/Lucas County Public Library)

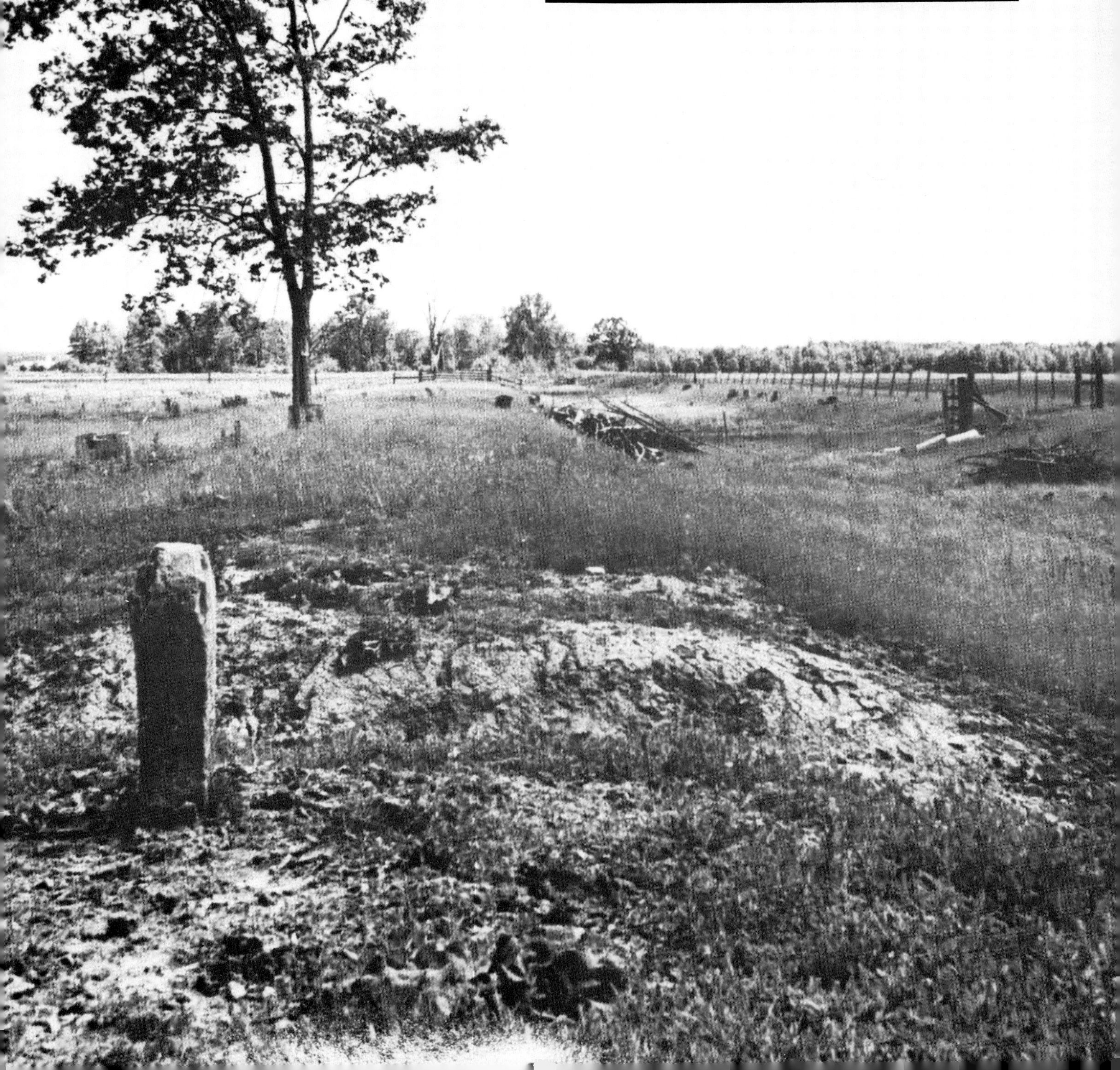

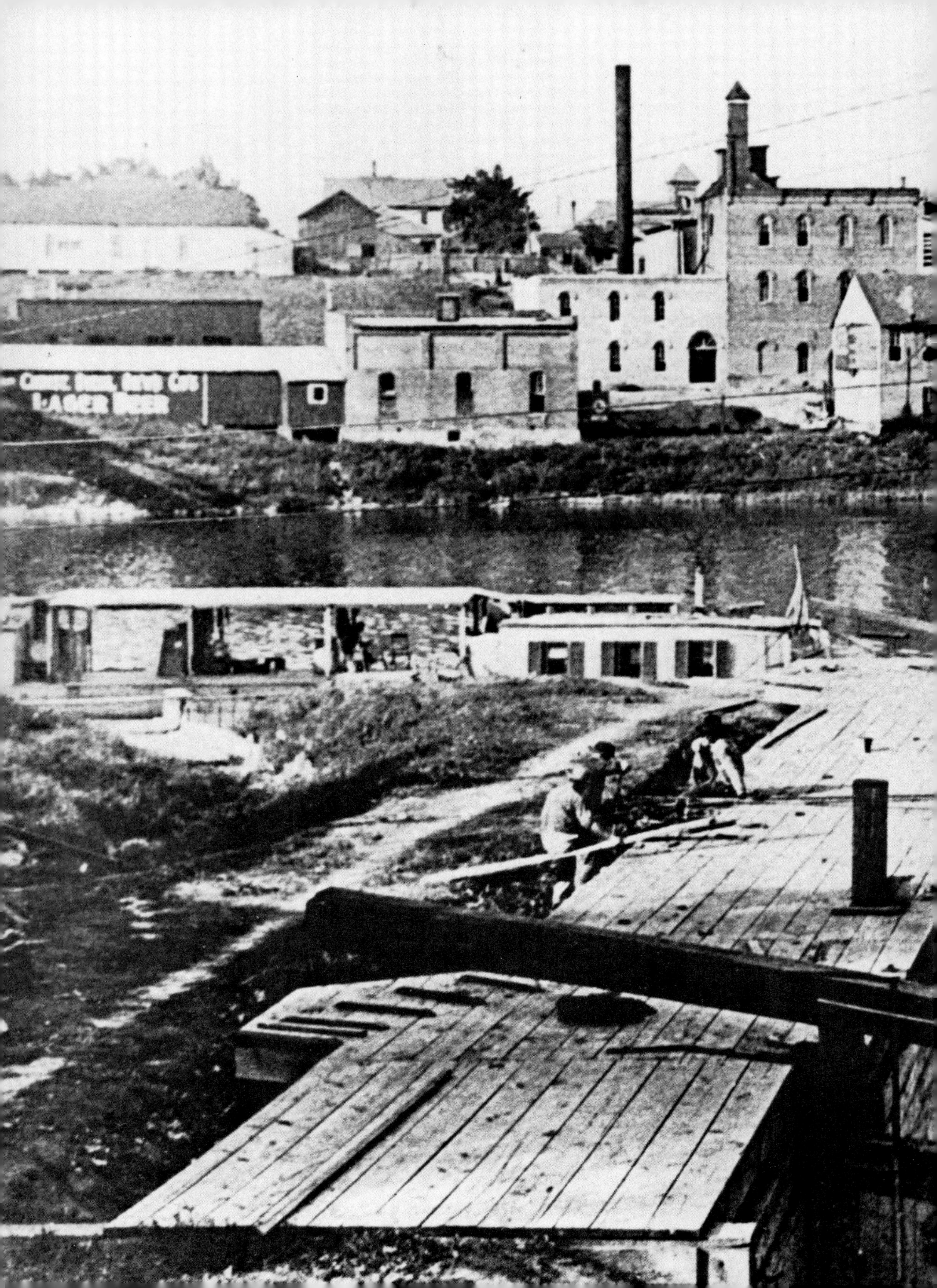
LAGER BEER

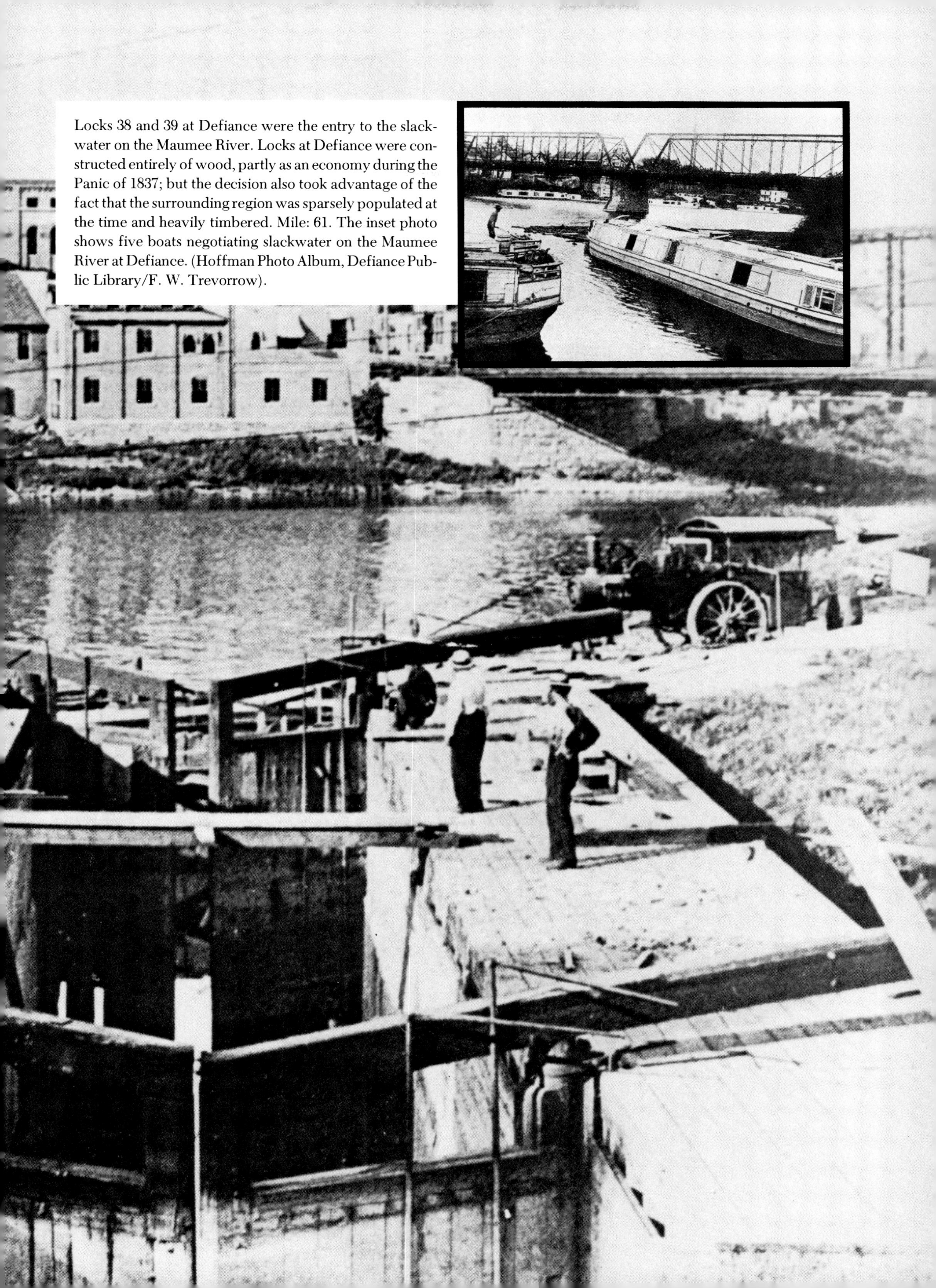

Locks 38 and 39 at Defiance were the entry to the slackwater on the Maumee River. Locks at Defiance were constructed entirely of wood, partly as an economy during the Panic of 1837; but the decision also took advantage of the fact that the surrounding region was sparsely populated at the time and heavily timbered. Mile: 61. The inset photo shows five boats negotiating slackwater on the Maumee River at Defiance. (Hoffman Photo Album, Defiance Public Library/F. W. Trevorrow).

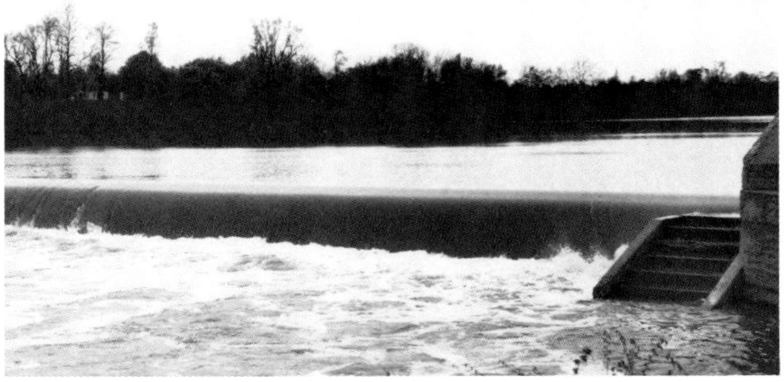

The slackwater was created by the dam at Independence, four miles downstream from Defiance. Mile: 57. (Photo by F. W. Trevorrow, 1973.)

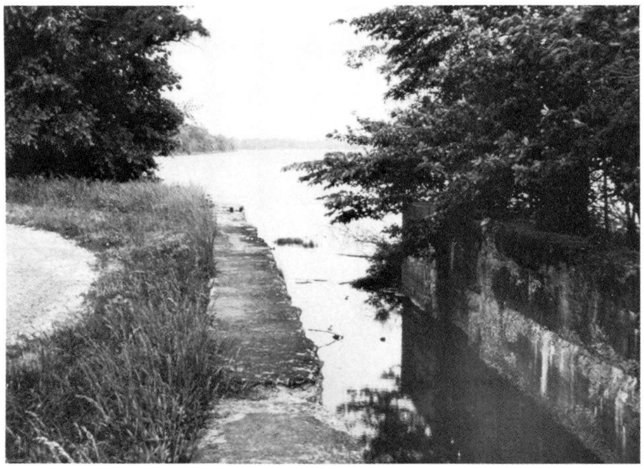

A guard lock at Independence through which boats reentered the canal from the river slackwater; 1978. Mile: 57. (From a color slide by Jack Gieck.)

The restored Lock 13, a lift lock in the canal channel at Independence Dam State Park. Mile: 57. (From a color slide by Jack Gieck.)

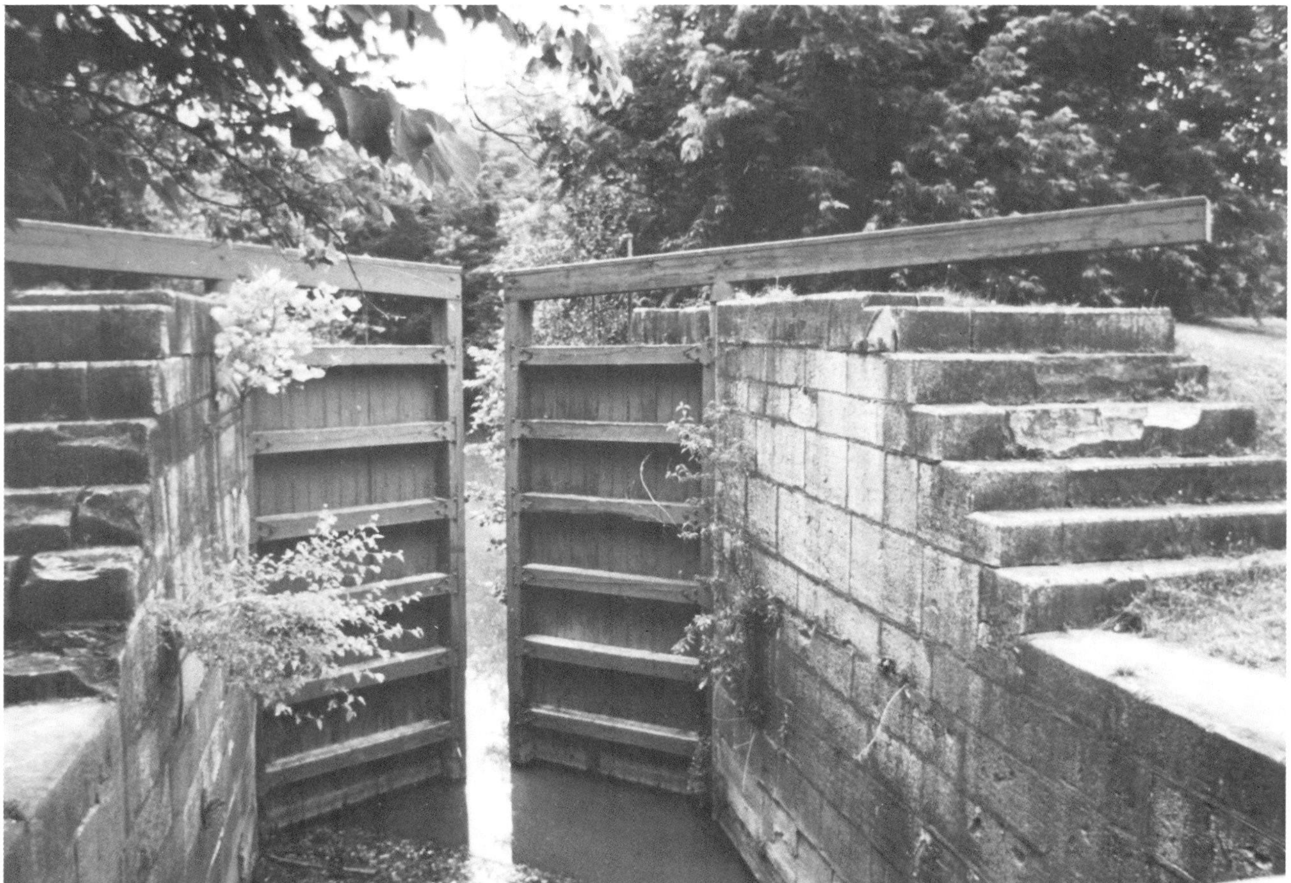

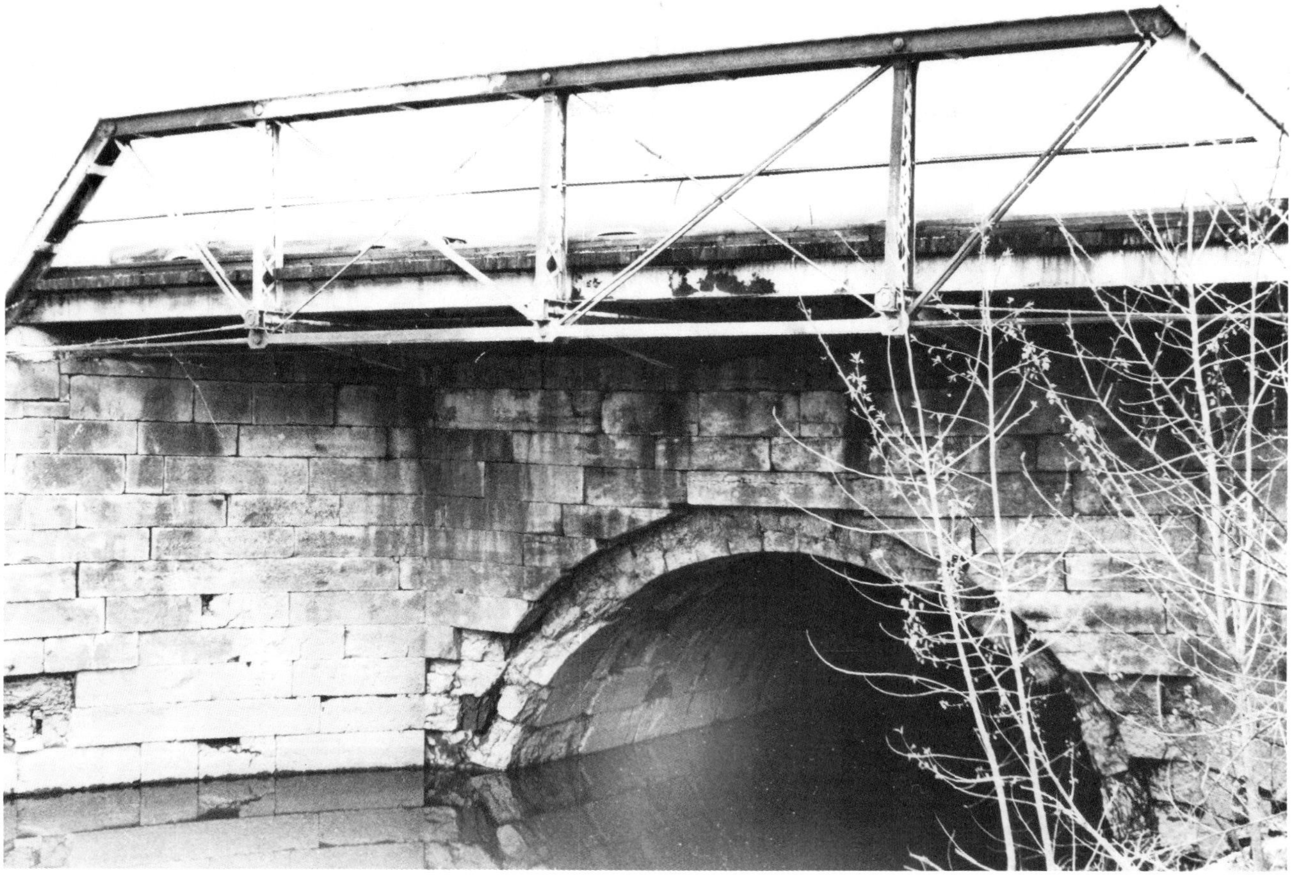

Oberhauser Creek Culvert, east of Napoleon, has become a bridge on U.S. Highway 24, the Anthony Wayne Parkway; 1973. Mile: 43. (Photo by F. W. Trevorrow.)

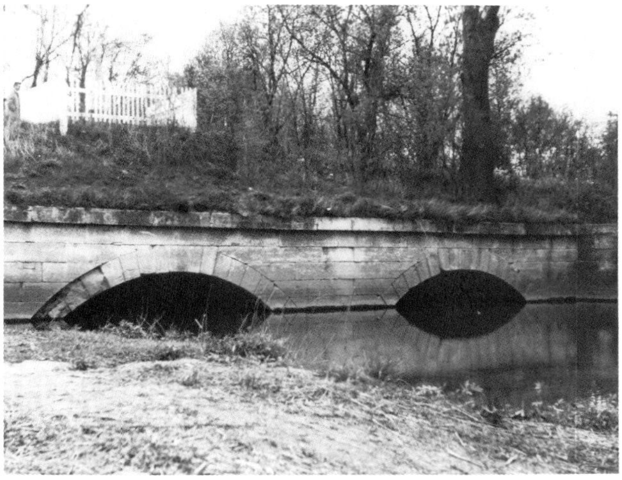

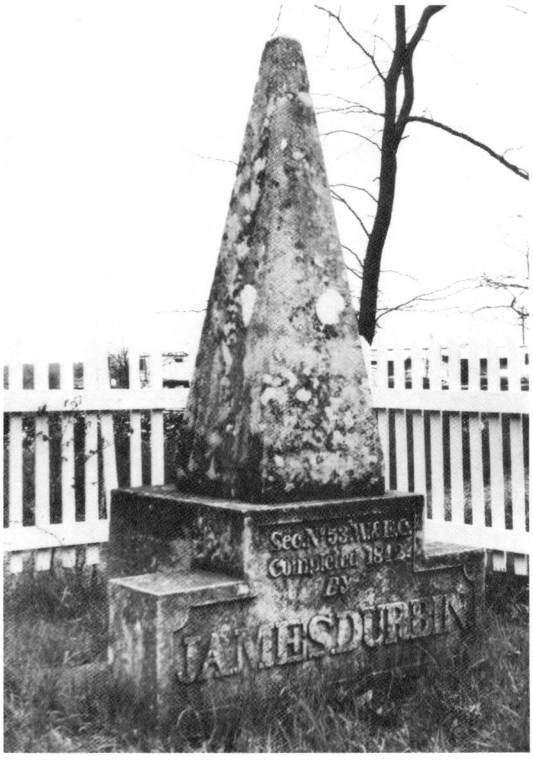

Above, Bad Creek Culvert, a double culvert construction at Texas. The white fence on the hill in background encloses a monument to the builder; 1973. Mile: 34. At right is contractor James Durbin's monument to himself for having built the Texas culvert in 1842. (Photos by F. W. Trevorrow.)

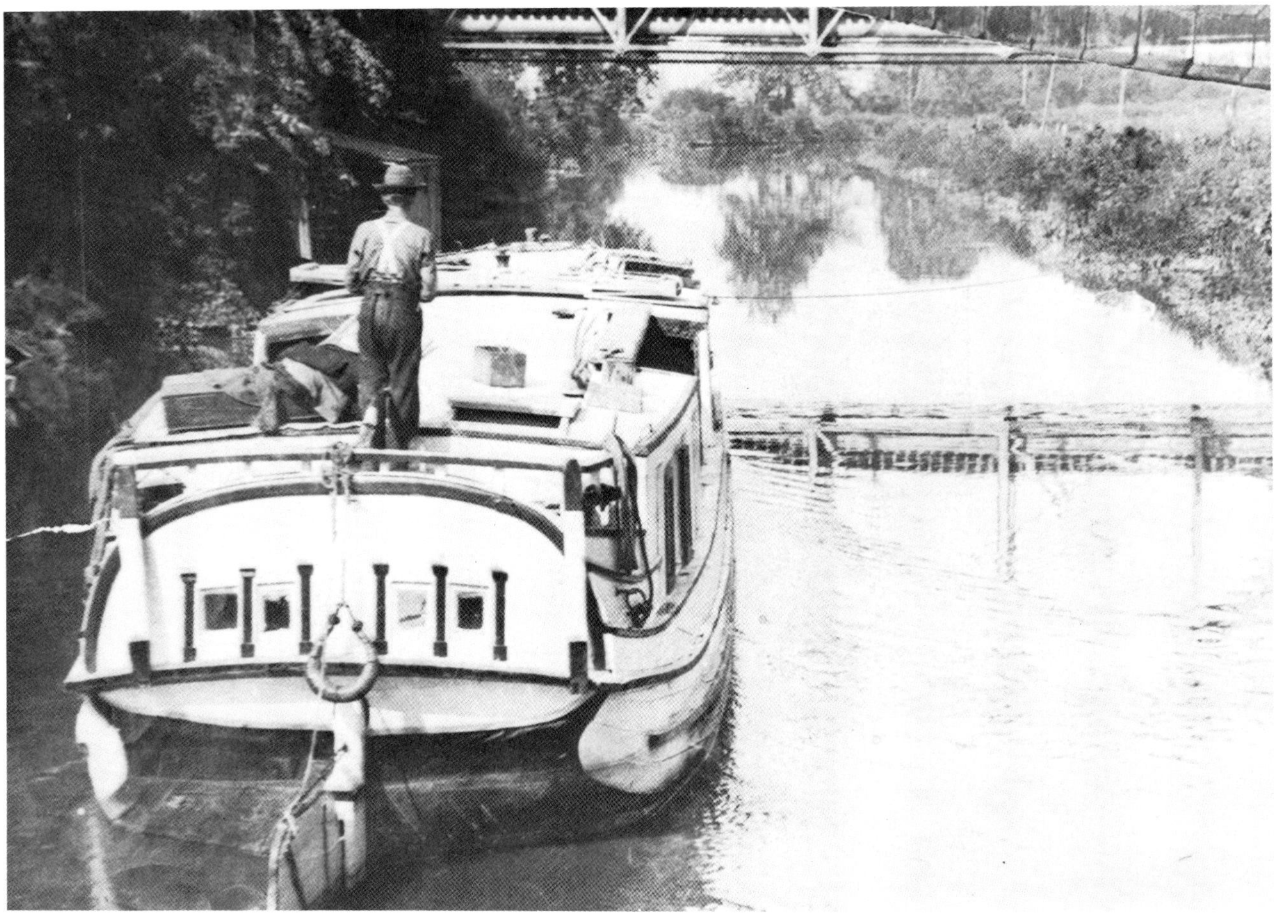

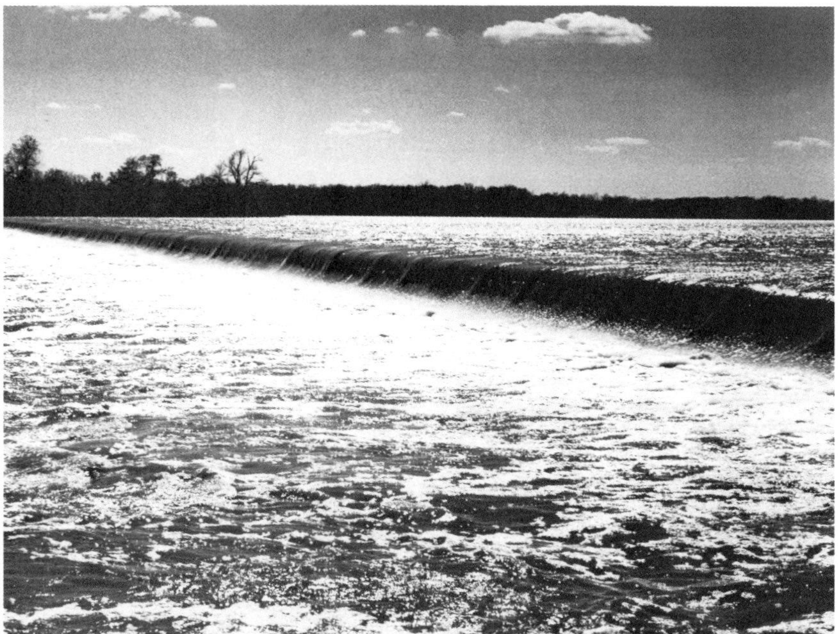

The captain of this boat (above) near Grand Rapids in 1898 steers the tiller with his knees. Mile: 31. (Toledo/Lucas County Public Library)

Providence Dam (left) on the Maumee River provided another Miami & Erie slackwater. Mile: 30. (Photo by F. W. Trevorrow, 1973.)

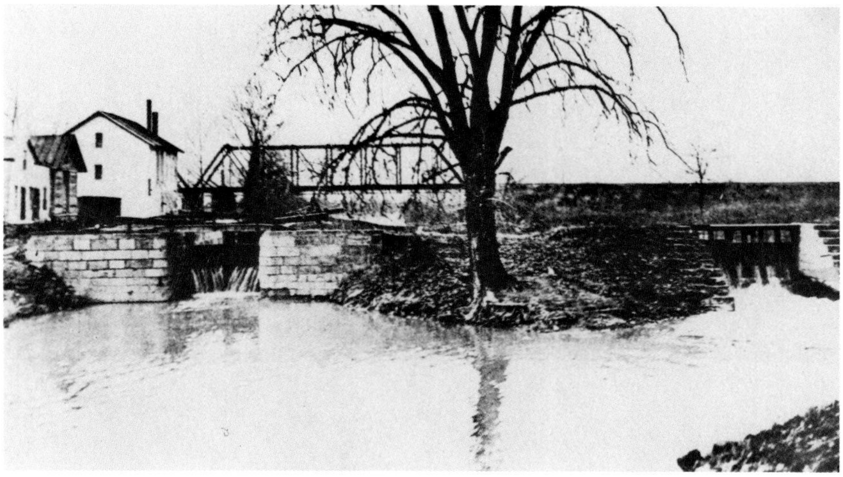

Ludwig's Grist Mill and Lock 44 in operating condition at Providence in 1904. Originally constructed in 1835, the mill was acquired by Augustine Pilliod in 1886, who equipped it to generate electric power. The mill supplied electricity to the city of Grand Rapids and surrounding communities until 1915, when Toledo Edison purchased the power rights. Mile: 30. (Toledo/Lucas County Public Library)

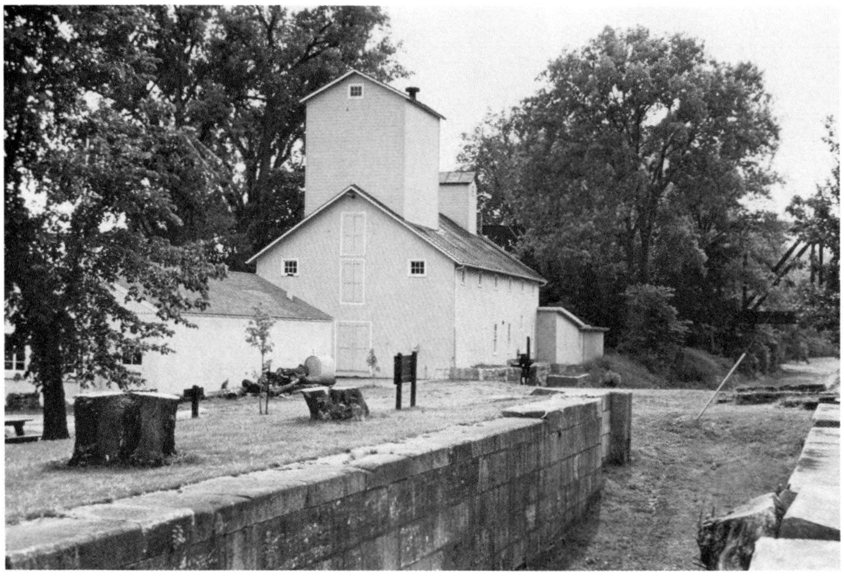

Ludwig's mill in 1978 (Lock 44 in foreground). Still operating on canal waterpower, the facility is maintained by the Toledo Metropolitan Park Board as an educational tourist attraction demonstrating both sawmill and grist mill operations having nineteenth-century flatbelt drive mechanisms. Visitors can purchase corn meal and other products milled here. In 1940, while owned by Frank Heising, the mill sustained a major fire which destroyed its upper story. Restored as a shorter structure, the mill has a "head house" on the gable to enclose the belts and shafts of mechanical power transmission equipment. Mile: 30. (From a color slide by Jack Gieck, 1978.)

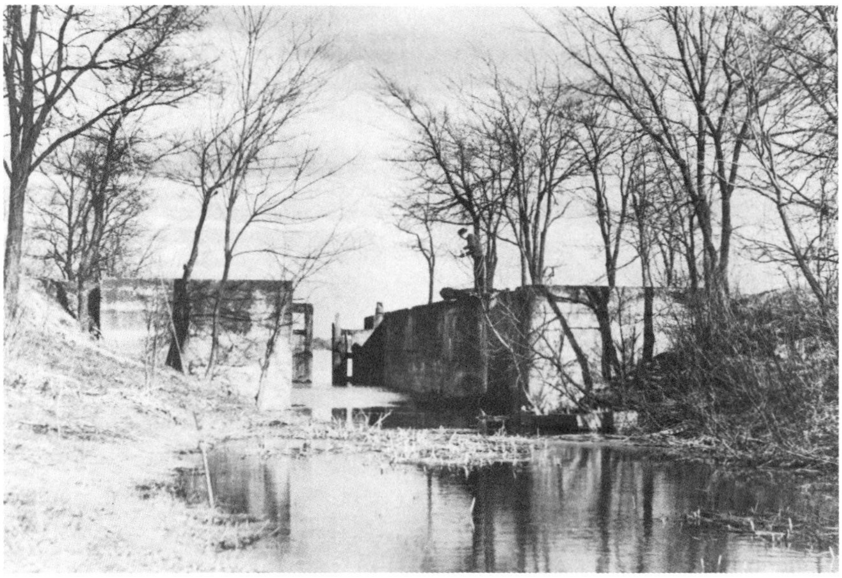

A guard lock from the slackwater at Providence Dam into the Gilead Canal on the opposite side of the Maumee River from Providence. The short feeder canal provided access to the Miami & Erie for the town of Gilead, now Grand Rapids; 1937. (Toledo/Lucas County Public Library)

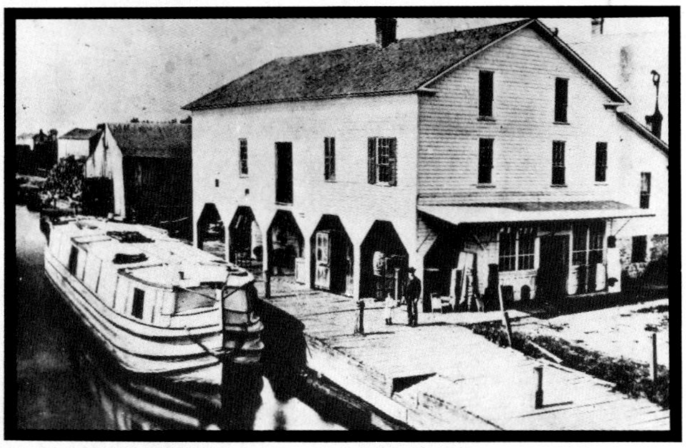

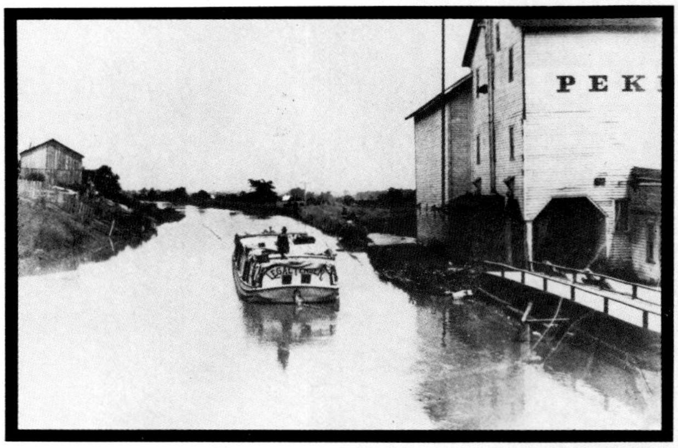

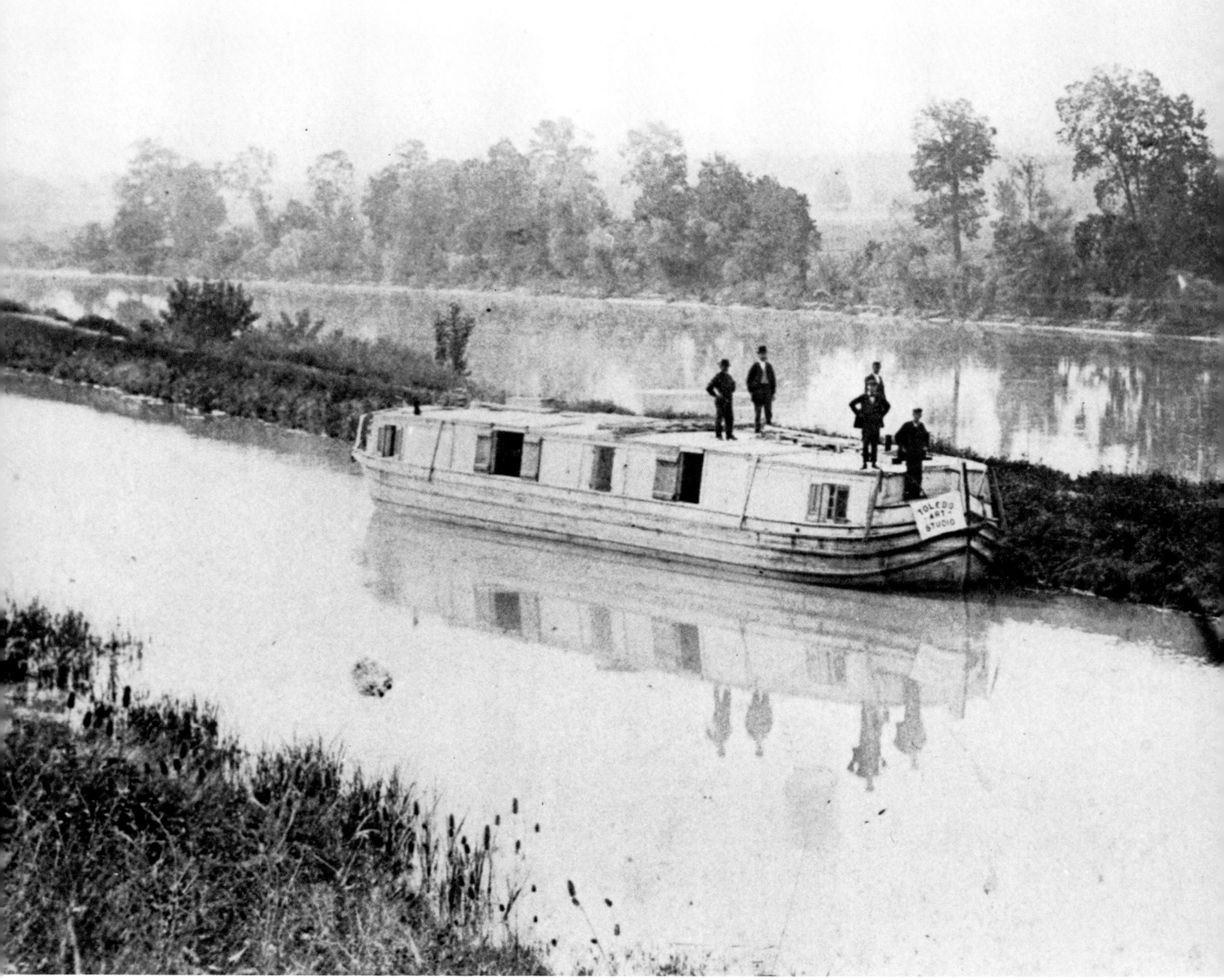

A boat and mule team on the canal parallel to the Maumee River northeast of Otsego. The sign on bow advertises "Toledo Art Studio." Mile: 22. (Toledo/Lucas County Public Library)

Inset, Two views from Waterville. A canal boat (left) is docked at a warehouse; 1869 (CSO). The *Legal Tender* (right) glides by the Pekin Mill in 1900. Mile: 18. (Toledo/Lucas County Public Library)

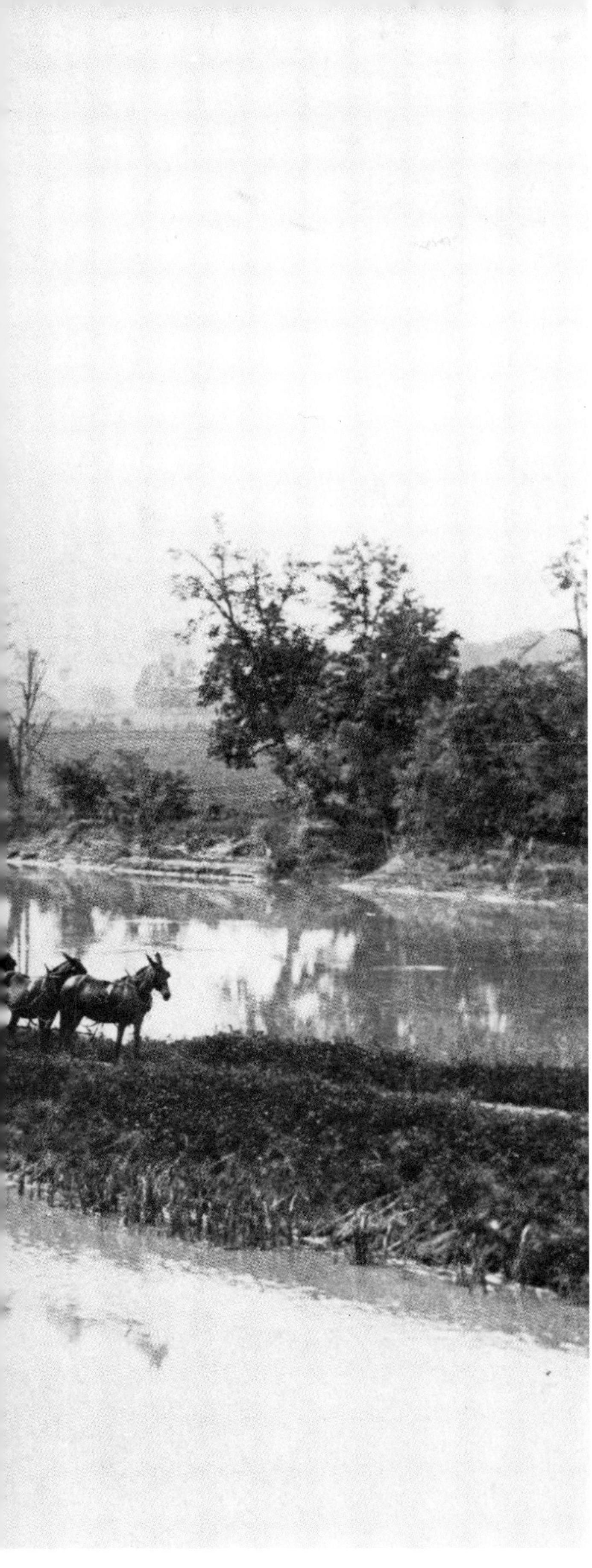

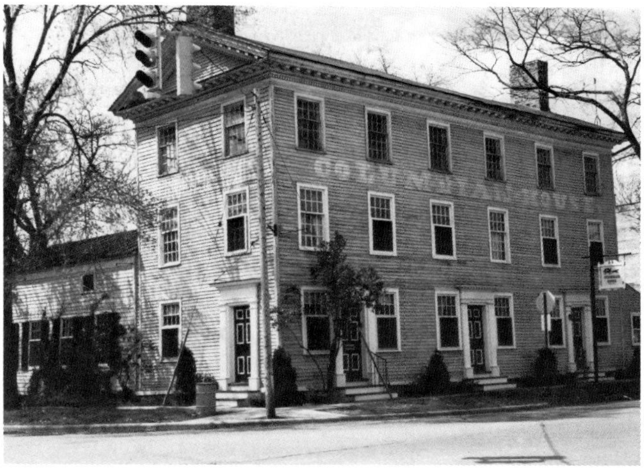

Built in 1828, the Columbian House, an inn that catered to canal travelers throughout the canal era, continues operation as a restaurant today. (Photo by F. W. Trevorrow, 1973.)

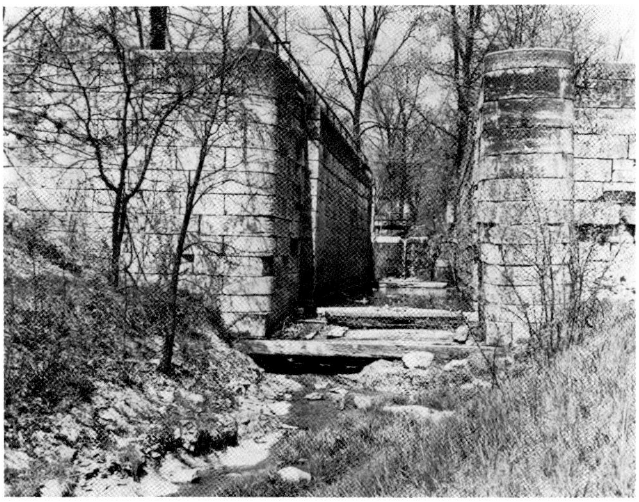

Sidecut Park (below), in Maumee, is operated by the Toledo Metropolitan Park Board. Maumee Sidecut was intended to provide direct access from the canal to Lake Erie traffic in the Maumee River; 1978. A lift lock in Sidecut Park (left), one of five preserved in this Toledo Metro Park; 1939. The lock remains in Sidecut Park today in a similar state of preservation. Mile: 14. (From color slides by Jack Gieck.)

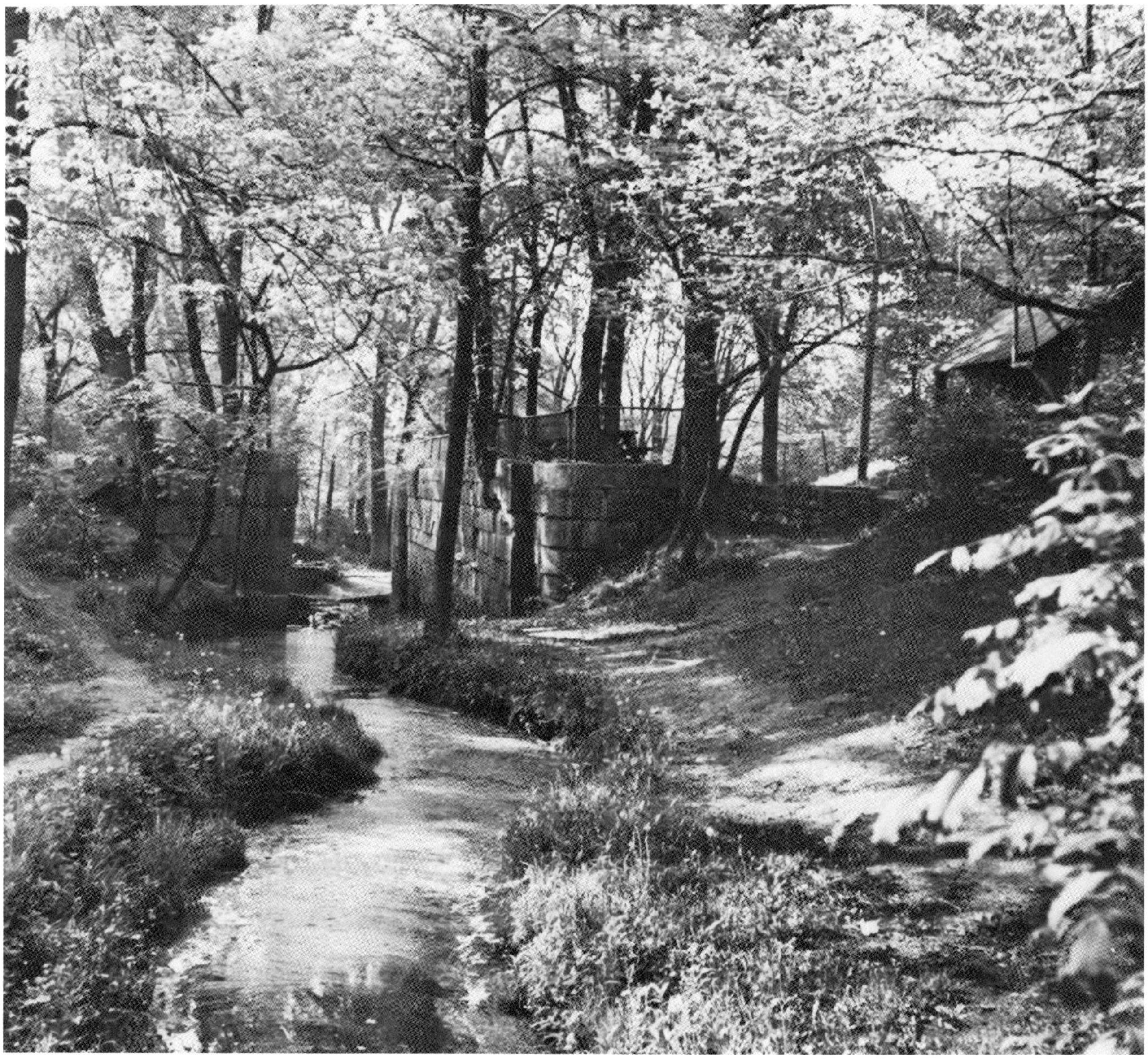

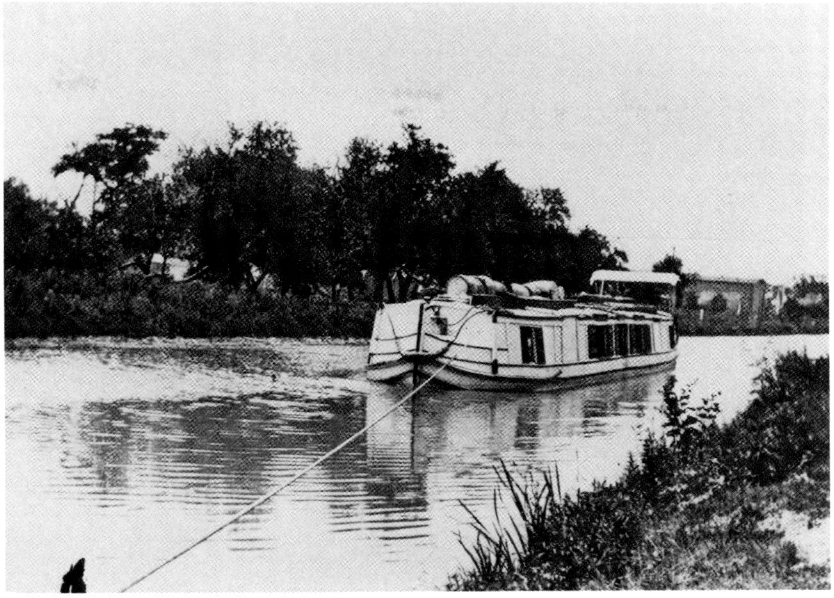

The *Turtle* at Maumee in 1898. Mile: 13. (Toledo/Lucas County Public Library)

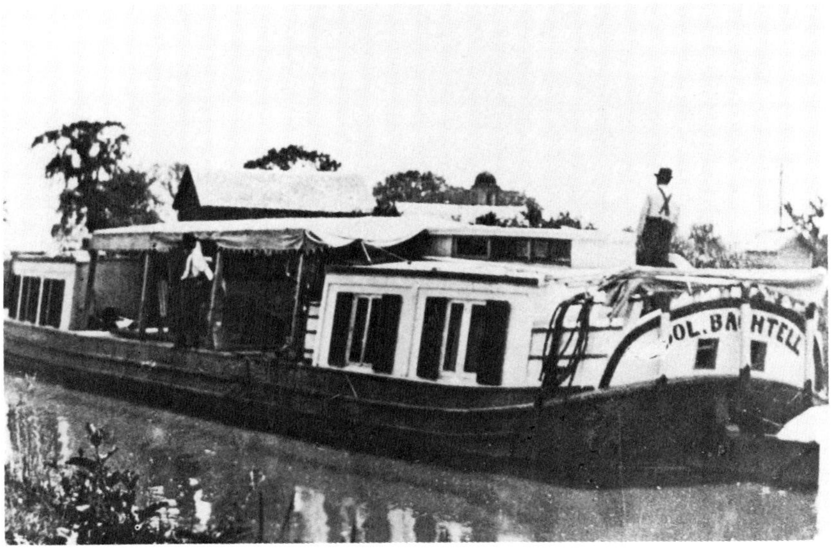

The *Col. Bachtell* between White and Elizabeth Streets at Maumee; 1898. Mile: 13. (Toledo/Lucas County Public Library)

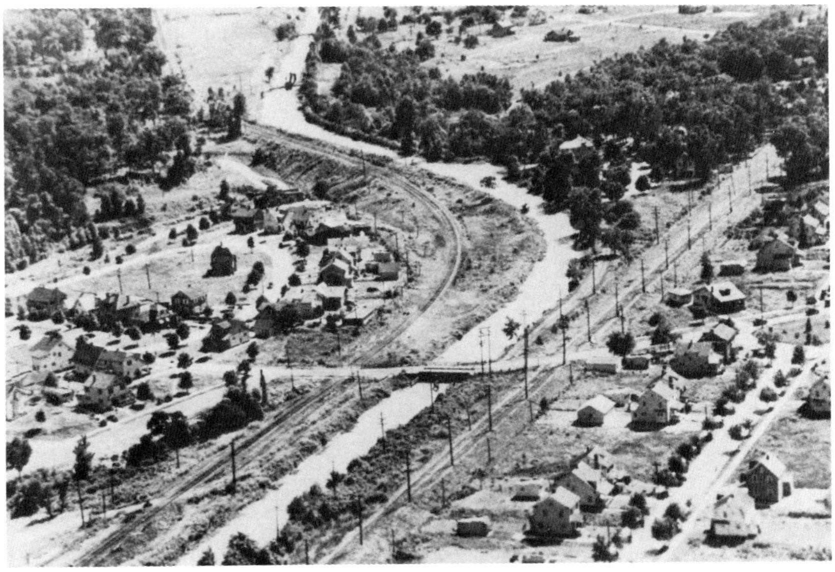

A 1921 aerial photograph of the canal in Glendale, near Toledo, with Socket Bridge and Jackknife Bridge (background). Mile: 8. (Toledo/Lucas County Public Library)

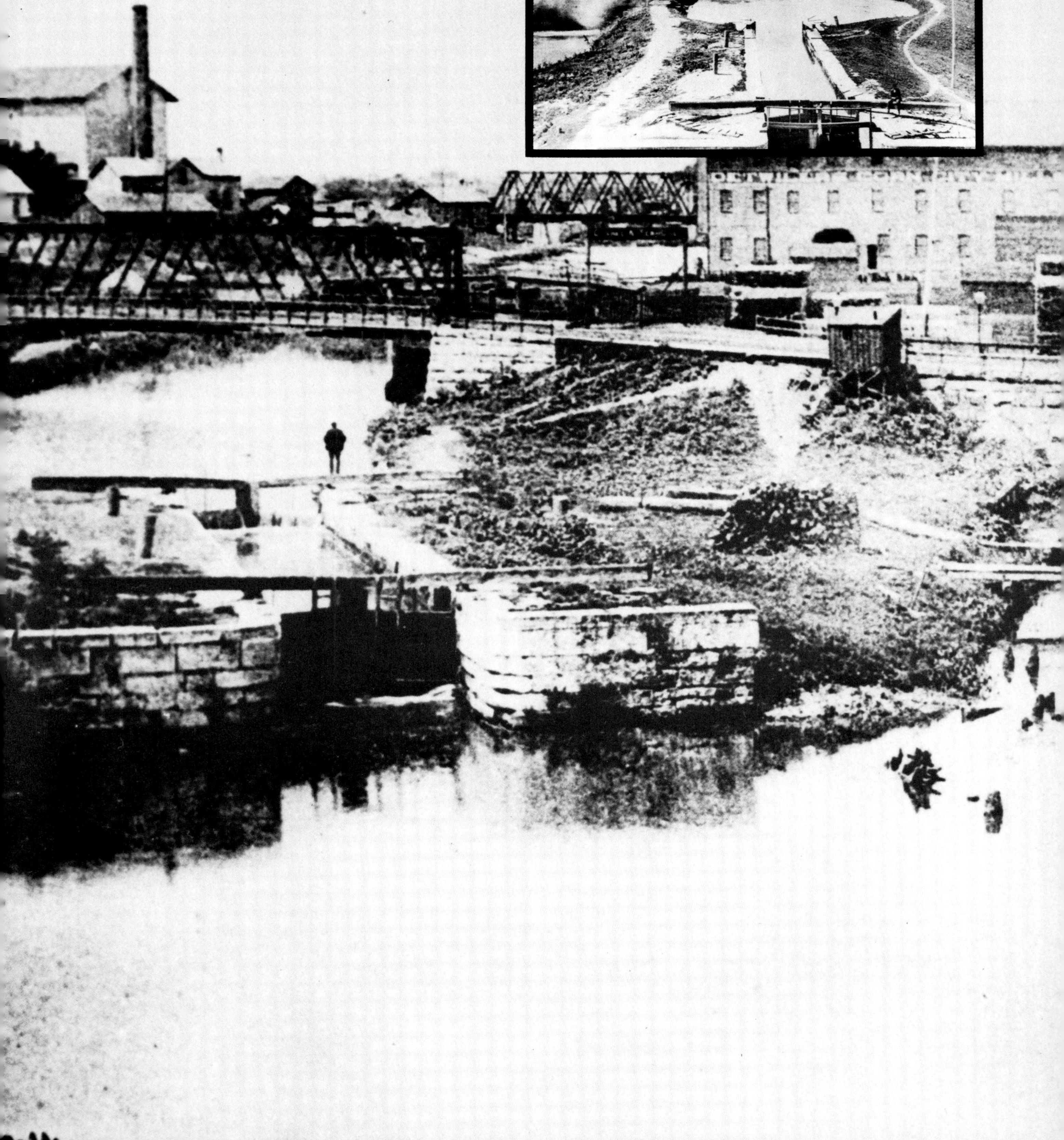

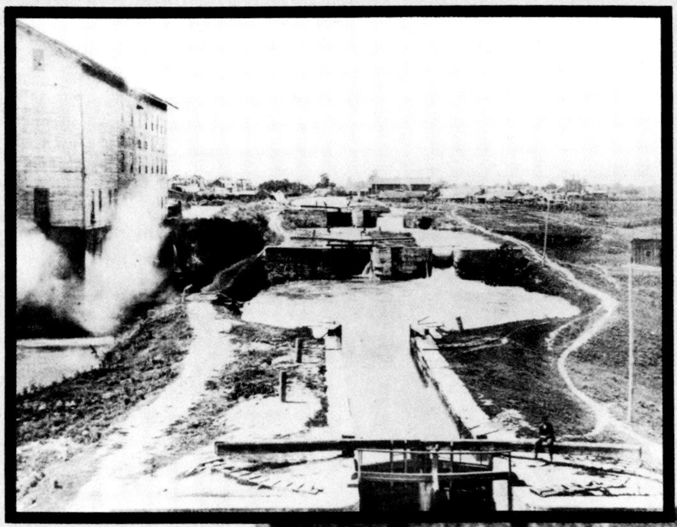

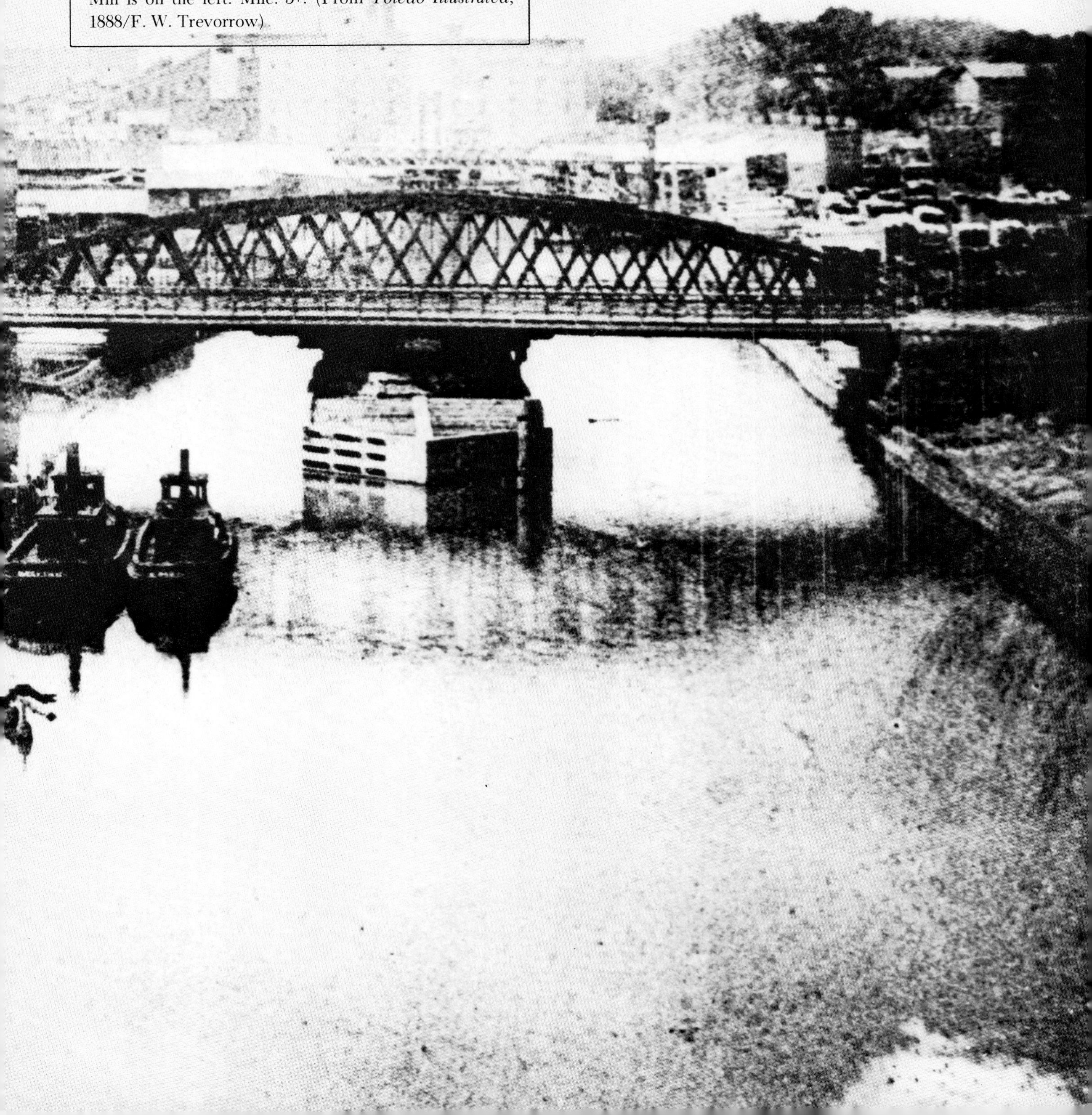

The river lock and entrance to the Maumee River from Swan Creek Sidecut, also called Toledo Sidecut. After 1864 this was the northern terminus of the canal, when the remaining channel through Toledo to Manhattan was abandoned. The building on the horizon at left is the Armada Mill. Because of the difficulty of controlling canal boats in the wide Maumee River, most cargo was transshipped from the last lock on the sidecut. Mile: 5. *Inset,* Locks 48, 49, and 50 (foreground) on the main trunk of the Miami & Erie as it approaches Swan Creek Sidecut, which went off to the left in front of the foreground gate. Armada Mill is on the left. Mile: 5+. (From *Toledo Illustrated,* 1888/F. W. Trevorrow)

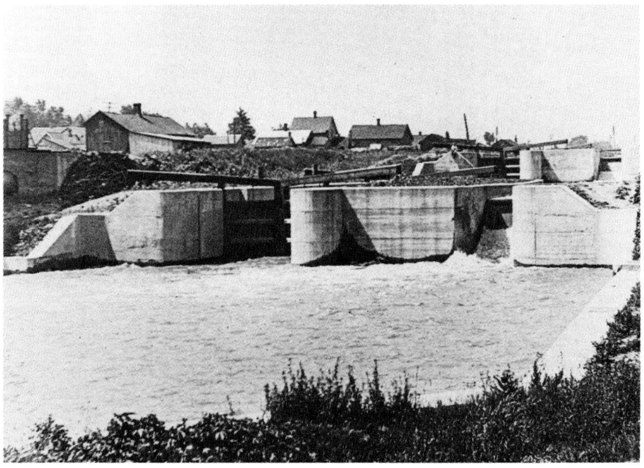

Harrison Street Locks, Toledo, after refurbishing of the canal after 1900. Mile: 5. (Toledo/Lucas County Public Library)

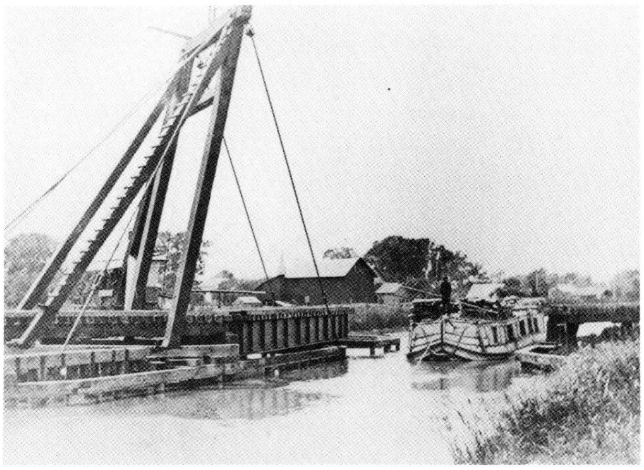

The *Ella May* and dredge in Toledo at St. Louis Railroad Bridge; 1898. Mile: 4. (Toledo/Lucas County Public Library)

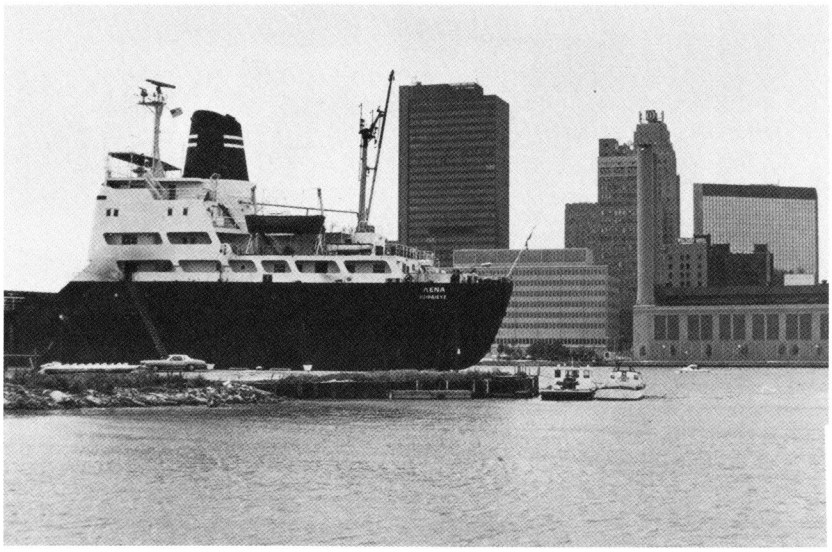

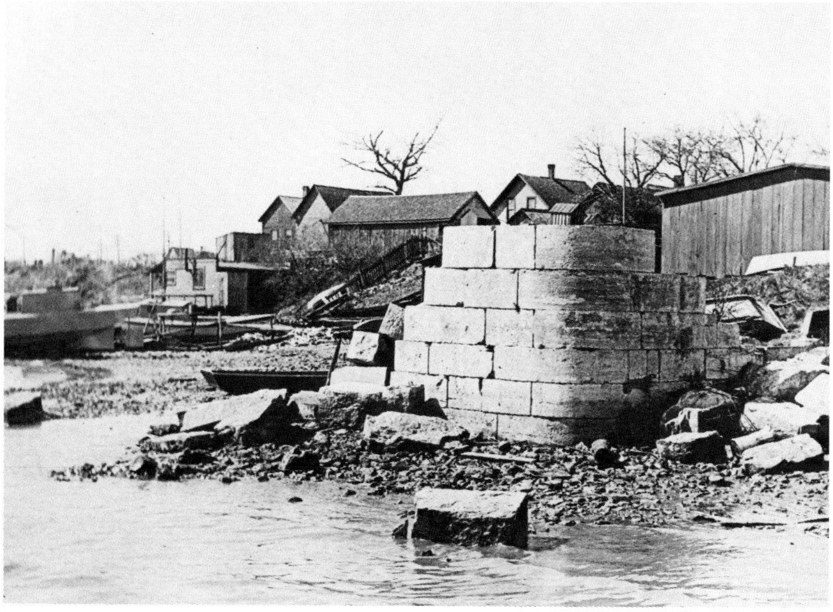

The original River Lock at Manhattan was near the opposite shore of the Maumee River shown here (above) in 1978 (from a color slide by Jack Gieck). Ruins of Lock One, River Lock (below) at Manhattan are near Lapier and Lasalle Streets in Toledo; 1911. Mile: 0. (Toledo/Lucas County Public Library)

The lake schooner *Hunter Savage* (top) at Toledo in 1895. Commerce with such vessels was the objective of merchants and boat captains on the Miami & Erie Canal, as well as those on the Wabash & Erie. Lake steamers like the *Pastime* (center) brought passengers from Buffalo to Toledo. Many immigrants embarked on the Wabash & Erie here to settle in Indiana; 1894. The *Idlewild* (bottom) was a Detroit-Toledo packet, which also carried immigrants to Michigan in its time; 1894. (Photos from the Vanderlip Collection, courtesy Toledo/Lucas County Public Library.)

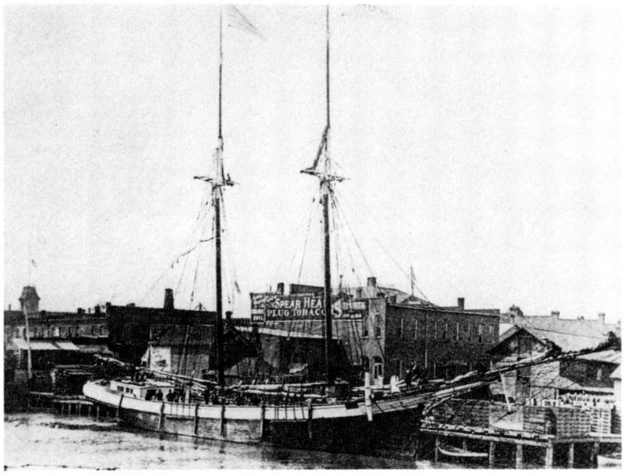

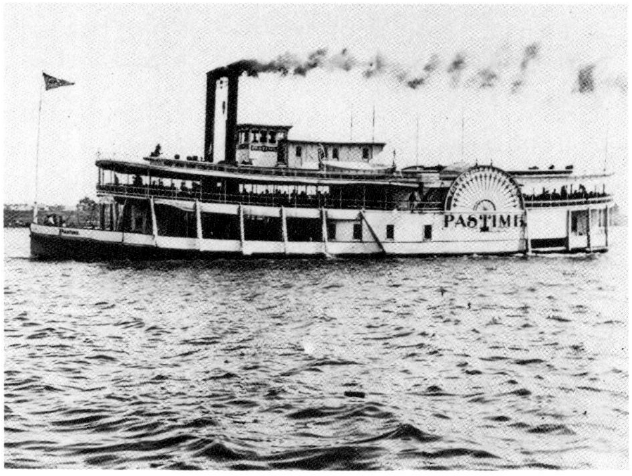

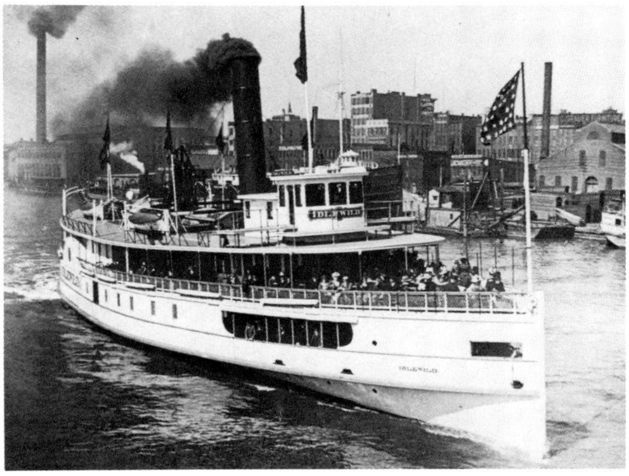

"Travelers on Ohio's canals could board steamboats at either Lake Erie or the Ohio River. These romantic vessels, fighting lake gales or river hazards, were important links in the transportation chain. The *Orleans* had amazed spectators along the Ohio in 1811. Only seven years later, the *Walk-in-the-Water,* the first steamboat on Lake Erie, churned from Buffalo to Detroit in three days with stops at Cleveland and Sandusky. By mid-century the colorful steamboat era produced great boats which competed with one another for supremacy even if high-pressure boilers exploded in the process. This free-wheeling spirit coupled with abundant natural hazards made steamboating a risky business. But who could resist the allure of the lighted ships with their grand salons, crystal chandeliers, and gay music as they navigated the Ohio or the lakes with their engines throbbing with a powerful beat? No less romantic were the stern or side wheelers, manned by roustabouts and deck hands, many of whom were free Negroes, which churned from port to port carrying passengers and cargo short distances, though on occasion they went all the way to Buffalo on the lakes or to Cairo or even New Orleans on the rivers. Ship-building became an important industry for Cincinnati, Marietta, and several other Ohio River ports, while along Lake Erie nearly every town or sizeable river mouth was the scene of construction activity." (George W. Knepper, *An Ohio Portrait,* Ohio Historical Society, 1976)

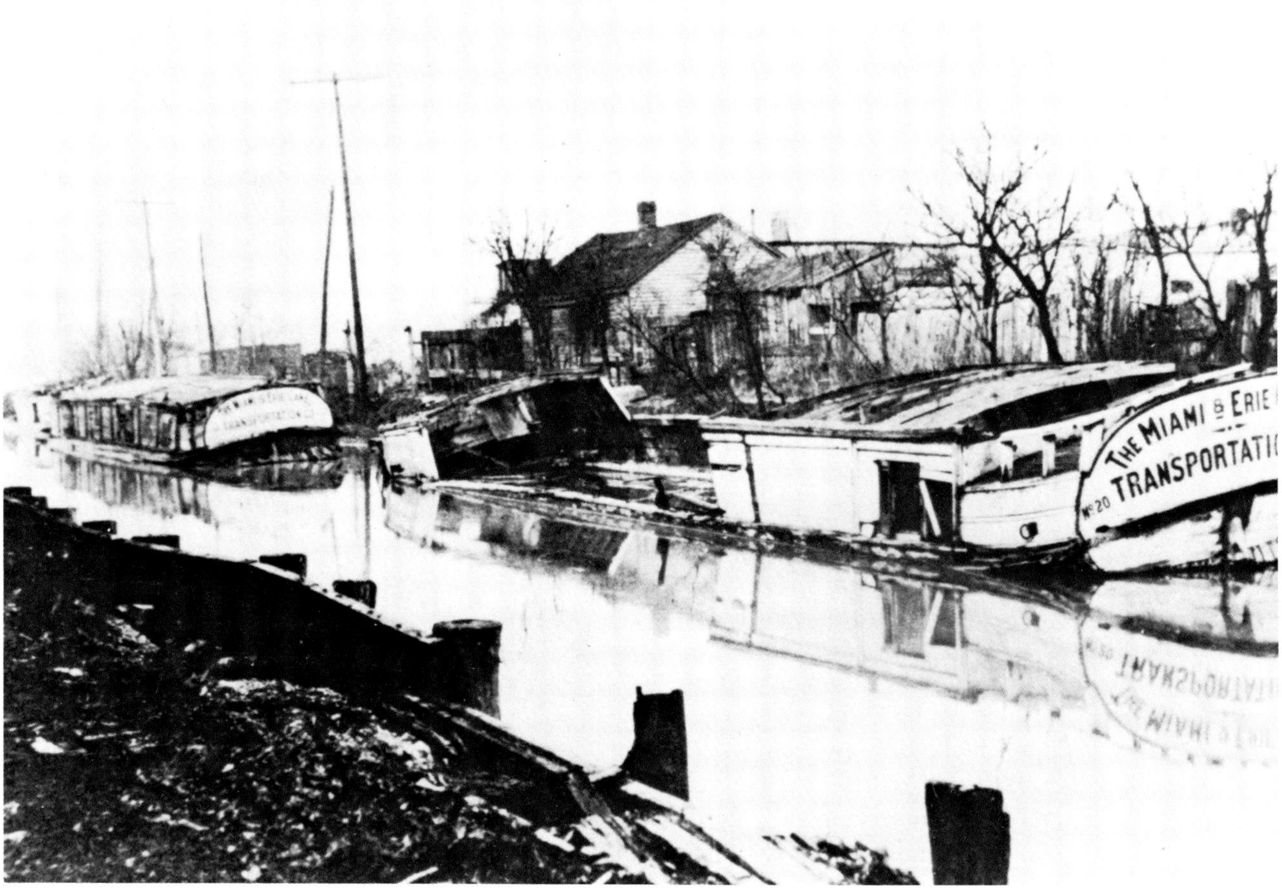

Derelict canal boats of the Miami & Erie Transportation Company at the end of an era. Ever the subject of inaugural festivities, the Miami & Erie Canal was formally *closed* with celebratory speeches and a giant parade in 1929 in Middletown. (Herbert F. Verity)

Feeder Canals

8

Although the Ohio & Erie and the Miami & Erie constituted the two main trunks of Ohio's canal system, there were many feeder canals that extended the useful range served by the state's inland waterway transportation system. Some of these branch canals, like the twelve-mile Columbus Feeder running north from Lockbourne, simply connected an important city, an industry, or other business-interest group with the main trunk. Others, especially along the Maumee River near Toledo, were sidecuts between the canal channel and a navigable river. To serve southeastern communities, the Muskingum River Improvement canalized the river with a system of dams and locks, and connected with the Ohio & Erie via the Dresden Sidecut—providing the Ohio & Erie with a second Ohio River outlet at Marietta. Still other branches were hydraulic canals serving local industries or providing water to one of the main trunks, although the hydraulics were also occasionally navigated by canal boats. There were, in addition, important local canals in Ohio that were not directly connected with either main trunk.

During Ohio's canal era, the state boasted nearly thirty distinct or separately named canals (including some that were abandoned before completion and others that were later absorbed by the main trunk at the Miami & Erie). The following list, grouped more or less geographically, was compiled by canal historian Terry Woods for the Indexing Committee of the American Canal Society:

EASTERN OHIO

Ohio City Canal
Ohio & Erie Canal
Pennsylvania & Ohio Canal
Middlebury Branch Canal
Cuyahoga Feeder
South Feeder
Sandy & Beaver Canal
Nimishillen & Sandy Canal
Zoar Feeder & Sidecut
Trenton Feeder
Walhonding Canal
Dresden Sidecut
Muskingum River Improvement
Granville Feeder
Hocking Canal
Columbus Feeder

WESTERN OHIO

Miami Canal
Miami Extension Canal
Wabash & Erie Canal
Miami & Erie Canal
Gilead Sidecut
Toledo (Swan Creek) Sidecut
Maumee Sidecut
Grand Reservoir Feeder
Sidney Feeder
Hamilton Sidecut
Warren County Canal
Cincinnati & Whitewater Canal

CENTRAL OHIO

Milan Canal

"Twenty five miles to nowhere" is how one contemporary writer dismissed the Walhonding Canal, which terminated at the village of Rochester—a town no longer in existence. The characterization was accurate for the Walhonding as well as some of the others on the foregoing list—often because plans for extension did not materialize. Others, like the Nimishillen & Sandy Canal, were never completed.

Because it provided a market outlet for the coal mines and extensive salt deposits in the Hocking Valley, the Hocking Canal maintained a high level of traffic for at least a decade after other Ohio canals were in eclipse. Even so, its receipts never equalled its cost of operation.

But some feeders were a resounding success and had a corresponding economic impact on their section of the state. One such success was the Pennsylvania & Ohio Canal, an indirect product of James Geddes's farsighted vision when he ran a survey along the Cuyahoga and Mahoning valleys in 1823, concluding that it would be feasible to lead a canal through those two valleys to connect with Pennsylvania's canal system. After years of community meetings and political negotiations, a private company (organized by Ohioans Eliakim Crosby, Leicester King, and William Ryan, in cooperation with a consortium of businessmen from Philadelphia and Beaver County, Pennsylvania) was chartered by the legislatures of both Ohio and Pennsylvania to construct the canal.

The Pennsylvania & Ohio was eighty-two miles long—sixty-seven miles of it in Ohio. Its western terminus was in Akron, where it connected with the Ohio & Erie Canal just below Lock One in the Lower Basin adjacent to property where the B. F. Goodrich Company would locate almost half a century later. From here it went north, going up the middle of Akron's Main Street and preempting most of the thoroughfare's width, entering a waterway at Mill Street known as Crosby's Ditch. Before the age of steam this mill race had diverted water from the Little Cuyahoga River to power Eliakim Crosby's Stone Mill at Lock Five and other industries in the Cascade, North Akron's canal valley. A system of sixteen locks and several slackwaters led the canal through Cuyahoga Falls, Monroe Falls, and Franklin Mills (now Kent) to Ravenna, overcoming a 112-foot rise to the summit level—which was two and a half miles long, and included a deep cut that was nearly five thousand feet long and over forty feet deep in places.

From Ravenna the canal entered the valley created by the East Branch of the Mahoning River, going through Campbellsport, Newport, McClintocksburg, and Newton Falls, where it crossed the East Branch on a stone masonry aqueduct of three fifty-foot spans. After crossing the Mahoning on another slackwater, the canal route led through Niles, Girard, Youngstown, Campbell, and Lowellville before entering the state of Pennsylvania—ultimately joining the Beaver Division of the Pennsylvania Canal System at Western Reserve Basin, about seven miles south of New Castle.

In its eighty-two miles, the Pennsylvania & Ohio required fifty-four locks to overcome 424 feet of rise and fall. In addition, the canal contained two aqueducts, nine dams, and fifty-seven road bridges. The canal was obviously a major engineering challenge, primarily the work of military engineers borrowed by the Pennsylvania & Ohio Canal Company from the state Corps of Engineers, the U.S. Topography Engineers, and the U.S. Army. Actual construction was under the supervision of two principal engineers, Colonel Sebried Dodge, who had been with the project since its initial surveys in 1827, and James Dunlop Harris, a civilian.

The Pennsylvania & Ohio Canal enjoyed a decade of freedom from rail competition and brought prosperity to the string of communities it served. The prosperity it would engender seems to have been anticipated on August 4, 1840, when four brightly decorated packets left New Castle with a party of Pennsylvania dignitaries, signalling the start of a gala that would pick up momentum in Youngstown, Warren, Ravenna, and Franklin Mills. At Cuyahoga Falls, after yet another reception on August 6, two more boats carrying the Akron delegation joined the flotilla.

By the time the boats reached Akron, where a reception had been prepared at the American House, the canal banks were lined with cheering throngs, brass bands, and loaded cannons. In addition to Pennsylvania's Governor David Wittenhouse Porter, the Pennsylvania fleet carried a cargo of sherry and Madeira wines, as well as imported champagne. The ensuing celebration is said to have virtually paralyzed Akron business for several days.

•

Only thirty-five miles away, success managed to completely elude a parallel feeder project. As conceived by the residents of New Lisbon, Ohio, seat of Columbiana County, the Sandy & Beaver would tie their community to both the Ohio & Erie Canal and to Pennsylvania markets, much like the Pennsylvania & Ohio—perhaps even better, since their canal would have access to the Ohio River.

By 1822 New Lisbon's Rebecca Furnace, a charcoal blast furnace named after owner Gideon

Hughes's wife, had expanded to include a rolling mill and nail making machinery. As scarce as nails were in the frontier (some settlers actually burned their eastern homes to salvage the nails before heading west), access to the market afforded by new home construction along the Ohio & Erie to the west, and to Ohio River markets and Pennsylvania coal to the east suggested that New Lisbon (now Lisbon) could become a future Pittsburgh. There was also Jacob Hostetter's brick kiln and later a pottery works that utilized the region's fine clay. New Lisbon even had a clock and watch works. The idea of a canal was very attractive to the community, and canal fever quickly caught on after the 1825 canal act was passed.

The Sandy & Beaver Canal seemed to enjoy other natural advantages. Unlike other American canals it would require few bridges, since there were so few recognized roads in the area. It would be a canal through an undeveloped area connecting two well-developed areas. It would, thus, open up the new West to Philadelphia merchants. Stock in the privately financed Sandy & Beaver Canal Company was easy to sell. The company's president, Benjamin Hanna, reported on the results of the first survey on February 1, 1830.

As originally laid out by Major D. B. Douglas, a professional engineer on loan from the U.S. Military Academy at West Point, the canal would be just over ninety miles in length, with 100 locks, seven aqueducts, and with three feet of extra depth at the summit to act as a reservoir. At its western terminus it would connect with the Ohio & Erie after crossing an aqueduct over the Tuscarawas River at Bolivar. At its eastern extremity it would run along the north bank of the Ohio River from Glasgow, Pennsylvania, to Beaver, where it would connect with the Pennsylvania canal system at the mouth of the Big Beaver River.

Douglas's plan also called for a unique, 2,700-foot tunnel to take the canal through the ridge dividing the Big Sandy and Little Beaver watersheds.

Innovative as it was, Douglas's design was rejected by investors associated with Philadelphia's Board of Trade, whose experts concluded that the summit reservoir would not provide sufficient water to support heavy traffic during the dry season. Worse, the Philadelphia group threw its support behind the concurrent Pennsylvania & Ohio Canal project.

In September 1834 two new engineers were hired who had demonstrated experience in hydraulic engineering. Edward H. Gill from the Schuykill Navigation Company and Hother Hage from the Susquehanna Division of the Pennsylvania system were directed to reexamine the route with particular attention to improving the water supply at the summit. They succeeded in shortening the route by seventeen miles, and improved the water supply situation by lowering the summit level through the creation of a second tunnel; and they lengthened Douglas's Big Tunnel to 1,060 yards. They also designed additional reservoirs for the system. Douglas's recommendation of seven aqueducts was eventually reduced to one.

The investors thus satisfied, construction began at once, with Gill as chief engineer. The tunnels received special attention. They would be about fifteen feet wide (the width of a lock) and seventeen to eighteen feet high. The Big Tunnel would begin with two vertical excavations that would become air shafts, and digging could proceed in both directions from each, as well as from the ends, thus providing work stations for six crews simultaneously. At its deepest point the Big Tunnel was eighty feet below the surface. To avoid the added expense of drilling and blasting a towpath through the solid rock, an endless chain, driven by a stationary mule-powered treadmill, would be installed to drag the boats through, with the boat's own draft animals riding aboard. No such mechanism was provided in the Little Tunnel, and it is likely that the boats were poled through.

The 600-yard Little Tunnel provided an unexpected serendipity. During its excavation east of Dungannon, workers discovered a twenty-inch horizontal vein of high-grade bituminous coal ten feet above the channel level. This coal was mined and used to operate the steam engines that hauled away the rock spoil from the Big Tunnel via the air shafts. Some of it may also have been used to heat workers' homes in the village Little Tunnel, which had sprung up on top of the hill above. Just below the coal vein was a deposit of excellent limestone that was later burned to provide lime for mortaring lock masonry.

Edward Gill was a stern boss. Contractors were warned that work was done at their risk until their job was completed and approved. And his specifications were demanding: A typical example was his detailed stipulation for building a dam across a stream. The contractor was ordered to begin with six parallel trenches in which white oak timbers more than sixty feet long would be laid. Row upon row of timbers would be bored and pinned together on top of these, course after course, until a barrier more than twelve feet above the foundation had been fabricated. After facing the piling with white oak planks fastened by wrought iron spikes, ice guards would be placed at the top and stone abutments would be meticulously constructed at the sides.

Gill was absolutely intolerant of "ardent spirits," and an intoxicated laborer was to be automatically discharged—this in an age when alcohol was often a lubricant for the hardships of frontier life (his prohibition, however, was more often honored in the breach, since there were no less than twelve saloons in Dungannon—a town fortuitously located between the two tunnels).

Gill and his contractors estimated that the Big Tunnel could be completed by October 1837. Their estimate proved to be off by more than ten years! In fairness, they cannot be blamed. Cash flow problems besieged the canal company beginning in 1836, and some contractors were never paid. With the arrival of the panic of 1837 came a 90 percent layoff—reducing the labor force from 2,000 to less than 200. As the depression continued, all work was shut down and the value of the canal stock collapsed. Edward Gill remained in the area for about a year, but then left for greener pastures.

The Sandy & Beaver Canal lay dormant for seven years. But in 1844, the economy having improved, new stock issues were promoted. The Canal Society of Ohio's Sesquicentennial issue of *Towpaths* describes the canal's new, if altered, lease on life:

"W. Milnor Roberts was hired as Chief Engineer. The departure of perfectionist Gill and the need for economy meant that adequate, but 'cheap' engineering was the guideline for construction in the 40's. Locks of rough stone lined with plank replaced the magnificent stone structures, such as Lusk's lock, which were characteristic of Gill.

"Roberts made a few changes in the route, some as economic measures and some because, in the 7 years the canal lay dormant, a number of mills had been erected along the Big Sandy [Creek] and the route was altered to avoid ruining these properties. . . .

"As completed, the Sandy & Beaver left the Ohio river through a stone gate near Glasgow, Pennsylvania (a proposed extension up the west bank of the Ohio river to Beaver Falls was surveyed, but never constructed), and followed the Little Beaver east for 18 miles, to a point just opposite Fredricktown. It then followed the Middle Fork for approximately 19 miles to the beginning of the summit level, some 2 miles above New Lisbon. The eastern division was 27 miles long (17 miles of it in slackwater). 57 locks overcame 465 feet; the last 120 feet required 15 locks in a little over a mile at Furnace Hollow. This division touched the towns of Jamestown, Martinsburgh, Sprucevale, Williamsport, Middle Beaver, Elkton, New Lisbon and Lockbridge.

"The summit level began at the 'paper village' of Lockbridge and wound its way for 14 miles along the valleys of Cold Run, West Fork, Willard's Run, and Sandy Creek. A reservoir on Cold Run and one on the West Fork of the Little Beaver (now Guilford Lake) supplied the summit level with water. The communities of Guilford, Dungannon and Hanover lay along this stretch. . . .

"The western division began at Kensington and ran 32 miles to Bolivar, 5 miles in slackwater, while passing through the towns of Kensington, Lynchburg, East Rochester, Minerva, Pekin, Oneida, Malvern, Waynesburg, Magnolia, Sandyville and East Bolivar. This division contained one aqueduct (across the Tuscarawas at Bolivar), 2 small reservoirs near Kensington, 10 dams, and it required 33 locks to descend 220 feet."[1]

The Little Tunnel was completed in 1846; and, although the eastern and western divisions could be opened, the Big Tunnel was not finished until New Year's Day 1848—nearly eight years after completion of the rival Pennsylvania & Ohio Canal. The first boat made it through the tunnel a few days later (see p. 223). Unfortunately, the boat's passage proved that the summit level was not yet ready to carry regular traffic; and it took two more years before it was.

By this time railroads were beginning to cobweb the country. Nevertheless, contrary to the myth that only one boat ever travelled its length, the Sandy & Beaver operated for about two years, albeit unprofitably. Then, on April 12, 1852, the Cold Run Reservoir gave way. Only the alertness of seventy-three-year-old Judge John J. Bowman and his Paul Revere-like ride into the valley below saved the people living there. The banks of the canal were

1. *Towpaths*, Sesquicentennial Issue No. 3 (1975).

washed out in several places; dams and bridges were swept away. Mills and other industry were wiped out, and local sentiment turned against the canal.

Nearly dry, the summit level was never rebuilt. The canal company failed in 1853, and much canal land was sold off to satisfy overdue judgments. Although the eastern and western ends were used for several years, the dry summer of 1854 stranded a number of boats near the summit; they never moved again.

The story of the Sandy & Beaver is incomplete without a final note about Edward Hall Gill. For all his sternness, Gill is said to have had a tender side. People tell how he cherished his only daughter, Gretchen, especially after his wife's death and burial at sea during their voyage from Europe while emigrating to the United States. According to the legend, when Gretchen died of malaria during construction of a lock above Sprucevale, Gill had a crypt built into the wall of the lock; his daughter was entombed there for several years, until after Gill left Ohio. When he decided to return to Ireland, he had her body removed from the crypt and took it with him for burial in the town of her birthplace. But Gill's ship encountered a storm on the return voyage and all aboard were lost.

As author Max Gard puts it, "E. H. Gill and little Gretchen joined their wife and mother in the waters of the Atlantic. Little Gretchen's monument is the lock near Sprucevale where she was entombed. Everyone who knows of this lock knows it as 'Gretchen's Lock.' The pilgrims who wish to make the trip . . . will be rewarded with a view of the open crypt where Gretchen slept for a while."[2]

Alas, subsequent research by historian T. Gibson Hobbs, Jr., of Lynchburg, Virginia, reveals that Gill died in Virginia after serving as superintendent of several railroads in that state.[3] He had, in fact, arrived in the United States as an unmarried teenager and later married a woman from Pennsylvania, but the Gills had no children. Doing crass, objective violence to such colorful romanticism almost calls for an apology. That Gretchen never lived somehow seems sadder than the legend of her death.

•

There were other feeder canals that turned out to be resounding successes. About halfway between the Ohio & Erie and the Miami & Erie canals, near the shore of Lake Erie, the shortest of Ohio's canals was constructed. And for a time it was also the busiest. Only three miles long, the Milan Canal transformed the village of Milan into a major lake port.

Not technically a feeder, the canal connected Milan's basin with Fries Landing—a point on the Huron River from which the river was navigable out into Lake Erie. The basin was formed by damming the Huron River, which also provided the water necessary to operate the canal's two locks, as it crossed the swampy bottomland to the head of navigation at Fries Landing. The Milan Canal not only carried goods down to the river by canal boat. It actually permitted lake schooners to sail right into Milan Basin to load their cargoes.

By 1847 (the year Thomas Edison was born in the village), Milan had become second only to the Russian port of Odessa in the shipment of wheat. In May of that year, some six hundred wagons, waiting to unload their grain, made up a line extending from Milan's town square to the Blue Fly Inn—a distance of six miles. And 35,000 bushels of wheat were handled in a single day. Although wounded by the railroads, the Milan Canal survived until the flood of 1868.

•

Lore of the kind recounted on the feeders described above abounds in all of the byways of Ohio's canal system. Some of it has been preserved by a few dedicated researchers, several of whom have recorded oral histories before it was too late. But this is a photo album.

2. R. Max Gard and William H. Vodrey, Jr., *The Sandy and Beaver Canal* (East Liverpool Historical Society, 1952).

3. T. Gibson Hobbs, private communication.

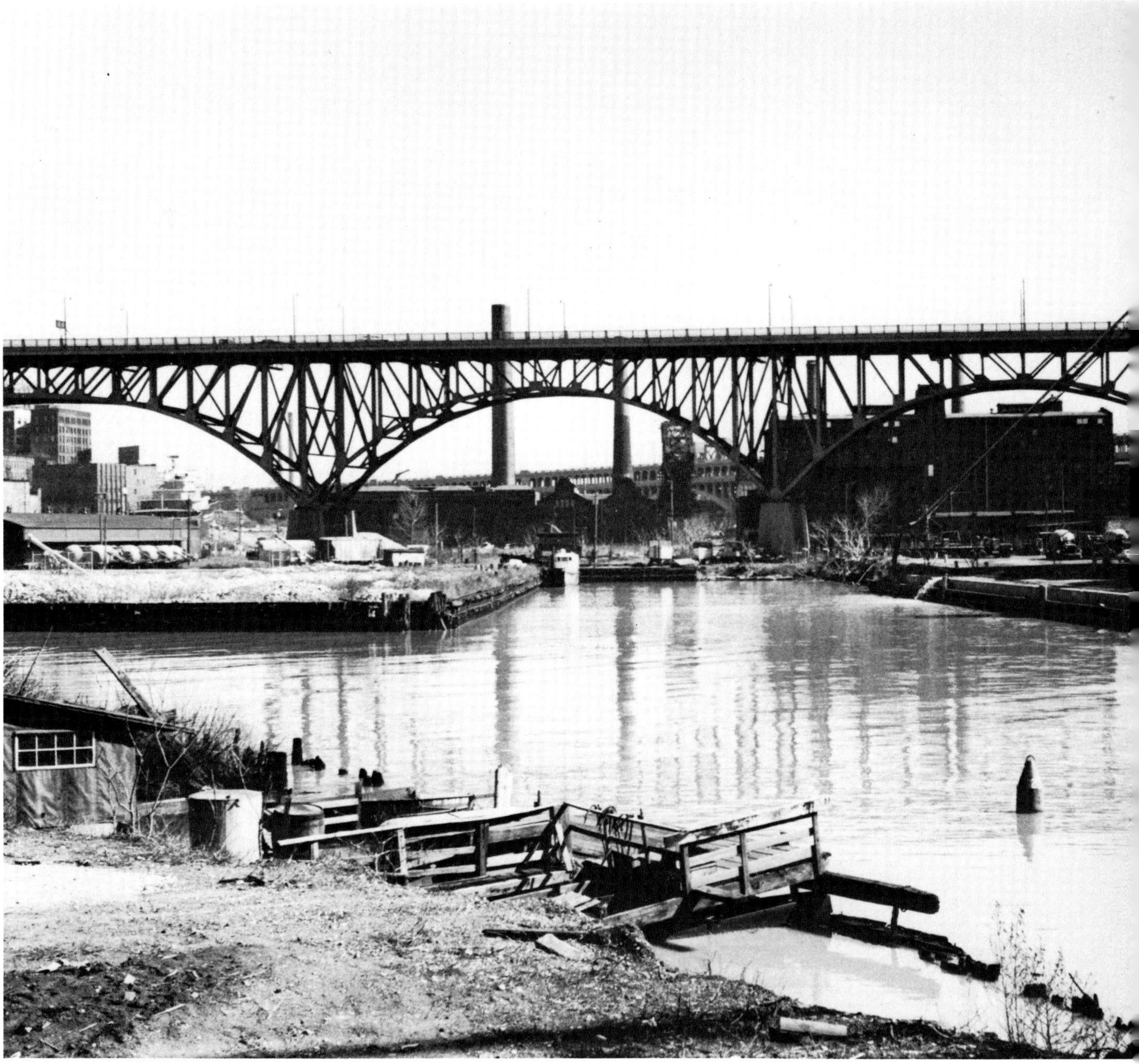

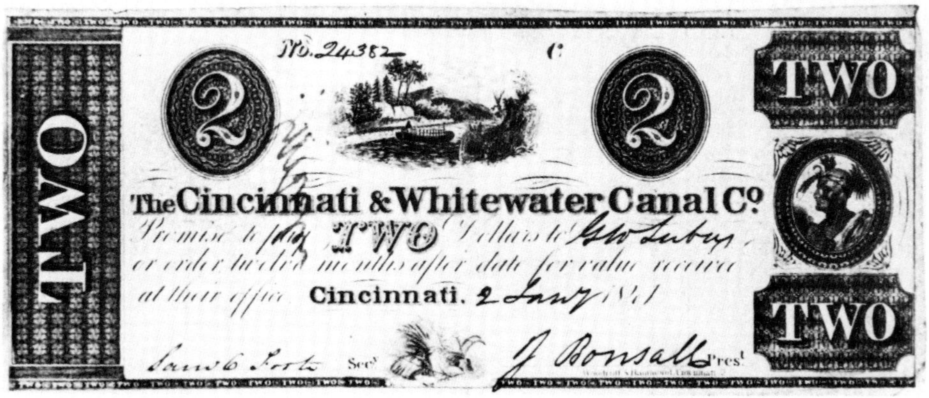

Organized in 1837, just in time for the economic panic of the same year, the Cincinnati & Whitewater Canal Company resorted to the use of scrip to pay its workers and contractors; at left is a two-dollar certificate.

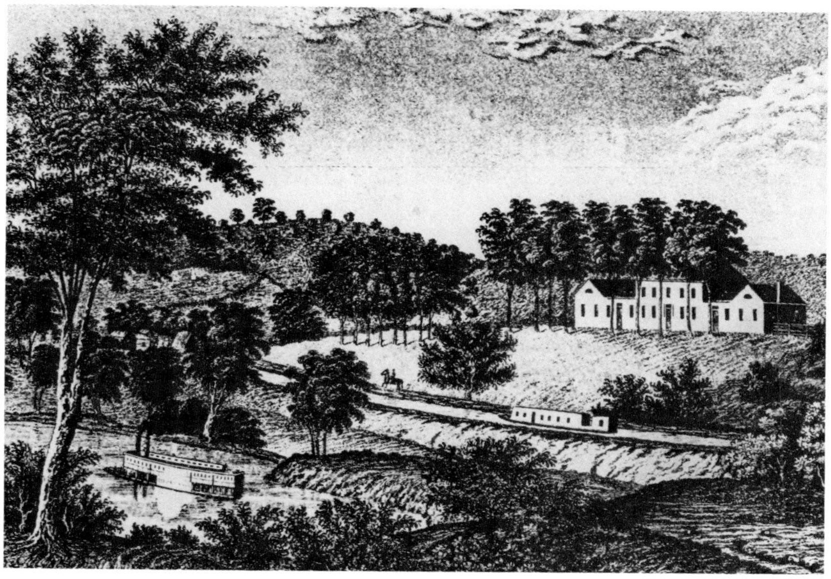

The engraving above shows the Cincinnati & Whitewater Canal paralleling the Ohio River (foreground, with steamboat) at North Bend, with the residence of William Henry Harrison in the background; c. 1842 (*Ladies Repository*/CSO). At North Bend, this twenty-five-mile canal turned northward through a 990-foot, brick-lined tunnel (whose masonry arch extended an additional 800 feet through open cuts to keep them from filling with debris) to pierce the ridge that separates the Great Miami and Ohio River valleys. Above the community of Cleves the canal crossed the Great Miami River on an aqueduct, ultimately joining Indiana's Whitewater Canal below Harrison in a slackwater pool formed by a dam across the Whitewater River.

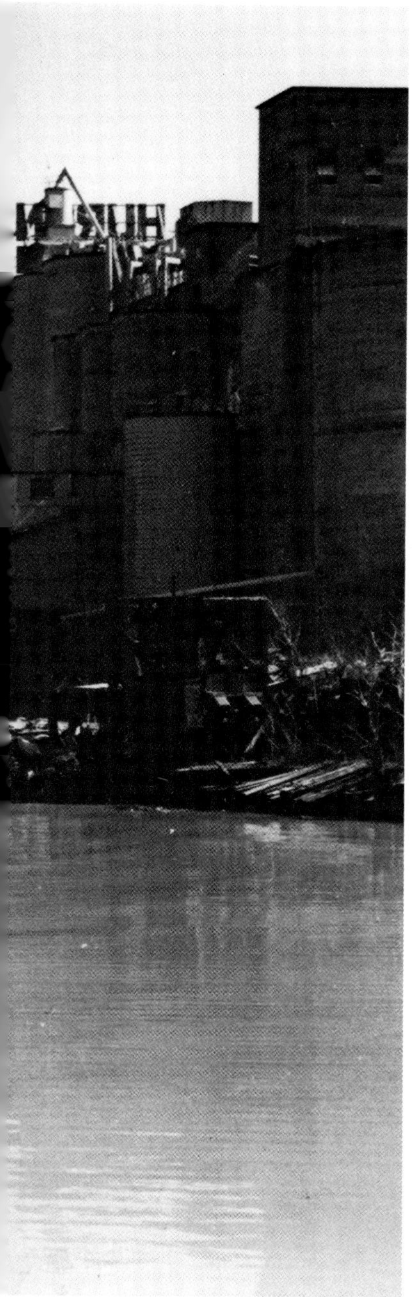

A surviving remnant of the Ohio City Canal is in Cleveland's flats where the canal entered the Cuyahoga River opposite the outlet of the Ohio & Erie. Soon after its incorporation in 1836, the village of Ohio City, on the west bank of the Cuyahoga River, authorized the digging of this branch canal, thus attracting traffic to warehouses and docks on its side of the river. (Photo by F. W. Trevorrow, 1966.)

The Columbus Feeder Canal in the city, looking northeast. The line boat *Wave* is moored in the foreground, with a freight boat behind it. Water flowed to the right (south); c. 1880s. *Inset*, An earlier scene a few hundred yards to the north on the Columbus Feeder. Although the young muleskinner is taking a rest, the mules have the towline taut and are ready to go as soon as the picture-taking is over.

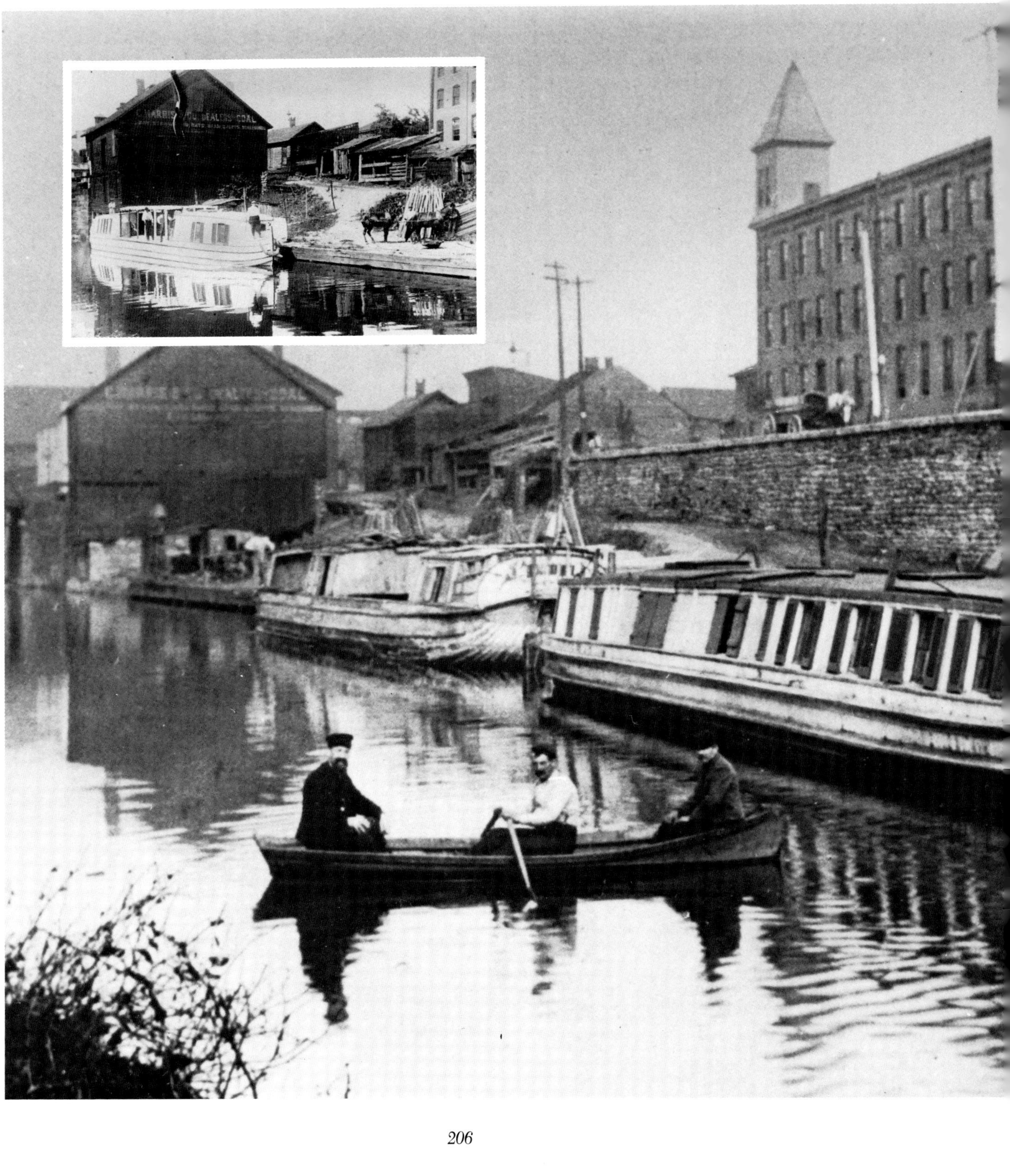

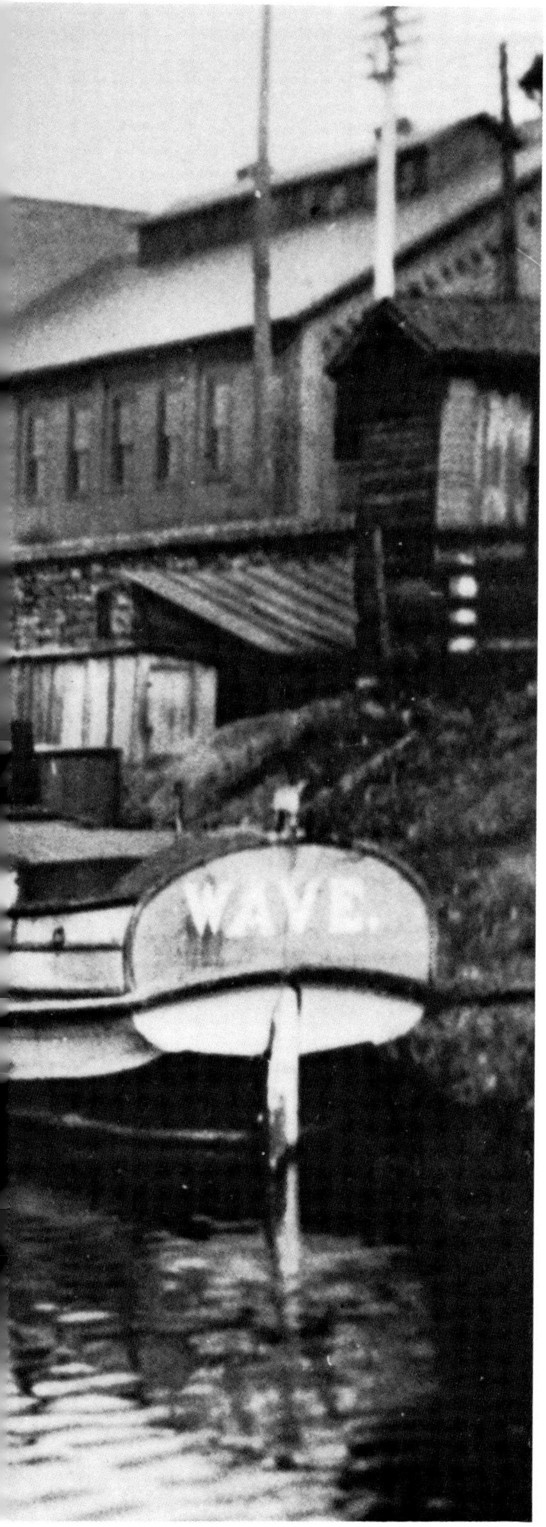

Below is an 1832 engraving of the Columbus Feeder, or Lateral Canal, a twelve-mile link that provided the capital city of Columbus with access to the main trunk of the Ohio & Erie Canal. The fourteen-foot difference in elevation between Columbus and the junction at Lockbourne was overcome by two locks, one shown here. Nathaniel Medbury was chief engineer on this project (at an annual salary of $720, plus a subsistence of $156 and another $52 for his horse). (New York Public Library)

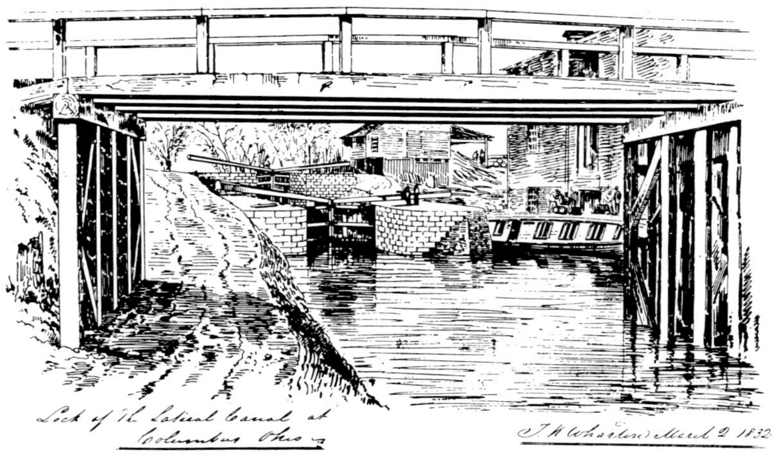

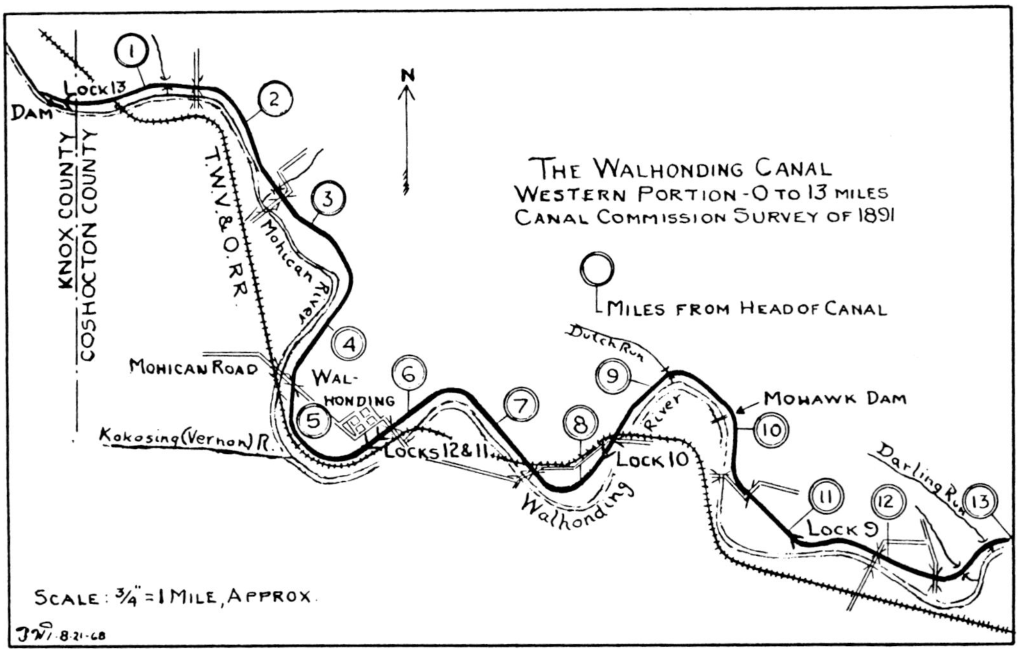

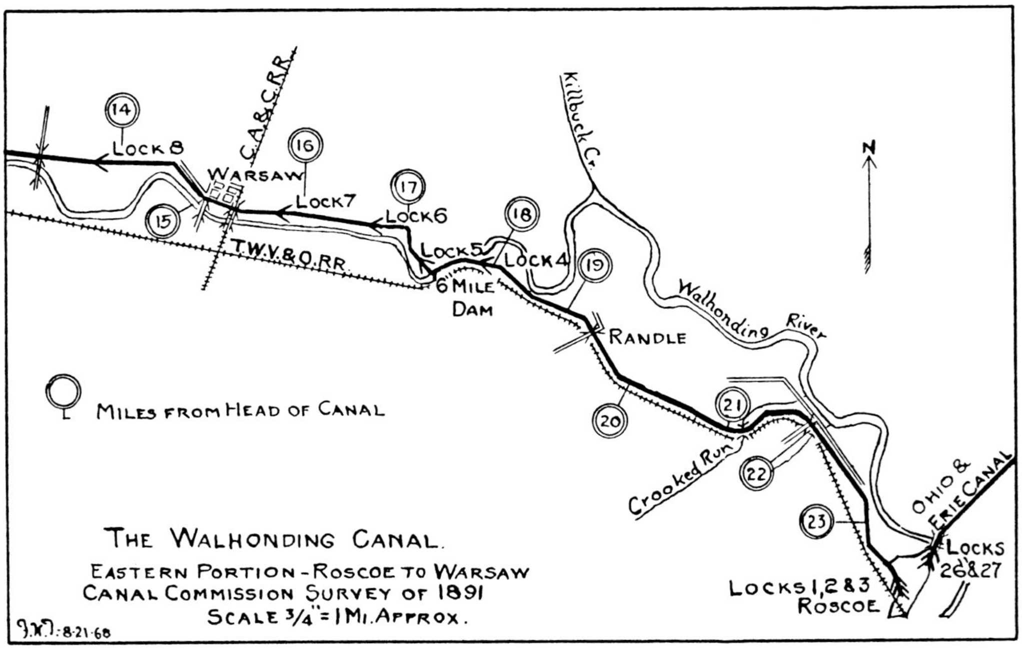

Above, top to bottom: These maps of the eastern and western portions of the Walhonding Canal show the three locks at Roscoe and its junction with the Ohio & Erie Canal in Roscoe's Lower Basin. The Ohio & Erie is lifted from the basin through two locks to the level of an aqueduct spanning the Walhonding River. (Drawing by F. W. Trevorrow, 1968, from *Ohio's Canals* by F. W. Trevorrow, 1973.)

The Triple Locks at Roscoe, which brought the Walhonding Canal down to the level of the town's Lower Basin; 1986. (Photo by Jack Gieck.) *Inset,* A state canal boat which has expired in the Triple Locks, 1937. The boat's tailgate later became a part of Pearl Nye's nearby Camp Charming. (RVF/CSO)

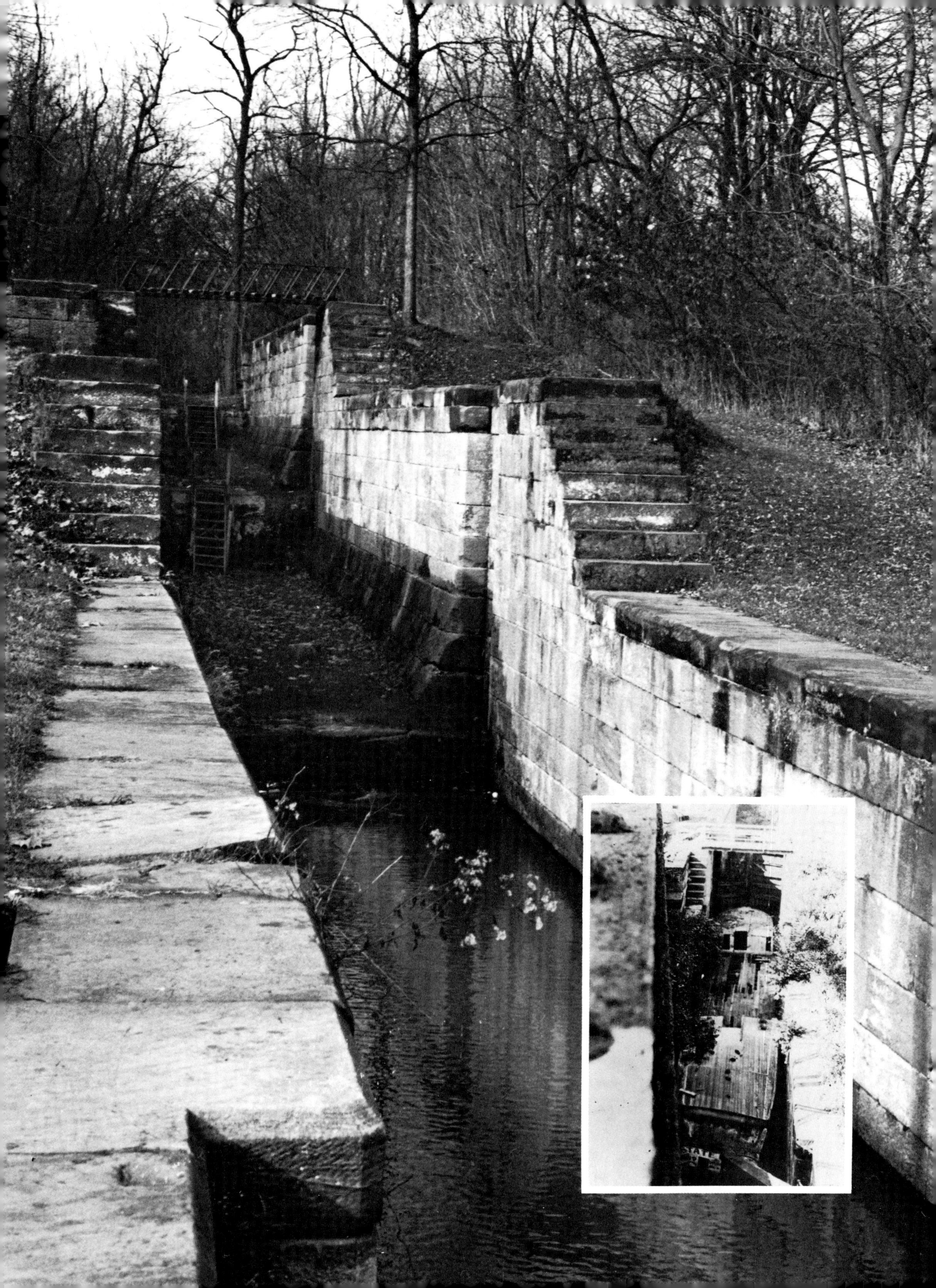

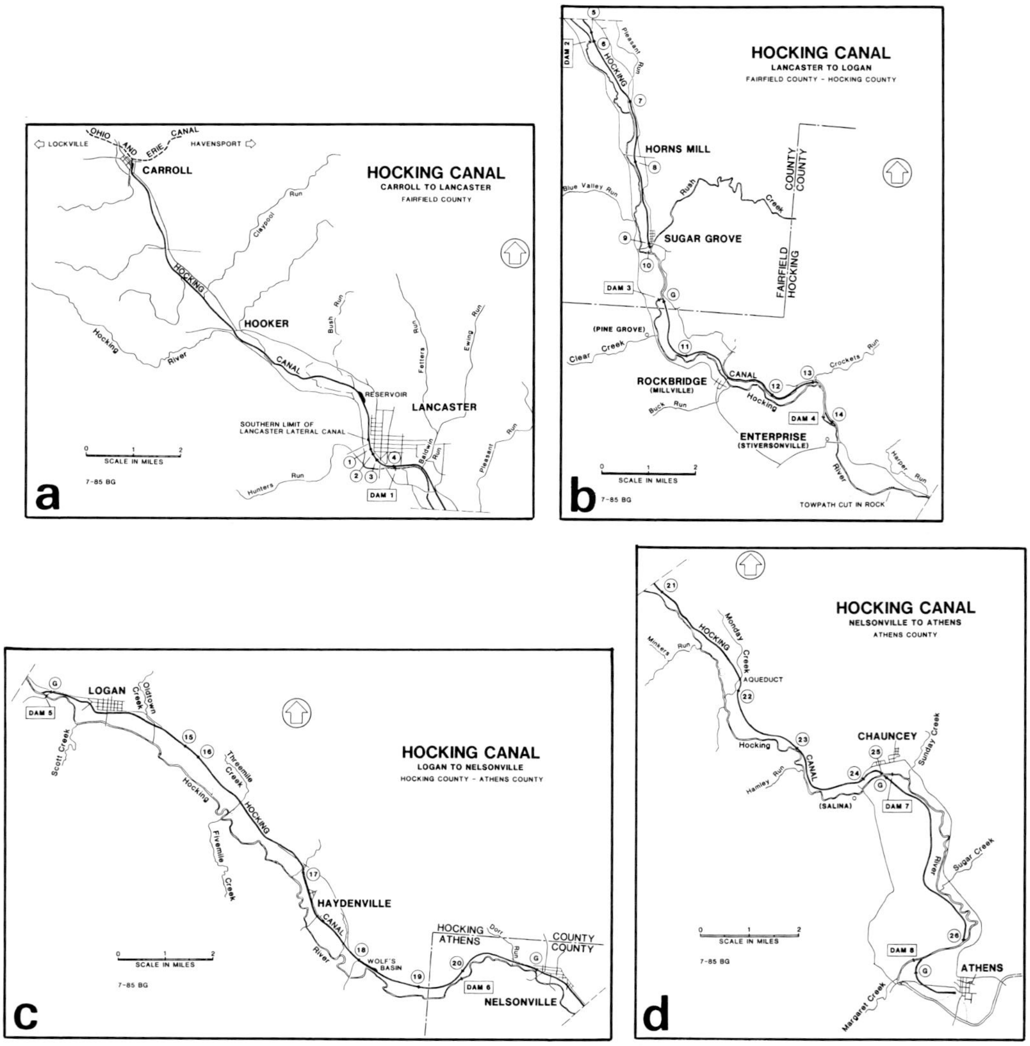

Above, a through d: Beginning at Carroll, where it joined the Ohio & Erie, the northern portion of the Hocking Canal absorbed the earlier Lancaster Lateral as it ran southeast through Sugar Grove to Nelsonville. (Maps by Barnett L. Golding.)

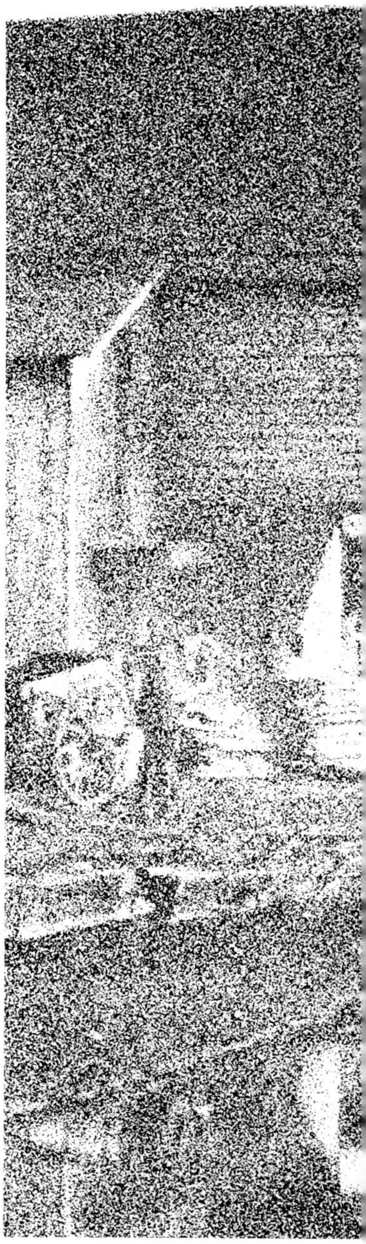

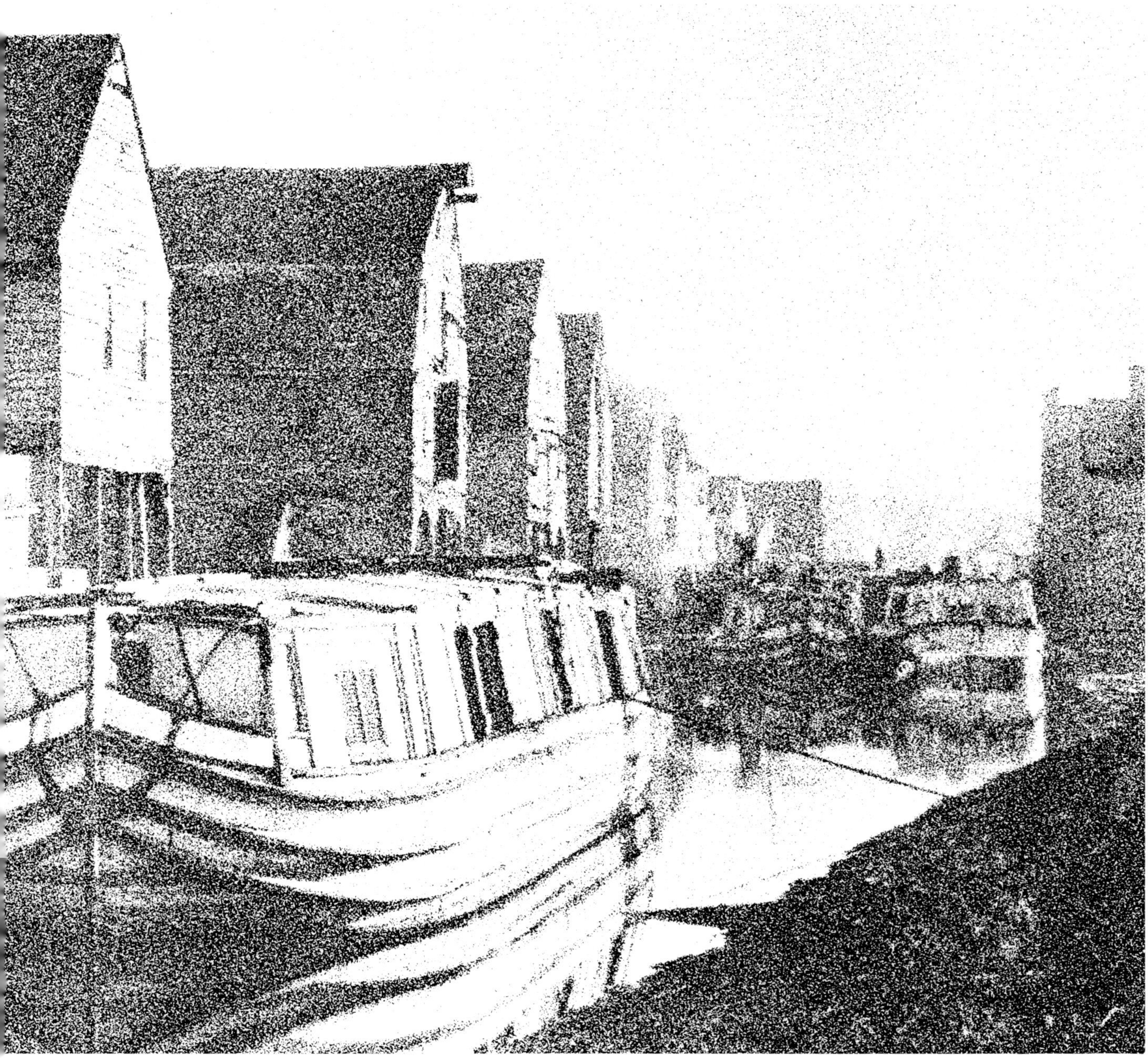

Canal boats at Lancaster, on the Hocking Canal. (OHS)

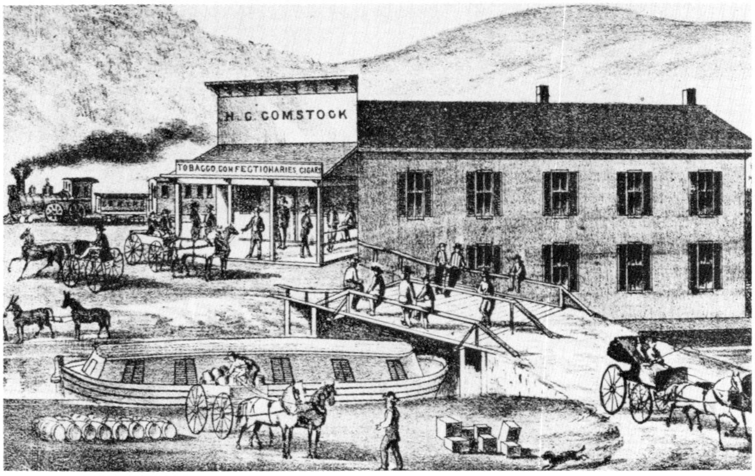

Goods are unloaded from a Hocking canal boat at the H. C. Comstock General Store. (OHS)

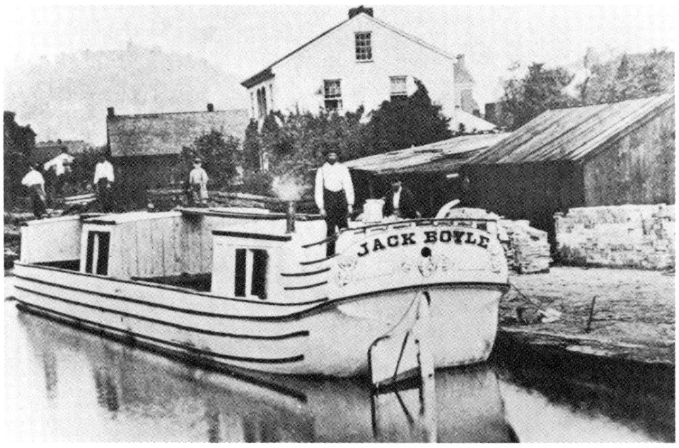

The *Jack Boyle* at Sugar Grove on the Hocking Canal; 1893. (Pockrandt Collection, SCHS)

An engraving of Salina Salt Works, Athens County, along the Hocking Canal. Salt was one of the mineral commodities of the Hocking River that justified the canal. (1875 *Athens County Atlas*, courtesy WRHS.)

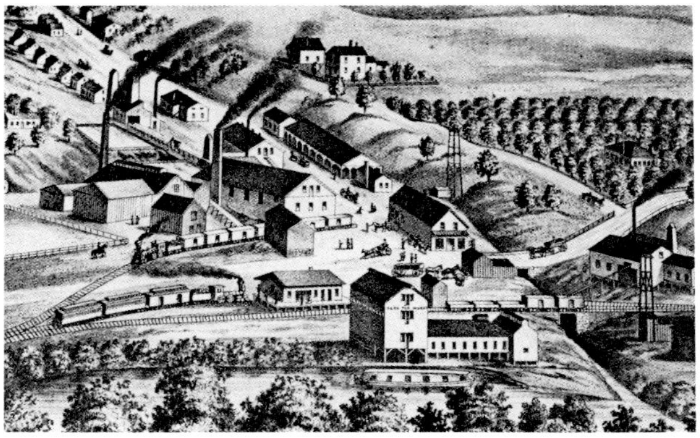

A panoramic engraving of "Farm Houses and Coal Land" in Athens County along the Hocking Canal (foreground). The W. B. Brooks & Son Coal Works operated an underground mine here, together with a narrow gauge railway that hauled coal from the mine shaft to a loading ramp. (1875 *Athens County Atlas*, courtesy WRHS.)

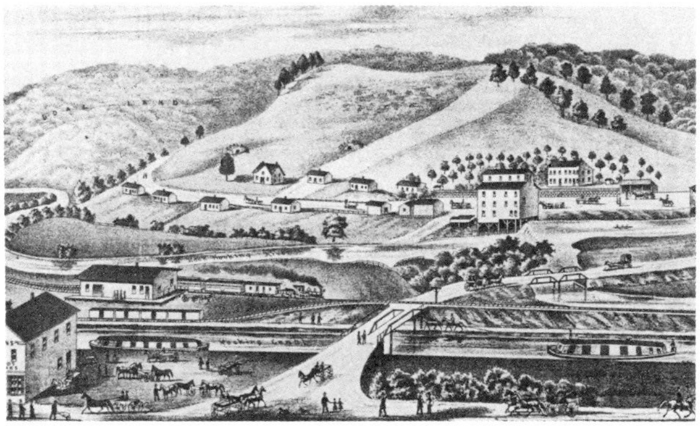

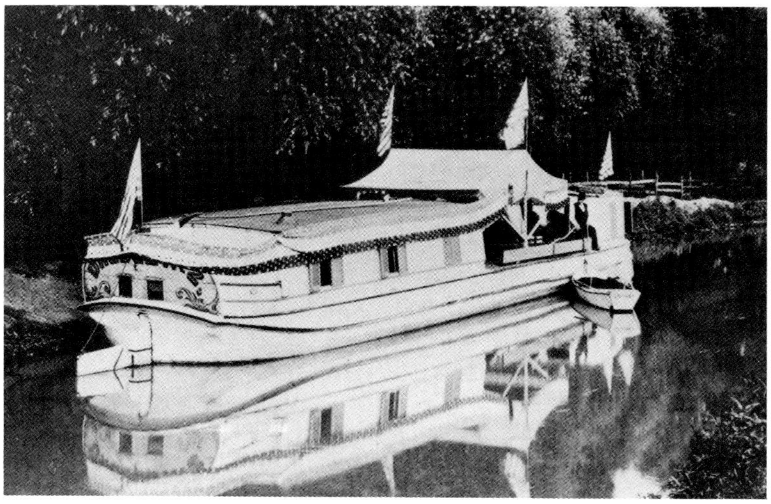

A very patriotic state boat on the Hocking Canal at Logan. (Baus Collection, RVF)

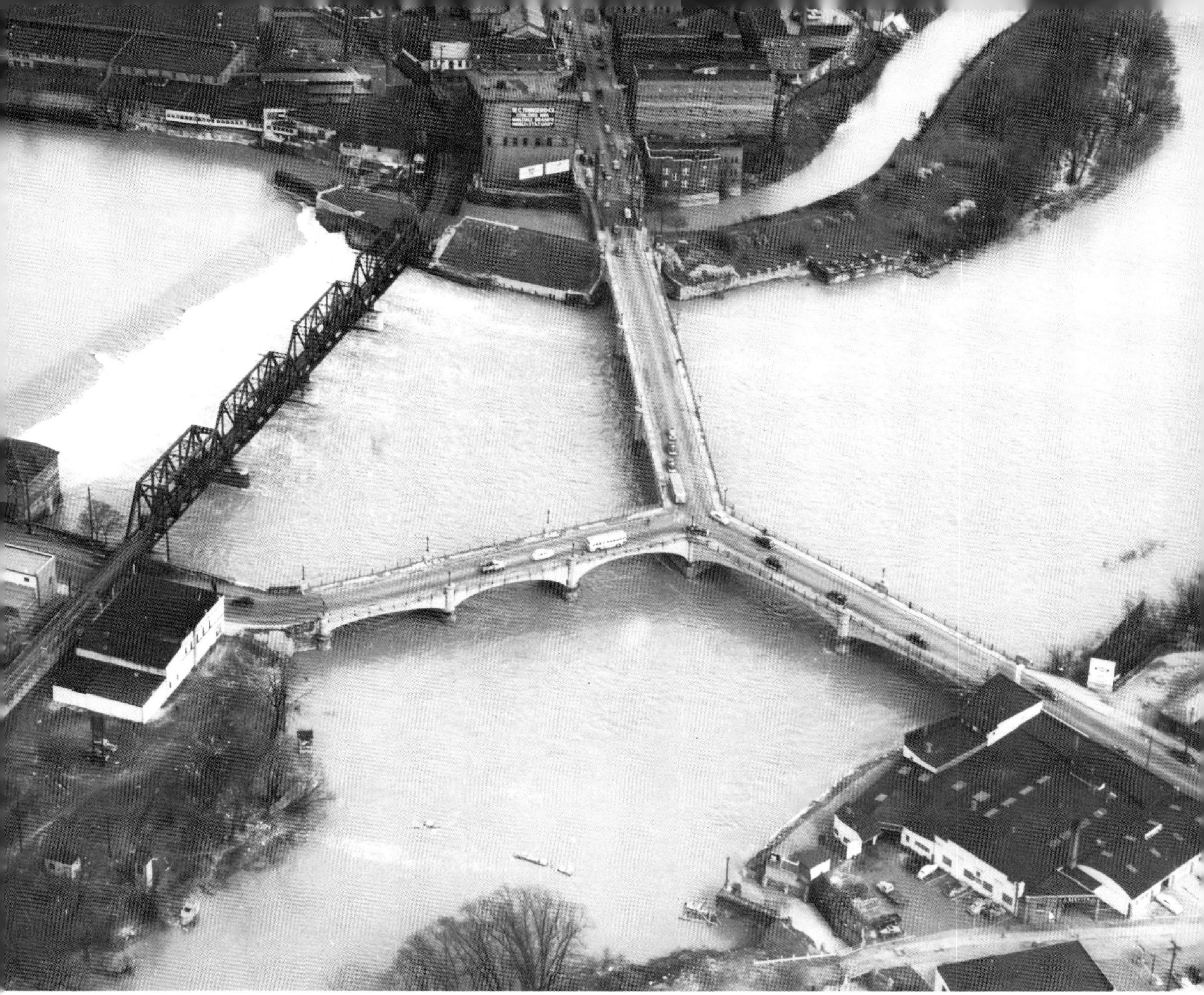

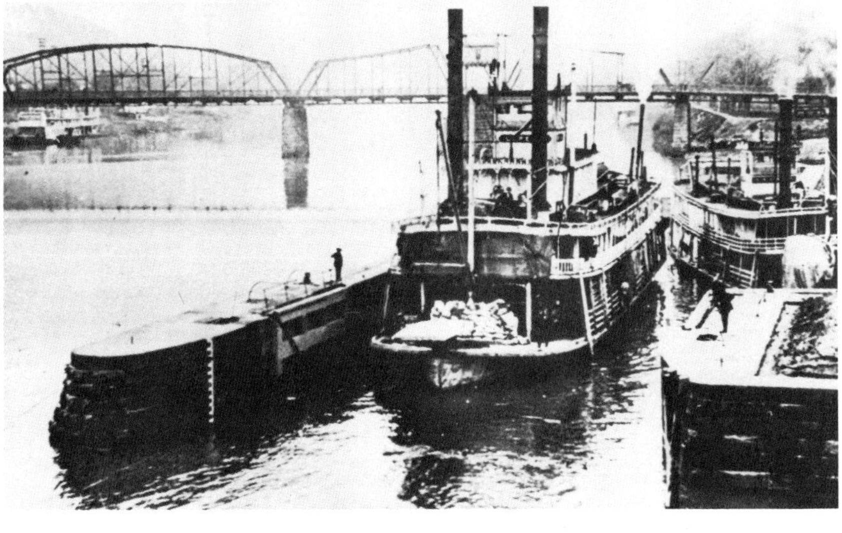

An aerial view of Zanesville (above) shows canalization around the dam on the far side of the city's famous Y-bridge, since dismantled. Licking River enters here from lower left; c. 1949. (*Zanesville Times Recorder*)

The steamboat *Lorena* (left) at Malta, on Muskingum River Improvement; c. 1870. (CSO)

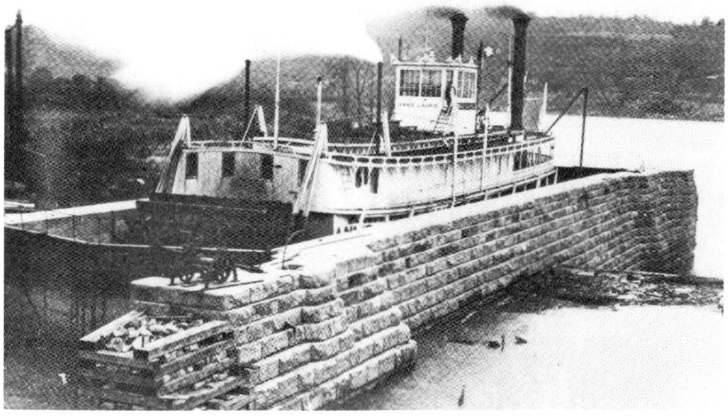

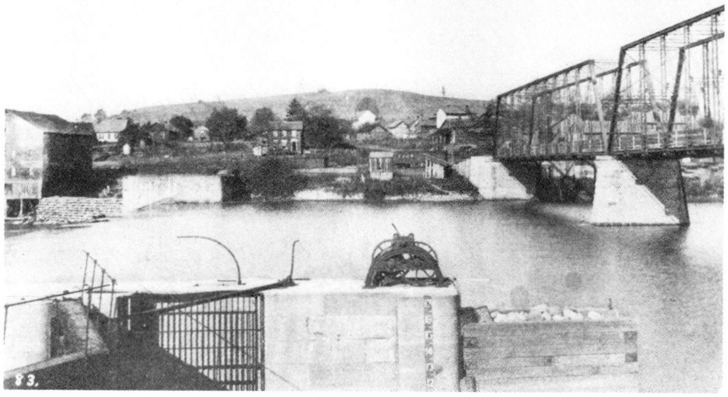

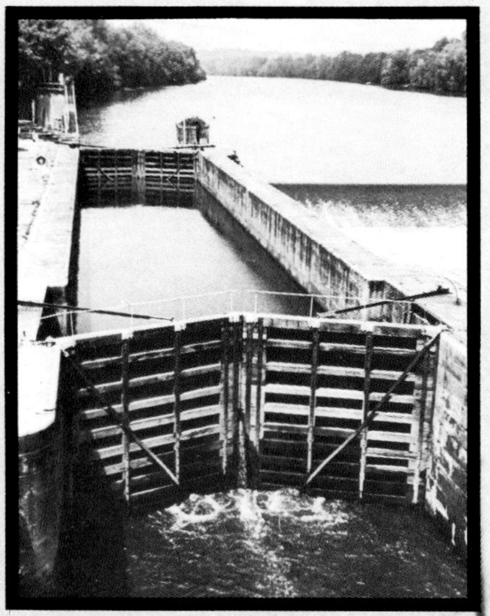

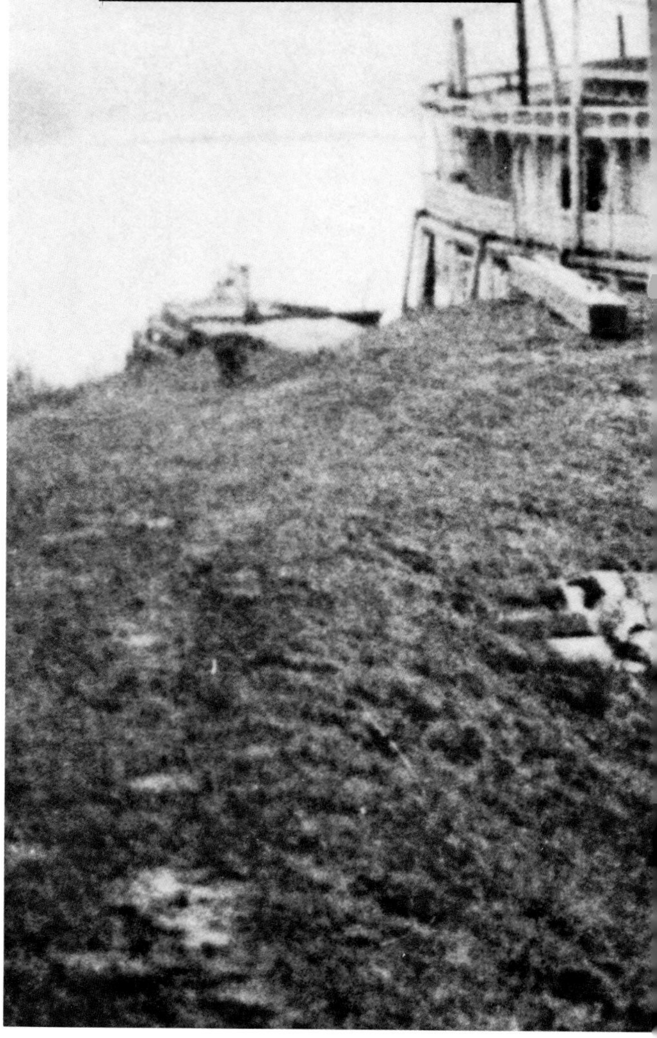

The steamboat *Annie Laurie* (top) in lock at Stockport on the Muskingum River Improvement. Lock wall has tapered stone construction, thicker at the base, which provides an integral buttress in the structure of the lock wall unsupported by earth. (CSO)

Looking across Muskingum River, refurbished Stockport Lock (bottom) with concrete facing is in the foreground. The mill at the left of the town of Stockport is the only surviving mill on the Muskingum River. (CSO)

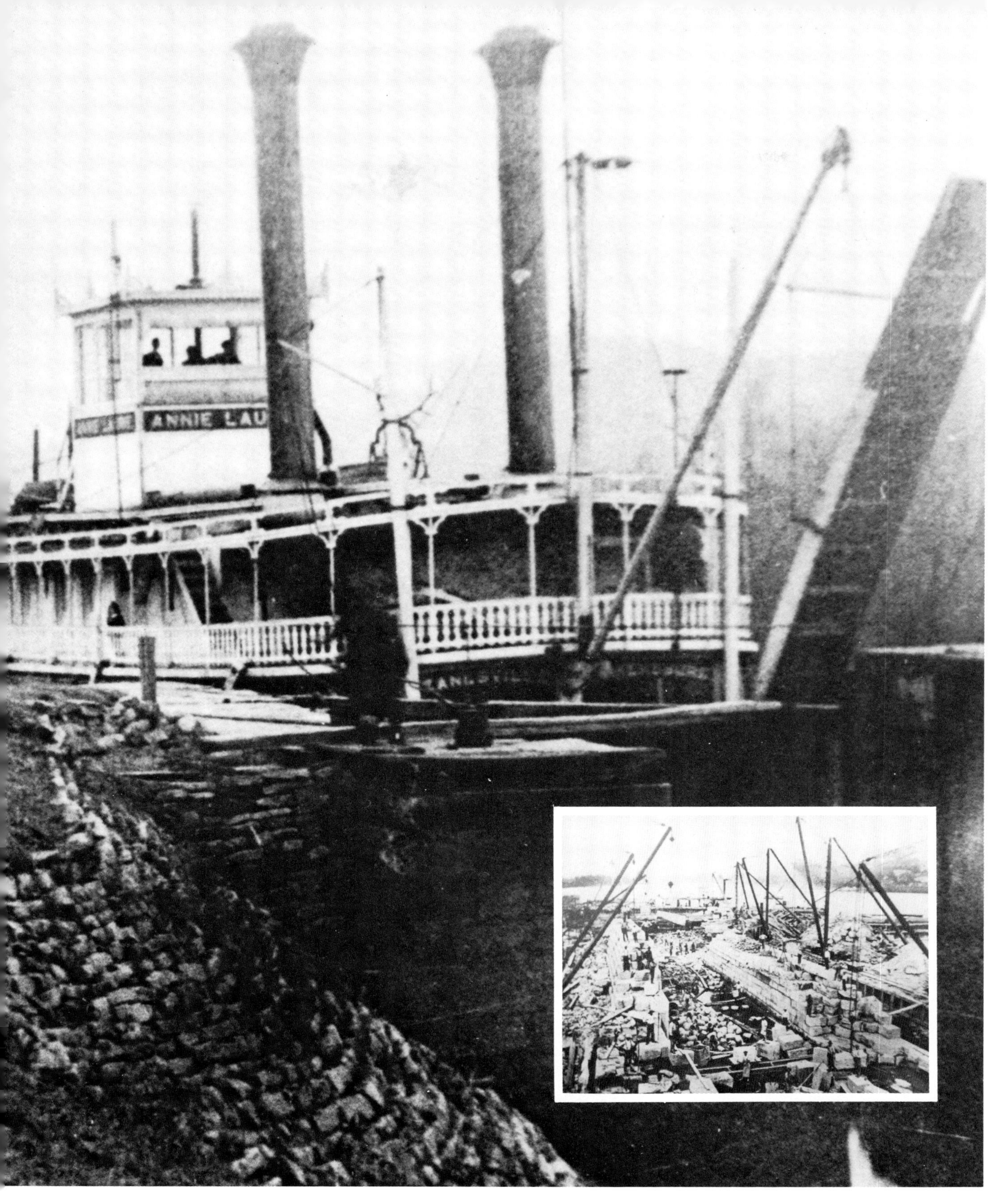

The steamboat *Annie Laurie* negotiates Philo Lock (Lock 9) on Muskingum Improvement. *Inset above*, Lock 9 and dam many years later. Although lock dimensions were standardized in 1838, most did not measure up to the specified 175 by 36 feet. *Right*, The rebuilding of the stone Lock 9. (CSO)

Right, A Summit County Surveyor's Office map shows the route of the Pennsylvania & Ohio Canal from Akron's Lower Basin, where the canal joined the Ohio & Erie, north through Cuyahoga Falls, then turned east toward Kent (formerly Franklin Mills). (SCHS)

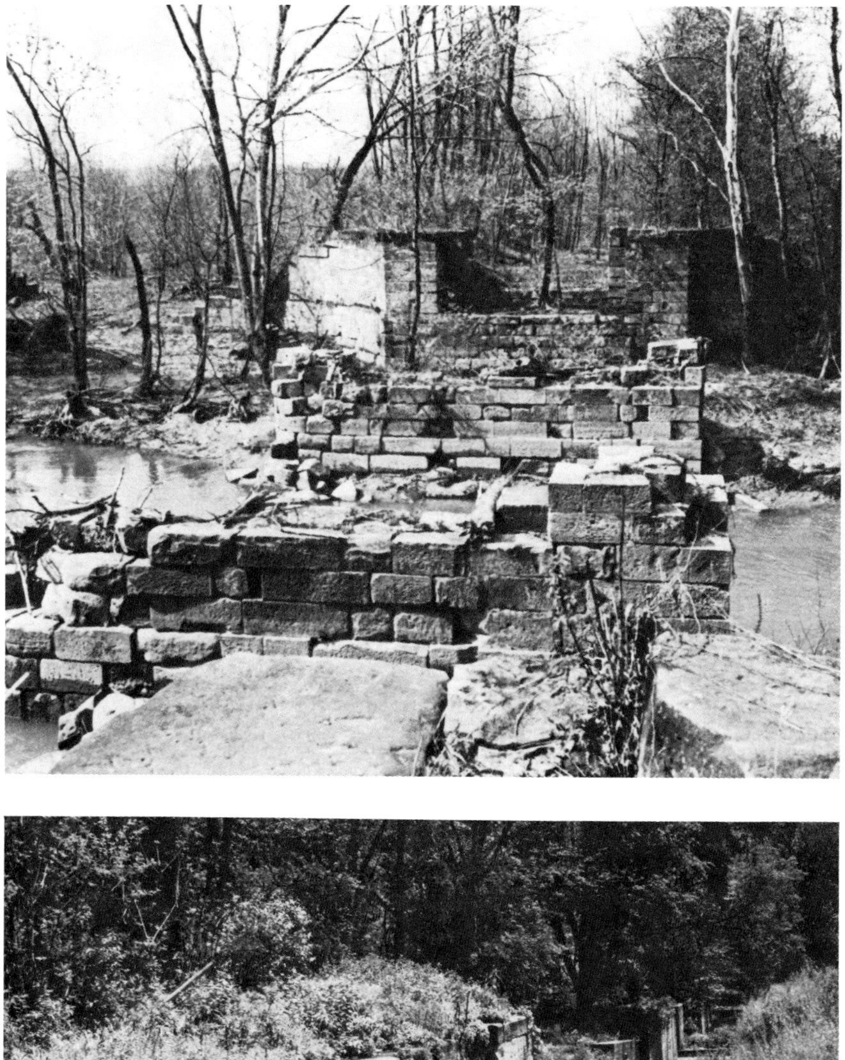

Connecting the Muskingum River Improvement with the Ohio & Erie Canal at Dresden was the Dresden Sidecut, two miles long. These are ruins of the Dresden Sidecut's aqueduct over the Wakatomica Creek. (CSO)

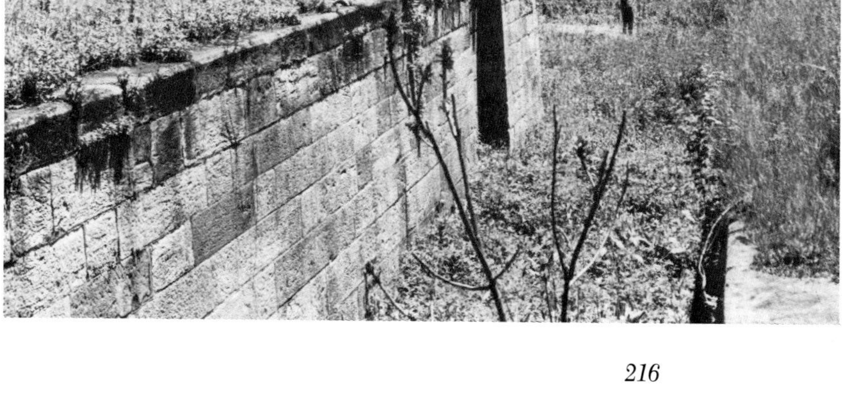

The descent of the Dresden Sidecut to the Muskingum River: Locks 31, 32, and 33, the river lock. (Photo by F. W. Trevorrow, 1966.)

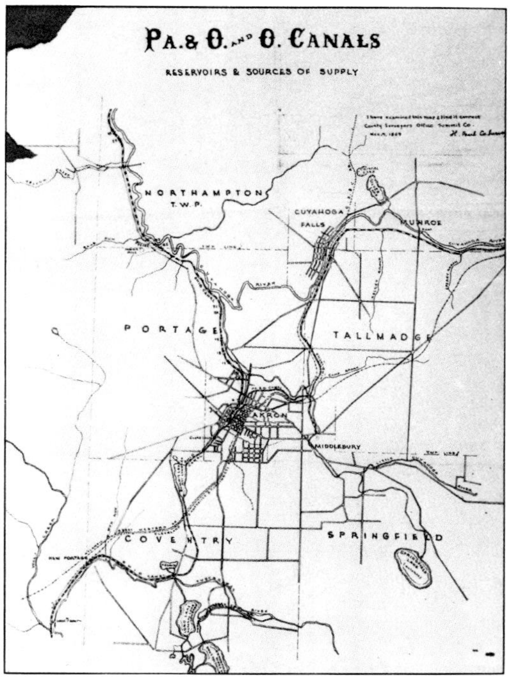

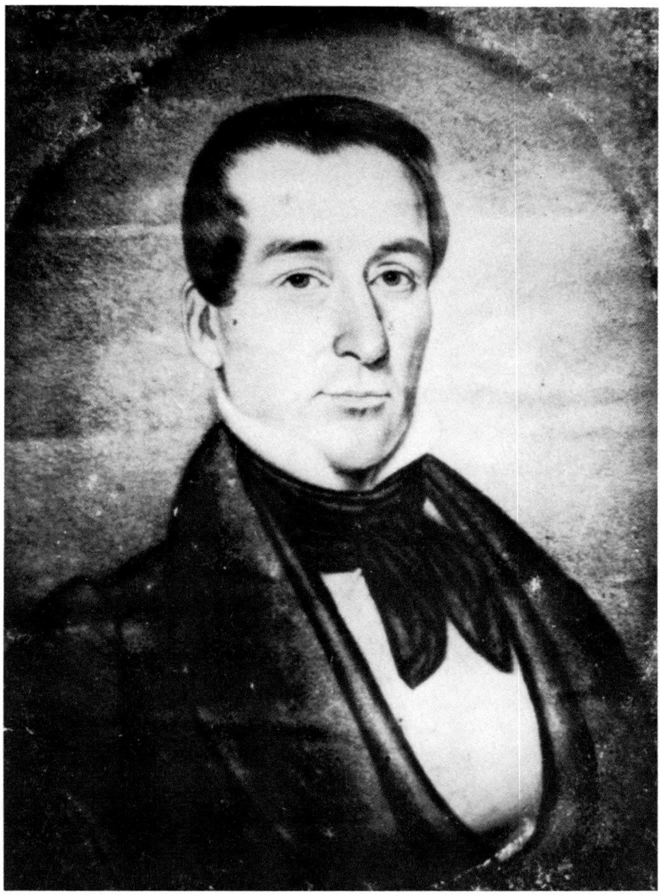

Right, Colonel Sebried Dodge of the state corps of engineers, a lawyer, surveyor, and canal commissioner who was also principal engineer on the Pennsylvania & Ohio Canal. From an 1845 (possibly retouched) daguerreotype. (J. Roy Dodge/F. W. Trevorrow)

Below, Just north of its junction with the Ohio & Erie in Akron's Lower Basin, the Pennsylvania & Ohio Canal preempted most of the width of the town's Main Street in 1856; several buggies and wagons may be seen on the remaining narrow lane on the foreground side of the canal, which also served as the canal towpath. (SCHS)

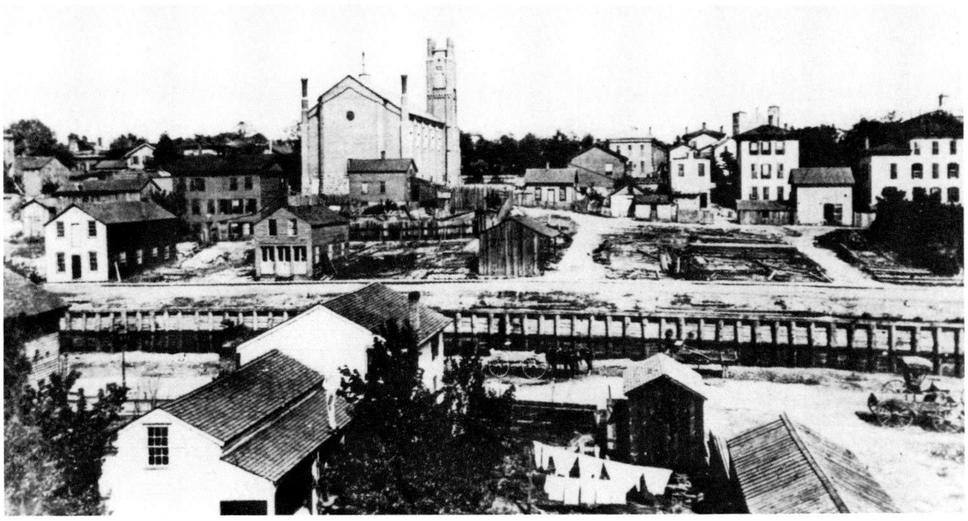

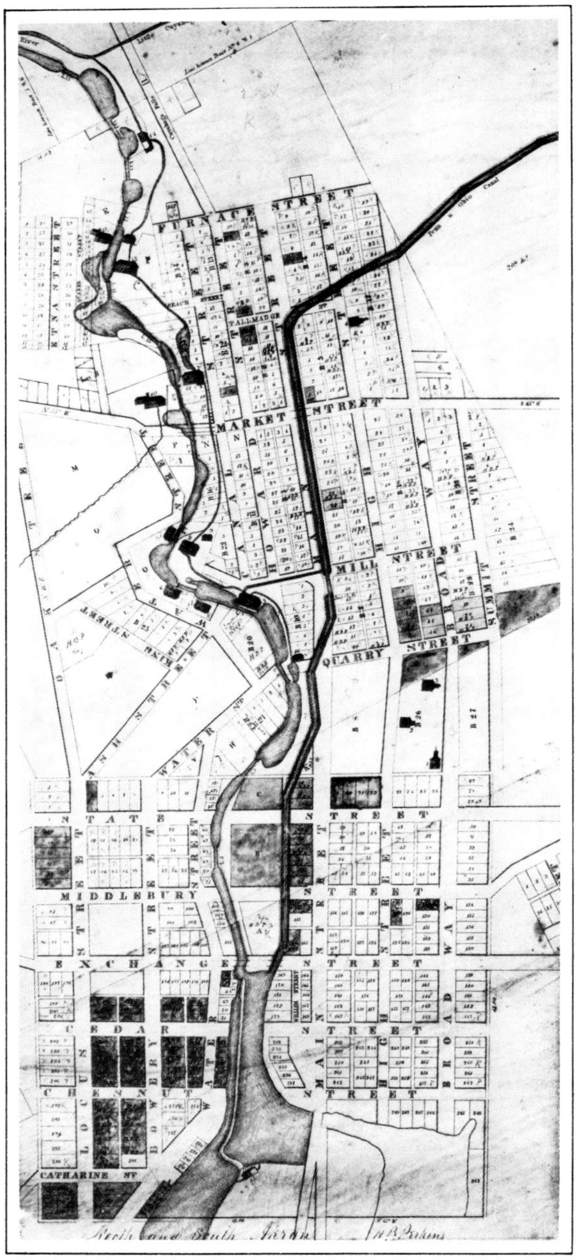

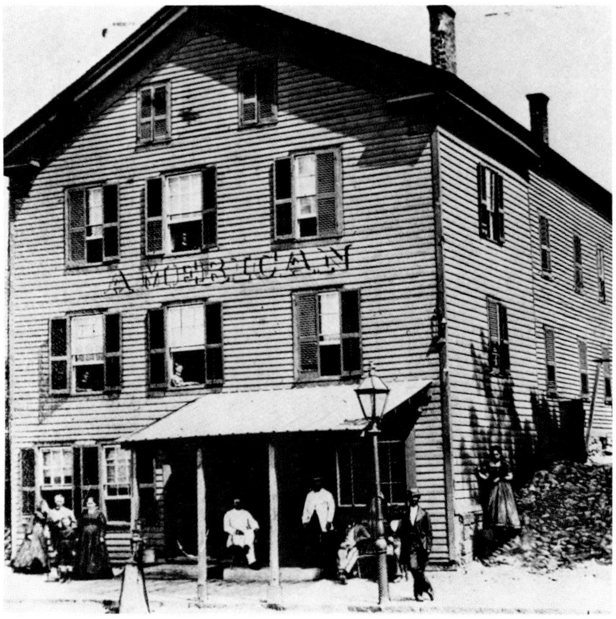

The American House in Akron, the hotel at which a gala reception was held when governors and other dignitaries from both Pennsylvania and Ohio arrived on the flotilla of six boats celebrating the opening of the Pennsylvania & Ohio Canal on August 6, 1840. After ample refreshments, the party reboarded the boats and, accompanied by a martial band and the uniformed Summit Guards, were cheered down the Main Street waterway by crowds lining both banks. Upon arrival in the Lower Basin, the brightly decorated boats were greeted by booming cannon and the roars of an even larger crowd. The flotilla's cargo of sherry and Madeira wines added significantly to the occasion.

An 1837 planning map at left shows how the Pennsylvania & Ohio Canal (the dark, wide waterway shown) is to come into Akron from the northeast, paralleling the Ohio & Erie with its many locks and basins. The two canals connect in the Lower Basin (bottom of map). The Pennsylvania & Ohio was completed in August 1840. (SCHS)

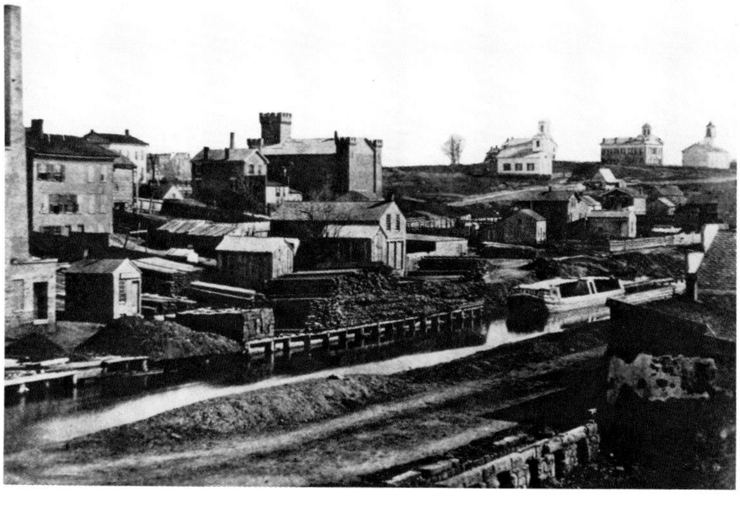

The Pennsylvania & Ohio Canal at the north end of Akron's Main Street in 1858. The ruins in lower right are on the present site of the Federal Building on the southwest corner of Main and West Market Streets; it is likely that the picture was taken from an upper story of the Empire House across West Market.

A dam on the Cuyahoga River at Franklin Mills, now Kent (right). Above this point the Pennsylvania & Ohio Canal was a slackwater for more than a mile, since the river to the north flowed through a rocky gorge. A towpath was blasted in the rock face of the gorge to accommodate the mules. The lock at lower right raised and lowered the boats between the slackwater pool and canal level; the canal channel proceeded along the south bank of the Cuyahoga, paralleling the river. The same scene a century later (below) shows vestiges of the lock next to the dam on the Cuyahoga River, which together with the flooded canal bed (lower left) may still be identified in Kent; 1963. (CSO)

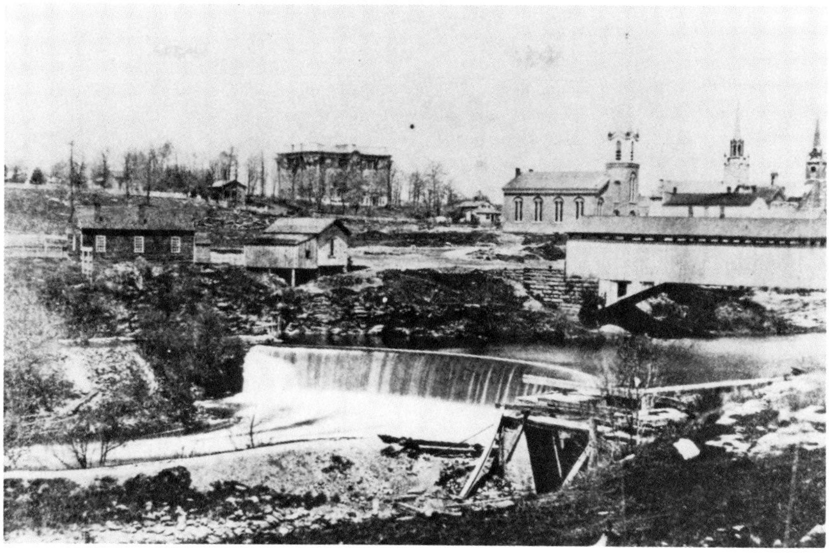

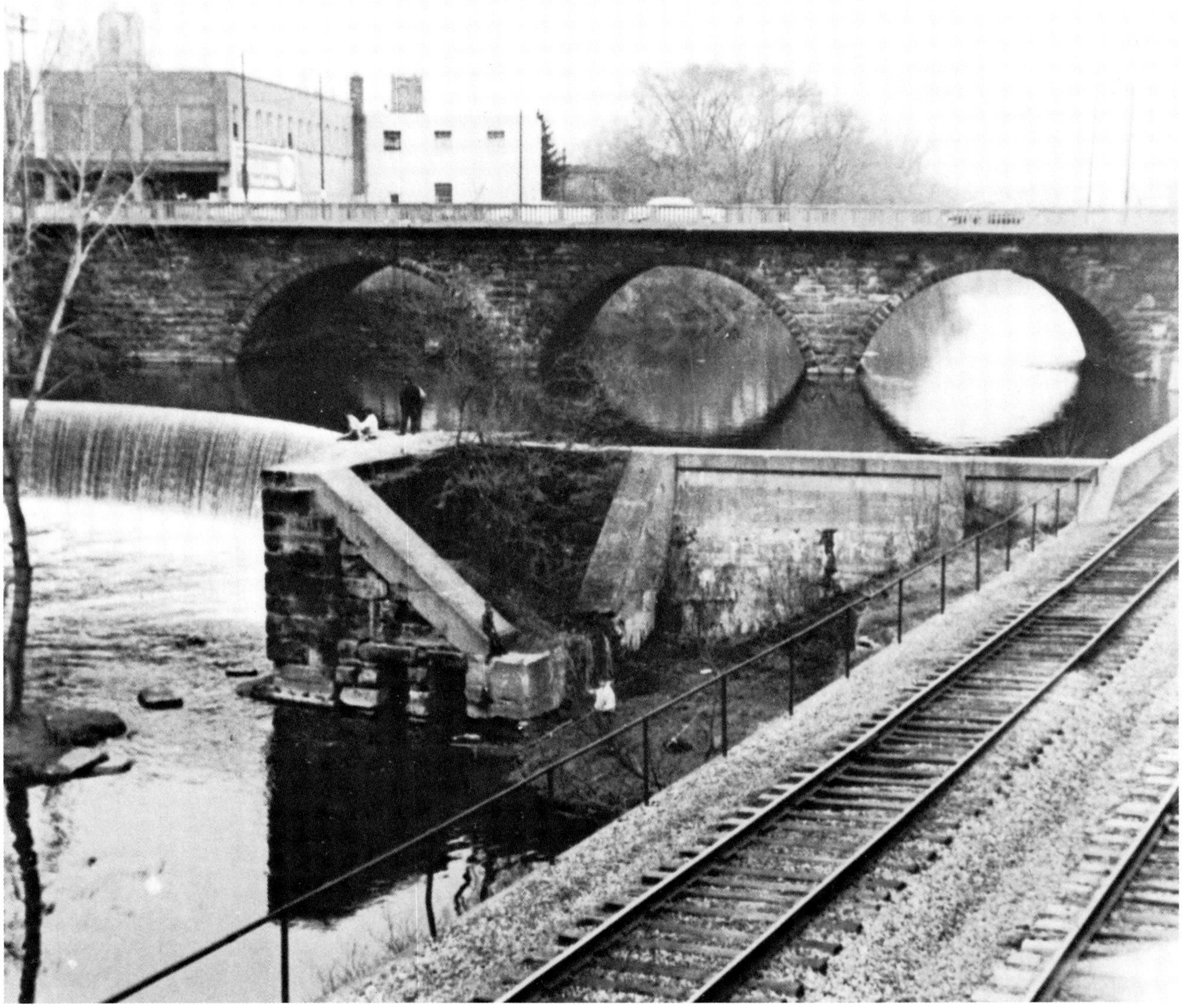

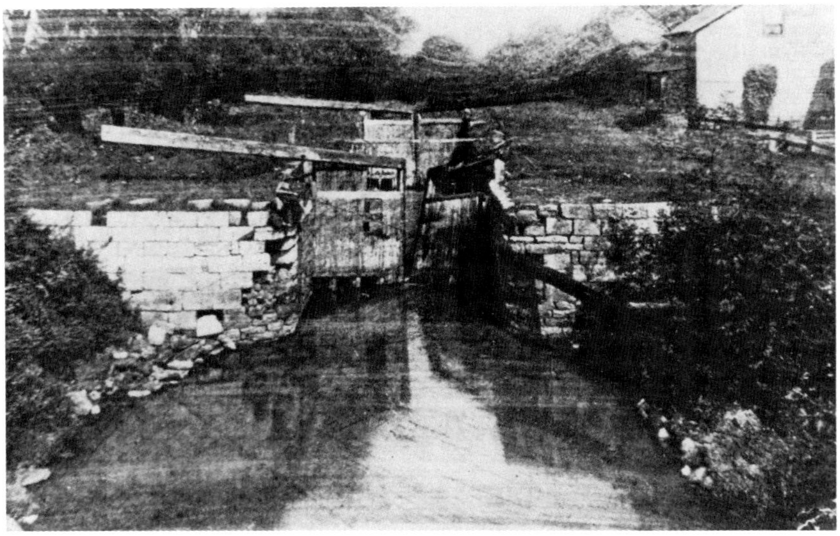

Lock 10 at New Campbellsport, on the Pennsylvania & Ohio Canal; 1870. This picture represents one of the few photographic relics remaining from the very brief era of the Pennsylvania & Ohio. (CSO)

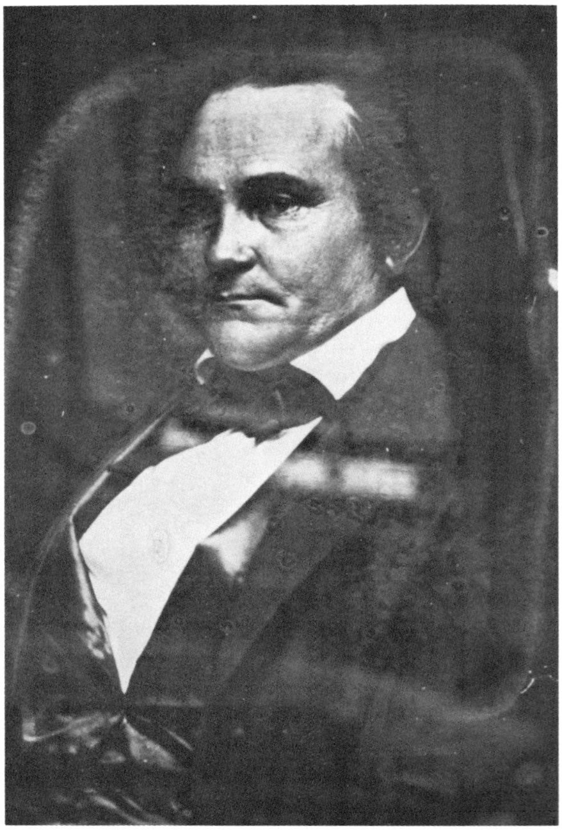

An 1842 (possibly earlier) daguerreotype of Edward H. Gill, an Irish immigrant who became chief engineer of the Sandy & Beaver Canal. Gill's 1834 engineering plan included the construction of the daring Big Tunnel, which shortened the original route by seventeen miles. With a reputation for unswerving excellence in design and workmanship, Gill supervised construction of the canal until 1837, when work was halted because financing evaporated in the economic panic of that year. The print suffered water damage in a fire. (VHS/T. Gibson Hobbs, Jr.)

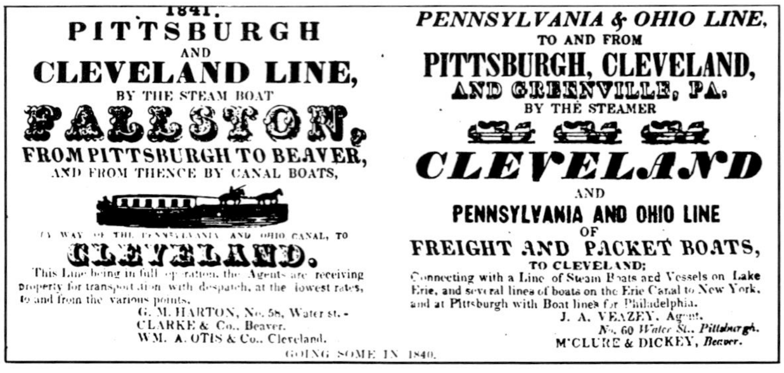

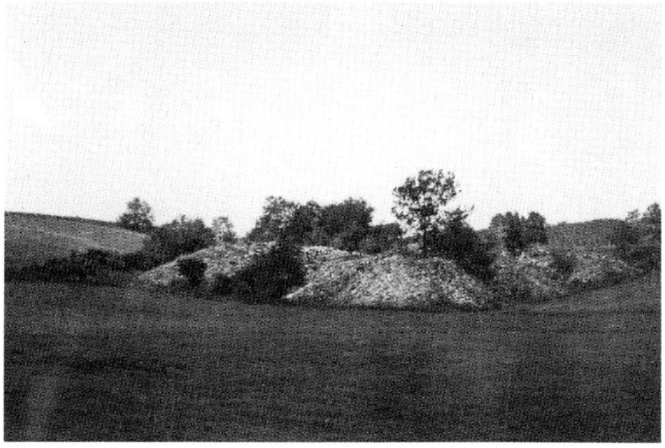

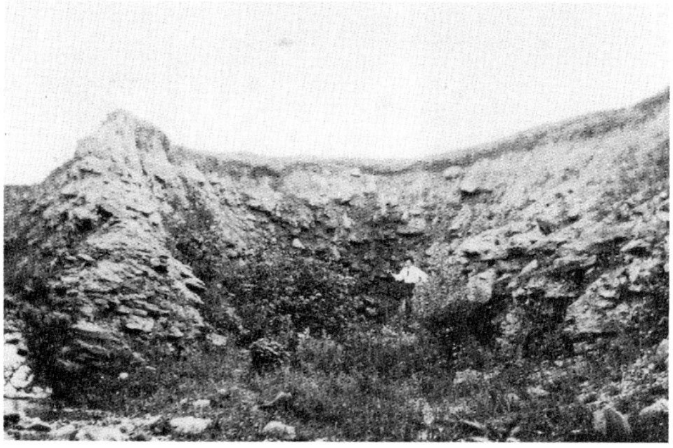

Topside view (above) of one of the two "air shafts" vertically excavated to the depth of the Sandy & Beaver Canal's Big Tunnel so that tunneling could proceed in both directions from each shaft, as well as from both ends simultaneously. Rock spoil, visible in the picture, was brought up the shaft on a conveyor mechanism driven by a steam engine fueled by a vein of coal discovered in the path of the Little Tunnel; c. 1900. (Tom Pike)

The collapsed entrance to the Big Tunnel on the Sandy & Beaver's Middle Divison. The tunnel was more than half a mile long and eighty feet below the surface in places. It was blasted through solid rock. To avoid the added expense of an internal towpath, an endless chain powered by mules on a treadmill pulled the boats through. Mules rode aboard the boat during the trip through the tunnel. (Tom Pike)

The east approach to the Sandy & Beaver's Little Tunnel (also in the Middle Division), which was 600 yards long. In digging this tunnel, workers found a twenty-inch horizontal vein of high-grade bituminous coal that heated workers' houses and powered the steam engine above. The mining operation also yielded limestone, which was burned to make lime mortar for tuckpointing locks; c. 1900. Inset is a view of the Sandy & Beaver Canal at Dungannon, midway between the two tunnels; c. 1900. (Tom Pike)

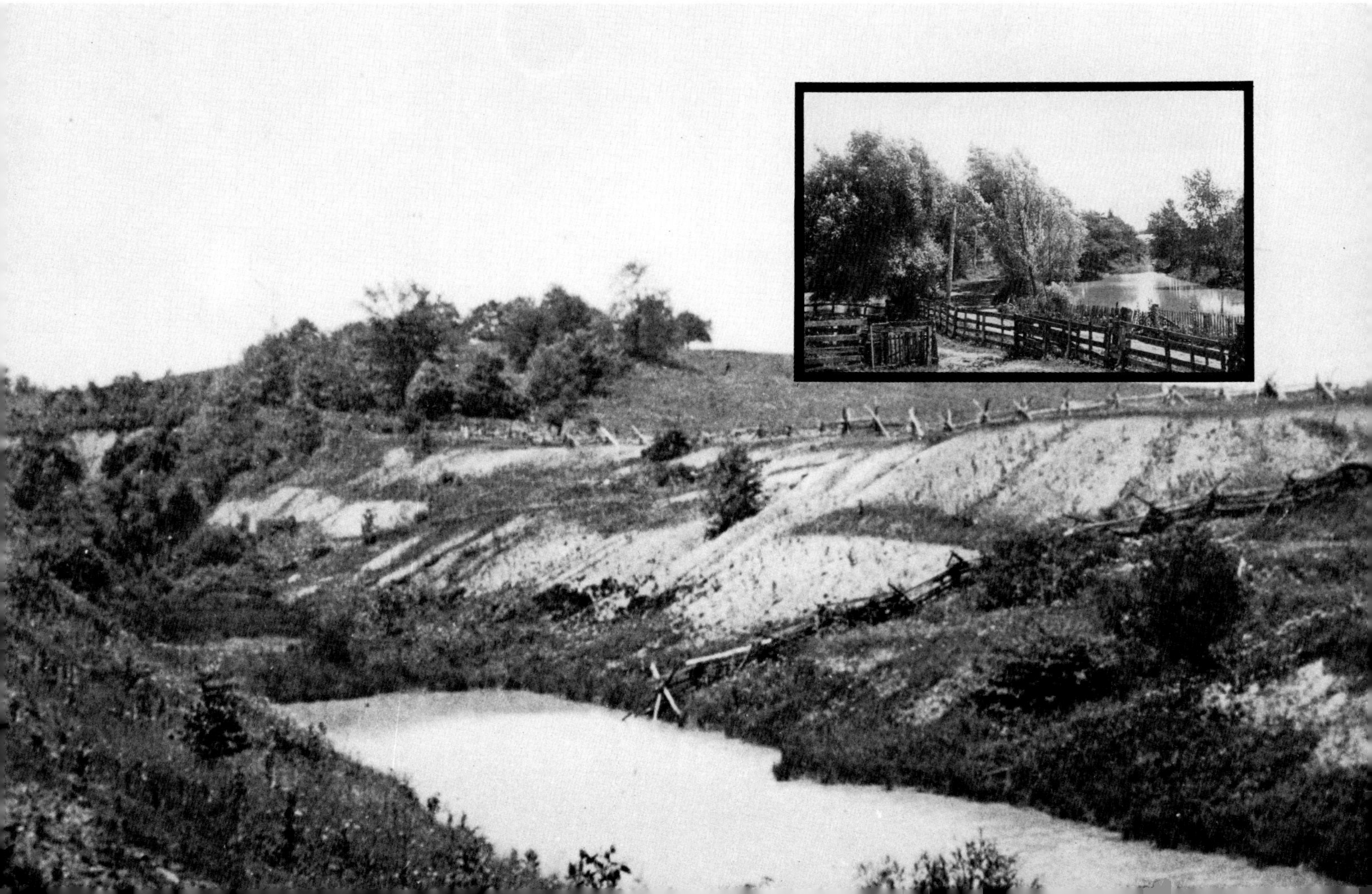

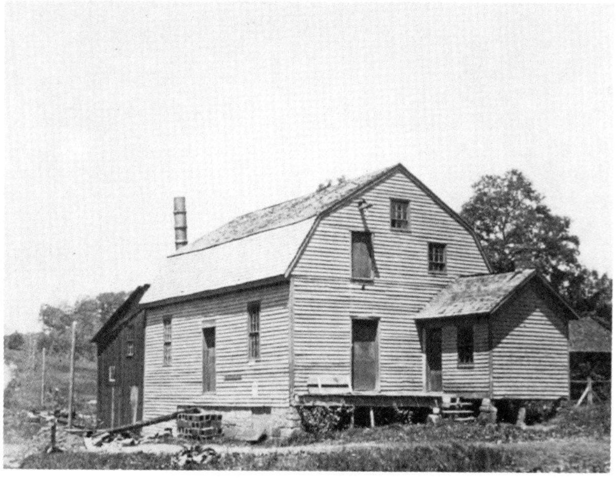

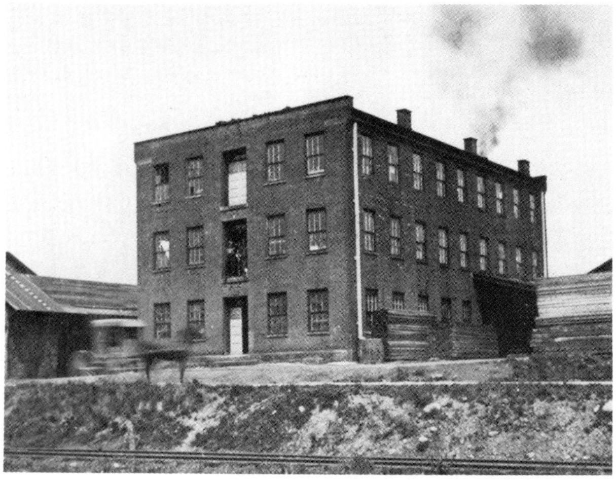

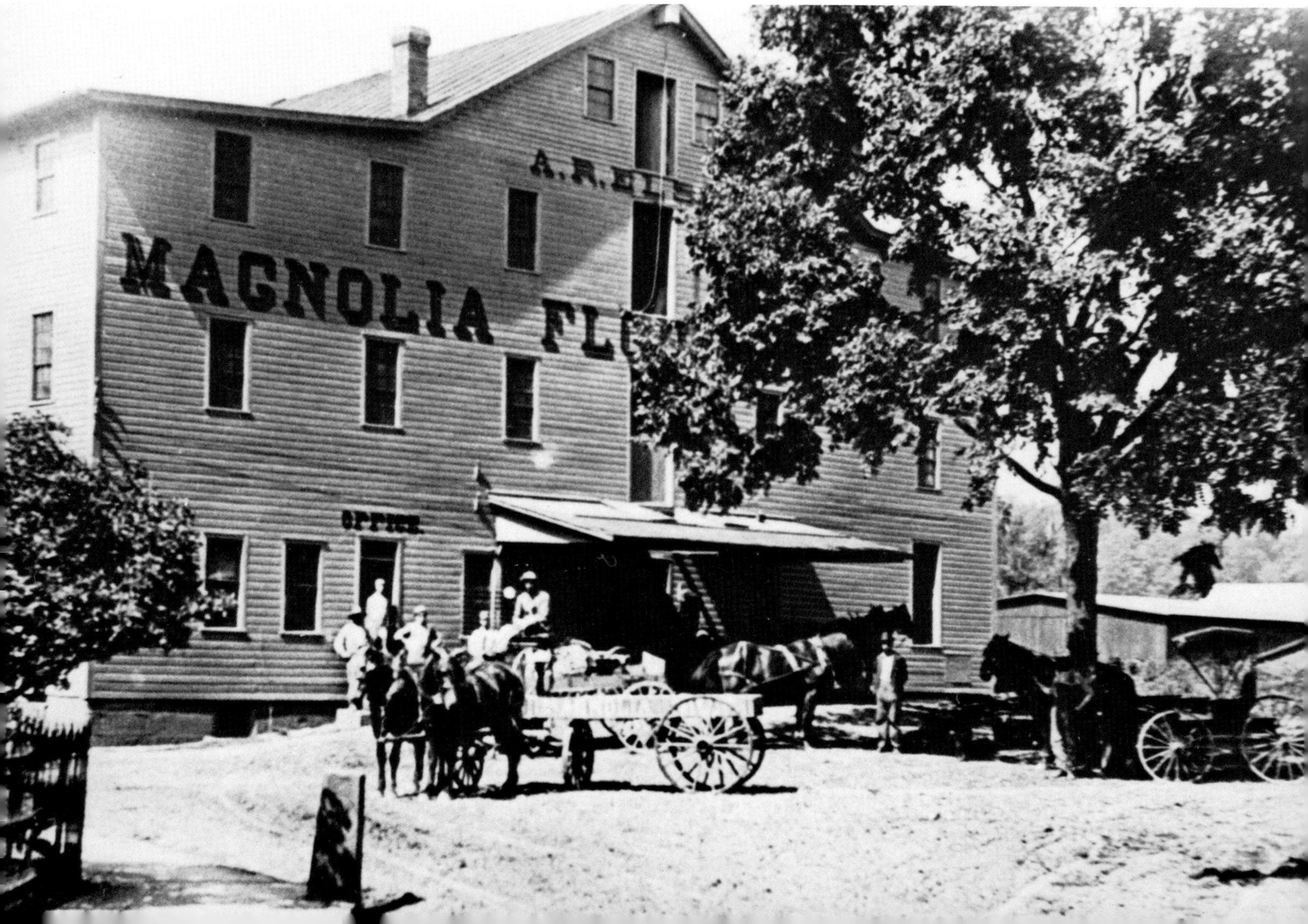
MAGNOLIA
OFFICE

Left to right, Judge Bowman's mill, near the Sandy & Beaver Canal; c. 1900. Woolen mill, built in 1836 on the bank of the canal (foreground ditch with railroad tracks) in anticipation of access to markets in both Ohio and Pennsylvania. But the owners had to wait a dozen years before the canal was fully operational. By that time railroads were taking over Ohio transportation needs. Note that the horse and buggy were too fast for the photographer's shutter speed; c. 1900. (Tom Pike)

"The tunnel was completed January 1, 1848. . . . On January 7th, the boat started on its way through the locks at New Lisbon and into the slackwater pool backed up by Dam No. 1 at 'Rebecca Furnace.' The canal crossed the creek at this pool and then went up 164 feet into the hills of the summit. . . . On went the boat past the old Miller Farm and the old Miller family cemetery. . . .

"Everything was going fine when the boat passed below the Hanover Road into the channel. . . . It became stuck when it came to the Frost Mill where it was necessary to cross an elevation of three feet. Here they were grounded and spent the night of January 7th.

"A messenger was sent to Hanover to get the Big Tunnel contractor, Morris Miller, to bring his seven yokes of oxen to drag the boat across the hump where it was stuck. Early on the morning of January 8th, Morris Miller arrived, bringing his oxen and an added feature, the Hanover Brass Band. The band had pledged themselves to accompany the boat and furnish music on the way till they tied up at the dock of the Arter and Nicholas Warehouse at Hanover. . . .

"On through the Little Tunnel went the boat, then through Dungannon amid deafening cheers. At Dungannon, a large parade of children, young and old, tagged along behind the boat, intent upon following it to the Big Tunnel, then walk across the Tunnel Hill to watch it come out the west end, then on to Hanover where a great celebration was planned.

"The boat entered the Big Tunnel where Robert Hall was in charge, hooked onto the endless chain, the mules started the chain going and the boat was on its way through the tunnel. The band struck up a lively tune inside the tunnel, almost deafening the passengers, and rocking the whole hill with its resonance. A few yards ahead of the boat a big rock fell from the tunnel roof into the channel before them.

"The band music was blamed for the rock falling, and was the only concert ever given inside the Big Tunnel; into the water they went with hammers, drills and wedges, breaking the stone up so they could move it out of the channel.

"It took several hours, and Edward Sinclair had a very important date to be married in Hanover at three o'clock in the afternoon on January 8, 1848. He waited until a few minutes before three o'clock, then shouted 'Boys, my time's up.' Into the water he jumped, swimming, splashing and wading past the workmen dispatching the rock, and arrived in Hanover in a thoroughly soaked band uniform, half frozen, ready to be married for better or for worse. It must have been better; nothing could be worse.

"Before sunset, the tunnel had been cleared and the boat passed through, docking at the Lower Arter & Nicholas Warehouse, amid a celebration never to be forgotten." (R. Max Gard and William H. Vodrey, Jr., *The Sandy and Beaver Canal,* 1952, East Liverpool Historical Society.)

Surviving to the present day, the A. R. Elson Flouring Mill at Magnolia, Ohio, on the Western Division of the Sandy & Beaver Canal utilized canal water power for its operations until the 1960s, when it converted to electricity to improve reliability; c. 1900. (CSO)

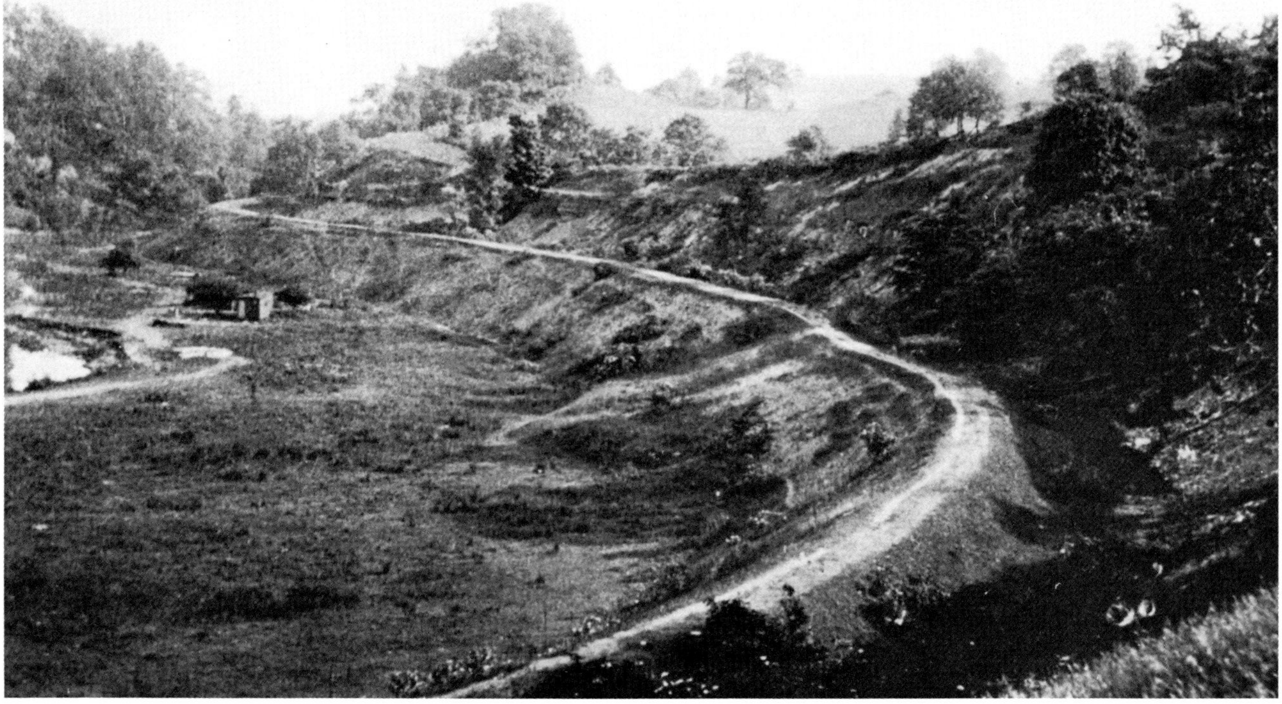

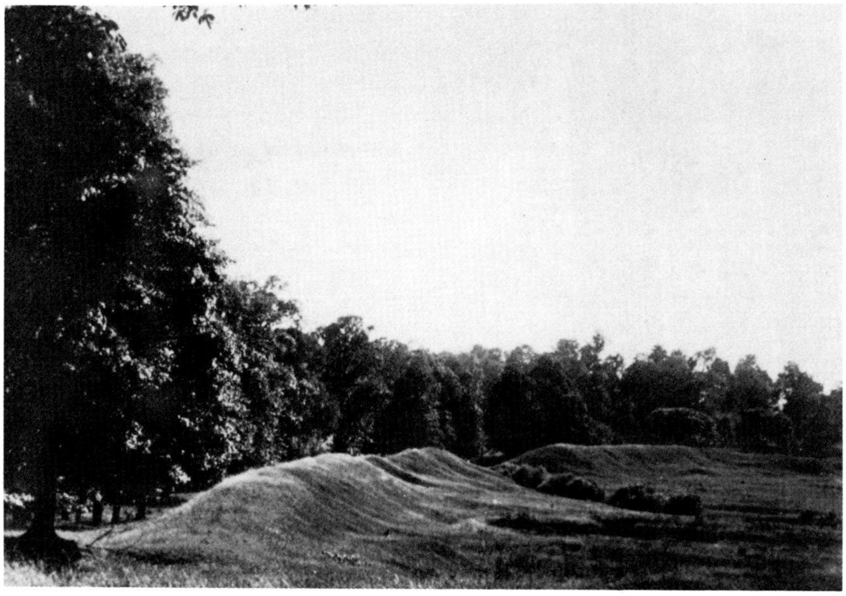

The Sandy & Beaver Canal bed and towpath (above) clinging to the hillside along West Fork, Pine Hill; c. 1900. (Tom Pike)

The Sandy & Beaver's Little Reservoir (right) also called Cold Run, east of Guilford on the Middle Division, was created by an earthen dam forty feet high. A washout here in 1852, still visible as a breach in the embankment half a century later (lower center of picture), flooded adjacent countryside. Many lives were saved as 73-year-old Judge John J. Bowman sped into the valley on horseback to spread the alarm; c. 1900. (Tom Pike)

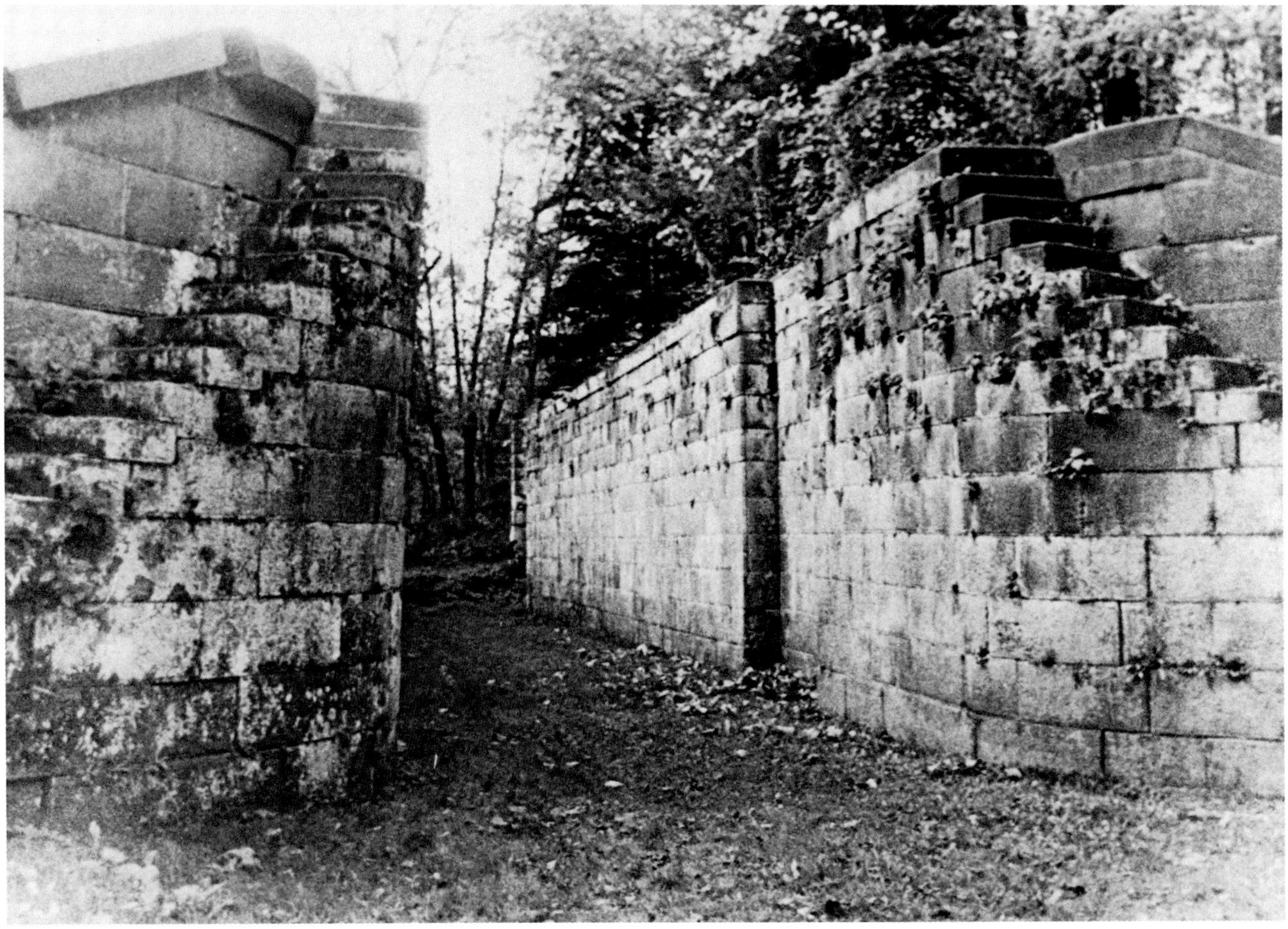

Edward Gill's magnificent structural masonry stands up well to the ravages of time at Lusk's Lock, Lock 27 on the Eastern Division of the Sandy & Beaver. Named for one of its contractors, Lusk's Lock had a lift of more than thirteen feet. Typical of Gill design perfection, the massive stone blocks were sawed and notched to achieve even height of the stairway risers. Near the top the steps were decorated with sculptured ends. The stairway was for the convenience of boatmen accessing the gate controls and sweeps to lock their boat through; 1985. (Photo by Larry McVan, courtesy Larry Turner.)

Lock 36 on the Eastern Division is the centerpiece of Beaver Creek State Park, near Gaston's Mill, a restored grist mill. (Photo by Larry McVan, courtesy Larry Turner.)

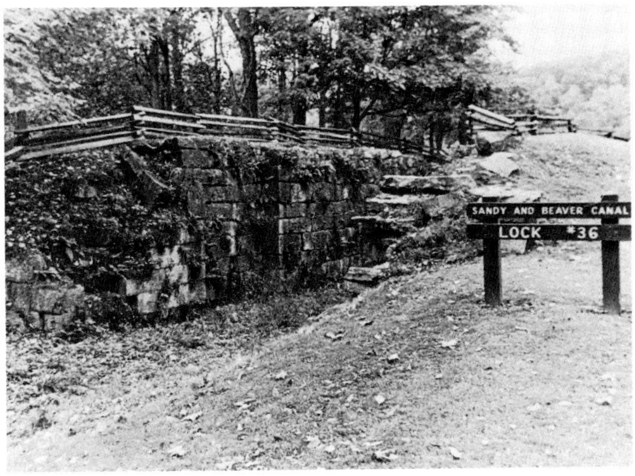

Lock 50, Lost Lock, on the Eastern Division of the Sandy & Beaver, was restored in 1971. (CSO, 1976 photo)

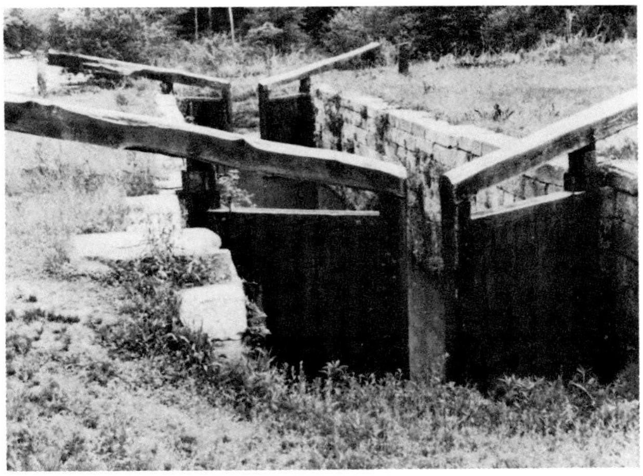

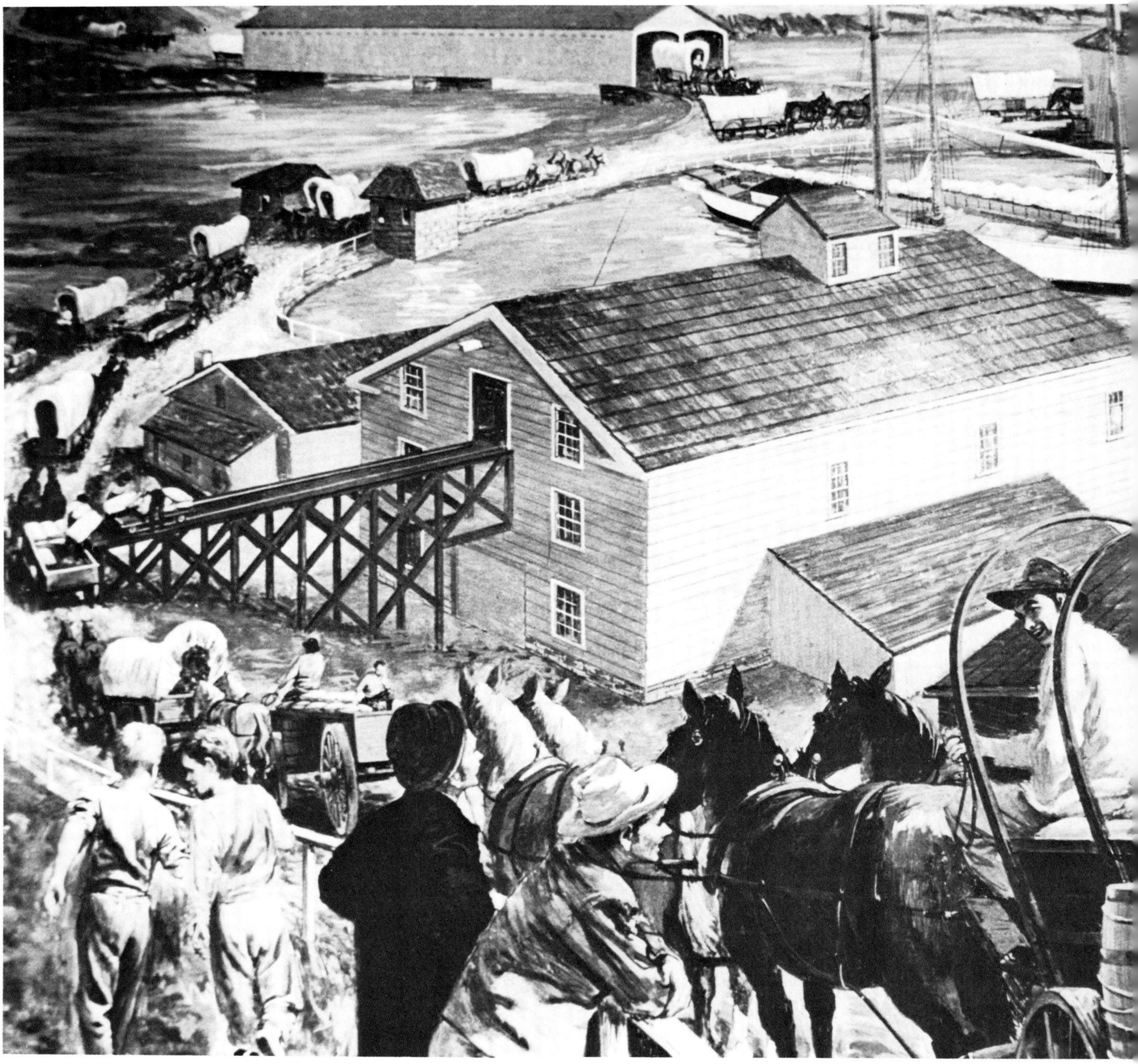

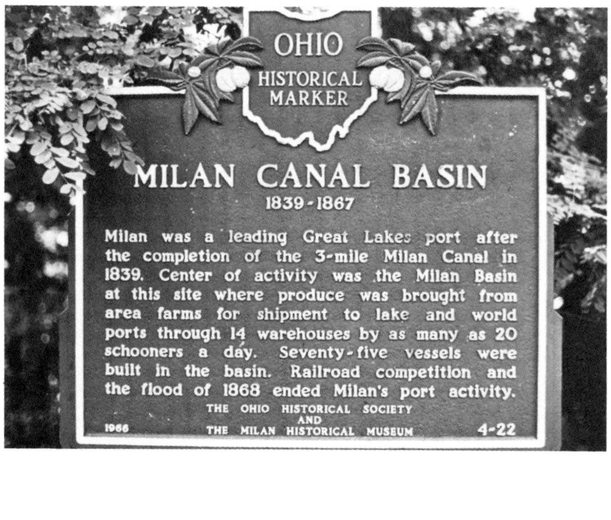

This painting illustrates a famous day in Milan's history. In May 1847, some six hundred wagons, waiting to unload their grain, made up a line extending from Milan's town square to the Blue Fly Inn—a distance of six miles. Thirty-five thousand bushels of wheat were handled that day. (Painting by T. Gorka, courtesy The Milan Historical Museum, Inc.)

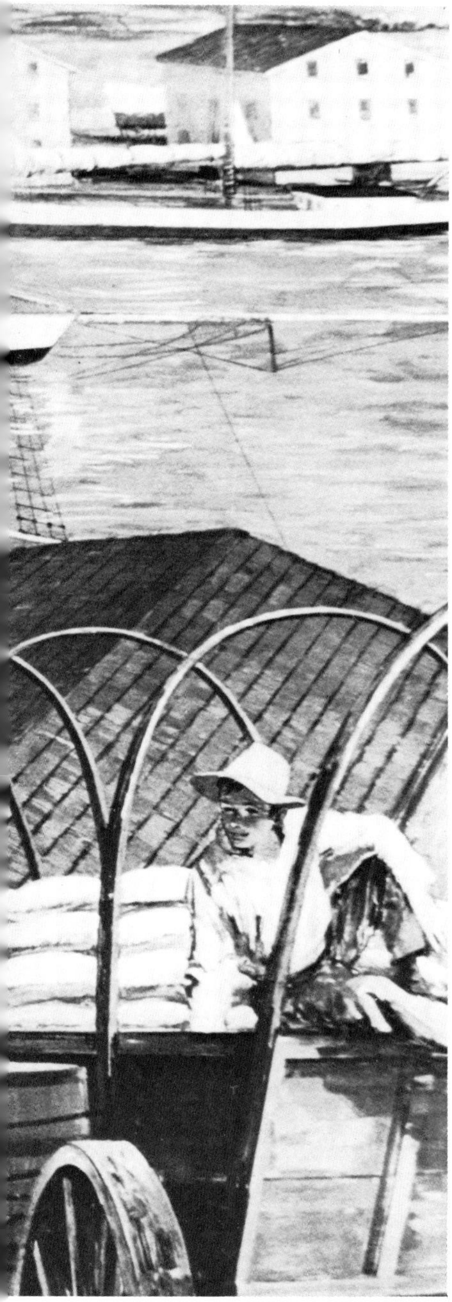

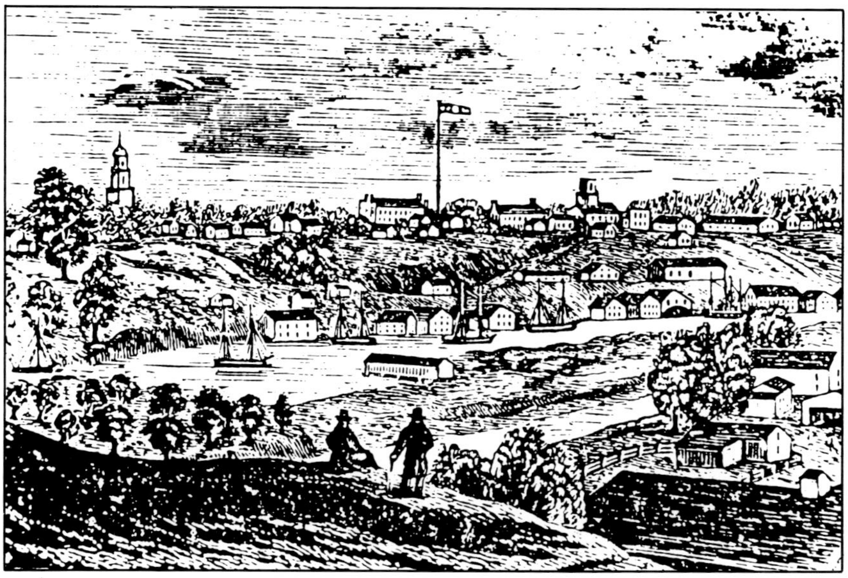

This old woodcut shows the Milan Basin and docks, with six lake schooners at anchor; as many as twenty schooners a day called at the port. The canal exited to the right toward Lake Erie.

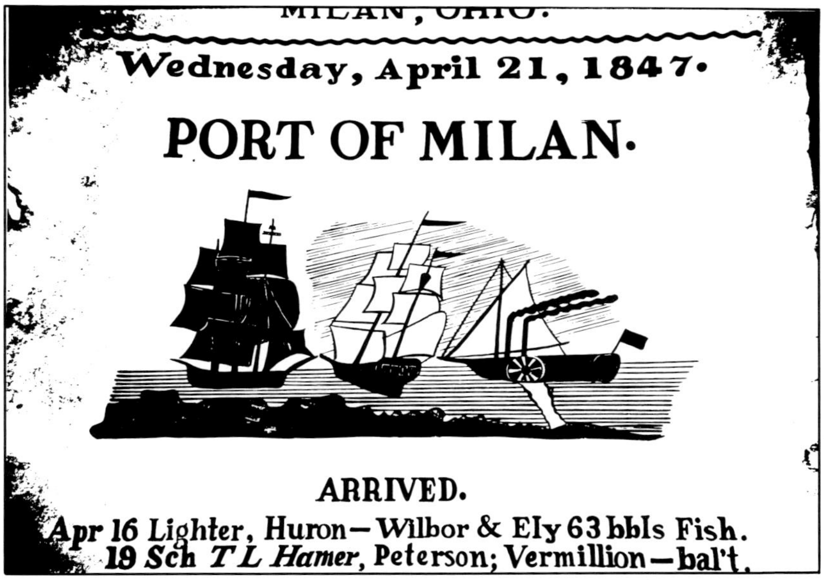

MILAN, OHIO.

Wednesday, April 21, 1847.

PORT OF MILAN.

ARRIVED.

Apr 16 Lighter, Huron—Wilbor & Ely 63 bbls Fish.
19 Sch *T L Hamer*, Peterson; Vermillion—bal't.

This broadside announced the arrival of Great Lakes vessels in the port of Milan in 1847, by which year Milan had become second only to the Russian port of Odessa in the shipment of wheat. (The Milan Inn)

9

The Canalers and Their Families

As noted earlier, a full crew complement for a freight boat on the canals consisted of a captain, a steersman, a bowsman, and a mule driver. The rank of captain automatically went to the boat's owner—which is why purists point out that the canal bard, "Captain" Pearl Nye, for example, was not technically a captain, since he operated a boat owned by his mother. Most captains affected a distinctive headgear—often a stovepipe hat. Some also had quasi-military dress uniforms with brass buttons that they wore on special occasions. But the dress of the rest of the crew was casual indeed.

Canalers had the reputation of being a rough lot, given to brawling on slight provocation. Actually, their fights weren't usually in genuine anger. At the plodding speed of four miles per hour or less, their job was essentially boring, and physical competition between boat crews helped to relieve the endless monotony. Right-of-way into locks, for example, was prescribed by state regulation: since the upstream boat was more difficult for the mules to restart, it had priority—according to the regulation, that is. But etiquette was not the canalers' long suit. In actual practice, when boats approached a lock (or an aqueduct or each other) from opposite directions, the decision on right-of-way was often determined by a free-for-all, or sometimes more formally by "duels" between rival drivers who used their switches as jousting weapons while balancing themselves on the top of the lock gates—the loser ending up dunked in the canal. When a parked boat was overtaken at night, canalers' idea of good clean fun dictated that its cookstove stack would probably be pulled down by the towline of the passing boat, together with other vertical projections, including flagstaffs, and any crew members who came out to investigate. Occasionally such competition got nasty, ending in a sort of armed truce, with neither party permitting the other to pass.

The saloons which sprang up at many locks afforded a facility for attitude adjustment—or, alternatively, a final definitive brawl to settle whatever subject was in contention. Some were hangouts for the gangs of toughs, or "canal bad men," who occasionally "captured" boats and rode along on them for days, terrorizing the passengers. In Canal Fulton, at the present intersection of State Routes 21 and 93, stood "Brimstone Corners." With a saloon on each of its four corners, the crossroads was the common meeting ground of canalers, farmers, and miners—and the unofficial final recourse of all disputes that could not be settled peaceably. There were several peace officers in the town, but it was generally conceded that their jurisdiction did not extend to the Corners—certainly not on Saturday night.

Life aboard the boats was something less than commodious. The owner-captain of the freight boat lived in the rear cabin, often with his wife and family. The tiny cabin had a cookstove and bunk beds, and exemplified the epitome in compact living. During the canals' prosperous salad days, the forward cabin was occupied by additional non-family hands. But on family boats members of the family constituted the entire crew. The last freight traffic on the Ohio & Erie, particularly after 1900, is said to have been almost exclusively the province of family boats, carrying coal from the mines of the Tuscarawas Valley to the mills in Akron or to the steamers moored at the Cleveland docks.

The cramped quarters aboard the boats did not seem to impede family proliferation. Captain Billy Smart of the Walhonding Canal and his wife had thirteen children, four of them born on the canal boat. Pearl Nye, also born on a canal boat, was one of eighteen children—accounting for his father's need to run two boats in tandem.

These canal families were a kind of subculture of the time. Their world was the canal. Although their boats were tied up for weeks or months in winter, most families lived aboard the boat year-round. And, except for those months when their home was immobilized in ice, some canal children rarely saw the inside of a school. But if it brought the hardship of school attendance to these Huck Finn types, winter also had its rewards. When frozen over, the canal provided a smooth, even skating rink that ran for miles; however, the creativity of canal kids found ways to turn such a winter idyll into an adventure, as recorded in this reminiscence by George Hilgeman of Cincinnati:

"Skating at times was too tame, and to liven things up we would cut holes in the ice, blocks of about four feet square and five inches thick, usually four or five blocks. Running across these was called 'playing ticklish.' After the surface of these blocks was washed over three or four times they sure were slippery, and I did it once too often and felt the icy chill of the canal at ten above zero."[1]

But summer provided the atmosphere most typical of life on the canals. At Canal Fulton, where Route 93 now crosses one of the canal bridges, village boys used to loll in the summertime, waiting for a boat to pass underneath. When one did come along they would drop onto the deck and, taking advantage of the good nature of the crew, would ride along to the first lock south of town. It was about a mile down to the lock and, what with berry picking and a short skinny-dip swim on the way back, the excursion was good for at least an hour's entertainment on a hot day.

If one could ignore the mosquitoes that bred in the quiet, algae-covered water—and was lucky enough not to contract malaria—being a canaler was a pleasant enough life.

1. Charles Ludwig, "Playmates of the Towpath: Happy Memories of The Canal Swimmers' Society," *Cincinnati Times-Star*, 1929.

At right is a remarkable group of contact prints from original glass plates made by Edwin B. Howe, a canal photographer who happened to be the grandson of Captain Richard Howe, one of the engineers who built the northern section of the Ohio & Erie and was later its resident engineer. It is apparent that Howe's wet plate collodion emulsion (see preface) is deteriorating beyond the edges of the printing frame. The group photo of what appears to be two families is earlier than the others by as much as ten years, perhaps making it circa 1878–80. The tiny rocker on the cabin roof behind the man seated next to the tiller suggests that one younger family member is absent; perhaps it is nap time.

An analysis of the images discloses a modest success story for one of these canal families. Possibly a former state boat, the weather-beaten, heavily-laden freighter in the top photo is of the short-haul type, which sometimes eliminated the center (stable) cabin housing the spare mule team. This configuration made it ideal for hauling lumber, which is, indeed, the cargo here. The somewhat haggard-looking man on the right—hat tipped back, corncob pipe in his left hand—stands next to his young wife, who wears a sack dress. We note that he wears trousers supported by a single suspender over his right shoulder, after the fashion of Al Capp's Li'l Abner.

The other two pictures, made some years later, reveal that the boat has since been in drydock, its hull and superstructure freshly painted, with decorative wood strips added to the sides of the aft deck in the process of refurbishing. The boat now has a name: the *Benjamin Harrison*. The stern is even decorated with a campaign poster for the new president, who was elected in 1888 (the same year, incidentally, that George Eastman introduced his flexible, dry-emulsion, photographic film, to which photographer Howe has obviously not yet converted).

Remarkable things have happened to the three principals in the later pictures. At bottom left, the boat's owner(?) has a new hat, a fuller mustache, has gained weight, and his countenance and body language now bespeak confidence; and his pants are held up by a pair of store-bought suspenders. His now-matronly wife has smart new clothes, including a ladylike hat. The other gentleman (seated on the left of the stern in this photo and tending the mules in the bottom right photo) appears to be the young man previously dandied up in western tie and galluses in the top photo; he now affects a conservative vest. It is apparent that none of them has been undernourished in recent years. Not bad for a canal family so late in the life of the Ohio & Erie.

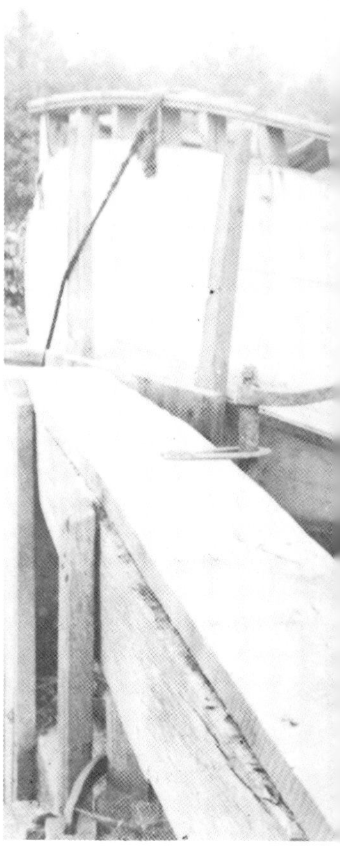

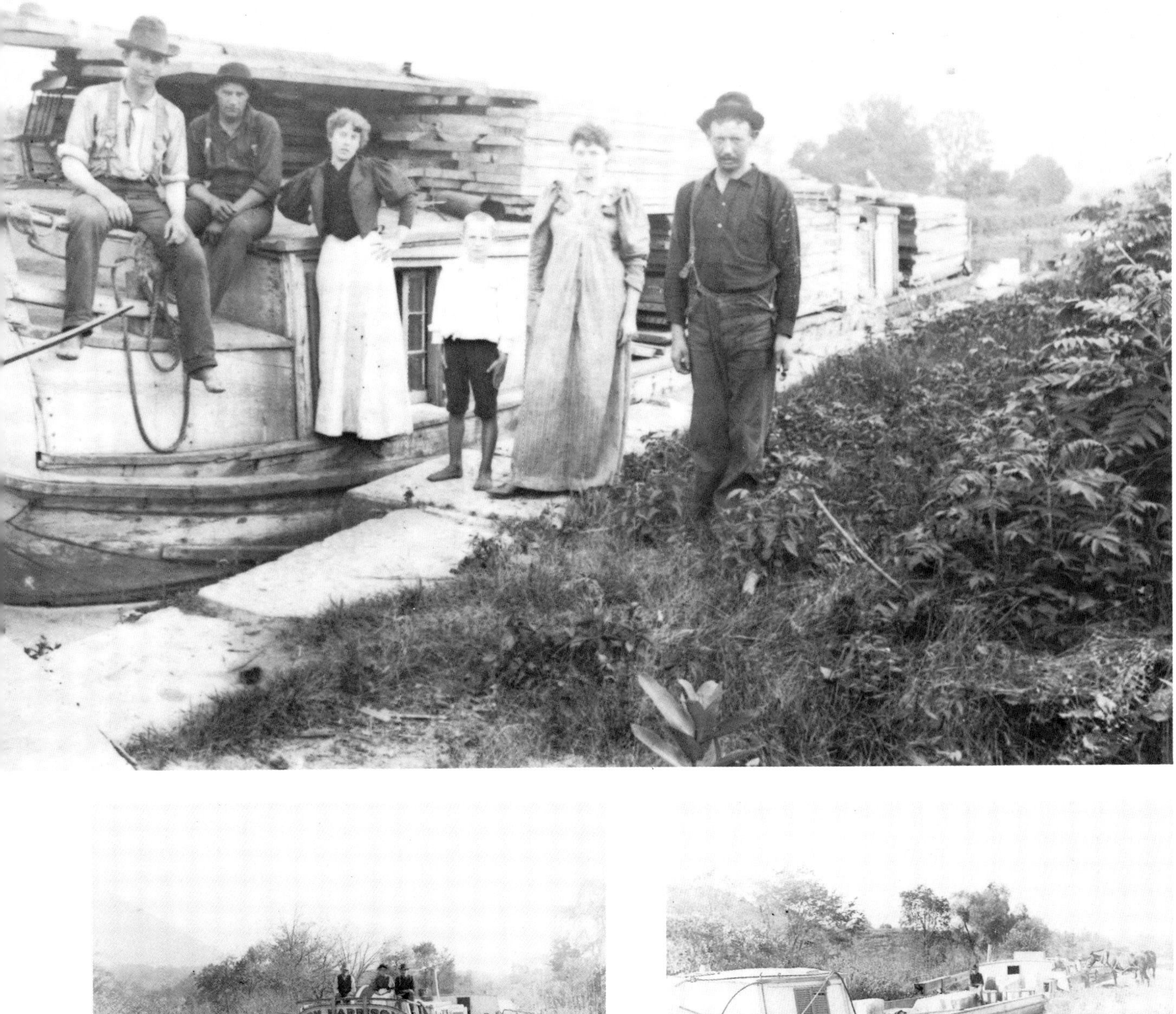

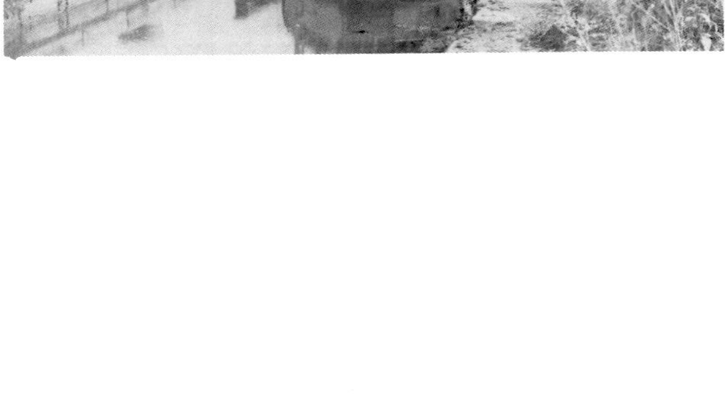
BEN. HARRISON.

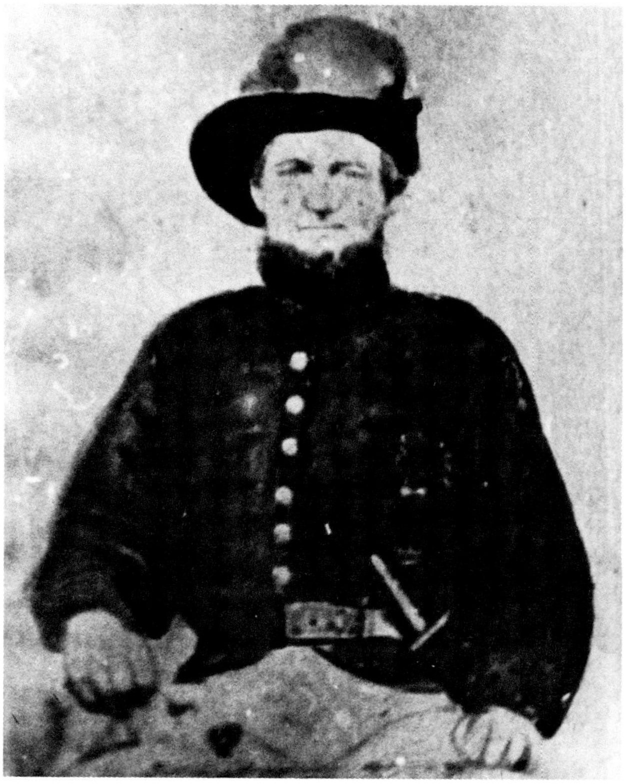

Above is an early portrait of an unidentified canal captain in his full dress uniform, including a militaresque jacket with a high collar into which he tucks his beard. Typically, he wears a distinctive headgear (perhaps a customized stovepipe hat); and this captain even bears a sword! The portrait has the grainy, indistinct image character of a calotype, or talbotype, a paper negative process popular in the late 1840s. Although the captain's head was undoubtedly clamped firmly in place for the requisite exposure of several minutes in direct sunlight, he obviously managed to move his right elbow during the process. The coarseness of calotypes was caused by reproduction of the fibers in the paper from which the negative was made. The process was invented by English scientist/mathematician William Henry Fox Talbot, who first published it in 1844. (SCHS)

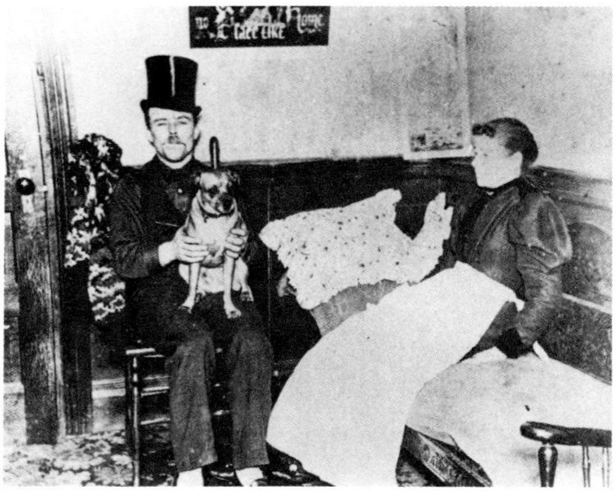

A Barberton captain models his "last captain's uniform" at home; undated. (Canal Visitors Center, CVNRA, courtesy Cloea Thomas.)

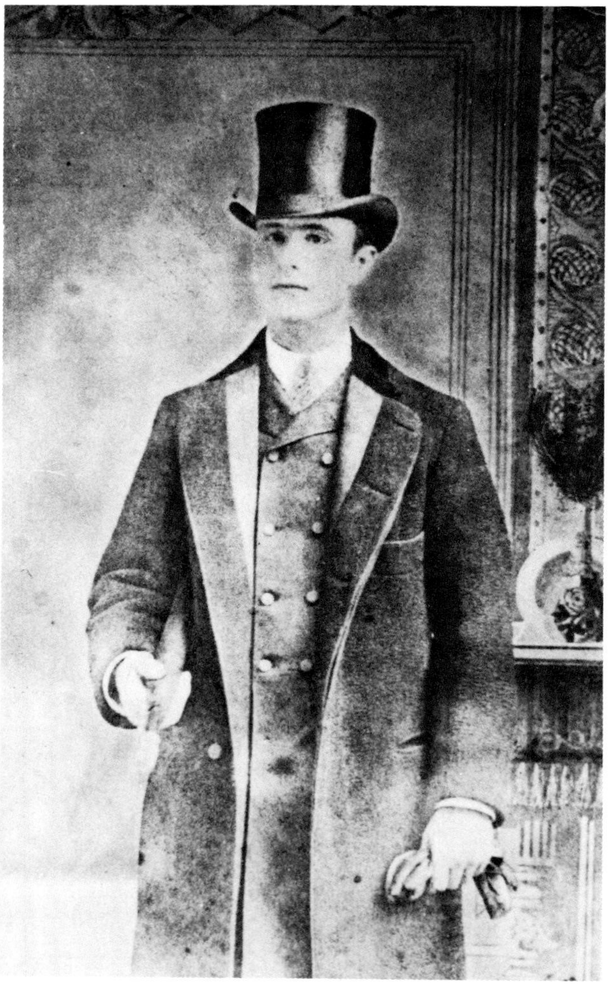

In this retouched daguerreotype, Captain Theodore Bennett's dress uniform is more genteel. Gloves held fashionably in his left hand (to facilitate shaking hands), and with a baton (umbrella?) tucked under his right elbow, Bennett is the essence of an antebellum fashion plate. His top hat of beaver or silk is unmodified and immaculate; c. 1860. (Pockrandt Collection, SCHS)

"Captain Brad Voshall was another type of 'canawler' entirely, a true town character in Walhonding during the 1920's. He would wander about the village dressed in a pair of tattered old trousers and a faded red long-sleeved undershirt, mooching meals and board wherever he could.

"By his own admission, Captain Brad was a 'peaceable man.' It is true that he once shot a fellow during his days on the canal, but it was only because, as he would say, 'I happened to have a gun.' He preferred to settle most of his arguments with his fists.

"Many times during his reign as the town character, he was invited in for a meal by a hospitable villager. The good captain would invariably extend these visits to marathon stays of two or three *weeks,* during which the harassed family would try everything to get rid of the captain. His uninvited appearance, clad in the inevitable trousers and undershirt, embarrassed many a formal dinner—as did his impromptu tales and rough talk. Still, he certainly added color and zest to many a meal and, perhaps, he paid many times over for his disconcerting, sudden and long visits with his stories and insights into a long ago era—the canal days of the Walhonding." (Terry K. Woods, *Twenty Five Miles To Nowhere*, Roscoe Village Foundation, 1978)

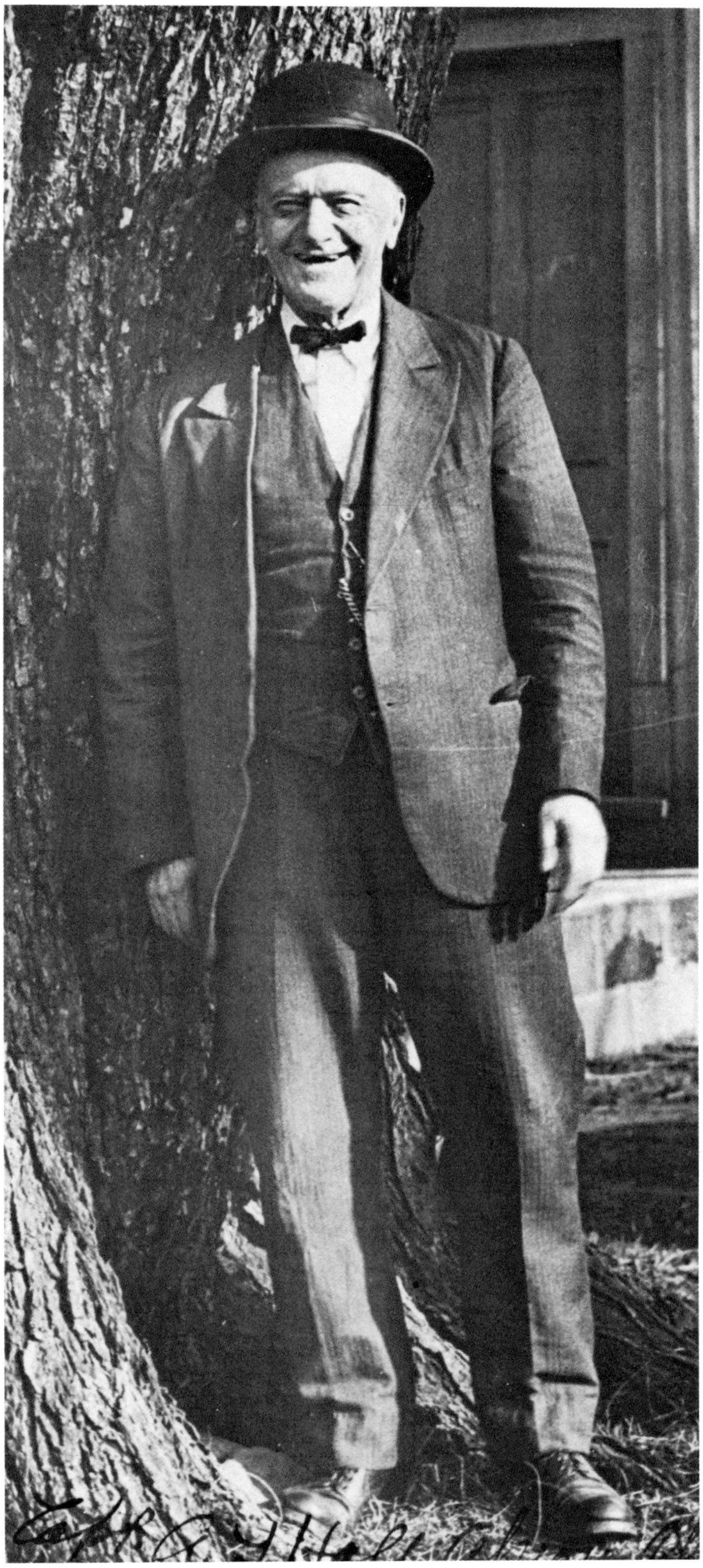

Master of an Akron boat, Captain A. L. Hall's special hat is a derby; c. 1900. (Pockrandt Collection, SCHS)

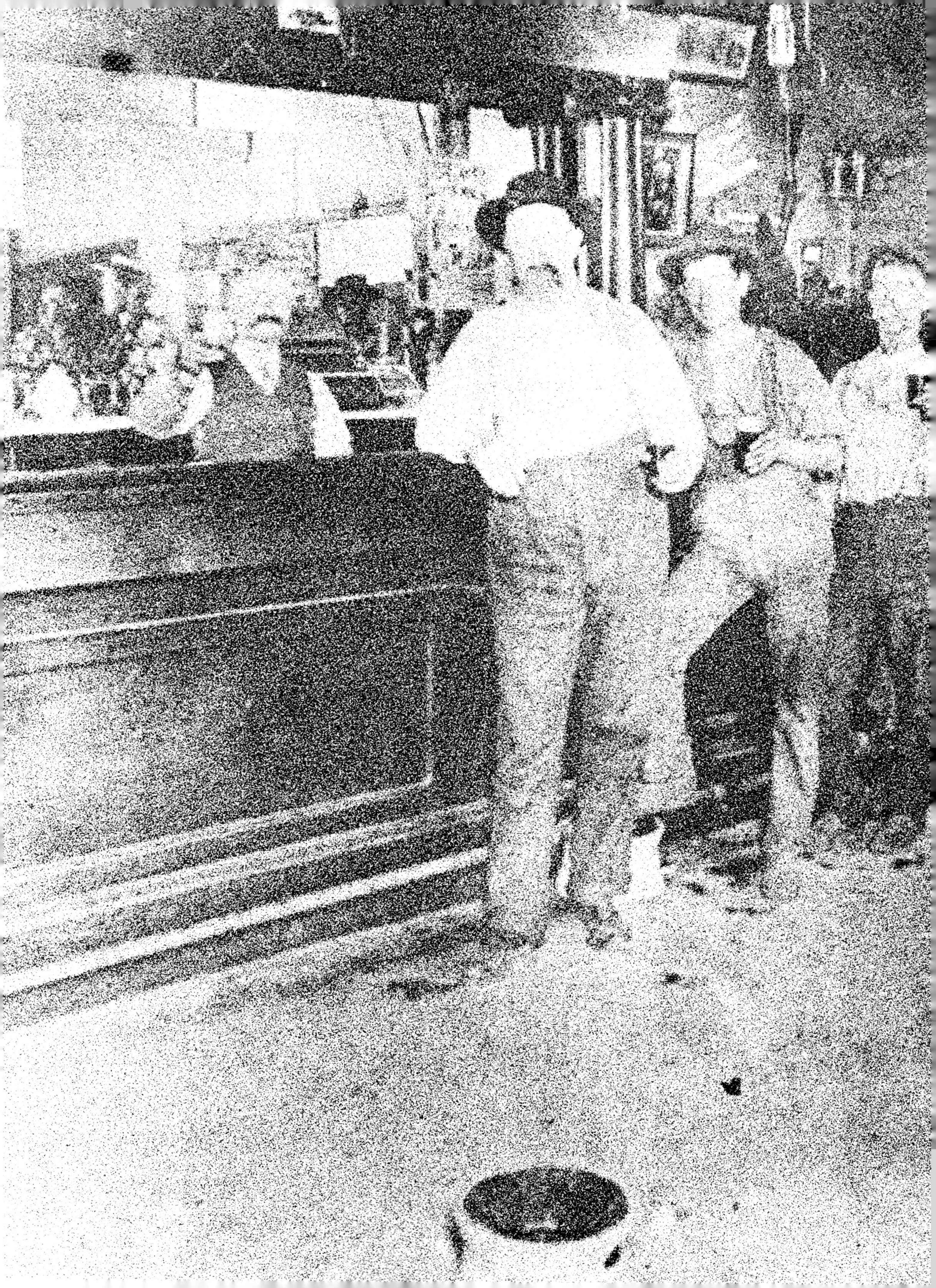

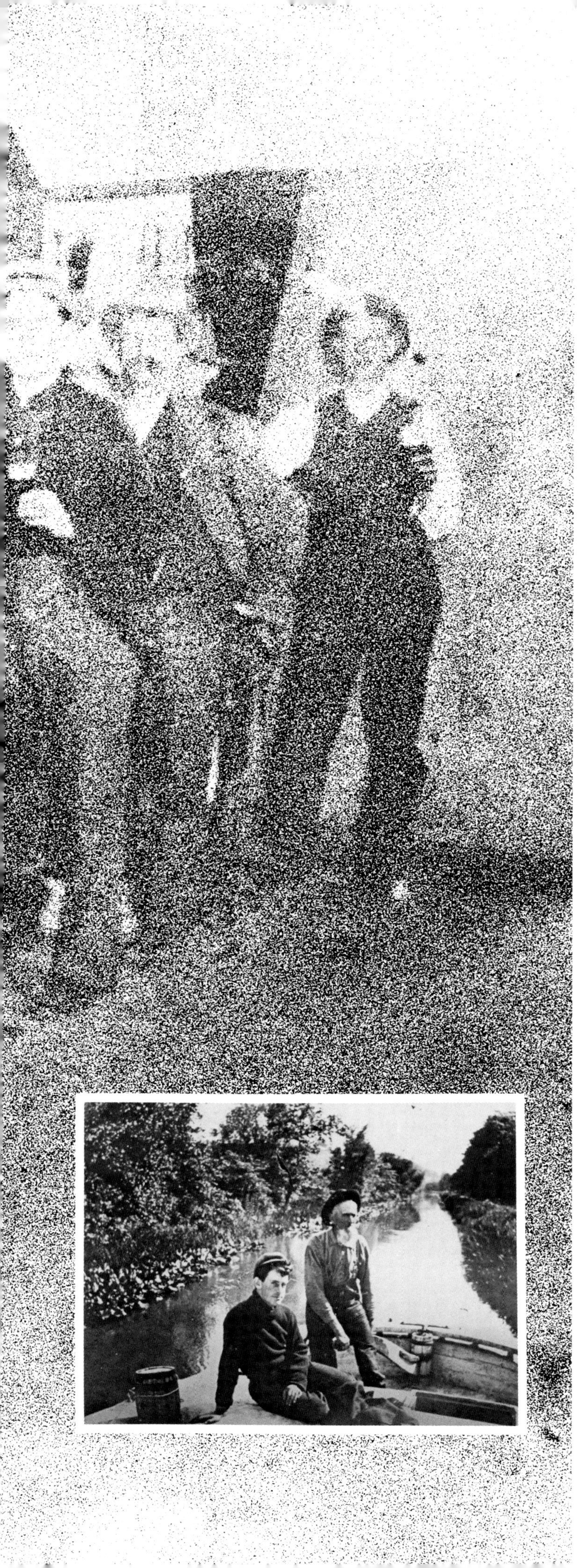

This canalers' saloon at New Portage, the head of navigation on the Tuscarawas River, was fitted with a mirrored mahogany bar, brass rail, Aladdin mantle gas light, and at least four cuspidors; c. 1890 (SCHS). *Inset,* South of New Philadelphia on the Ohio & Erie, Captain Neirmeyer takes a turn at the tiller for the benefit of the photographer, while his steersman takes a break; 1896. (Canal Visitors Center, CVNRA, courtesy Cloea Thomas.)

Mother Nye, boat owner, widow, and mother of eighteen Nye children (nine boys and nine girls), gets ready to prepare a meal on her boat. Inset is a photo of "Captain" Pearl R. Nye, who operated his mother's boat, and Mother Nye. Just how the two of them managed to eat the centers out of their watermelon slices (or why) is unknown; c. 1900. (Canal Visitors Center, CVNRA, courtesy Cloea Thomas.)

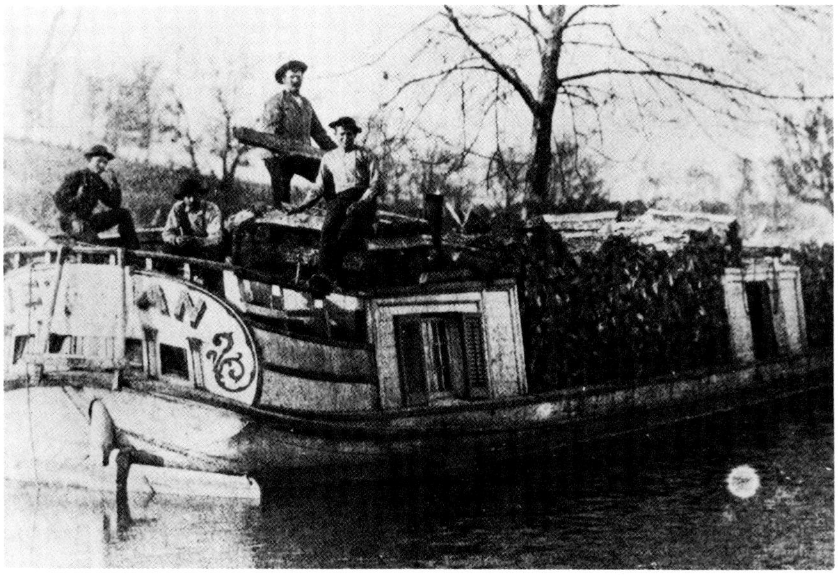

At Six Mile Lock near Zoarville, Captain Jim Perry and crew aboard the *Veteran* are obviously aware that their picture is being taken; 1888. (Canal Visitors Center, CVNRA, courtesy Cloea Thomas.)

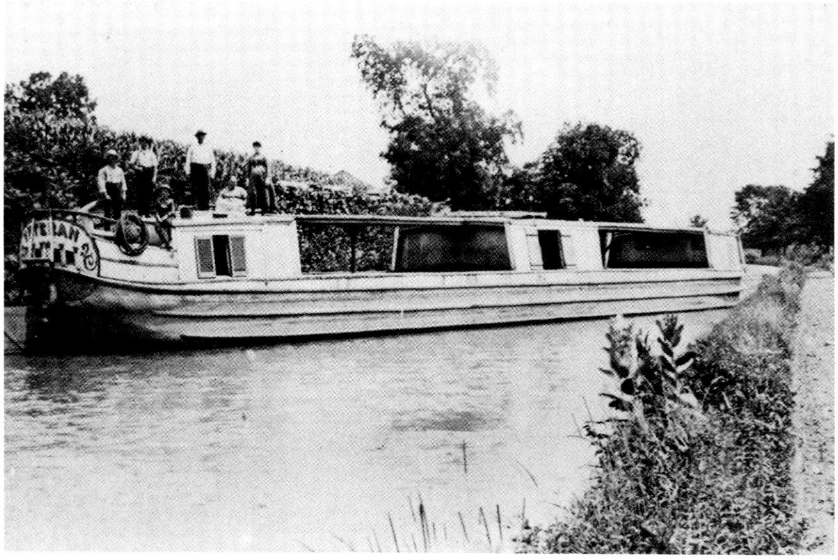

The *Veteran*'s cargo unloaded, Captain Perry's wife and mother join the cleaned-up crew for this picture, also near Zoarville. (SCHS)

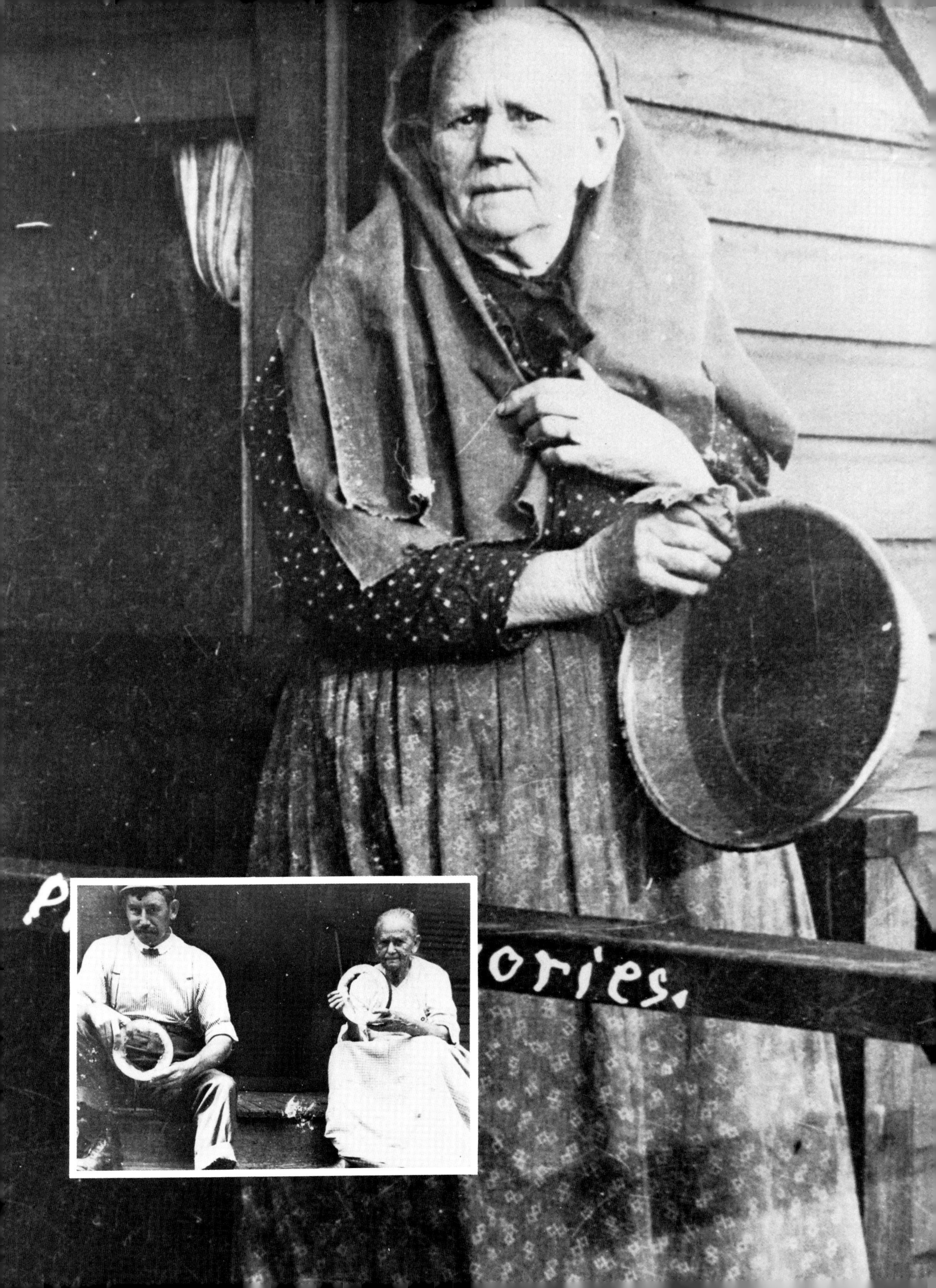
ories.

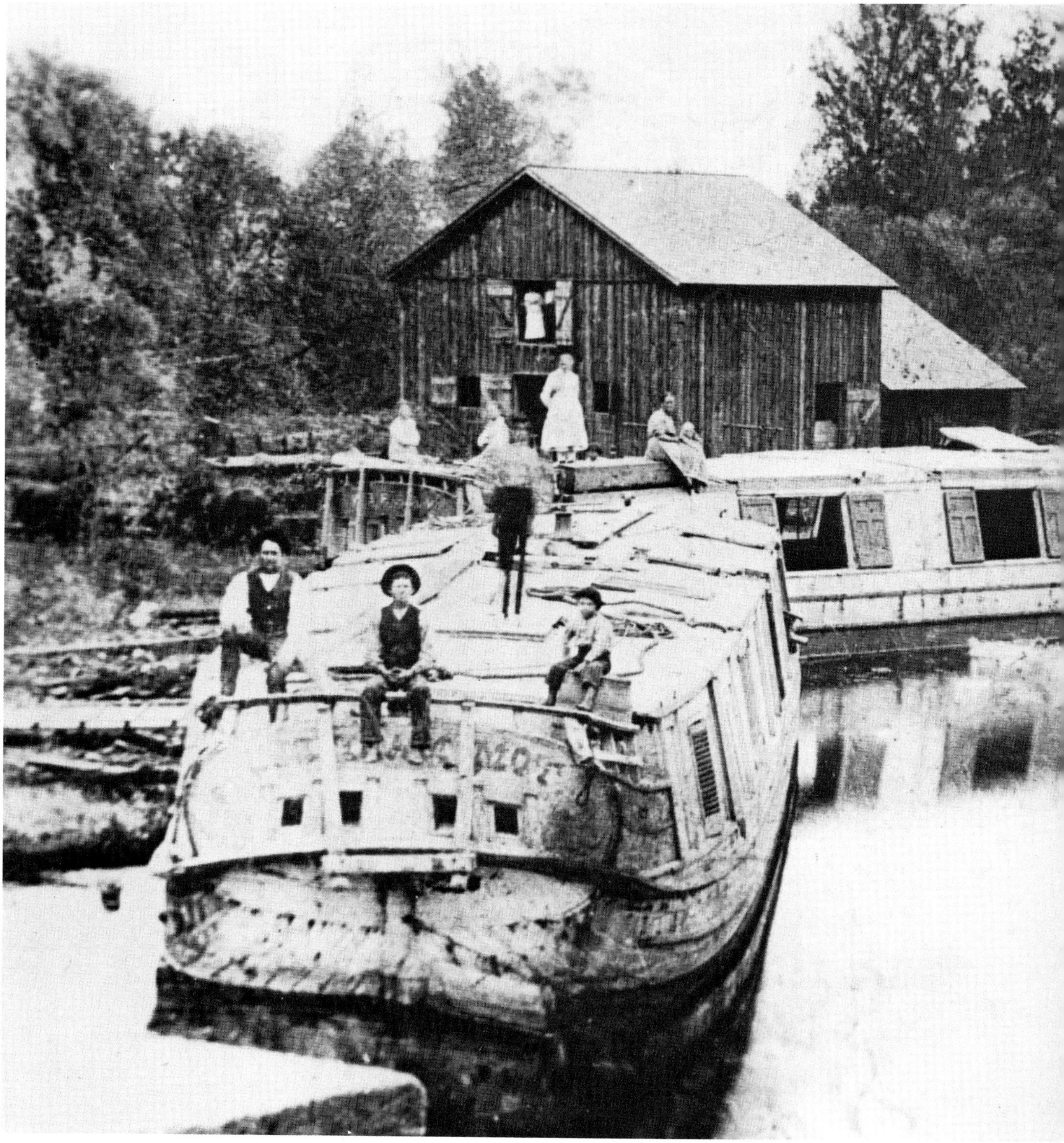

The pair of Nye family boats at Lockbourne. Father William Nye (at right holding the hand of his son Warren) had to operate two boats in tandem to accommodate his copious family. Mother Nye is seated at the rear of the foreground boat with their youngest child; 1884. (SCHS)

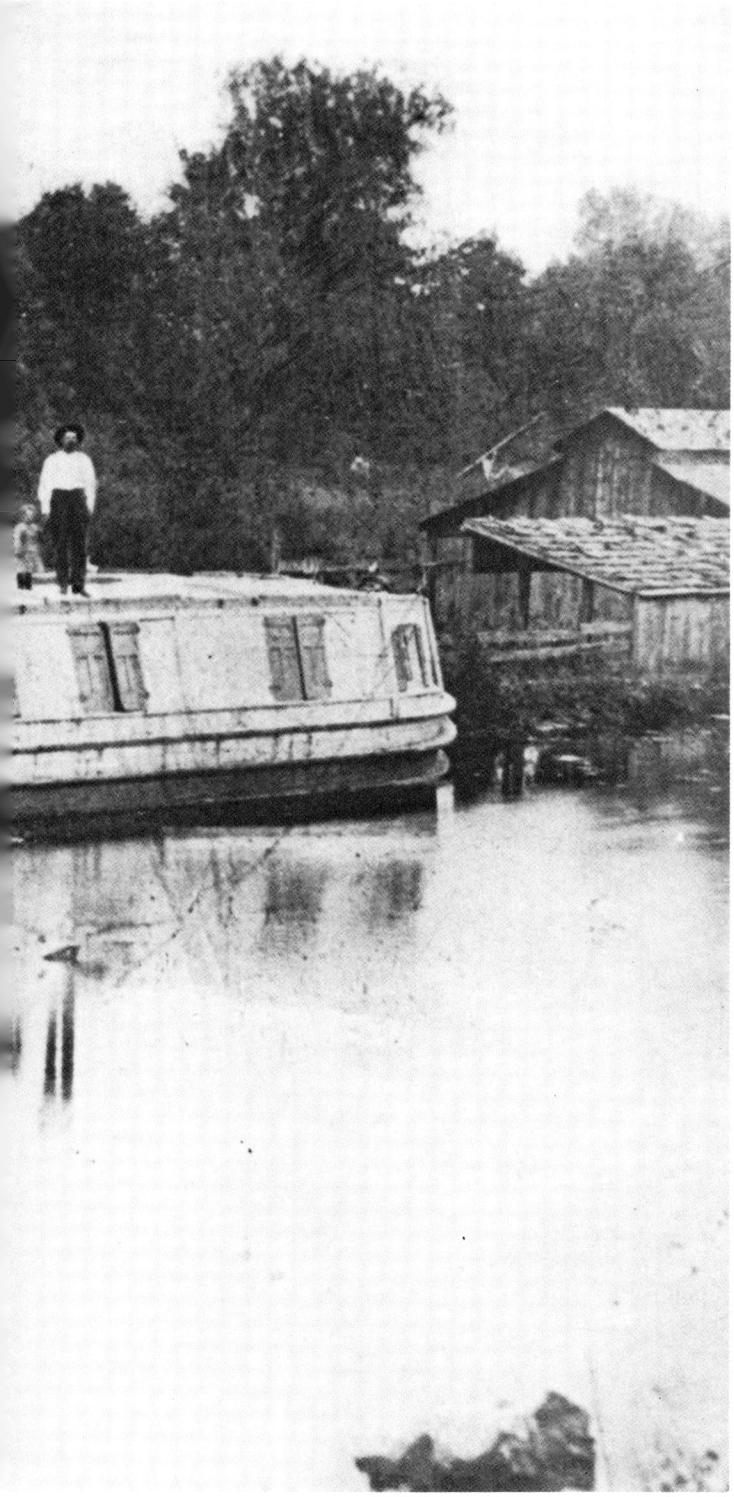

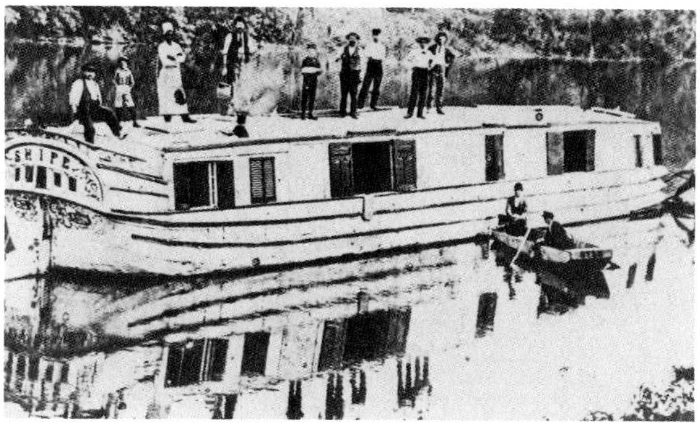

Crew members and friends gather for a fish fry at Jasper Basin, fifteen miles north of Portsmouth near Piketon, on the deck of the *Shife*, a Waverly boat. (SCHS)

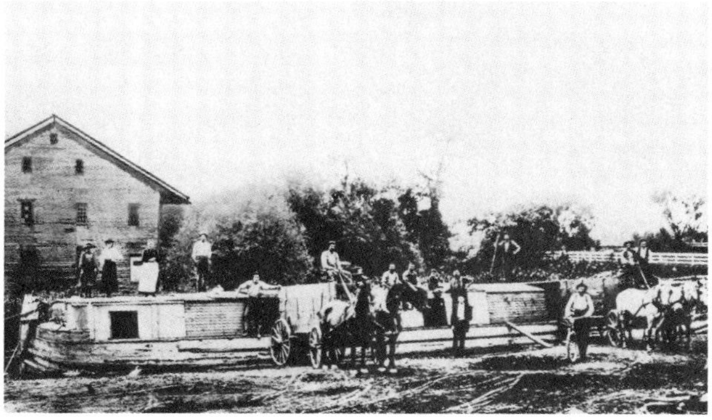

Captain John Hayes and family load grain aboard their boat at Circleville. They are probably on their way south to the Ohio River; 1890. (Pockrandt Collection, SCHS)

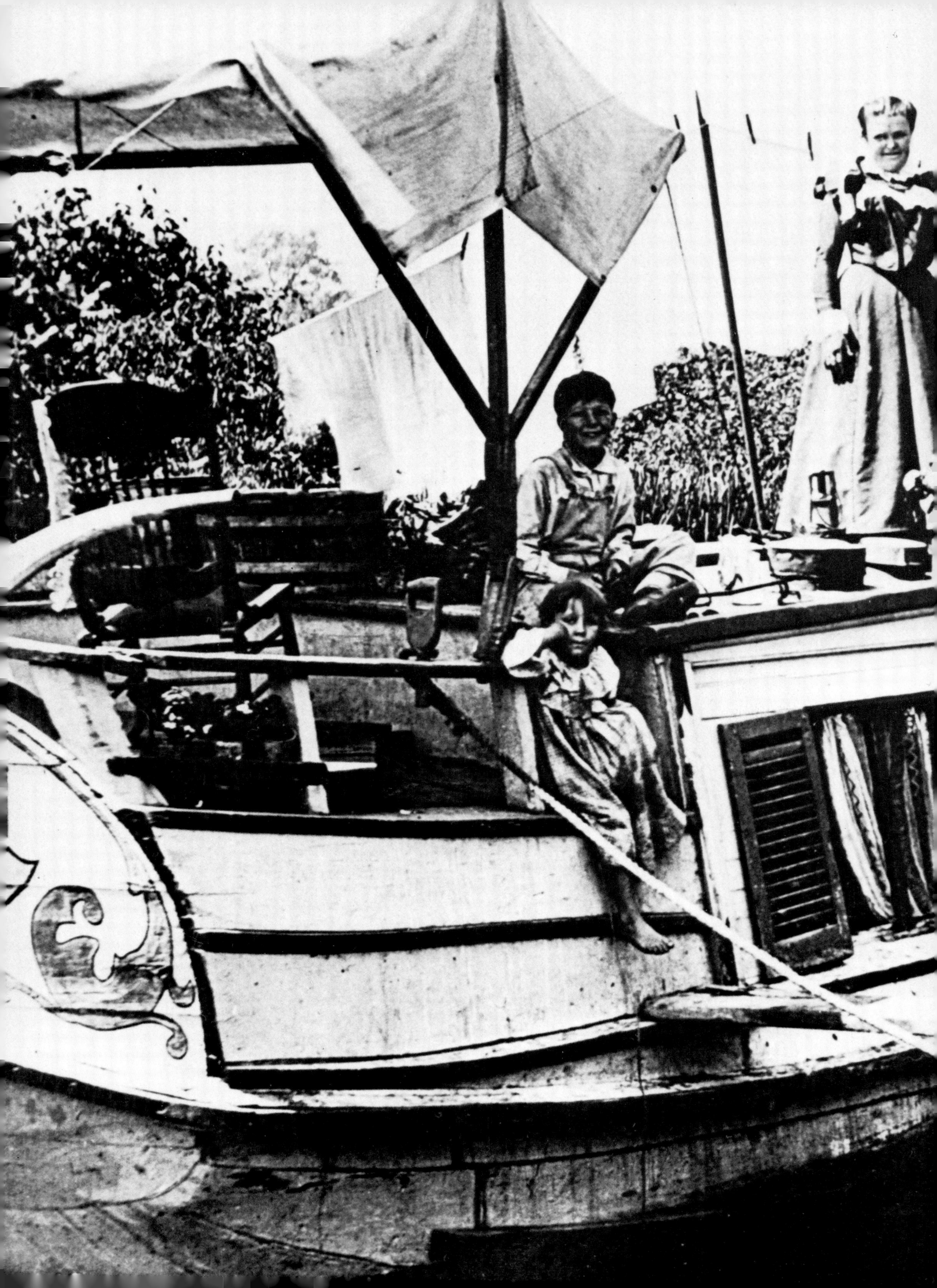

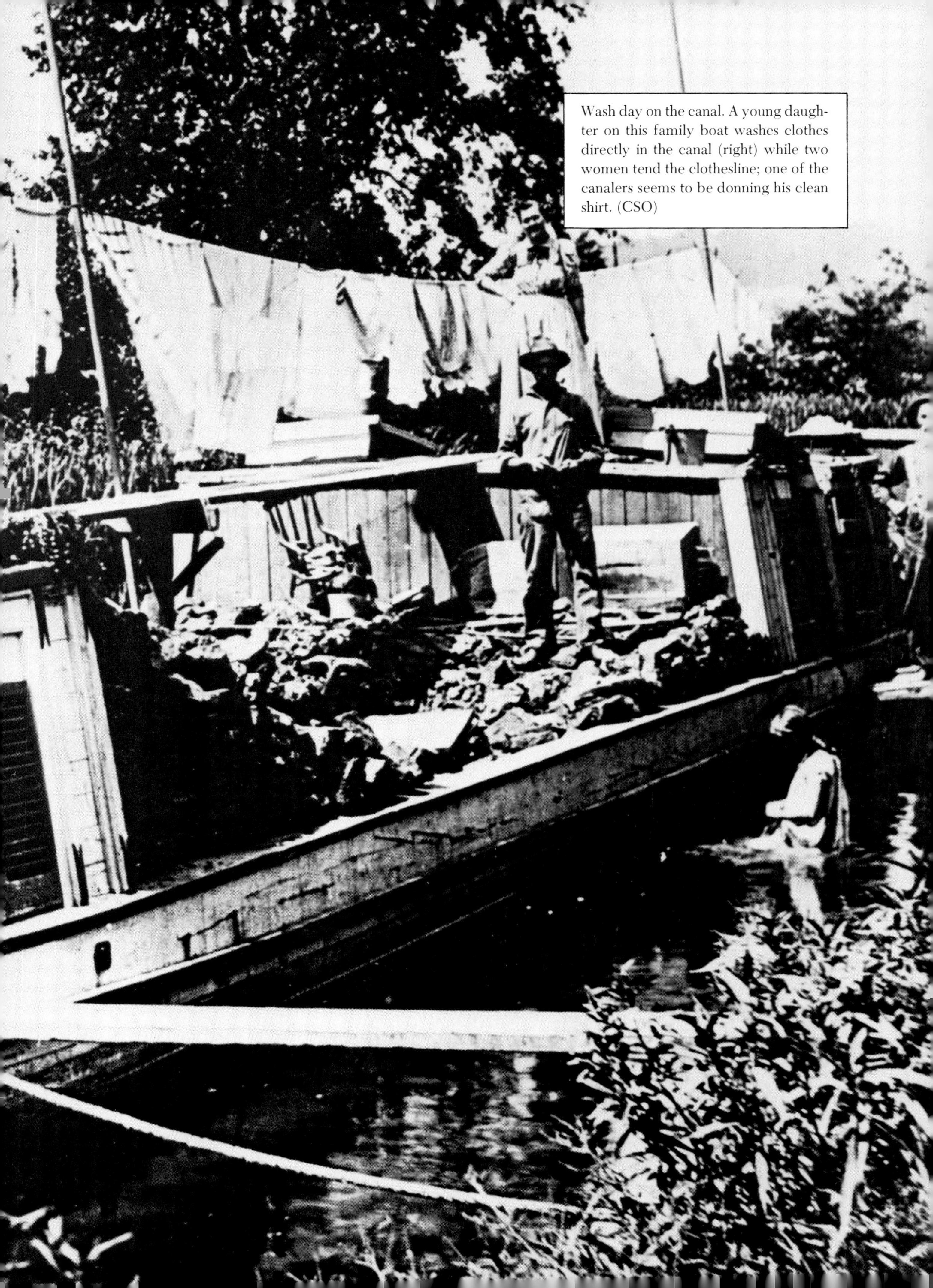

Wash day on the canal. A young daughter on this family boat washes clothes directly in the canal (right) while two women tend the clothesline; one of the canalers seems to be donning his clean shirt. (CSO)

Captain Johnny Reichert of Chillicothe gathers his family aboard the *E. Moore* for a picture at New Portage; c. 1880. (SCHS)

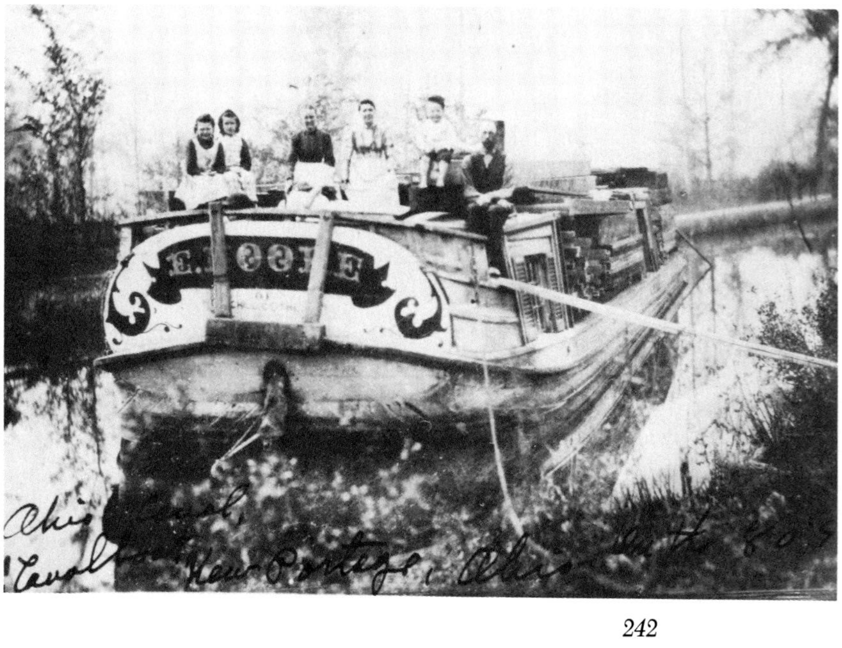

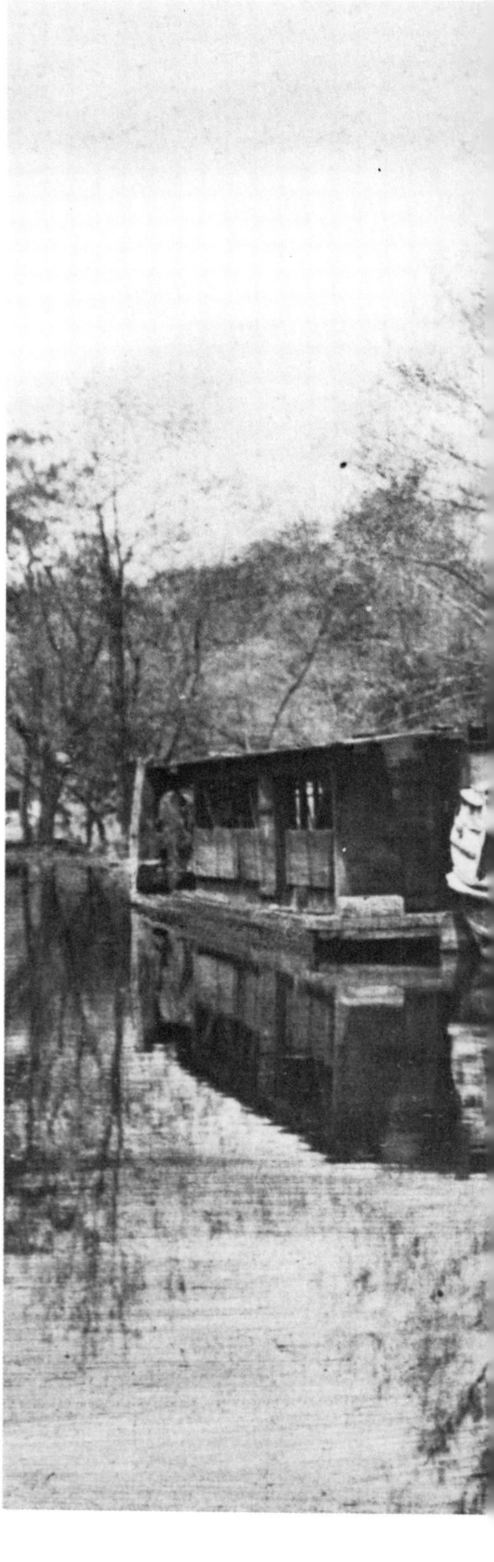

A state boat crew has invited family members aboard for a Sunday outing at 14 Mile Lock, Lock 37 on the Ohio & Erie in Valley View. Alexander's Mill (later Wilson's Mill) rises in the background; 1890. (Pockrandt Collection, SCHS)

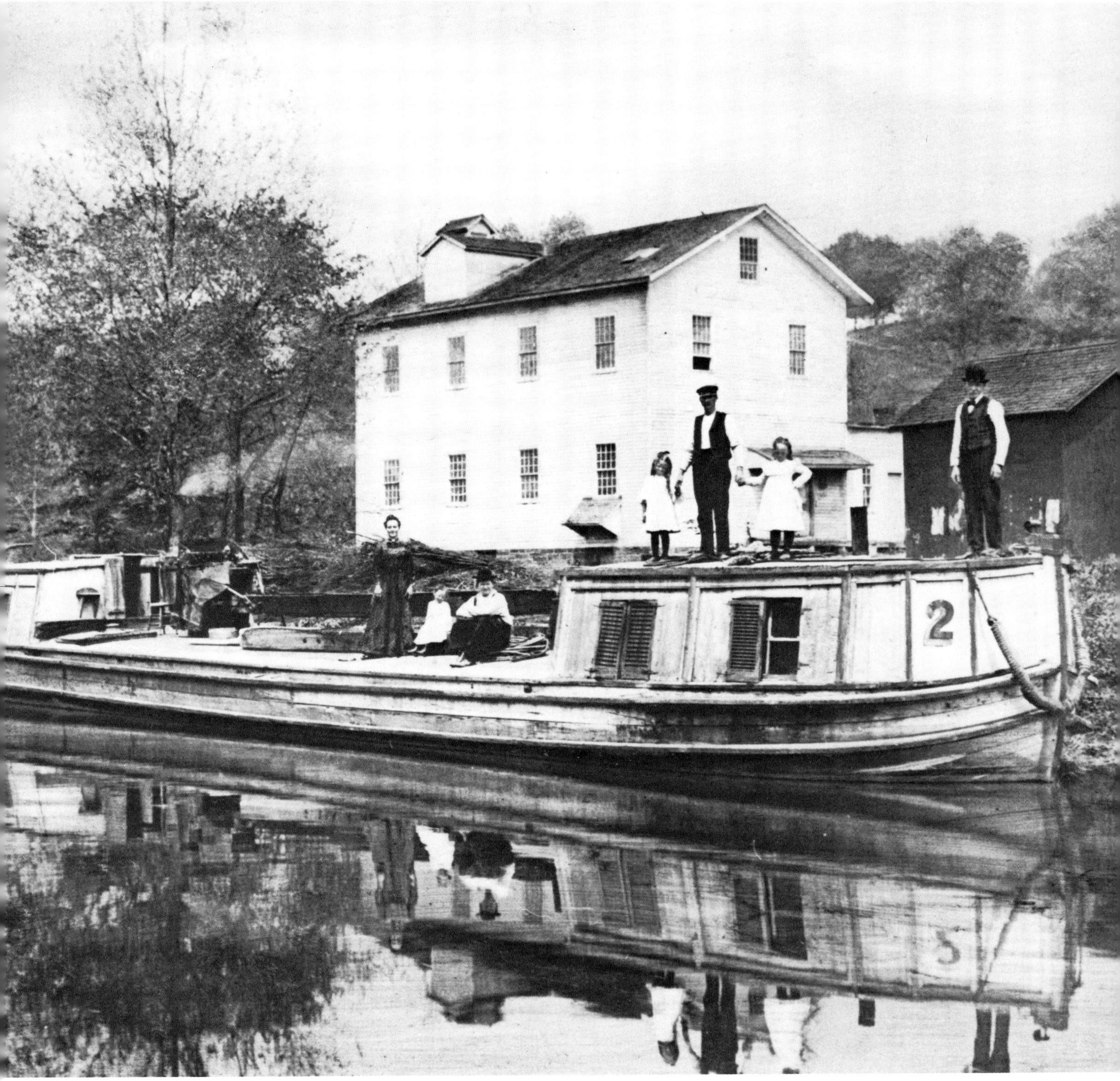

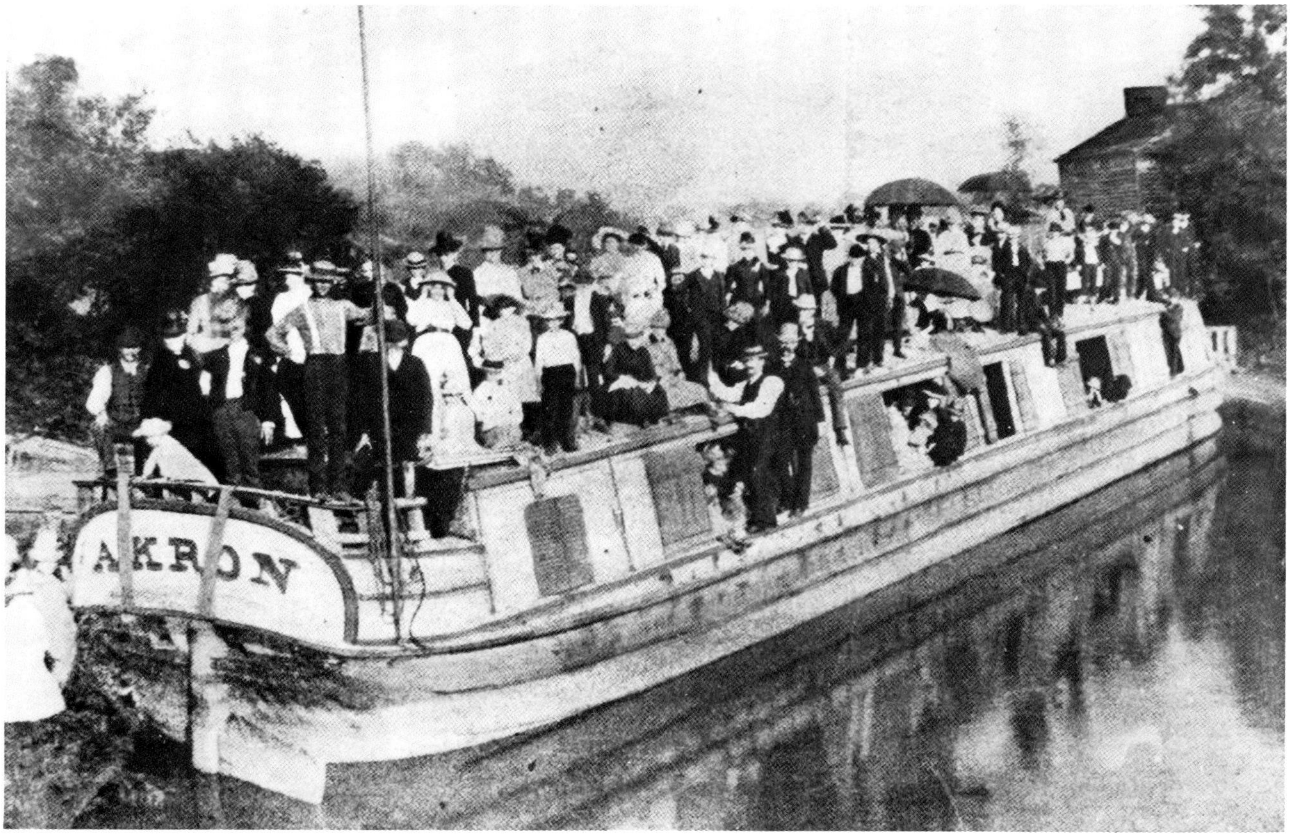

Captain Frank Lyons (above) of Newcomerstown hosts a large Methodist church picnic aboard his *Akron* on the Ohio & Erie near Cleveland; 1884 (Pockrandt Collection, SCHS). This gunpowder company boat (right), *Long Pinery* (signs on bow and forward cabin warn of dynamite), is host to a German Sunday school picnic that will include an excursion from 17 Mile Lock at Independence, near Pinery Dam and Feeder, to the weigh lock on South Dille Street in Cleveland. The mules at right are obviously ready to go when the picture taking is over. Umbrellas are to shield the ladies from the sun, primarily to prevent tanning of their fair complexions; 1886. (CSO)

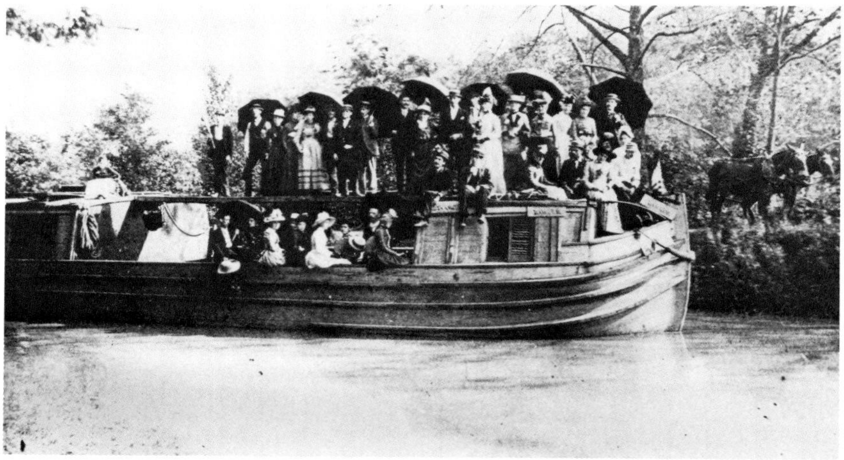

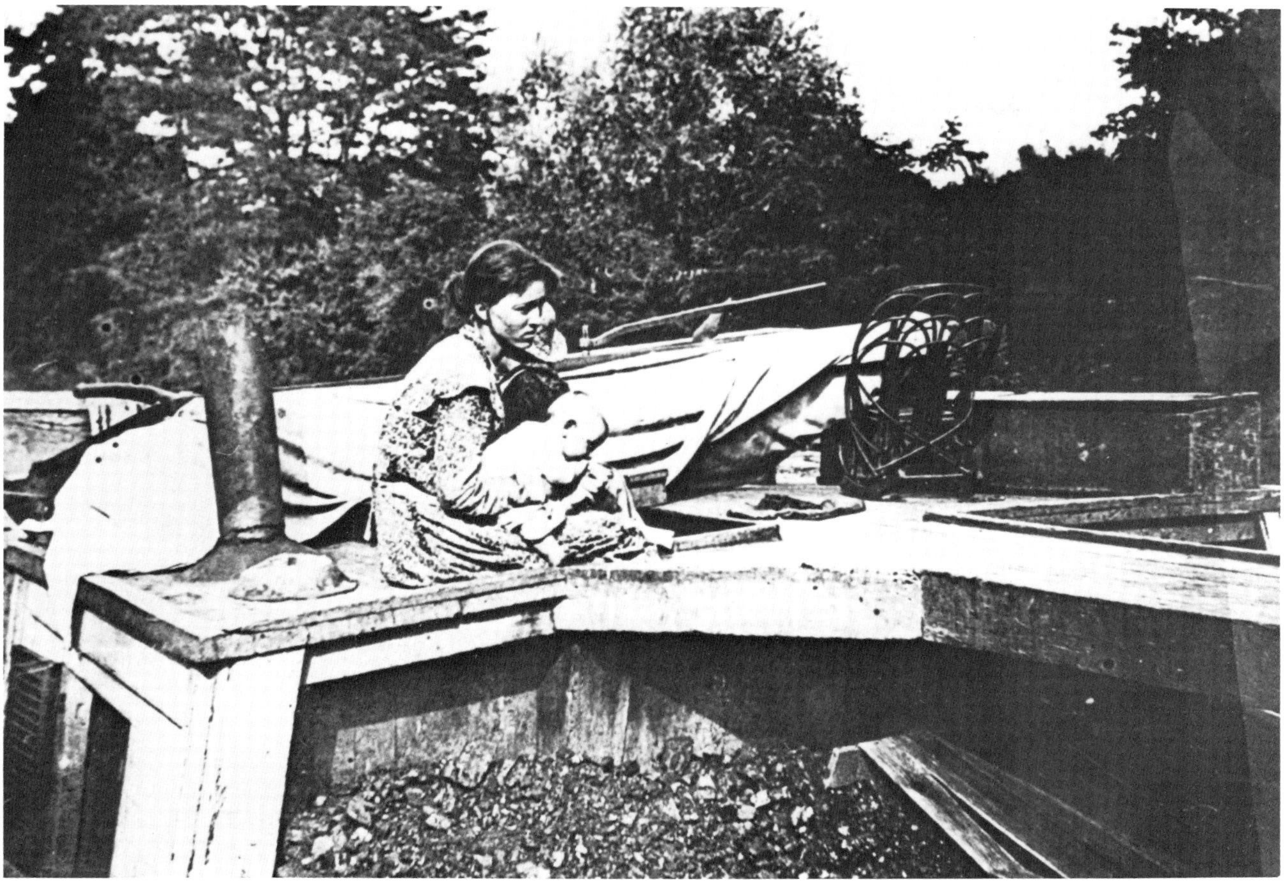

This poignant scene capturing life on a family coal boat at Navarre in 1896 needs no interpretation. (CSO)

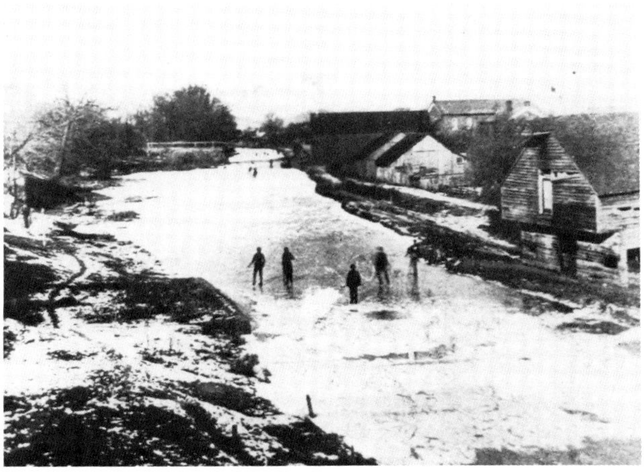

Ice skating on the canal near Trenton's coal mine was a small compensation for obligatory winter school attendance by canal children during the few weeks or months when their parents' boats were frozen in for the season. (CSO)

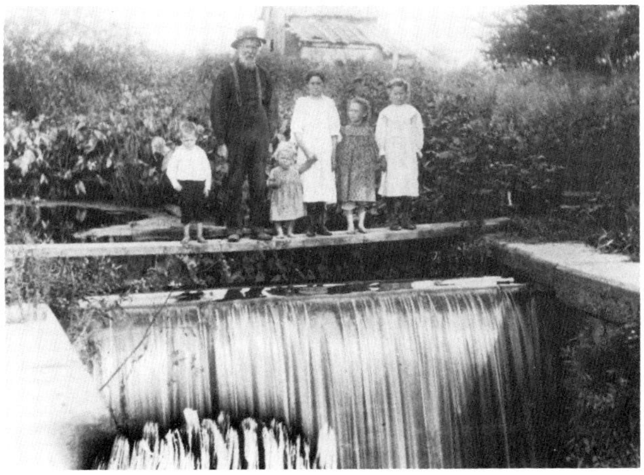

An old canaler and his grandchildren at the spillway of Lock 4 on the Columbus Feeder. (CSO)

10

Entrepreneurs, Opportunists, and Characters

Ohio has always had more than its share of innovators. It was the home of Charles Kettering, the Wright brothers, Thomas Edison, Harvey Firestone, the Seiberling brothers, and John D. Rockefeller, who began his career as a canal shipping clerk. In the early days of the automobile, hundreds of different makes were produced here, in such cities as Cleveland, Akron, Toledo, Sandusky, Cincinnati, and Dayton. The state is well known for its rubber and steel industries. Ohio contributed no less than seven native sons to the presidency.

Not surprisingly then, when the Canal Act was passed in 1825, there were a number of businessmen who correctly forecast that fortunes could be made from the social and economic changes it would bring. Of course, construction contractors (as well as farmers turned contractors) perceived their opportunities as they bid on the digging of canal sections, as well as on the building of locks, aqueducts, and other structures. But there were some entrepreneurs with greater vision—and broader ambitions.

One of these was General Simon Perkins of Warren. He had served as a surveyor and land agent for the Erie Land Company, controlled by Connecticut interests. In the course of this work, Perkins himself managed to become one of the largest landowners in Ohio. In a single transaction in 1807 (the same year he had run the abortive Cuyahoga and Muskingum Navigation Lottery) he acquired 1,003 acres by satisfying an unpaid 1806 tax bill for $4.01. Some of Perkins's holdings were in that portion of (then) Portage County, that is now the Akron–Cuyahoga Falls area.

As one of the first canal towns to develop directly as a result of imaginative scheming by local business interests, Akron exemplifies the kind of entrepreneurial drama that played out in communities all over Ohio. Historian Karl Grismer explains how Perkins's mind was working at the time:

"A man of vision and perspicacity, General Perkins realized that he owned land at a strategic point on the canal—that is, if the canal were built northward from Summit Lake . . . engineers decided that 41 locks would have to be constructed between [Summit Lake and Lake Erie]. Nearly half that number, 17, were needed in one short stretch, less than two miles in length . . . all in property which Perkins owned.

"Here was a place where canal boats would be held up in passing through the locks—several hours would be required for a boat to move from Lock 1 [at today's Exchange Street] to Lock 17 in the Little Cuyahoga valley. Passengers and boat captains would have plenty of time to stop and buy supplies, or loaf in taverns. This halting point was bound to become a business center, the general believed.

"Furthermore, two basins were to be constructed just south of Exchange Street. . . . Around these basins, the general believed, warehouses would be built and manufacturing plants started; boats would load and unload there; it would become a place where boats would be constructed and tied up in winter months; many boat captains and boat hands would live nearby.

"The general was certain that this was an ideal place to start a town. So he decided to become a town founder."[1]

But General Perkins had two problems: First,

1. Karl Hiram Grismer, *Akron and Summit County* (Akron, Ohio: Summit County Historical Society, 1952).

the route of the northern portion of the Ohio & Erie Canal was not only undecided, it was also the subject of bitter political battles raging in Columbus at the time; second, he did not control all of the strategic canal land in the area. Some of it was owned by Paul Williams, a native of Connecticut, who lived with his family in a log cabin at the present intersection of Akron's Buchtel Avenue and Broadway.

Fortunately for Perkins, he had a close friend and associate in Canal Commissioner Alfred Kelley. After fruitful discussions with both Williams and Kelley, Perkins pulled off a major coup to the considerable disappointment of such nearby competing communities as Middlebury. As George Knepper explains:

"Kelley asked Perkins and Williams to plat a town along this strategic part of the canal route. He further suggested that they donate lands for right-of-way and also town lots to be sold for the benefit of the canal project.

"Perkins and Williams employed Joshua Henshaw to make the town plat. His completed work was recorded in December 1825, at Ravenna, the county seat of Portage County."[2]

Since it would be on the summit level of the canal, they decided to call their town Akron, derived from a Greek word for "high." Williams owned approximately one-third of the platted site; Perkins owned the rest. Although he made a great deal of money from his town venture, General Perkins never lived there. But he built a classic Greek-revival stone mansion for his son, Colonel Simon Perkins, that is still called the Perkins Mansion. The city also has a Perkins School, and the little green park in the middle of Henshaw's plat is still called Perkins Square. Somehow, Paul Williams lost out in local perpetuitive honors. But, then, he never pretended to be an entrepreneur.

The canal warehouses predicted by Perkins began springing up almost as soon as the first section of the Ohio & Erie was opened to traffic between Akron and Cleveland in 1828. But for industry to develop, there had to be a substantial source of power available to operate heavy machinery—for this was before the age of steam.

In nearby Middlebury, Dr. Eliakim Crosby conceived a bold plan that would utilize both the topography of the area and the commerce wrought by the canal to expand General Perkins's town. In 1831 they joined forces to "make a river" that would provide the needed power for Akron industry.

From a diversion dam constructed on the Little Cuyahoga River at Middlebury, two miles east and a hundred feet higher, Crosby brought a stream of water down his Cascade Race that would flow south, down Akron's present Main Street, making an abrupt right turn at Mill Street, to deliver its torrent of waterpower to Crosby's other project simultaneously under construction: a five-story flour mill at Lock Five on the canal (at the present location of the Cascade Plaza Holiday Inn). With four sets of French burrstones, the mill was rated the finest in Ohio. From here to Lock Fifteen the descent was precipitous and the same water could be used over and over again to power two cupula iron furnaces, a textile machinery plant, and a distillery, all located in the canal valley below.

The result was a second hamlet called Cascade, a mile north of Akron's crossroads at Main and Exchange streets near Lock One. Cascade's main intersection would later bear the names of West Market and Howard streets. Here, Crosby persuaded James Baldwin to build a hotel, completed in 1832. Named the Cascade House, it was one of the first hotels to be built with strong links to the canal, which ran within feet of the establishment. Taking a cue from Perkins, Crosby and his fellow promoters donated choice lots to Baldwin and other businessmen as an inducement to build a downtown. By July 1833, Cascade boasted a population of 128; Akron had 329. A year later their populations were equal (at about 440 each), and they were known as North Akron and South Akron, respectively.

Riding past the complex in a canal boat in 1834, no less a personage than Prince Alexander Phillip Maximilian of Prussia recorded these impressions in his diary:

". . . Where the canal expands into a kind of small lake is Akron, a considerable town in a remarkable situation. . . . It already has an extensive trade, many neat wooden houses, stores, manufacturers, an iron foundry, and an establishment where, by means of a [water] wheel, bedsteads and other articles of furniture are turned."

Indeed, waterpower from the canal itself became a significant source of revenue for the state.

2. George W. Knepper, *Akron: City at the Summit* (Tulsa, Okla.: Continental Heritage Press, 1981).

Utilizing the excess water that would otherwise be wasted over the spillway, many lock mills were built on the canals. Their location also had the advantage of facilitating receipt of raw materials and shipment of product. Some grist mills sat astride the canal to facilitate delivery of the flour they ground directly into canal boats below.

Many other industries located on the banks of the canals: boatyards, stables, coal dealers, tanneries, gunpowder works, general stores; industries requiring process water, including many paper mills in the Dayton area (some of which had their own "hydraulic" canals), and rubber companies in Akron. Mostly on the Miami & Erie were a number of ice works, which harvested ice in winter by sawing it from the surface of the canal—or from adjacent "ice ponds" dug for the purpose—preserving it into the summer in ice houses, where it was buried in insulating sawdust.

A great deal of industry travelled on the canal itself. Most boat owners, captains of their own vessels, were self-employed. But there were also floating barber shops, store boats, sign-makers, print shops, and other artisans who served boat crews and canal families. There were even groups of itinerant actors, singers, and dancers who provided entertainment to communities along the route. Indeed, many canalers were collectors of songs and stories, and some also fancied themselves to be poets. The literary merit of most of these efforts is exemplified by the following quatrains from Pearl R. Nye's rhyming catalogue of every city, village, and point of interest along the Ohio & Erie from Cleveland to Portsmouth. The piece is meant to be sung to the tune of "Shortnin' Bread":

> Groveport has the longest straight,
> Jewel, beautiful and great.
> At Canal Winchester, much I saw,
> The bullfrogs dance with their mother-in-law.
>
> At Lockville was a whisky tank—
> The fish were drunk from bank to bank
> While Carroll, where they dance and play
> And make you think that all is day.
>
> Akron had all things in store,
> Which made us love her more and more;
> Old Portage Locks, yes, and Yellow Creek,
> A place your thirst could always slake.
>
> At Millersport, south of Black Hand,
> 'Twas a privilege great, exceeding grand,
> To see those gems of history shine,
> It always thrilled this heart of mine.

It goes on that way for forty-six stanzas—each one followed by a chorus. Such romanticism was not limited to canalers like Nye. The Ross County Historical Society recounts how one entrepreneur suddenly turned canal captain, was literally carried away with his zeal:

James Emmitt, the wealthiest man in Waverly, personally funded a party to welcome the water into Waverly in 1832. Water welcoming had been a major cause for celebration in a number of communities in the state. Emmitt was so happy at the thought of the canal being completed that he forgot his politics and fed everyone—Whigs, Democrats, and non-partisans alike.

The section of canal between Chillicothe and Waverly had been built through a gravelly loam, and the water took its time coming down. The celebration was delayed for hours. When the water finally arrived, with two boats right behind, Emmitt bought one of them on the spot for $600. He immediately loaded the *Thomas Worthington* with friends and cast off—vowing to be the first boat to reach Portsmouth, the canal's southern terminus.

Emmitt didn't quite make it; he ran out of water about two miles short of his goal. A valley legend insists that Emmitt's boat, responding to his urging, at times got ahead of the water. More likely, the water soaked into the dry, porous canal bed faster than the flow replenished it, leaving Emmitt high and dry. His adventure resulted in public censure by canal engineers for his use of the canal before it was officially opened. But it is likely that Emmitt would feel that his immortality in legend made it all worthwhile.

A prototype canal entrepreneur and real estate developer, General Simon Perkins persuaded the canal commissioners to route the northern section of the Ohio & Erie Canal through property owned by himself and a local settler, Paul Williams. To this end, Perkins had a town platted in which he and Williams donated lands for canal right-of-way, plus town lots to be sold for the benefit of the canal project, and, indirectly, themselves. Boats would be delayed at least half a day negotiating the 15 locks in a single mile at this location, giving them plenty of time for shopping and entertainment; Perkins also correctly predicted that canal warehouses and industry would grow and thrive in his village, which he called "Akron," from a Greek word meaning "high place," since it was at the summit level of the canal. (Painting c. 1832, by Waldo & Jewett, New York commercial portrait artists, courtesy SCHS.)

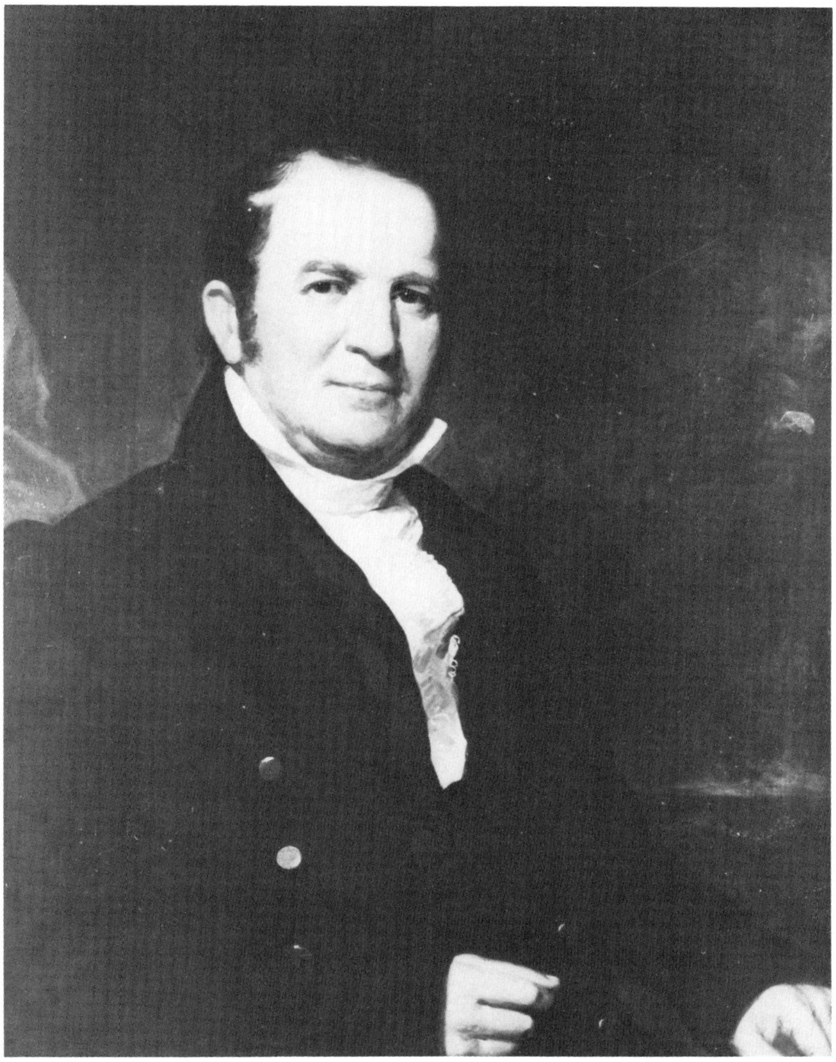

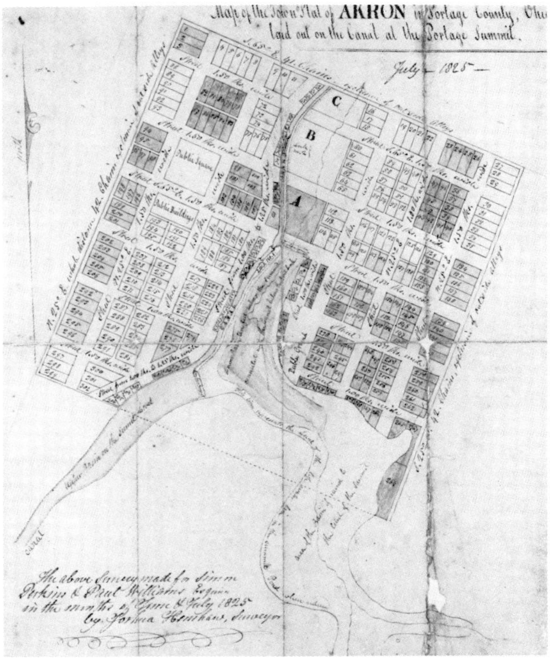

Akron's original town plat as laid out by Joshua Henshaw for Simon Perkins and Paul Williams, and recorded in Ravenna on December 6, 1825; the route of the canal is shown through the center of town. The village was located on the watershed ridge that is the north-south Continental Divide, accounting for the large number of locks (initially 41) required between Lake Erie and the Summit Level here. (From the original plat, courtesy SCHS.)

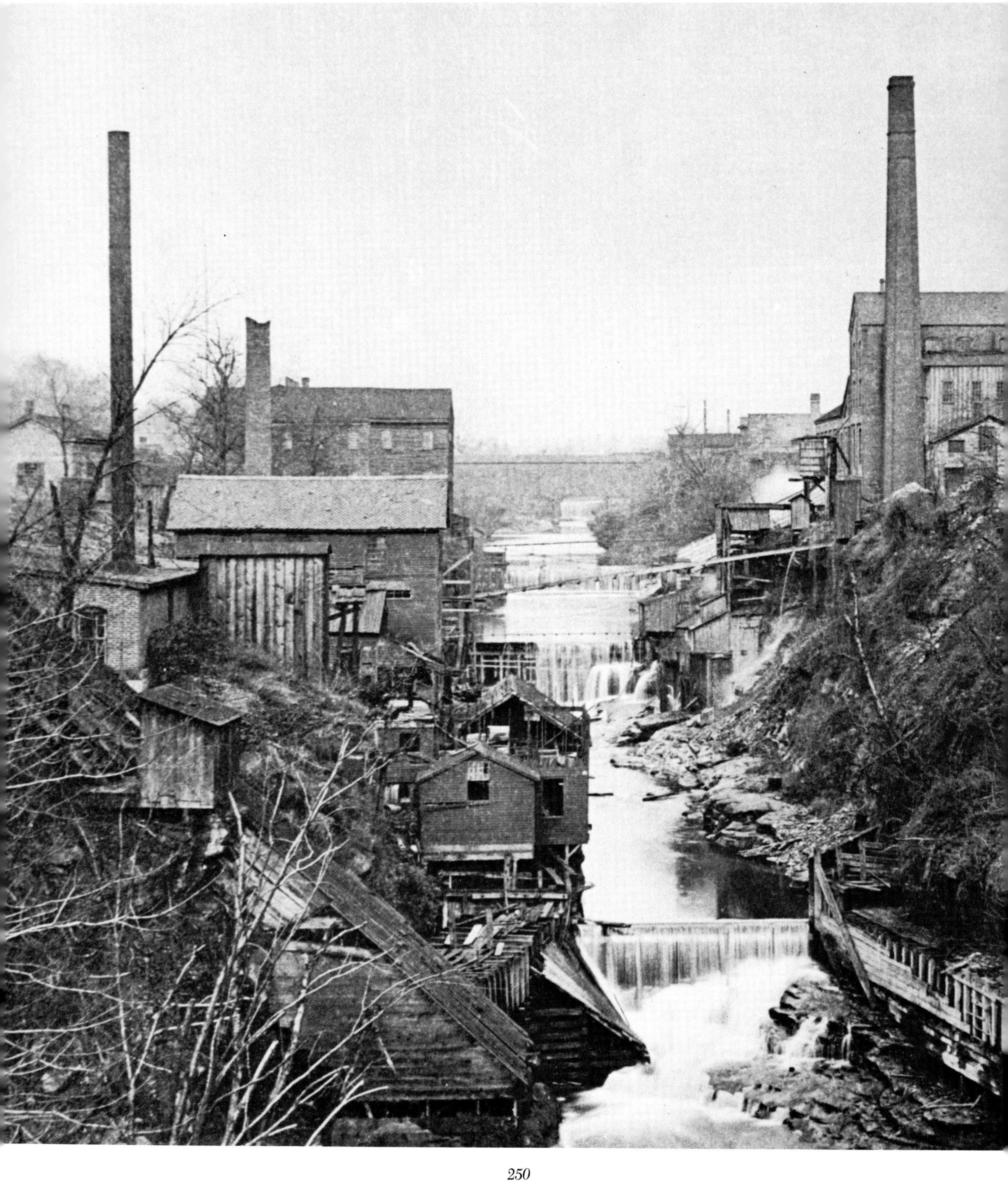

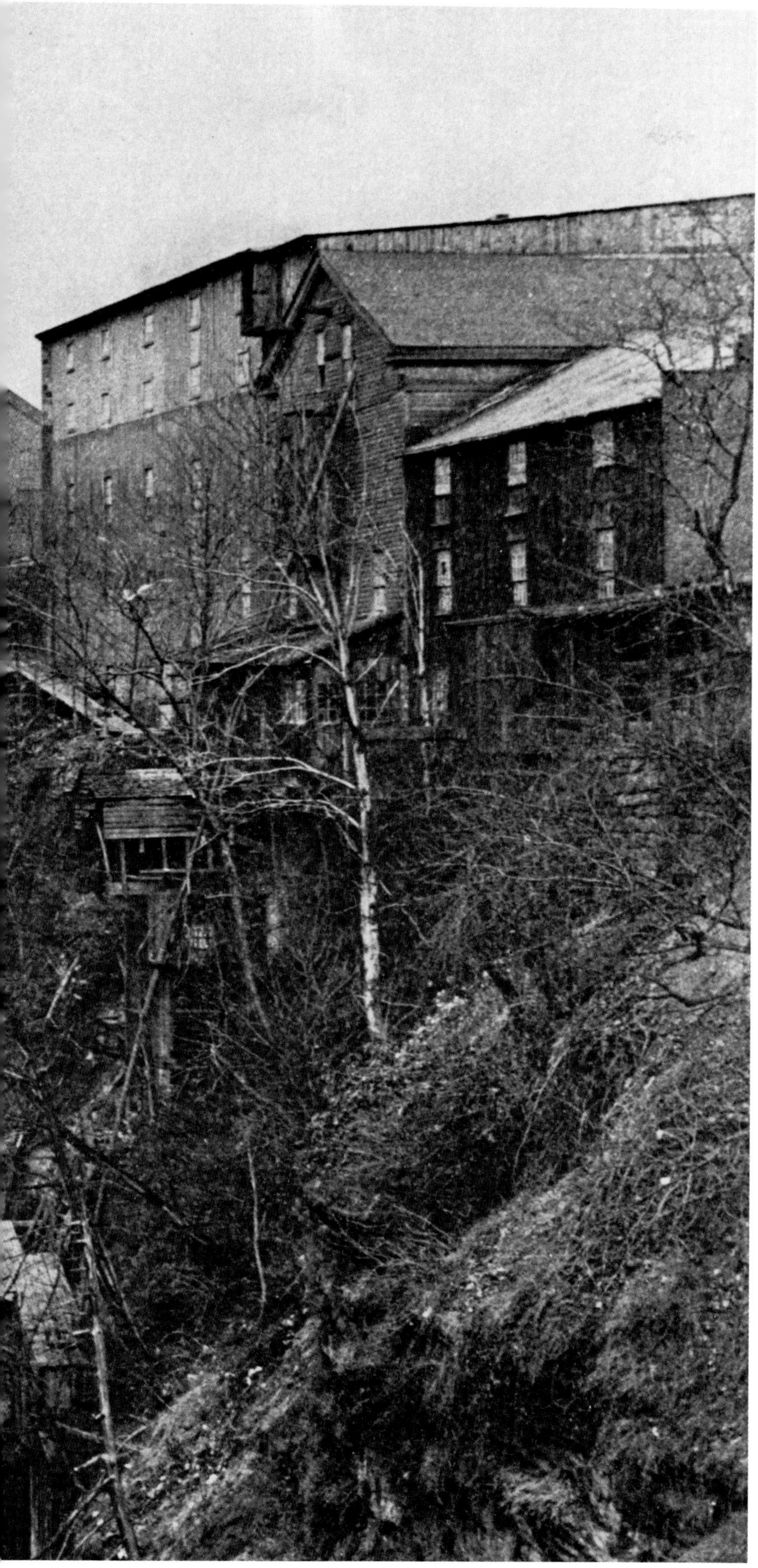

Below is an engraving of Dr. Eliakim Crosby, a physician turned entrepreneur, who conceived the bold plan of constructing a two-mile Cascade Race leading from a diversion dam on the Little Cuyahoga River at Middlebury (near the present site of Goodyear) to the canal valley in Akron, thereby making extensive waterpower available not only for his own flour mill, but for a number of additional industries that could use the same water at lower elevations to power their manufacturing equipment. (Karl H. Grismer, *Akron and Summit County*, SCHS, 1952.)

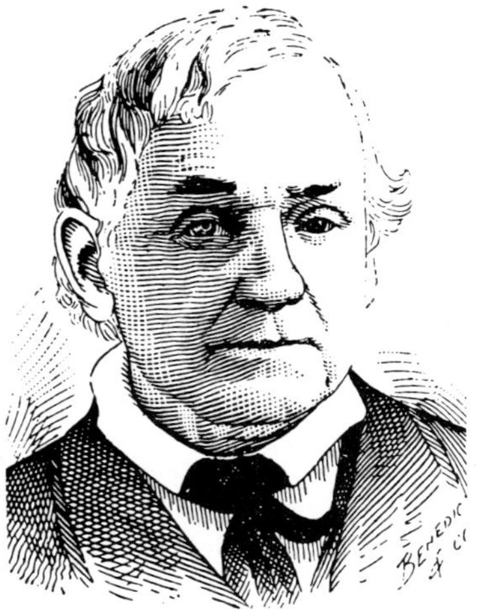

Below Crosby's mill in the canal valley, the water from the Cascade Race could be used over and over to power two iron furnaces, a textile machinery plant, a distillery, and the lathes of a furniture maker. Although no photographs of these industries survive, the scene shown at left in the nearby industrial valley of Cuyahoga Falls must look similar. (SCHS)

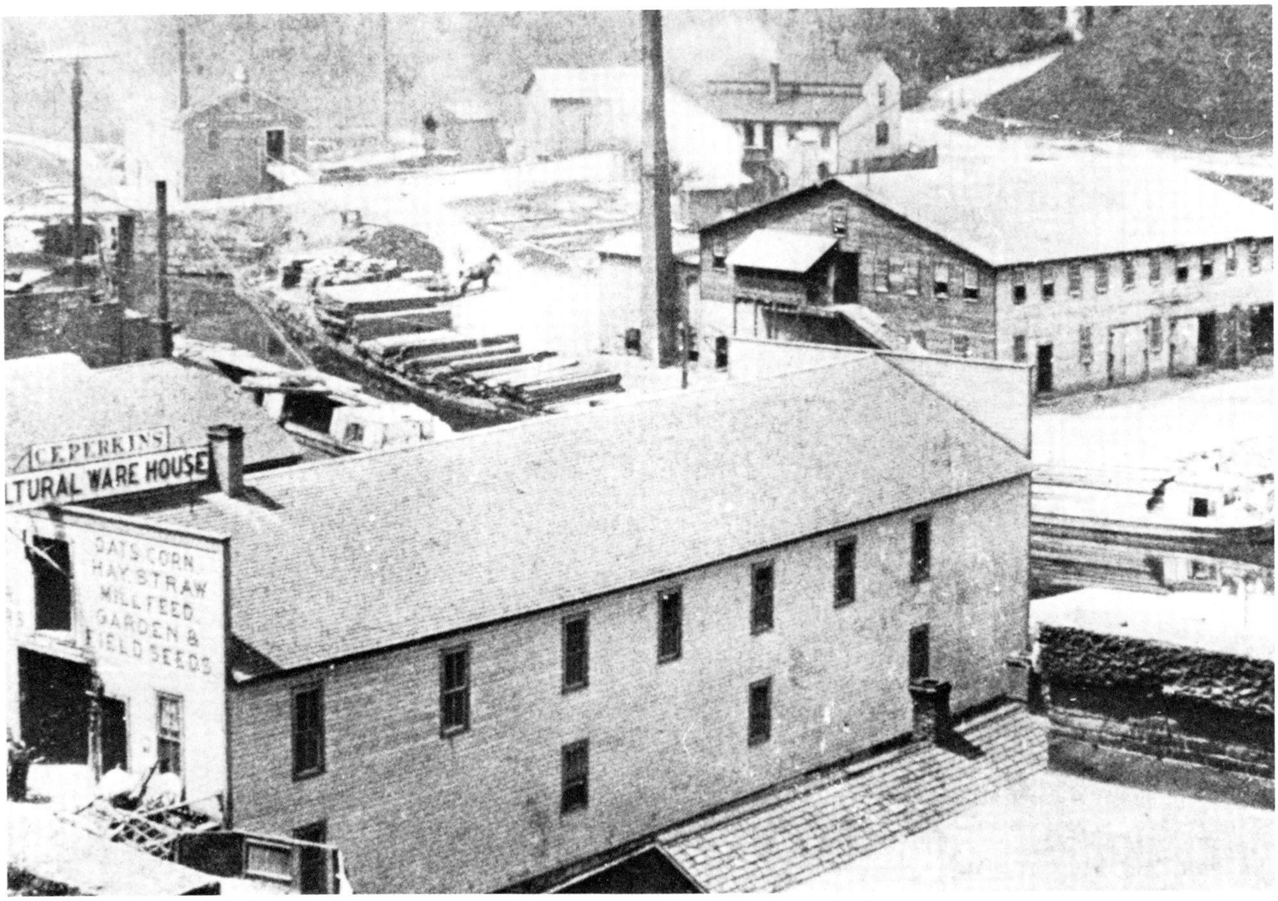

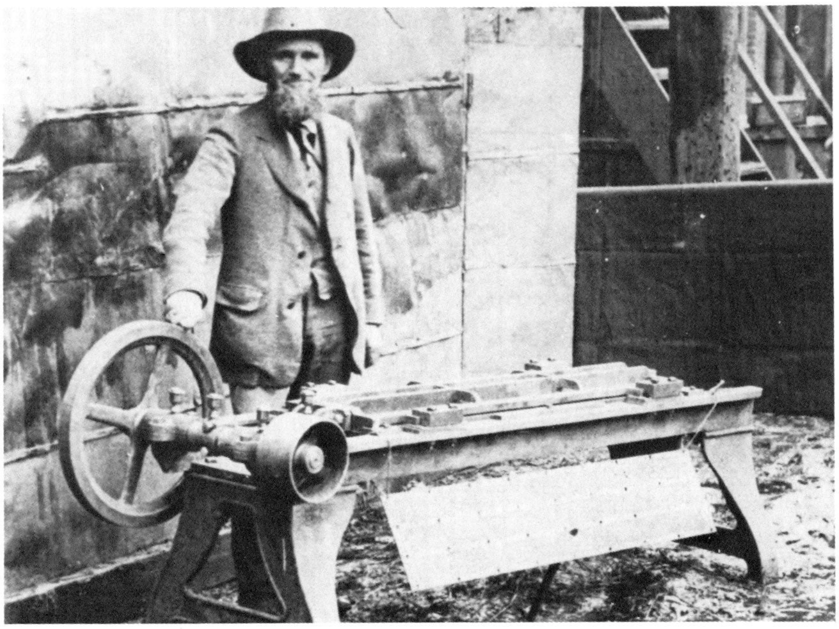

Canal warehouses lined both banks of the waterway in Akron (above). Canal boats are visible at the rear of the foreground warehouse to both the left and right. The business, it will be noted from the sign, is operated by Charles Ezra Perkins, grandson of General Simon Perkins. (SCHS)

Mill-owner Ferdinand Schumacher (left) and the machine he invented shortly after immigrating to this country in 1850 and settling in Akron. The device cracked the hulls of cereal grains to make rolled oats, the raw material for his oatmeal product. Schumacher built a chain of mills in Akron, several on the canal, and founded the American breakfast cereal industry that eventually became Quaker Oats. An ingenious innovator, Schumacher also invented pneumatic levitation, to blow his cereal products between his mill locations through miles of pipes. (Pockrandt Collection, SCHS)

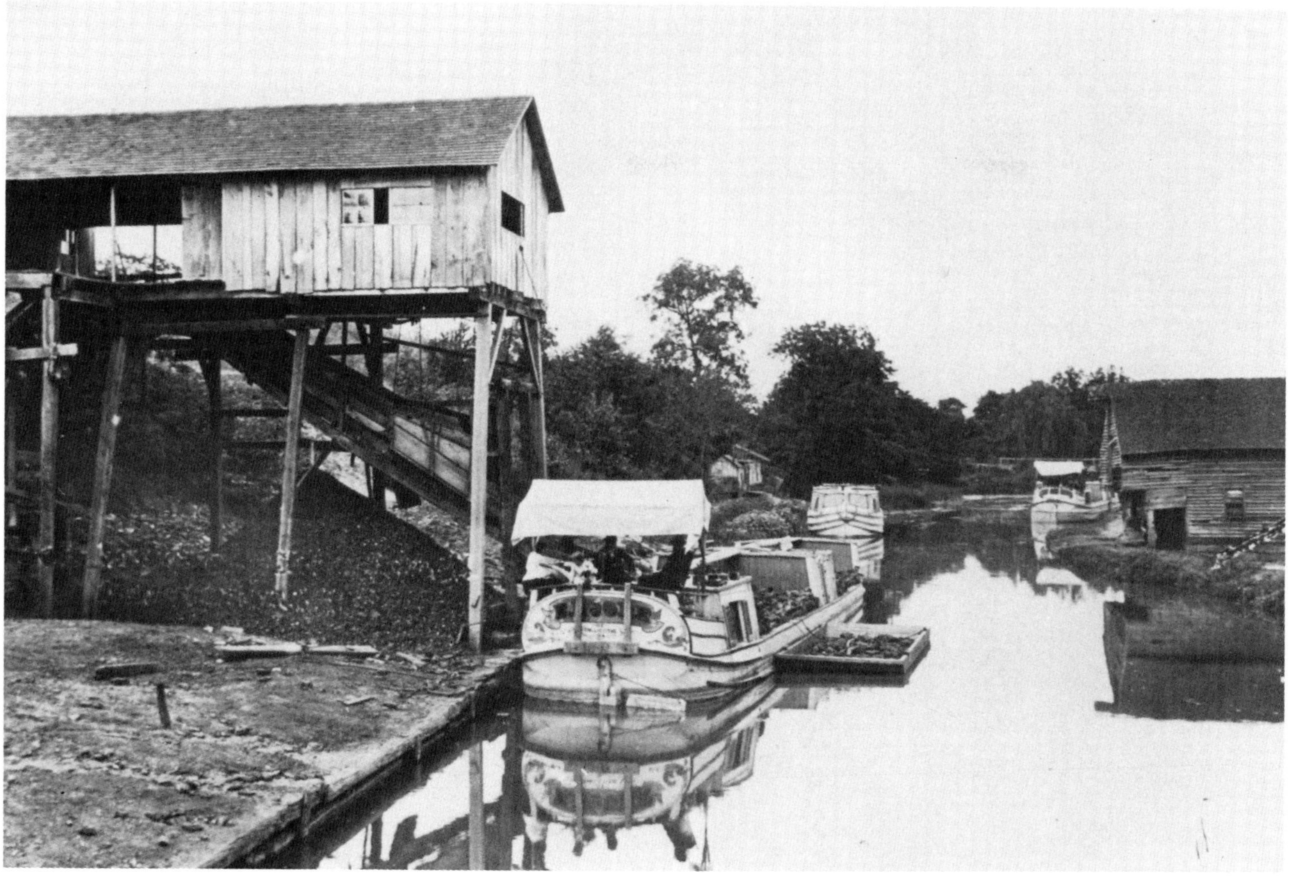

Hank Minnick's coal tipple, at Trenton, was built to directly load canal boats. Here, Captain Johnny Reichert of Chillicothe (under the awning at the boat's stern) has just taken on a load; 1902. (SCHS)

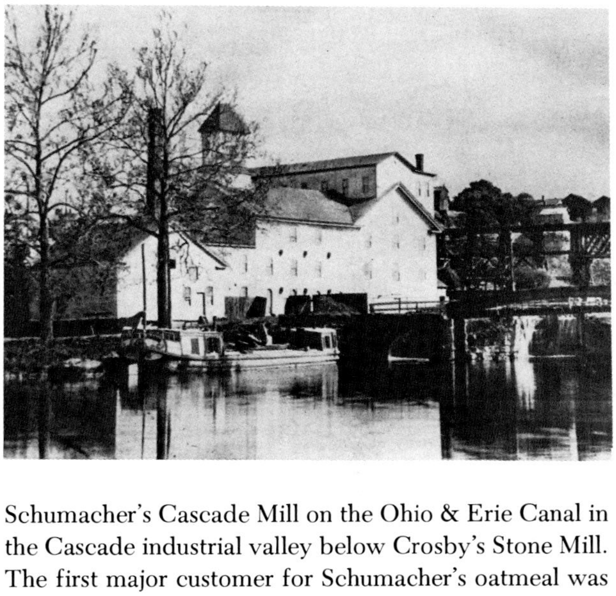

Schumacher's Cascade Mill on the Ohio & Erie Canal in the Cascade industrial valley below Crosby's Stone Mill. The first major customer for Schumacher's oatmeal was the Union Army during the Civil War. (Pockrandt Collection, SCHS)

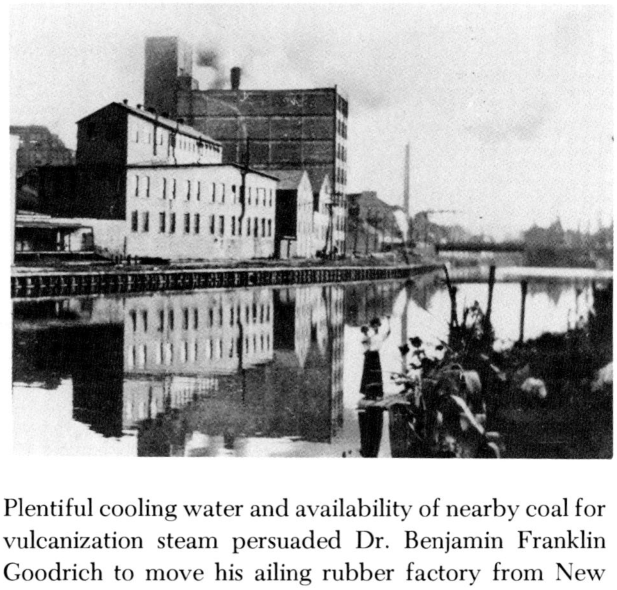

Plentiful cooling water and availability of nearby coal for vulcanization steam persuaded Dr. Benjamin Franklin Goodrich to move his ailing rubber factory from New York to the banks of the Ohio & Erie Canal in 1880, thereby founding Akron's rubber industry; 1916. (Pockrandt Collection, SCHS)

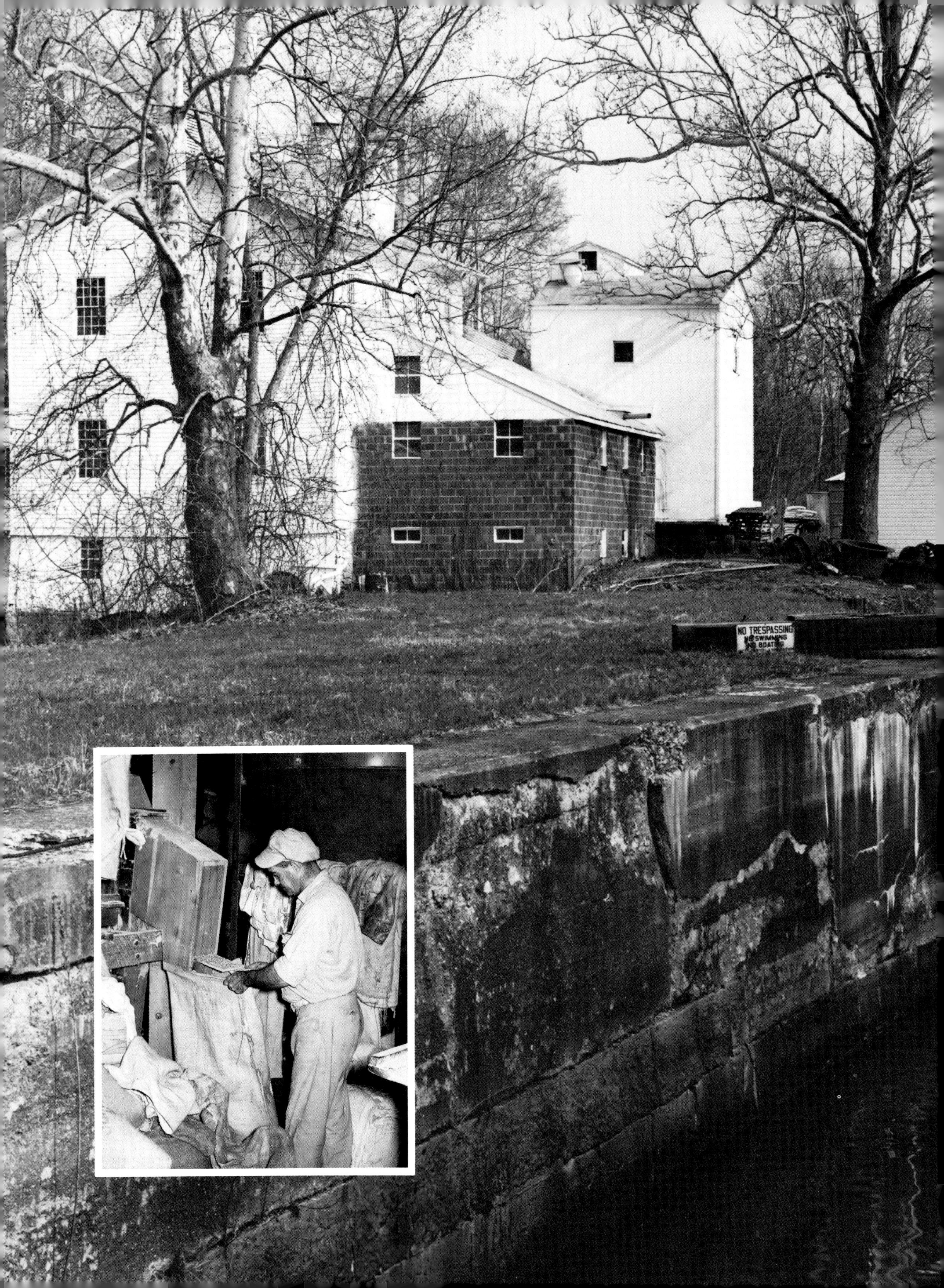
NO TRESPASSING

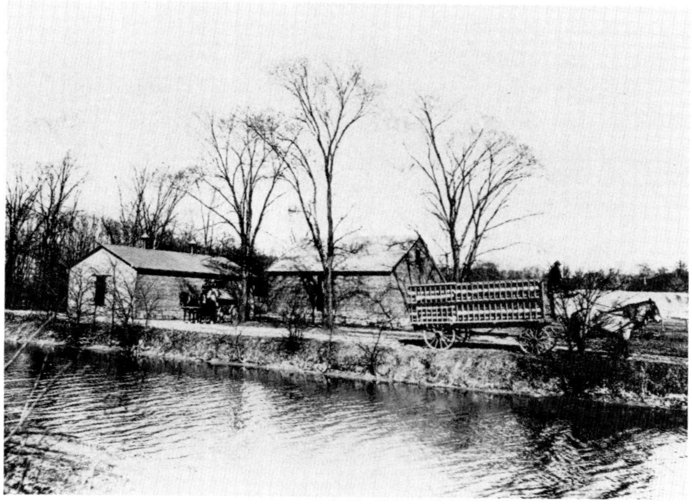

Another supplier to the Union Army was the Austin Powder Works, south of Cleveland. Here, a wagonload of black powder barrels leaves the powder mill, which brought in its raw materials of sulfur, charcoal, and saltpeter by canal. As a precaution against fire, the company's charcoal plant was a short distance down the canal at Lock 5. The Ohio & Erie is in the foreground; c. 1865. (CSO)

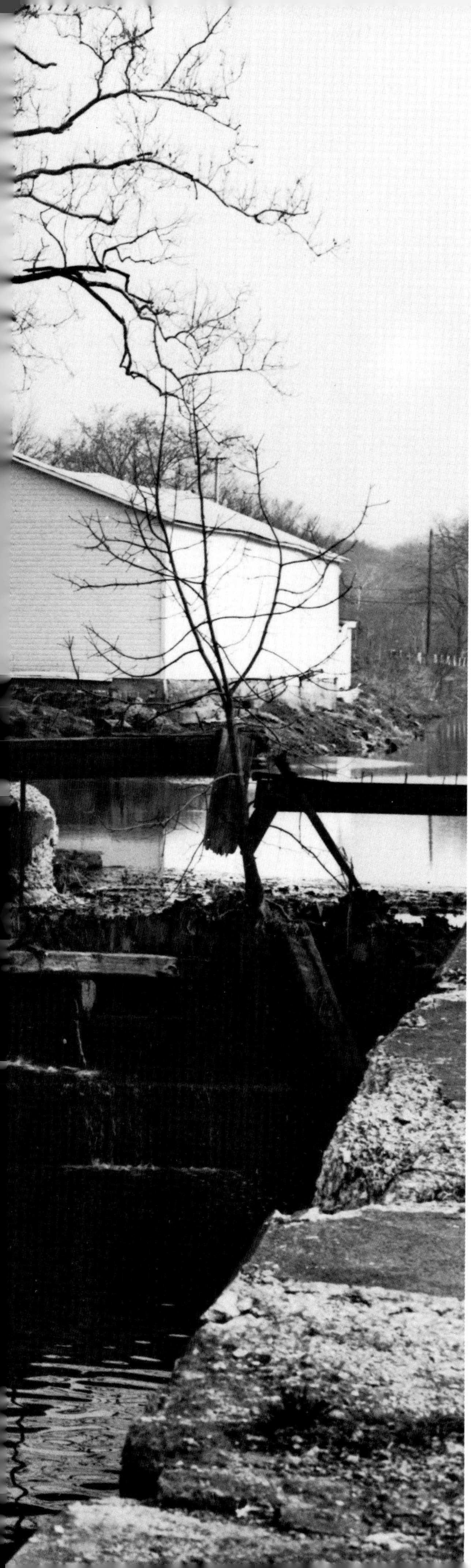

The waterpower available from the canal made "lock mills" a significant source of revenue for the state. Here, at Wilson's Mill on the Ohio & Erie at Valley View, the height of the fall for potential water power may be seen in the lock next to the Ohio & Erie's last operating lock mill (which utilized a turbine, not a waterwheel); 1986. (Photo by Jack Gieck.)

Inset, Fourth-generation miller Thomas Wilson demonstrates that some operations in Wilson's Mill were still carried out using available waterpower when this picture was shot in 1951. Wilson explained that preservation of the water-powered equipment was not practical, but that the family felt a kind of nostalgic responsibility as owners of the historic property. (Photo by Jack Gieck.)

Zoar Mills, at Zoar, Ohio, could load four boats at one time. Construction of the Ohio & Erie Canal first saved this communal society of German Separatists when they avoided foreclosure on their land by contracting to dig the canal through their property. Later, the community built up a sizable export trade in wheat, flour, and iron ore, and owned several boats of its own. (Canal Visitor Center, CVNRA, courtesy Cloea Thomas.)

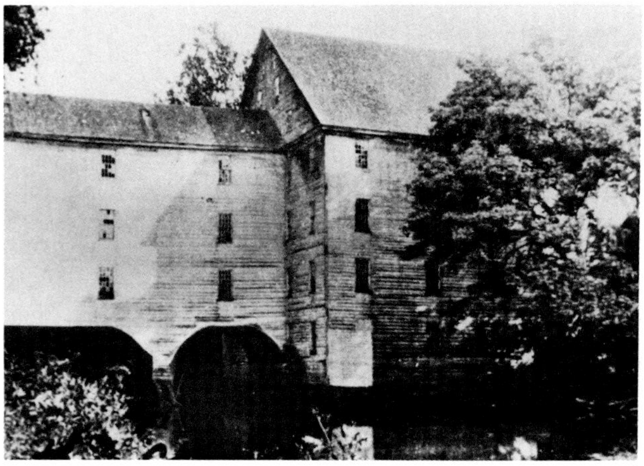

A close-up of this stone in the guard lock on Zoar Feeder clearly shows the date of its construction by the ambitious Zoarites. The canal was a mixed blessing for the community. In the first year of its operation it brought an epidemic of cholera to the town. Late in the century, the fashionable clothes and appearance of relative luxury enjoyed by canal passengers contributed to the dissolution of the communal society. (CSO)

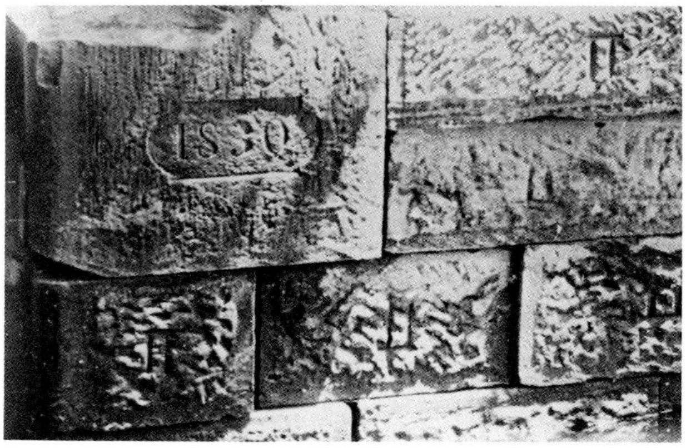

*"**Canal locks regulated** the size of the boats that went through. Their maximum size was 78 feet long and slightly over 14 feet wide. Often canalers built their own boats. If a person could build a boat and pay the toll . . . he could operate on the canal. Few were craftsmen, however, and boat operators would often have to pump out their boat at each stop.

"The cost of building [and equipping] a freight boat averaged $2100—including the mules for towing. The average individual boat owner had two boats. Boats were usually bought with a down payment and a mortgage on the boats for the balance . . . to be paid out of boat earnings. Boat owners usually had six head of stock (horses or mules), together with collars, harnesses, fly nets, and other necessary trappings for operation on the canal.

A boat owner's credit had to be good with drydock owners and canal boat suppliers in order to get the proper repair and supplies the operator needed to keep in business. . . . Accidents would occur on the canal, with boats occasionally having their cabins torn off by hitting low bridges—when the boats were lighter after depositing their freight in the previous town.

[Occasionally, an animal would fall into the canal.] Because the loss of a team, or even a single mule, was so devastating, escape holes were sometimes provided in the canal walls to save animals who might otherwise drown in the foul water. The animal escape hole was a ditch cut across the towpath to the water's edge. The opening in the towpath was planked over. A ramp was provided so that the animal could be led by the driver back up the side of the towpath to regain its place in the team. . . . (Terry D. Wright, *Miami Valley Weekend,* April 15, 1984)

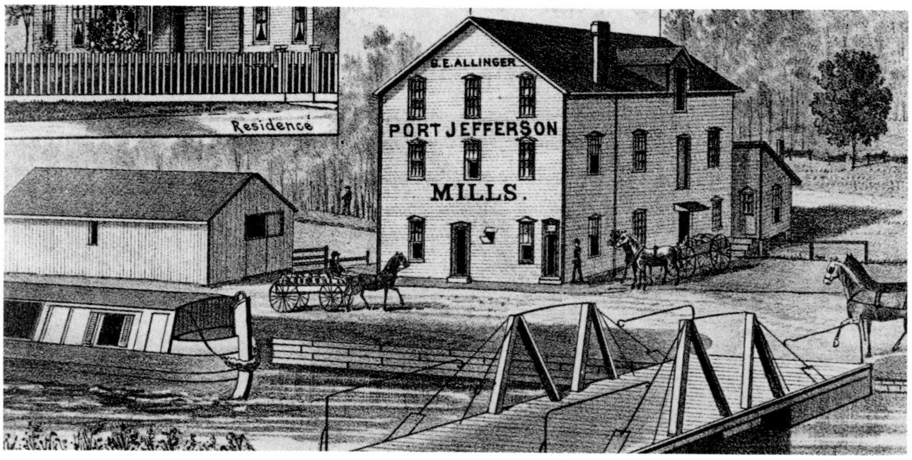

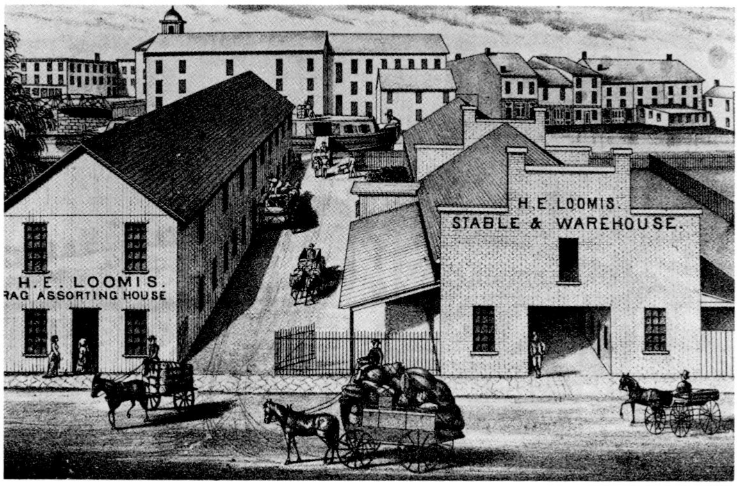

The engraving at top shows a prosperous lock mill on the Miami & Erie Canal's Sidney Feeder at Port Jefferson, with grain boat in left foreground. (*History of Shelby County, Ohio*, R. Sutton & Co., 1883/WRHS.)

The engraving immediately above shows warehouses along the Miami & Erie Canal (in background) at Piqua. (*Miami County Atlas*, 1875)

The engraving below shows a tannery on the bank of the Miami & Erie at Minster, with a line boat in foreground. Most tanneries had signs advertising "Cash Paid for Hides and Skins." (*Auglaize County Atlas*, 1880)

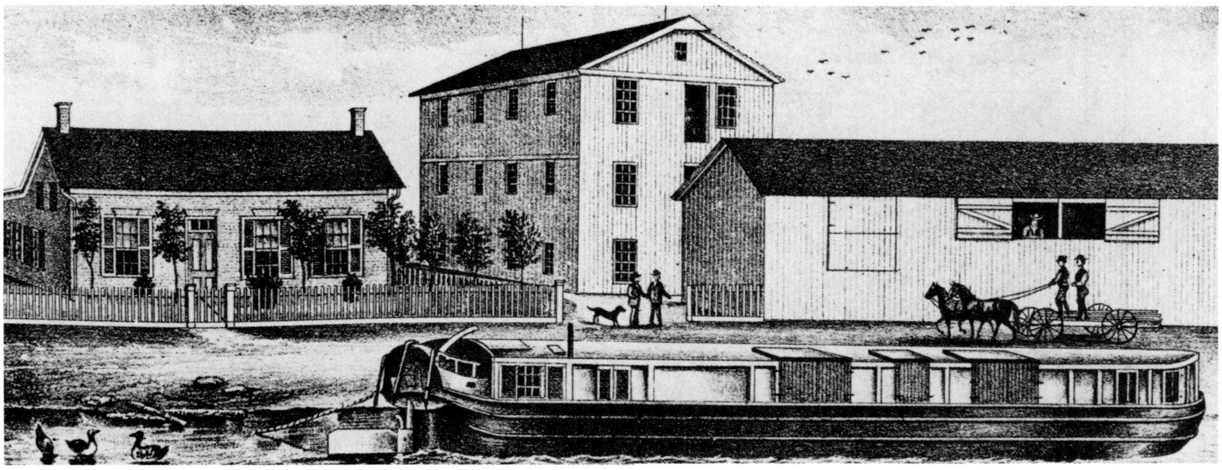

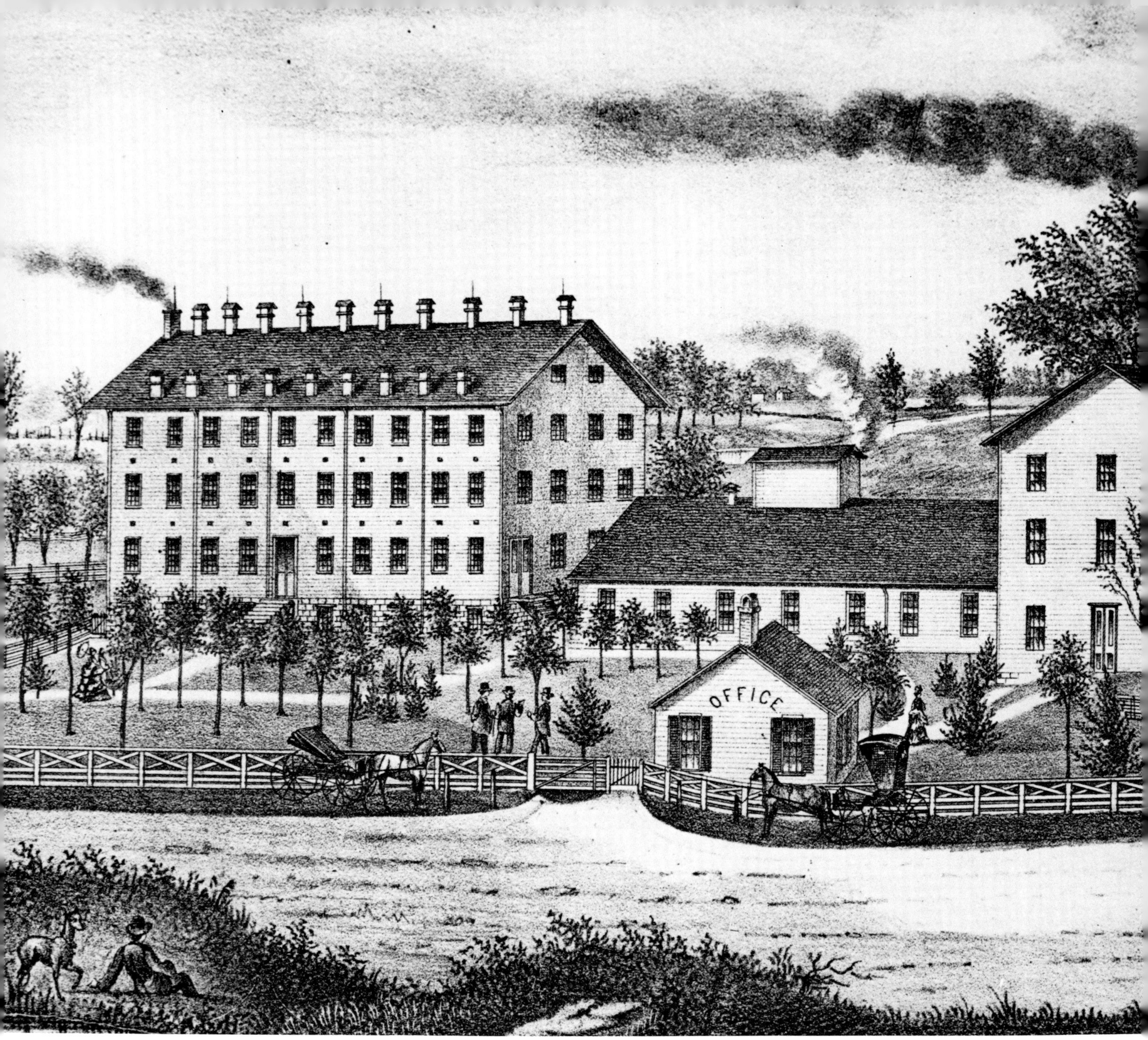
OFFICE

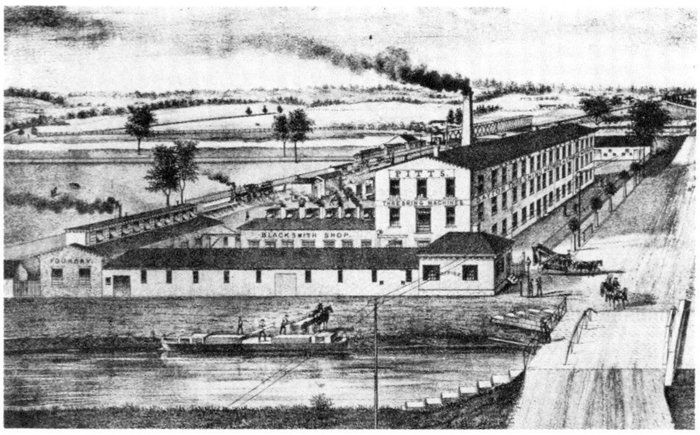
BLACKSMITH SHOP
FOUNDRY

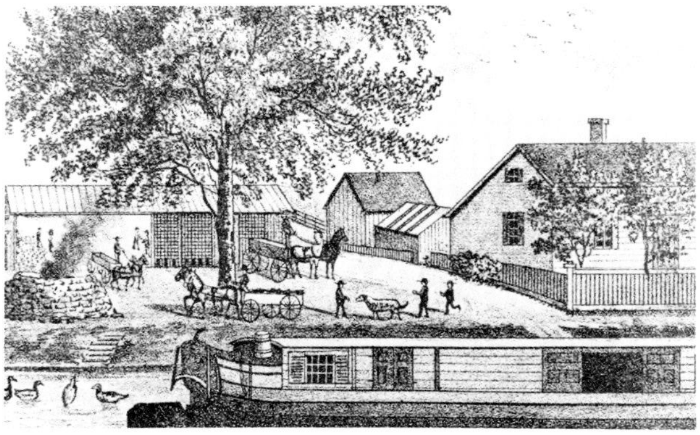

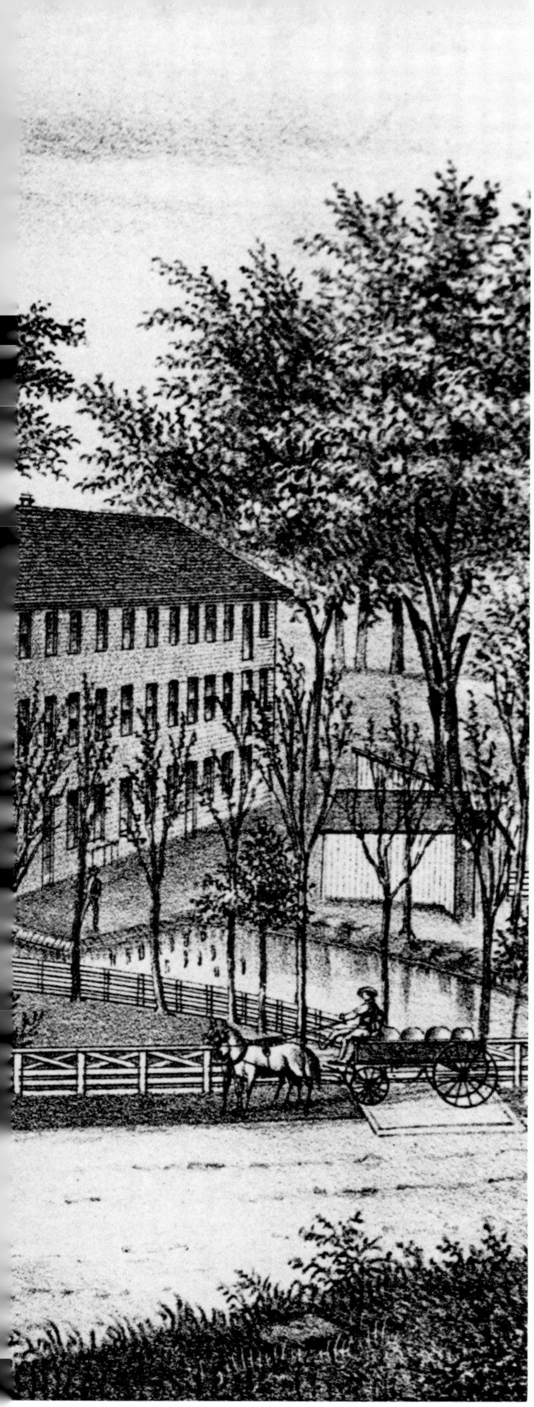

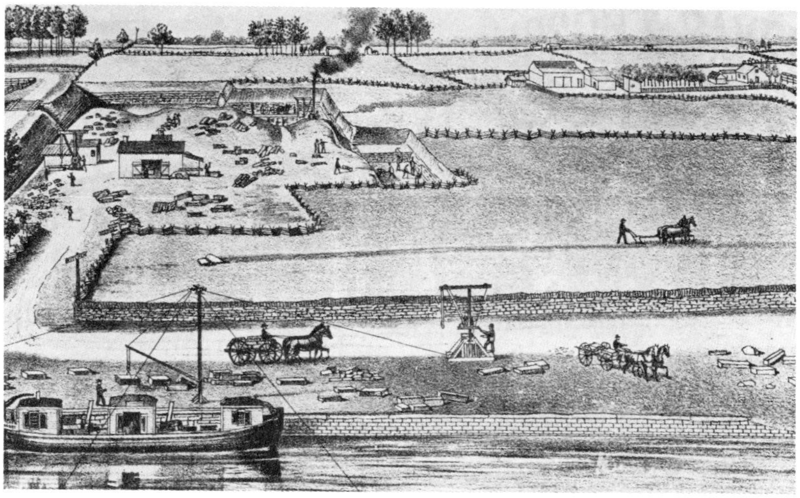

This stone quarry has an elaborate crane to load heavy stone onto canal boats. Taking obvious advantage of his location on the Miami & Erie Canal, the proprietor advertises "Special Attention Given to Orders from a Distance." (*Miami County Atlas*, 1875)

The Excello Paper Company at Middletown (which became an international firm in the twentieth century) had a mill race (right and background) on the hydraulic canal. Paper companies also used large volumes of process water, available from the canal, to make up their slurry of pulp and fibers. The size of the office structure implies an economy of administrative personnel. (*Butler County Atlas*, 1875)

A Miami & Erie canal boat (left) unloads raw materials for a Dayton threshing machine company that operates its own foundry. Telephone lines (two of them) were drawn in by the advertiser's artist at a time when the number of telephone exchanges in the nation numbered in the hundreds; real telephone wires do not appear in Ohio photographs until the 1880s. (*Montgomery County Atlas*, 1875)

If this brick yard (right), which also makes mortar in its lime kiln, were not located on the bank of the canal, its bricks would have little chance of finding a wide market because of the cost and difficulty of shipping the heavy, relatively cheap product by wagon. The canal system helped exploit Ohio's abundant natural resource of high-quality clay. (*Auglaize County Atlas*, 1880)

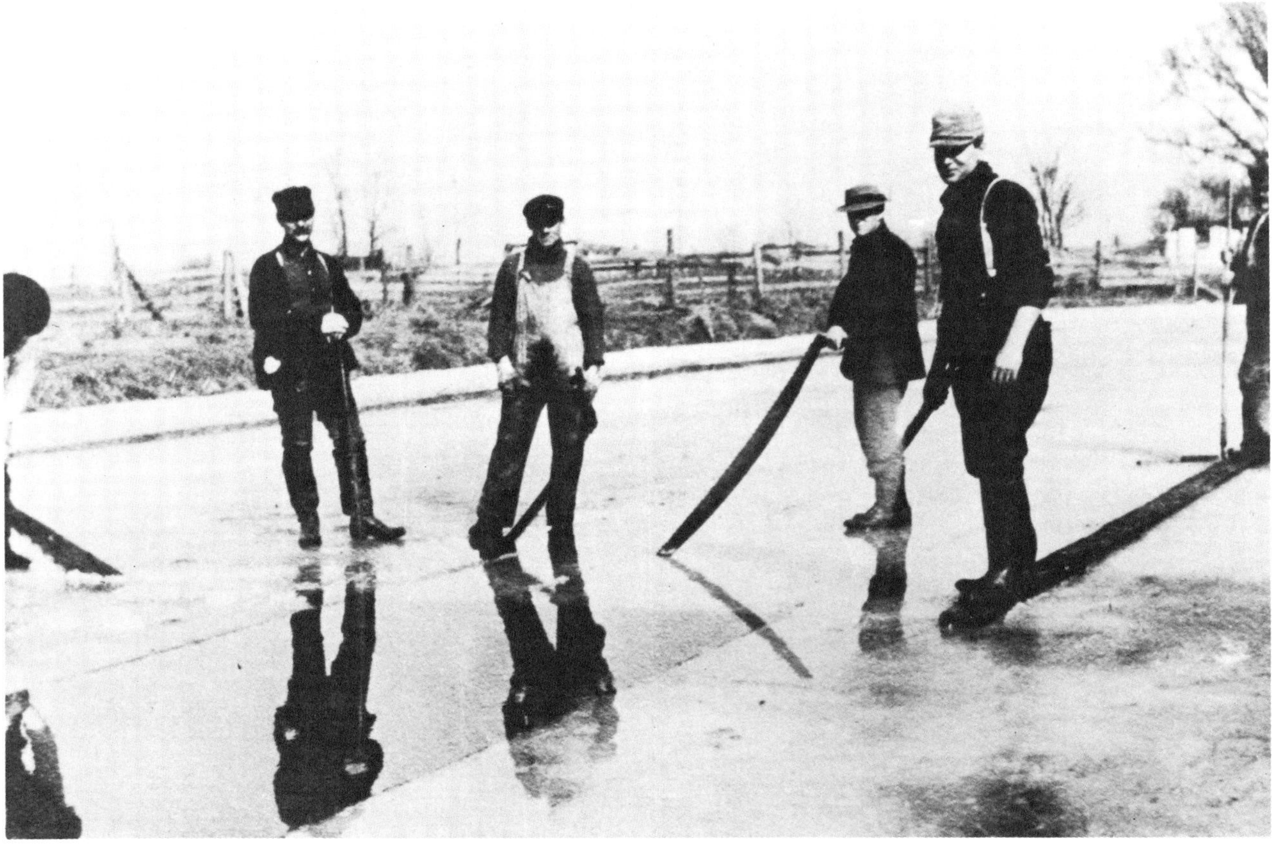

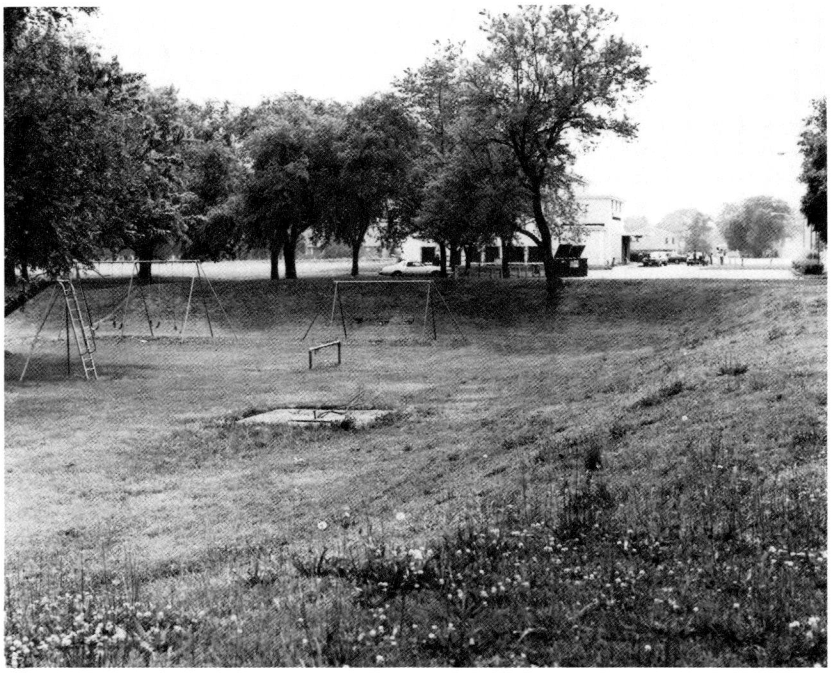

A quite different kind of industry on the canals was ice harvesting. Ice removed—either with hand saws or horse-drawn saws—from the frozen surface in winter was preserved into summer in warehouses, insulated under sawdust. The men in the photo above are harvesting ice from the surface of the Ohio & Erie at Groveport the old-fashioned way—with hand saws. Some ice-harvesting teams ate five meals a day. (RVF)

The remains of a nineteenth-century ice pond in Middletown (left) at the edge of the Miami & Erie Canal. The shallow area enclosed by the dyke was flooded with canal water for winter harvesting; 1986. Mile: 208. (Photo by Jack Gieck.)

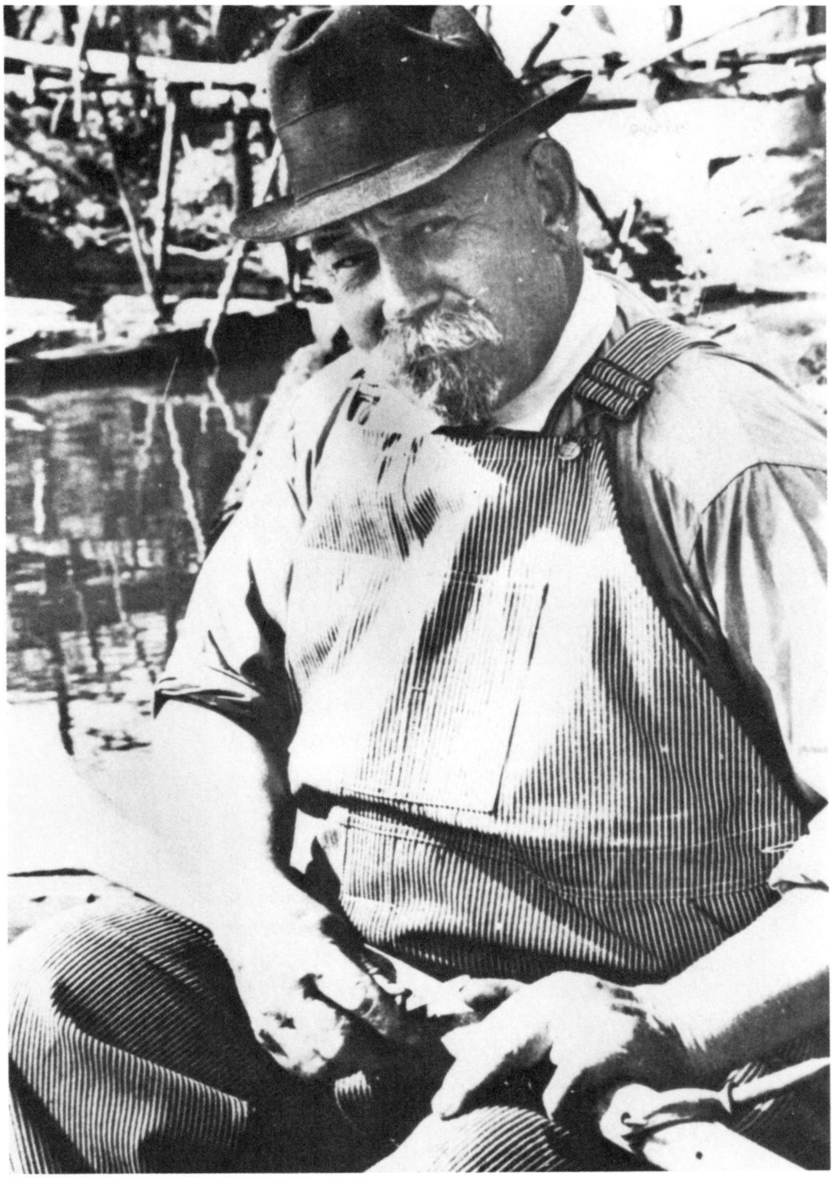

Pearl R. Nye was born (1872) on the canal boat *Reform* while it was docked in Chillicothe; fifteenth of eighteen Nye children, poet, canal historian and bard, collector of songs, and certified canal character, he preferred to be remembered as "the last of the canal captains." Actually, he never quite qualified as such, since, although he operated a boat, it belonged to his mother. His passion for collecting songs was shared by his mother, who accompanied him to camp meetings, where they learned gospel songs together; but his repertoire also included innumerable popular and traditional songs from the nineteenth century, ballads of American history, songs from minstrel shows, black and white dance tunes, and above all, canal songs—for many of which he made up additional verses (Richard H. Swain, "Captain Pearl R. Nye's Ohio Canal Songs," *The Gamut*, Cleveland State University, Fall, 1983). (Canal Visitor Center, CVNRA,, courtesy Cloea Thomas.)

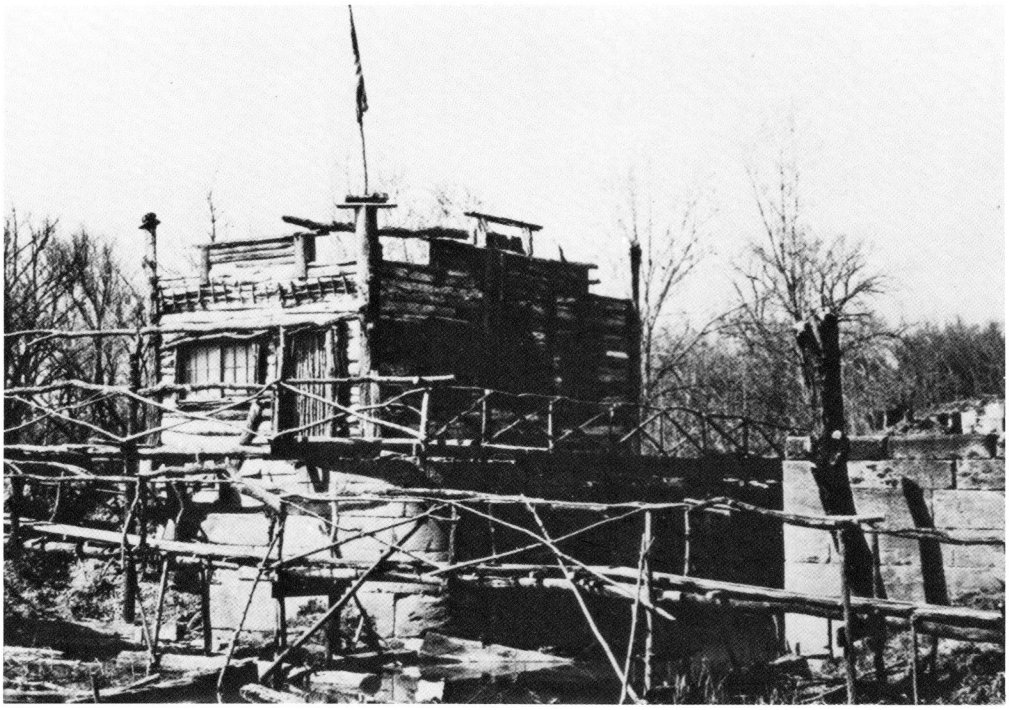

When the Ohio & Erie Canal came to an end in 1913, Nye was devastated for a time. In the late 1920s he built this cottage on an abandoned lock at Roscoe and called it Camp Charming. He lived there for almost two decades. (RVF)

Over the years, Camp Charming grew; building materials included fragments of canal boats and other memorabilia. Nye is seen at extreme left, where he has just caught a fish from the canal. During this period he wrote the Library of Congress, his congressman, and many others offering his services as a bard. In 1938, and again in 1942, he was invited to sing at the National Folk Festival in Washington, D.C.—two high points in his life (Swain, "Captain Pearl R. Nye's Ohio Canal Songs"). (Canal Visitor Center, CVNRA, courtesy Cloea Thomas.)

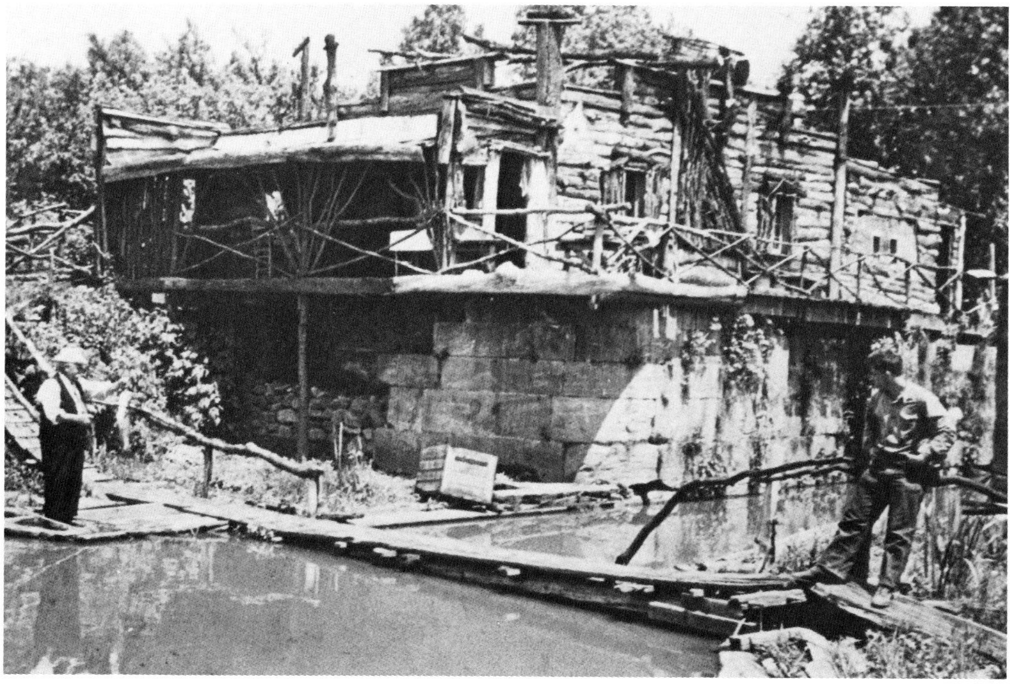

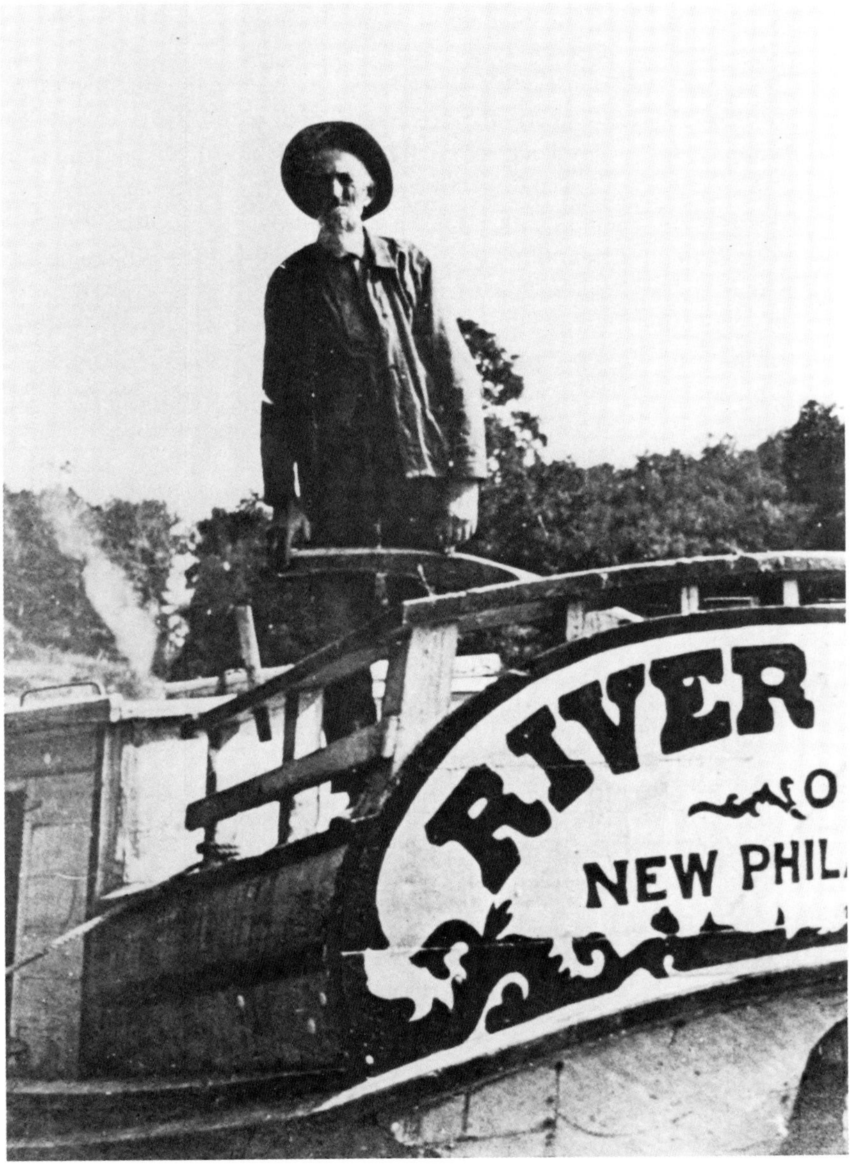

The true canal entrepreneur, exemplified by Captain Neirmeyer, of New Philadelphia, skipper of the *River Mills*, mans the tiller from the afterdeck of his boat to negotiate a lock. (Baus Collection, RVF)

11 Effects

The effects of the canals were immediate. The first beneficiaries were Ohio farmers, who could at last find a ready market for the fruits of their labors. Indeed, the canal system changed the way farmers utilized their land. As historian Harry Scheiber points out:

"Immediately after the Ohio Canal was opened to traffic in 1827, farmers in its newly populated northern region moved into commercial production of wheat for the first time. . . . Until the canal went into operation, the price of wheat near Akron had been twenty to thirty cents a bushel. But by 1833 the price was seventy-five cents; no longer was the farmer without cash markets. . . .

"Although the northward-flowing trade to Cleveland dominated the Ohio Canal's commerce, Portsmouth—the southern terminus—also attracted surplus produce from the interior. Exports destined for Portsmouth came mainly from the Scioto Valley counties on the canal's southern line. Farmers in the Scioto region shipped their wheat and flour north to the lake, but their pork and bacon (especially in bulk form, processed on the farm) they sent to the South, where dietary preferences sustained demand for such meat. Moreover, the southern river route was free of ice earlier than Lake Erie, an important advantage in the meat trade. Bulk pork and bacon thus comprised Portsmouth's principal export, with 200,000 to 700,000 pounds arriving there by canal each year in the late 1830's."[1]

As early as November 1830, the *Scioto Gazette* reported that "the Canal has reduced the price of salt from 87 to 50 cents a bushel and has reduced the carriage on every imported article in corresponding ratio. It has advanced the price of flour from three to four dollars per barrel and wheat from 40 to 65 cents per bushel. It has raised the price of real estate . . . and has increased the business and hustle of the town fifty per cent."

Before the advent of the canals, Buffalo merchants had purchased about a thousand barrels of flour per year in Cleveland. But in 1828, the first full year the northern section of the Ohio & Erie was open, they bought 250,000 barrels, suddenly transforming Cleveland from a sleepy little village into a major lake port.

That same year Henry Newberry discovered bituminous coal a few feet below the surface on his property in Tallmadge. Newberry shipped some of it by canal boat to Cleveland, where he loaded it into a wagon and drove around town all day trying to persuade merchants that the dirty-looking rocks would burn. He made only one sale—to Philo Scovill, manager of the Franklin House, who agreed to try it in his barroom stove. Soon, a few blacksmiths began using it. A few decades later, Cleveland's flats were the home of a new steel industry that used Ohio coal (in the form of coke) in its blast furnaces to smelt iron from its ore.

By 1838 more than four hundred steamers were operating in midwestern waters. The village of Milan became second only to the Russian port of Odessa in the shipment of wheat to world markets.

In the western part of the state, the relatively short Miami Canal from Dayton to Cincinnati helped that port earn the title of the Queen City as

1. Harry N. Scheiber, *The Ohio Canal Era: A Case Study of Government and the Economy, 1820–1861* (Athens, Ohio: Ohio University Press, 1969).

early as 1831. When this canal was extended north to Piqua in 1837, it encouraged new settlers to move into the area and clear the land for farming.

A comparison of Ohio market prices between 1826 and 1858 reveals that wheat rose from 50 cents a bushel to $1.08, while corn went from 25 to 70 cents during the same period. Meanwhile, imported sugar dropped from 9 to 7 cents a pound, while coffee fell from 15 cents to 12. Property values in the thirty-seven canal counties increased fourteen times between 1826 and 1859.

Settlers were able to outfit their households with imported hardware and merchandise that brought relative luxury to their frontier homesteads. The state became a magnet for easterners from New York and Pennsylvania. By 1850 Ohio's population ranked third in the nation.

The canal system resulted in industrial development in communities throughout the state. Entrepreneurs set up warehouses as well as mills and manufacturing enterprises along the canal, improving their access to raw materials and to profitable export of their products, and frequently also taking advantage of available waterpower or canal process water.

On a smaller scale, locks—the very essence of canal technology—produced significant local change. Travellers were delayed at least fifteen minutes at each lock, encouraging the establishment of saloons, hotels, general stores, and other sources of entertainment (including gambling halls and an occasional brothel). Locks, thus, caused towns; when more than a dozen locks occurred in a single mile, a major city was inevitable—as Simon Perkins rightly predicted when he formed his partnership with Paul Williams in 1825 and platted the town they named Akron.

In summary, Ohio's canal system delivered on its promise of a new era for the state. Harlan Hatcher put it in perspective:

". . . This trickle of water, conducted through the valley between its low, man-made banks, first connected the interior of the Western Reserve with New York, New Orleans, the interior of Europe, and the markets of the world. It expressed the vision and the hopes of our forefathers. It was in the future of their country and their magnificent gamble with their destiny. They likened their undertaking to 'the Revolution achieved by our fathers.' On its four feet of water floated the products of their toil. It made all the difference between a stagnant and isolated western frontier and a thriving, prosperous and growing region linked up with the rest of the world."[2]

It was, in short, the ditch that brought the world to the wilderness.

2. Hatcher, *The Western Reserve*.

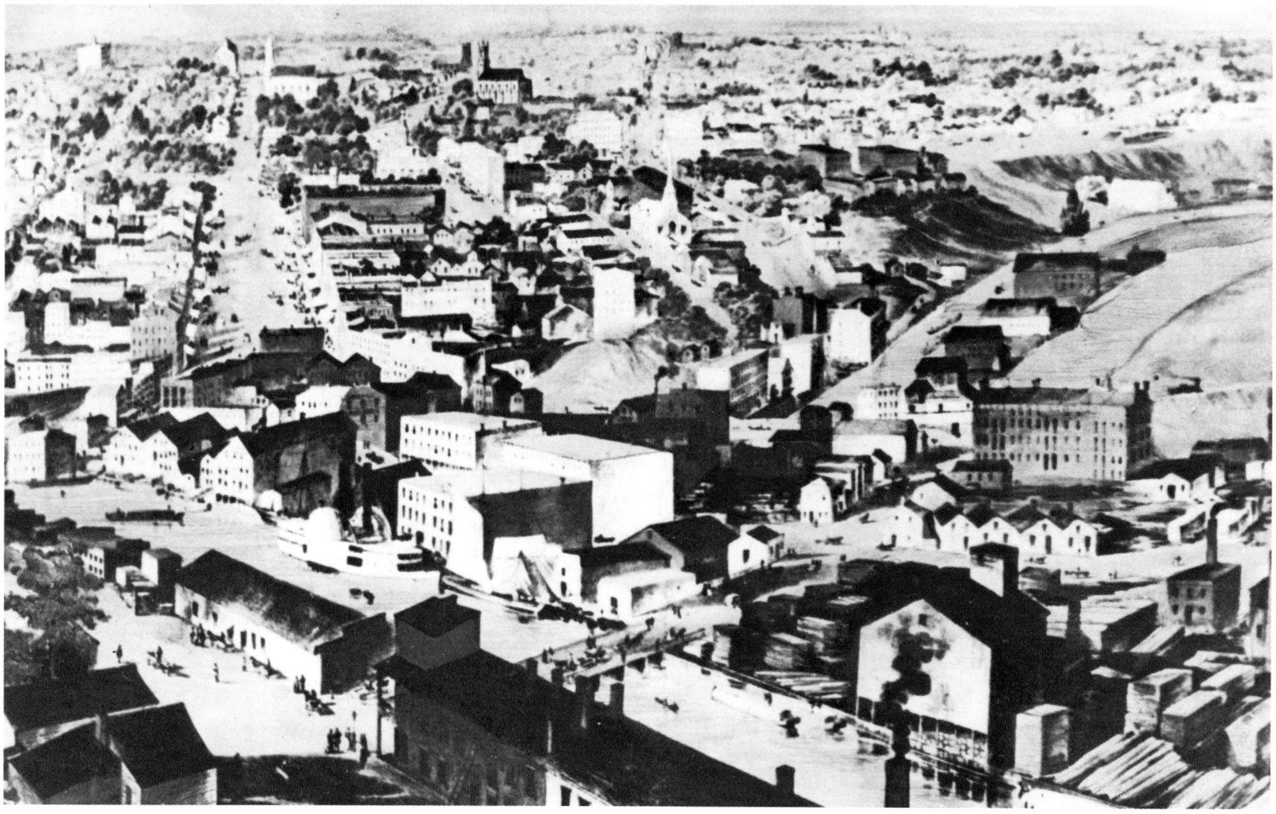

An unknown painter's bird's-eye view of Cleveland in 1860, shows the city after its expansion from a quiet little village to the dimensions of a major lake port, as a direct result of canal commerce. Forwarding firms in the city arranged for trans-shipment of freight between Ohio & Erie canal boats and Great Lakes schooners and steam vessels calling at the Port of Cleveland. In 1828, the first full year the northern section of the Ohio & Erie Canal was opened, Cleveland's exports to Buffalo jumped from 1,000 barrels of flour a year to 250,000 barrels. The Cuyahoga River appears as the diagonal waterway in the foreground. The Ohio & Erie Canal can be seen coming in from the upper right. (Pockrandt Collection, SCHS)

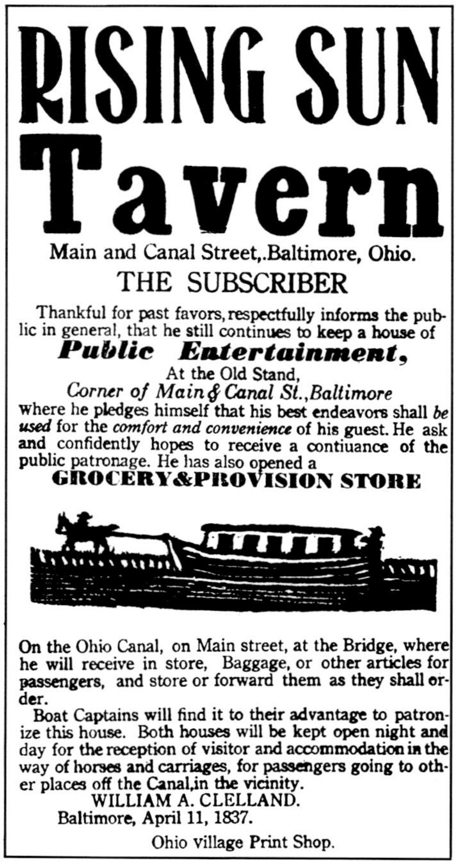

RISING SUN
Tavern
Main and Canal Street,.Baltimore, Ohio.
THE SUBSCRIBER
Thankful for past favors, respectfully informs the public in general, that he still continues to keep a house of
Public Entertainment,
At the Old Stand,
Corner of Main & Canal St.,Baltimore
where he pledges himself that his best endeavors shall *be used* for the *comfort and convenience* of his guest. He ask and confidently hopes to receive a contiuance of the public patronage. He has also opened a
GROCERY&PROVISION STORE

On the Ohio Canal, on Main street, at the Bridge, where he will receive in store, Baggage, or other articles for passengers, and store or forward them as they shall order.
Boat Captains will find it to their advantage to patronize this house. Both houses will be kept open night and day for the reception of visitor and accommodation in the way of horses and carriages, for passengers going to other places off the Canal,in the vicinity.
WILLIAM A. CLELLAND.
Baltimore, April 11, 1837.
Ohio village Print Shop.

This 1837 Baltimore, Ohio, tavern owner advertises "Groceries and Provisions" to his canal-captain customers. He does not specify the nature of his "Public Entertainment" but does assure that "his best endeavors shall *be used* for the *comfort and convenience* of his guests" (emphasis in original). He makes no secret of his policy of giving boat captains some kind of special accommodation for their help in building his business: "Boat captains will find it to their advantage to patronize this establishment. . . ." (A wood engraving with hand-set foundry type duplicating an 1855 broadside, by OHS artist Dale Star, who operates the OHS Print Shop in Ohio Village.)

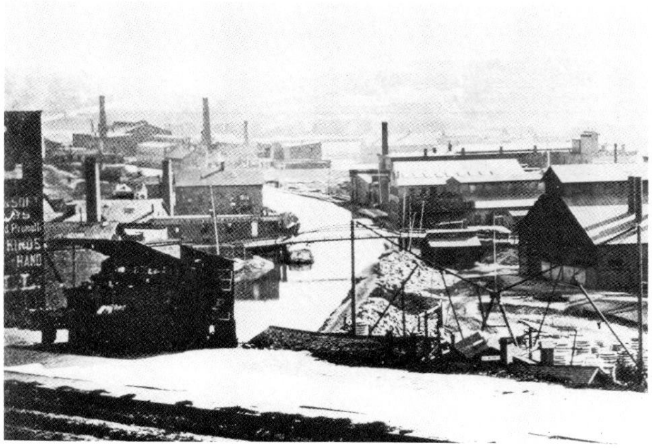

Early industry in Cleveland's Flats was fueled by coal and other raw materials brought in by canal boat. The canal descended to river level near this bend in the Cuyahoga River; 1870. (WRHS)

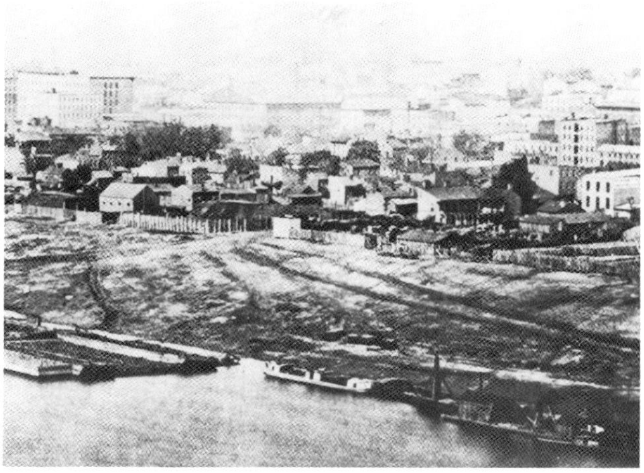

In 1862, at Cincinnati, the southern terminus of the Miami & Erie, canal boats braved the currents of the Ohio River to transfer their cargoes to river steamboats which carried freight as far south as New Orleans. The Miami Canal had helped earn Cincinnati the title of "Queen City" as early as 1831. (CSO)

Locks frequently created nuclei that caused towns to develop. In addition to a lock-master's house, most had a saloon, a grocery, and a general store. This is Eight Mile Lock near Cleveland; c. 1860. (Pockrandt Collection, SCHS)

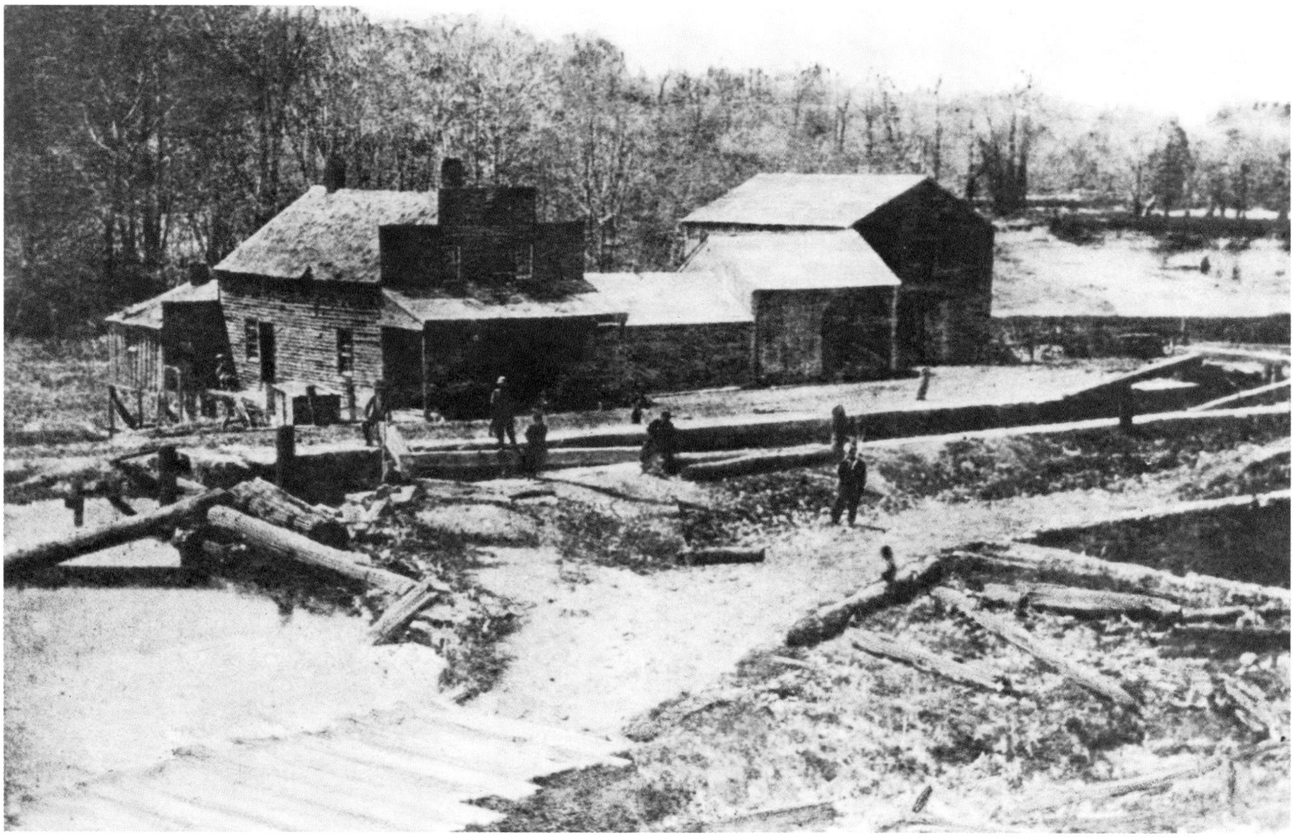

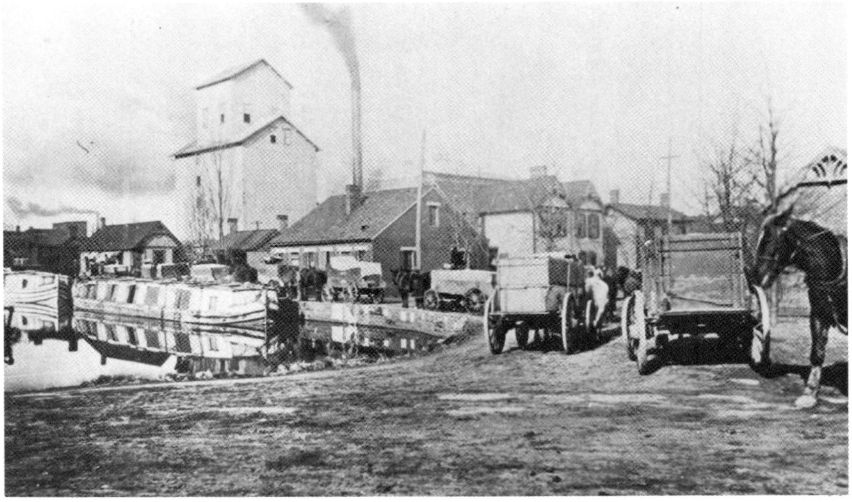

Farm wagons load canal boats at the basin in Chillicothe. By providing a market for farm produce, the canal system significantly changed land utilization in Ohio. (CSO)

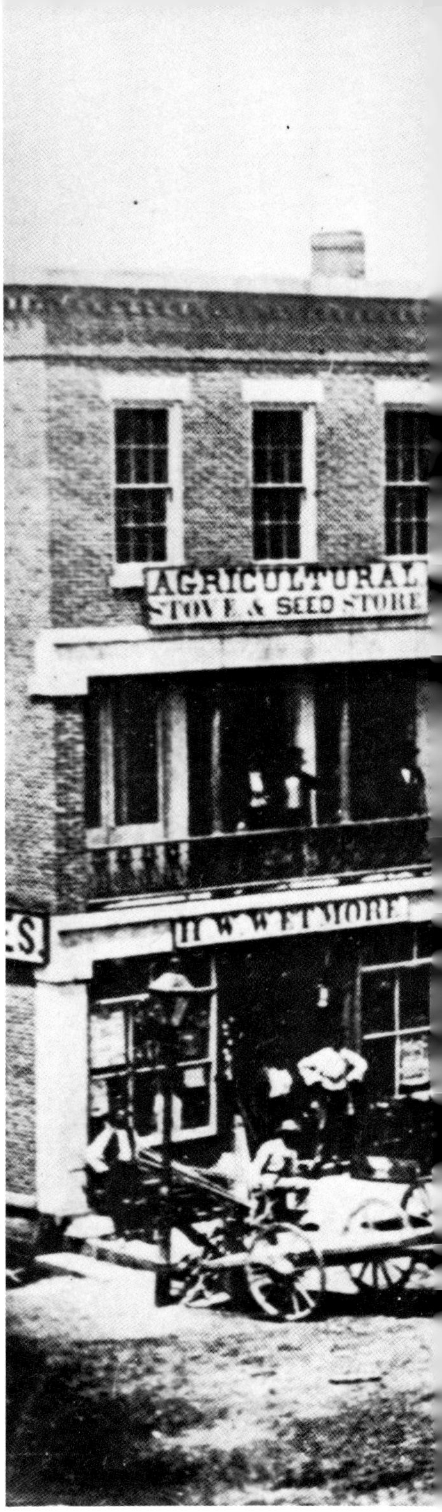

The village of Cascade, later North Akron, was located on the bank of the Ohio & Erie Canal an the edge of the industrial valley created by Eliakim Crosby's Cascade Race. (Ferdinand Schumacher's Aetna Mills on the canal may be glimpsed as the white structure in the distance between the buildings.) Only a block away was the Pennsylvania & Ohio Canal, assuring the prosperity of the settlement; 1858 (Pockrandt Collection, SCHS). Inset photo shows that more than a century later the same buildings are recognizable at Akron's Howard Street and West Market; 1967. The area has since been razed. (From a color slide by Jack Gieck.)

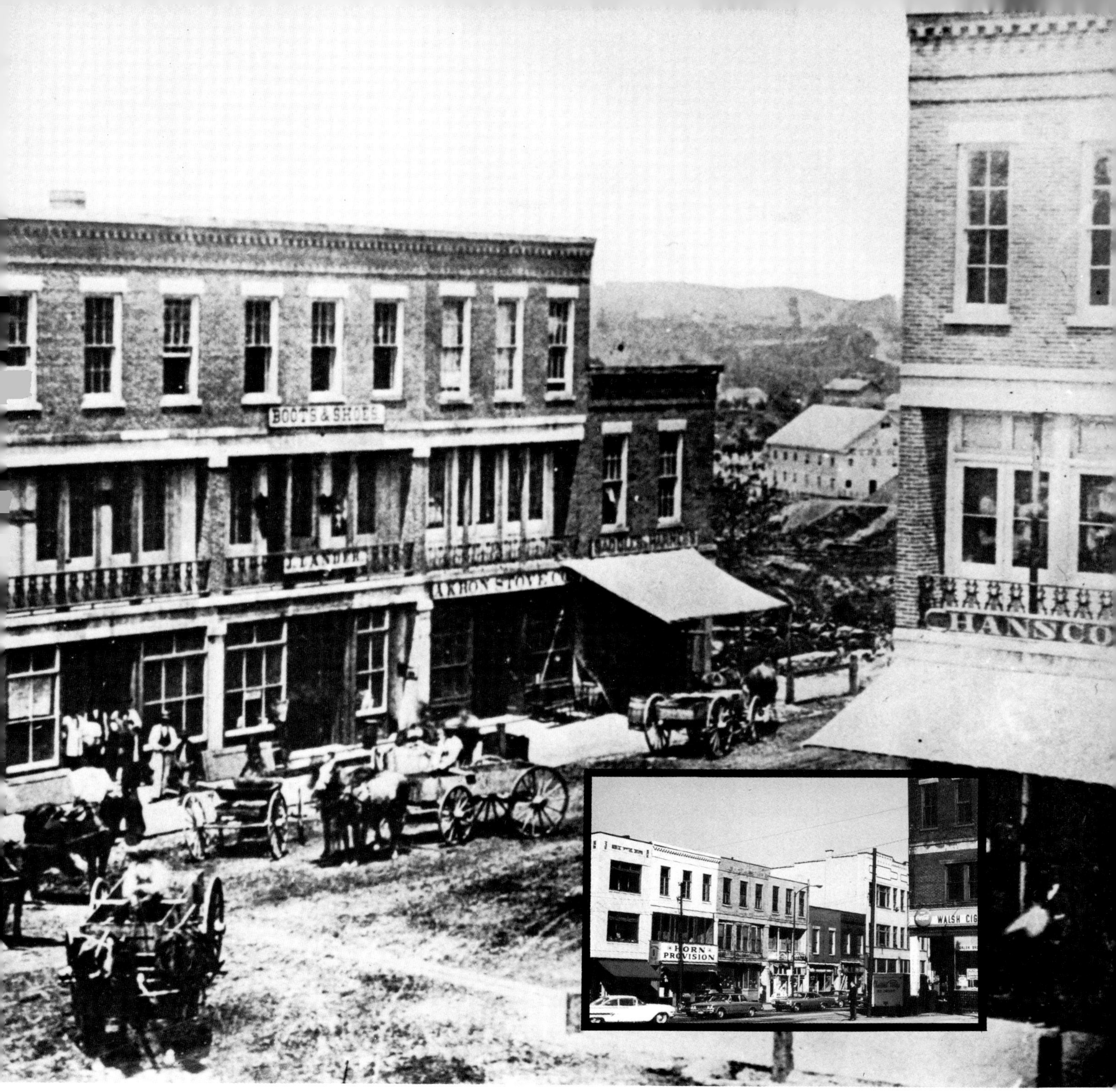

This engraving shows the prosperous downtown engendered by the Miami & Erie Canal at Berlin (canal boat in left foreground). (*History of Shelby County, Ohio*, R. Sutton & Co., Philadelphia, 1883.)

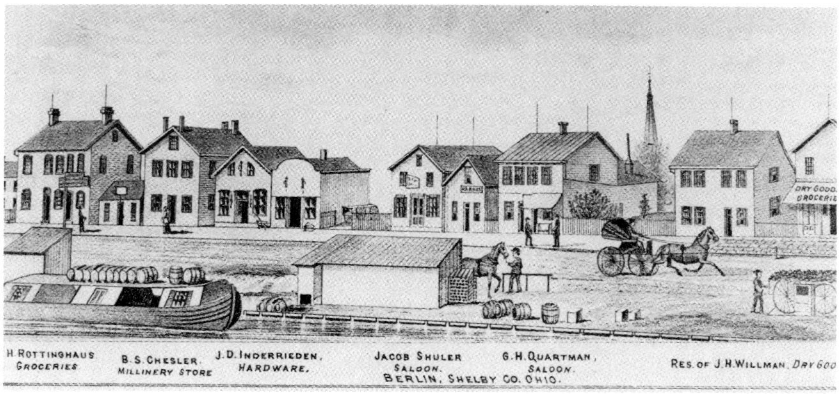

The Decline

12

Although profit had not been its objective, Ohio's state-owned canal system grossed over $14 million prior to the Civil War, representing a net profit of more than $7 million for the state (even though the Miami & Erie was not completed until 1845)—a considerable sum for the time. In benefits to the communities it served, its total effects defy calculation. By mid-century it had transformed Ohio from an agrarian frontier wilderness into a prosperous agricultural and industrial commonwealth.

But the fate of the canals was sealed while it was still under construction. In 1828, Charles Carroll (last surviving signer of the Declaration of Independence) drove the first spike for the Baltimore & Ohio Railroad. Within a generation, rail networks cobwebbed the state. Moreover, the iron horses ran year-round, while the canals were limited to nine or ten months of operation at best. As a result, net revenues from the canal system fell below zero for the first time in 1856. Private leasing (for a mere $20,000 per year) restored a scant margin of profit from 1861 to 1877. But the private companies failed to maintain the canals and the banks were permitted to wash out in a number of locations, flooding adjacent farmland. When it reverted to state control in 1877, the canal had gone into a decline from which it would never recover.

An attempt at restoration was made between 1904 and 1908, at considerable expense, but the renovated canal attracted little more than excursion trade.

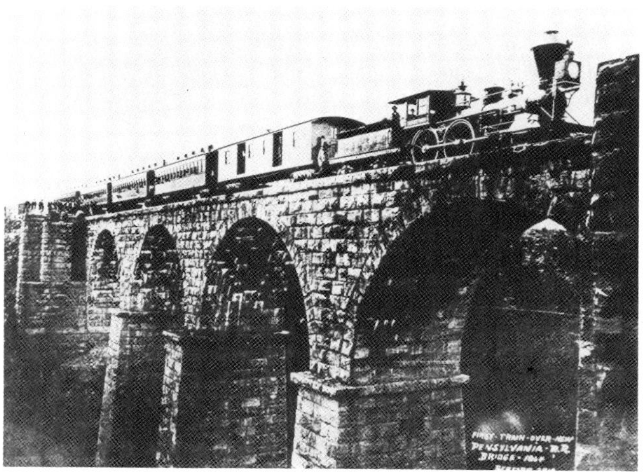

Harbinger of doom for Ohio's canal system, the first train to cross the new stone railroad bridge built by the Pennsylvania Railroad over Tinkers Creek Gorge lumbers into Bedford in 1864. (Canal Visitors Center, CVNRA)

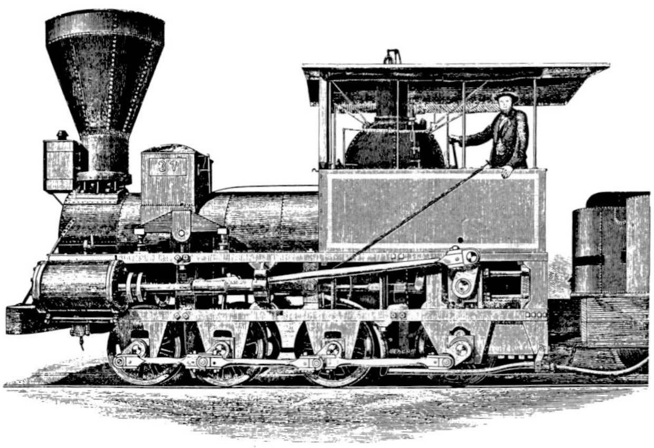

An engraving of an early Winans locomotive, the Cumberland, built for the Baltimore & Ohio Railroad in 1845. (Dover Publications)

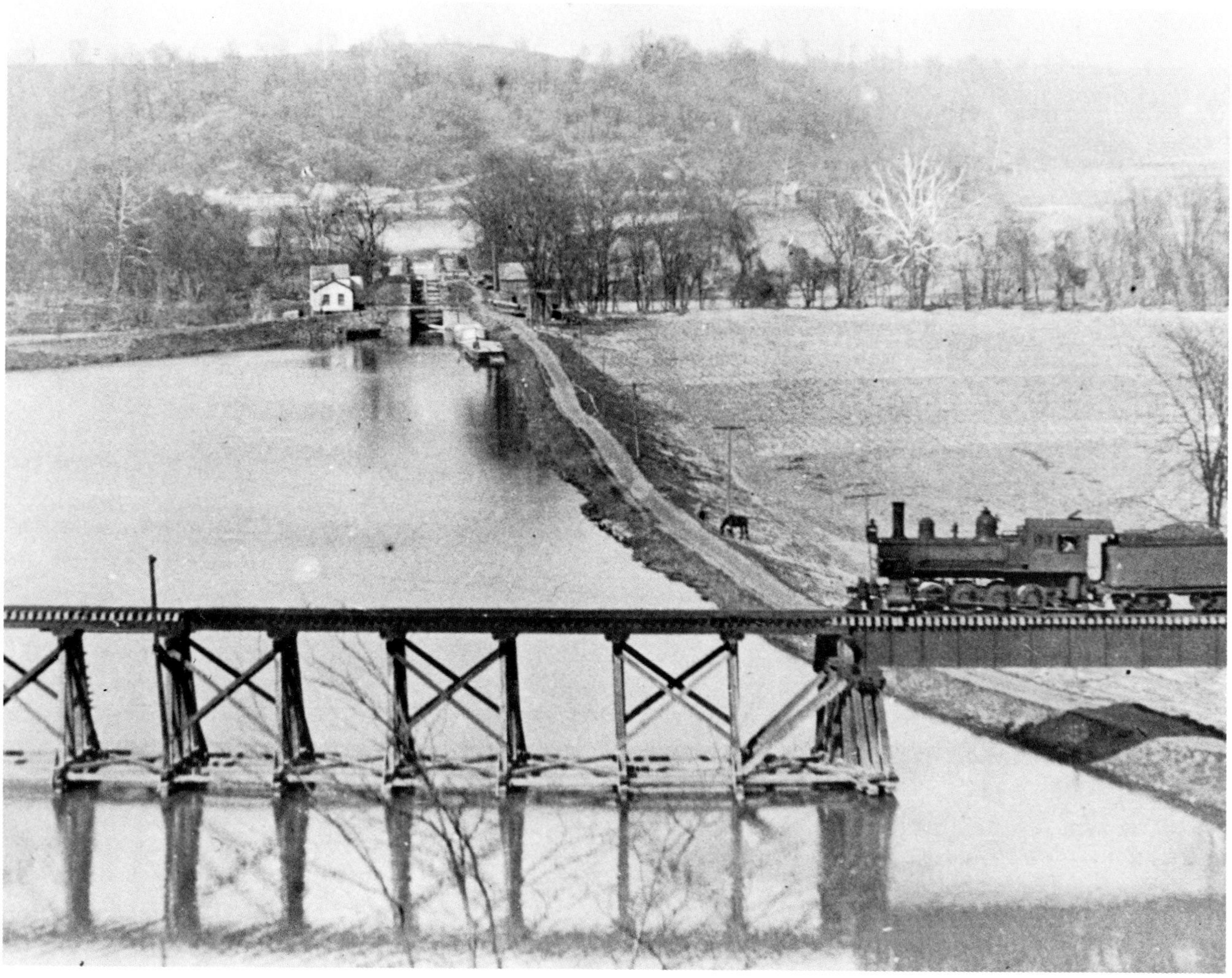

A locomotive crosses the canal at Roscoe's Lower Basin, while a boat stands idle at the double locks and a mule grazes next to the towpath. (RVF)

OHIO and PENNSYLVANIA

RAIL ROAD:

NEW ARRANGEMENT,

Commencing August 8,1853.

EXPRESS TRAIN

Leaves Pittsburgh at 6 o'clock, A.M.; stops only at the principal Stations and dines at Crestline at 1.30 P.M. arriving at Cincinnati in the evening, via Columbus and Xenia.

Fare from Pittsburgh to Cincinnati, $7.75: Columbus. $5.25

MAIL TRAIN

Leaves Pittsburgh at 8.30 A.M.; dines at Alliance; and reaches Crestline at 6.00 P.M connecting with the Night Train to Columbus and Cincinnati, and with the

BELLEFONTAINE & INDIANA RAIL ROAD

For Dayton, Indianapolis, Terre Haute, and towns in Indiana

Connections are made at Alliance by Rail Road with

RAVENNA, HUDSON, and CLEVELAND.

Fare from Pittsburgh to Cleveland, $4.00.

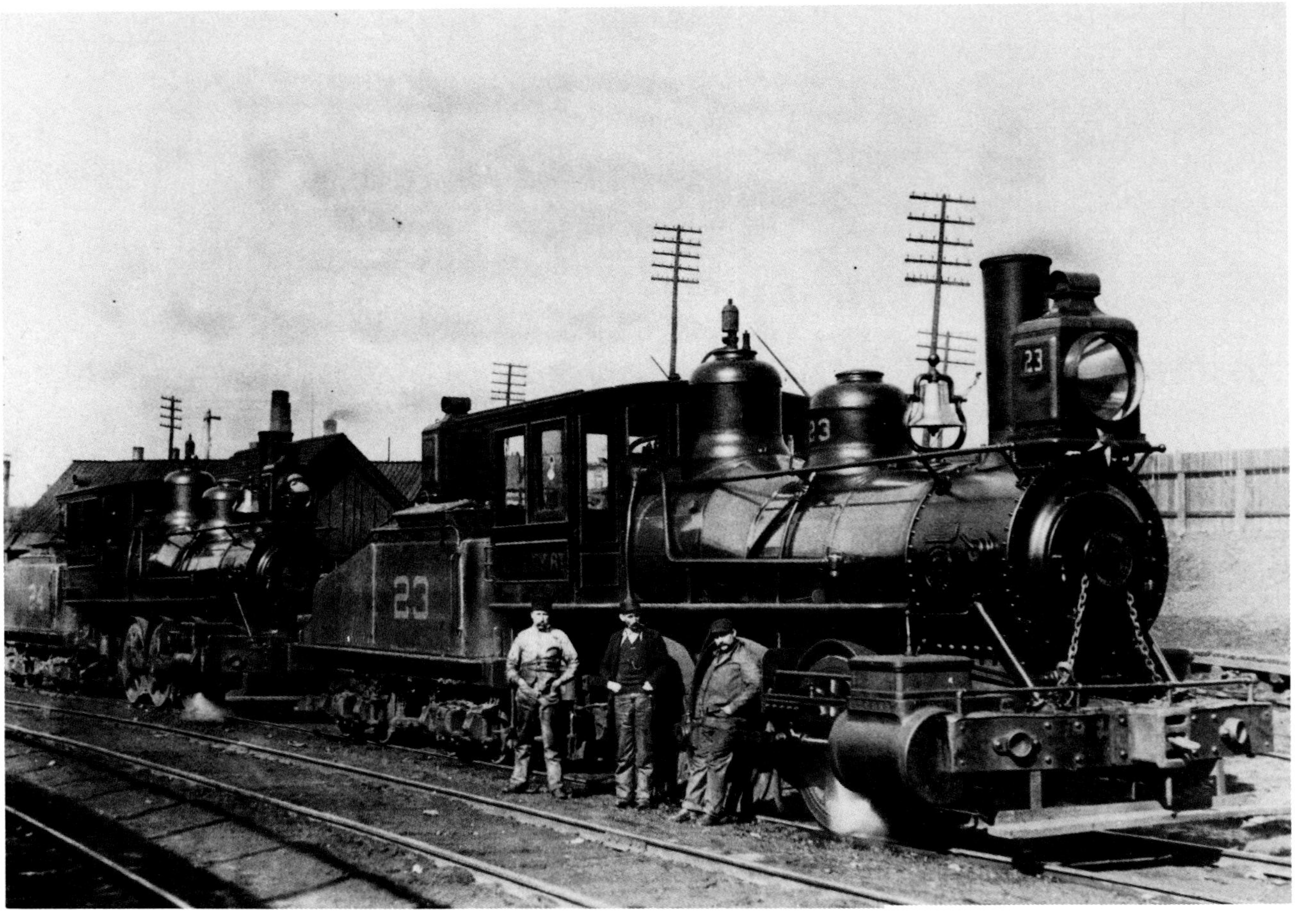

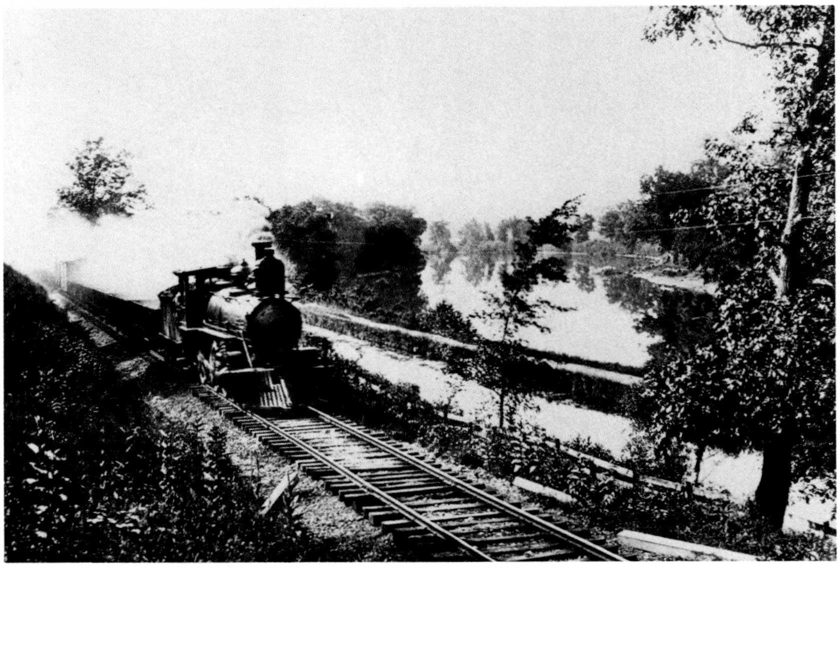

The crew proudly poses with a Valley Railway locomotive. The Valley Line, which operated through the Cuyahoga Valley, was taken over by Baltimore & Ohio in 1889; c. 1880. (SCHS)

A freight engine pulls its string of cars along its heel path right-of-way between Locks 9 and 10 in Tuscarawas County near Bolivar; the towpath is on far side of the canal. (Baus Collection, RVF)

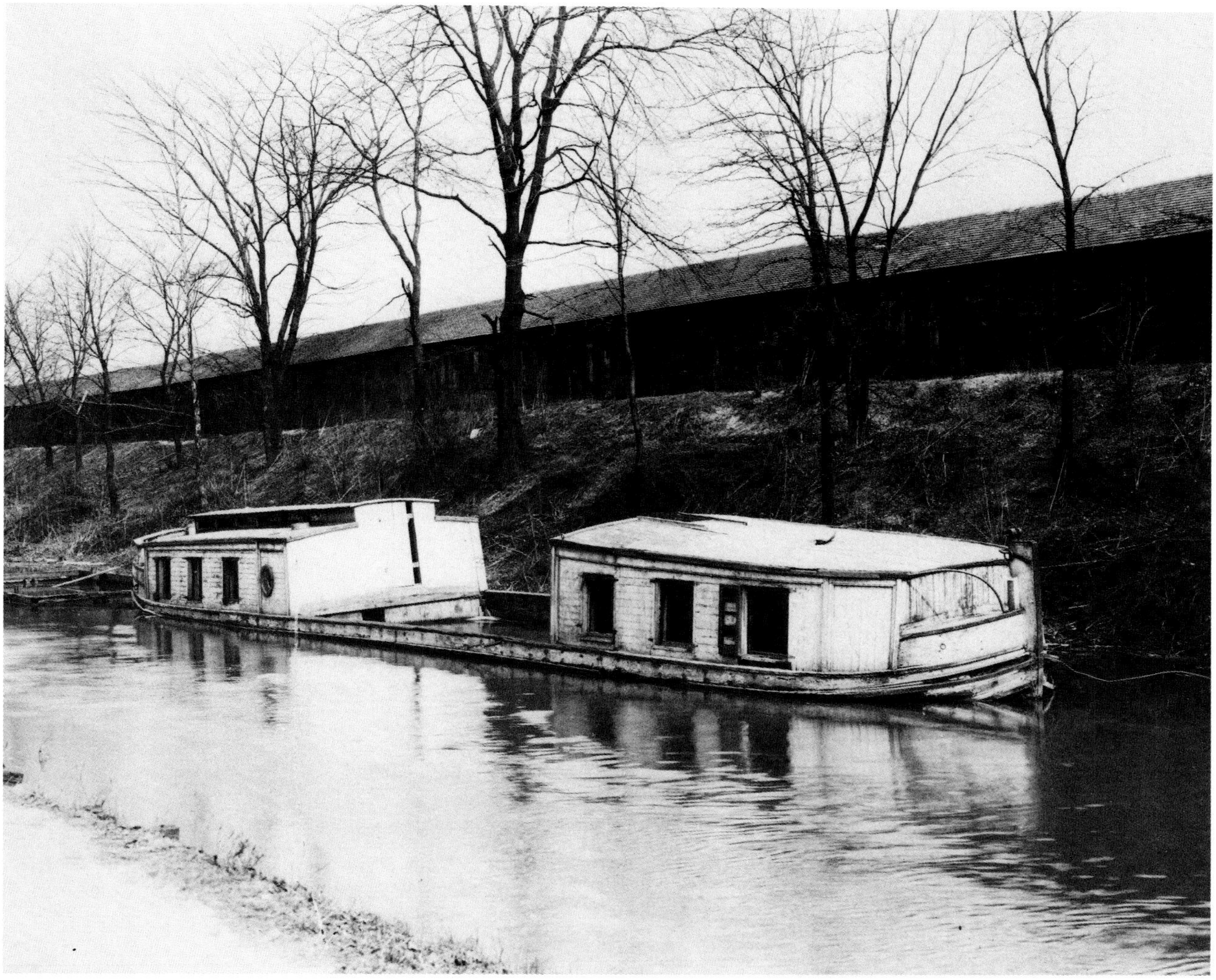

An abandoned boat sunk near Dayton, 1911. (RVF)

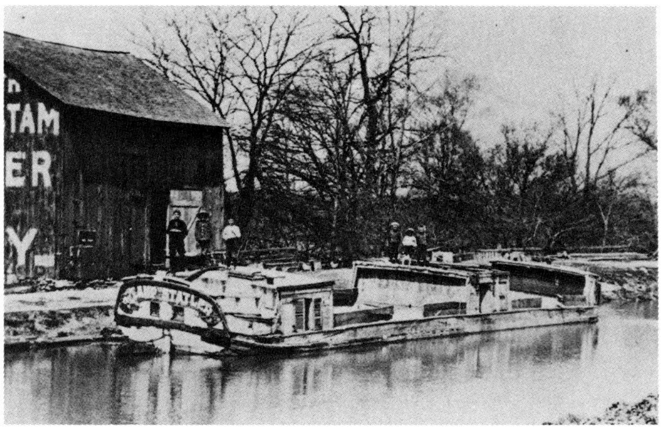

Children play on the abandoned *DeCamp Statler* on the Miami & Erie; c. 1900. (CSO)

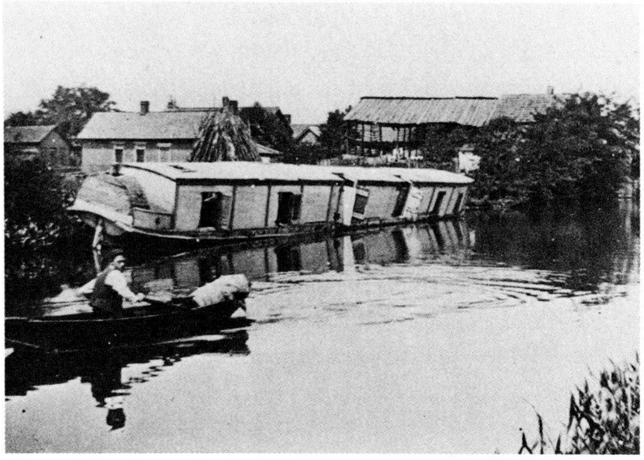

An abandoned boat near Park Street, Chillicothe. (CSO)

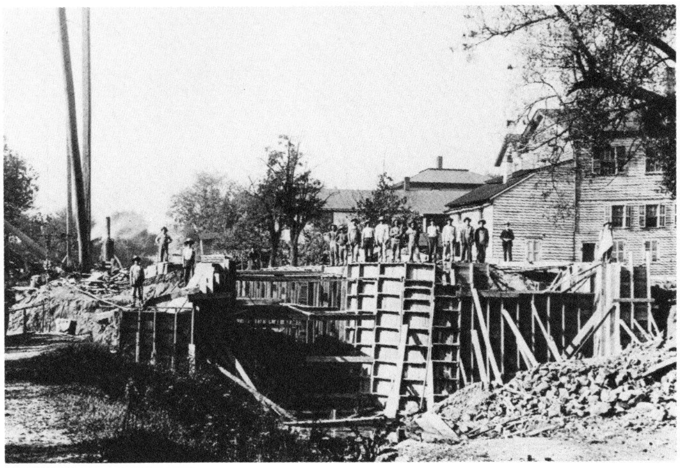

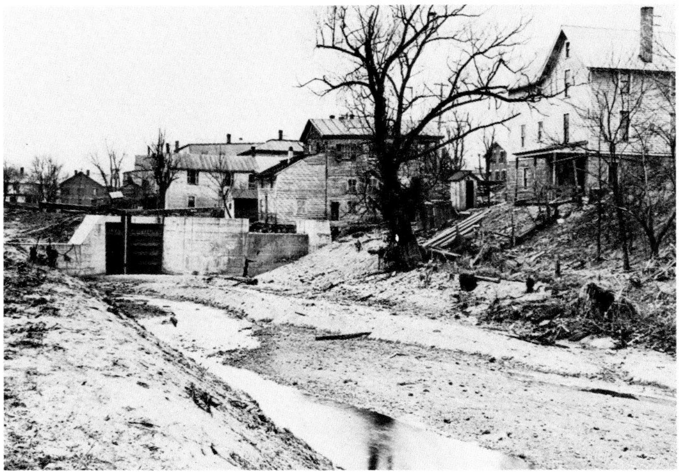

Typical of the state's 1904–8 canal restoration program was this well-documented reconstruction of Lock 6 at Navarre, seen here (top) in 1906. To waterproof and reinforce the lock, the Berea sandstone is being covered with a layer of concrete, then a relatively new, still-experimental building material; here, wooden forms are still in place. (Pockrandt Collection, SCHS)

The repaired Navarre Lock (bottom) before the rewatering of the canal. This lock has become part of the foundation for the Nickles Bakery loading dock. (Pockrandt Collection, SCHS)

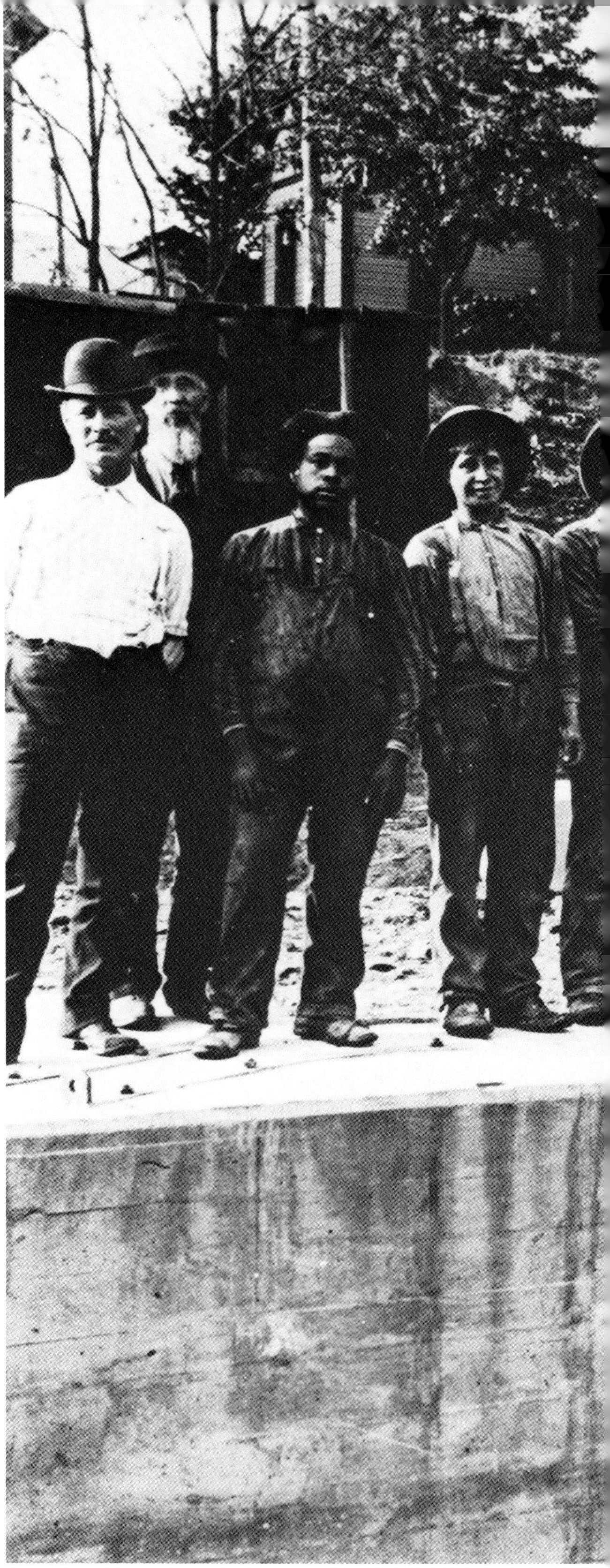

With obvious pride, the construction crew poses on top of the restored Navarre Lock in 1907 (right); the Daley Brothers Construction Company logo, cast into the new concrete wall face, may still be read on many surviving Ohio & Erie locks. (Pockrandt Collection, SCHS)

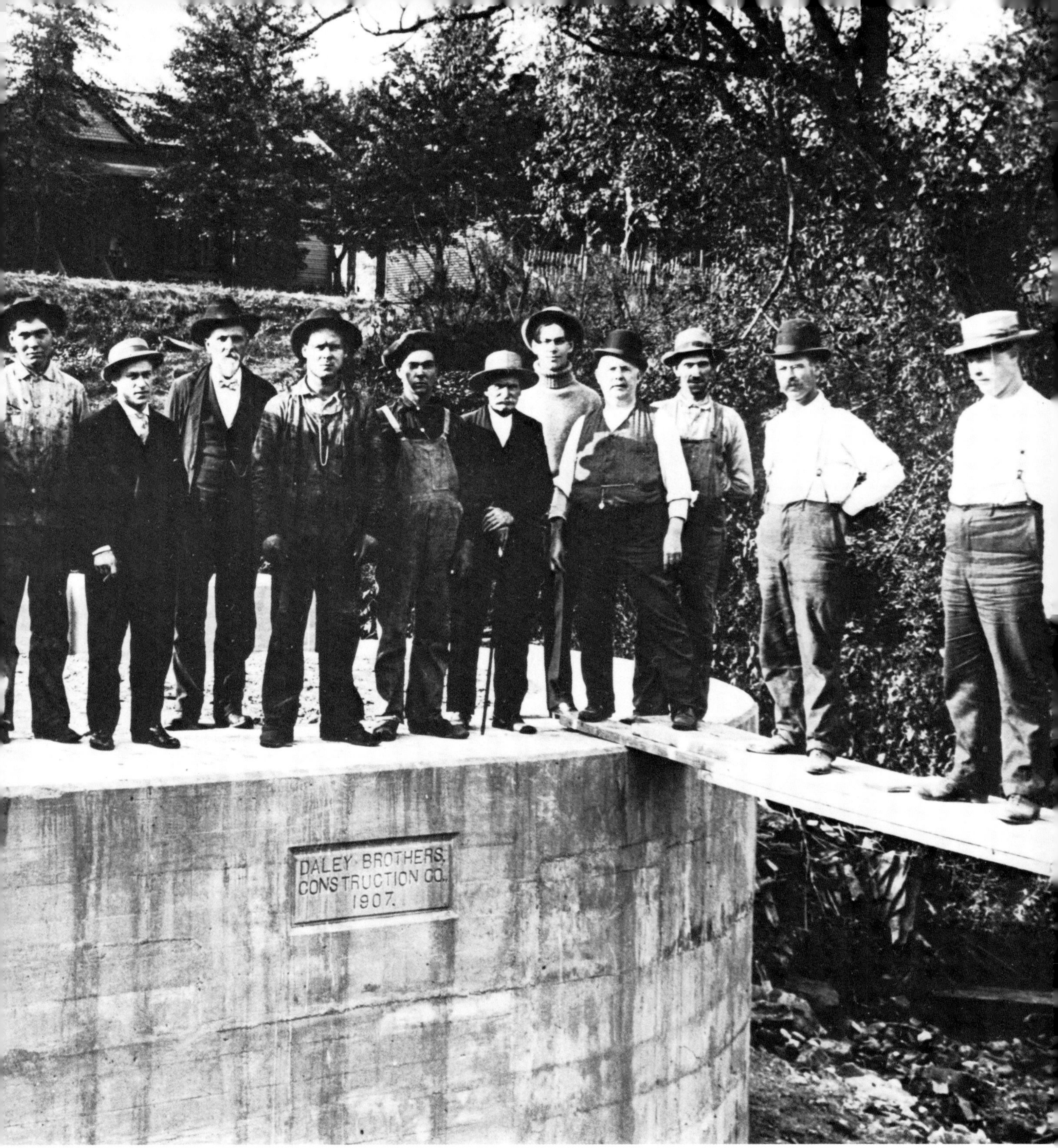
DALEY BROTHERS
CONSTRUCTION CO.
1907.

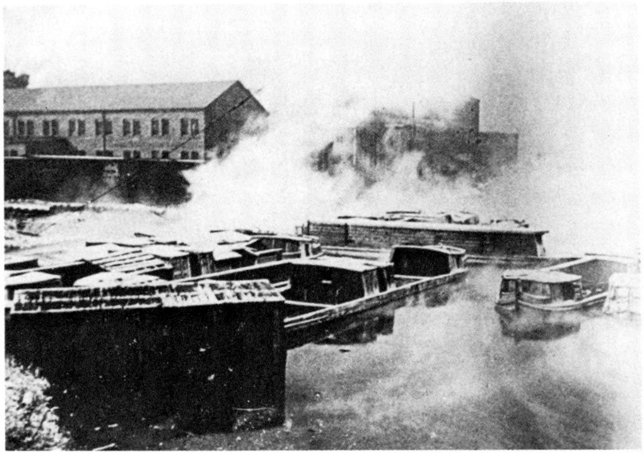

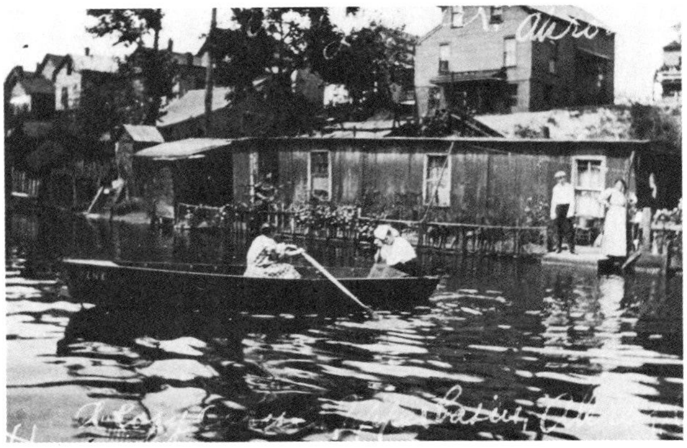

Many boats were destroyed when the canal boat "graveyard" (left) caught fire (CSO). The permanently moored houseboat, *Cozy Corner* (right), an Akron residence in Akron's Upper Basin; 1910 (Pockrandt Collection, SCHS). A dredge near Dayton's fairgrounds, abandoned and sunk, 1911 (below). (RVF)

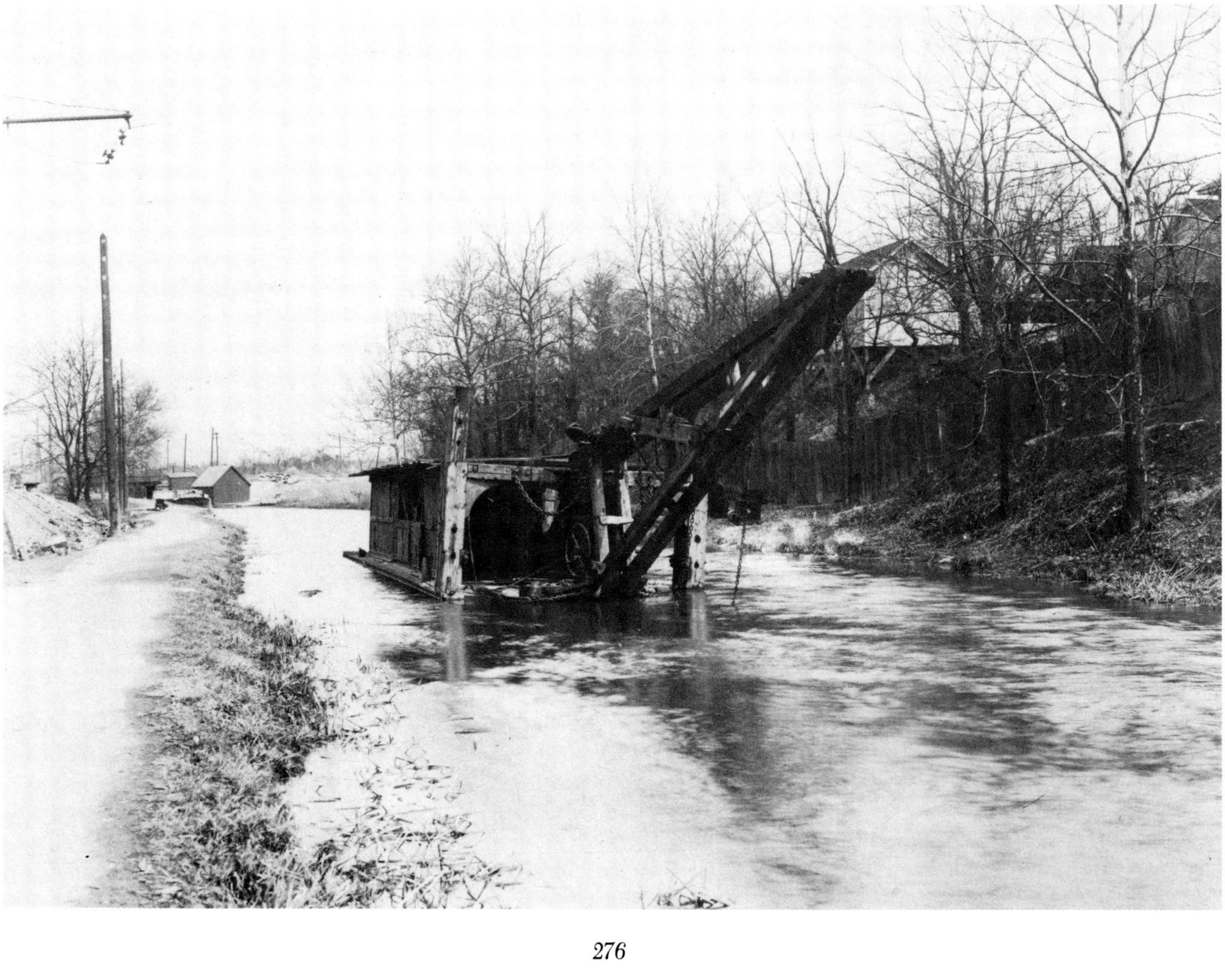

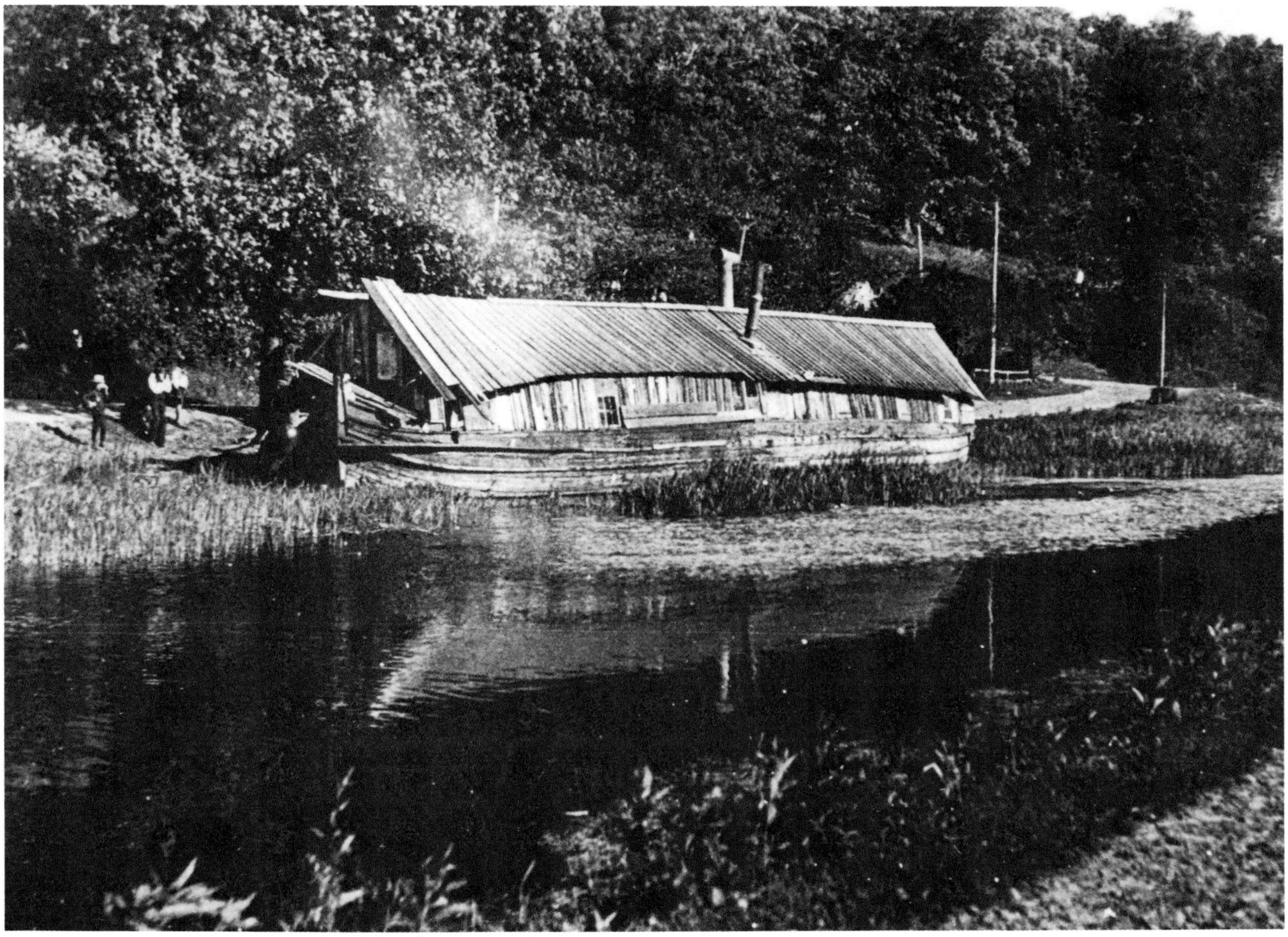

This Massillon residence (above) presents no illusions about its origins. (CSO)

Innovative carpentry cannot disguise this canal boat (right) converted into a residence in Bolivar. (SCHS)

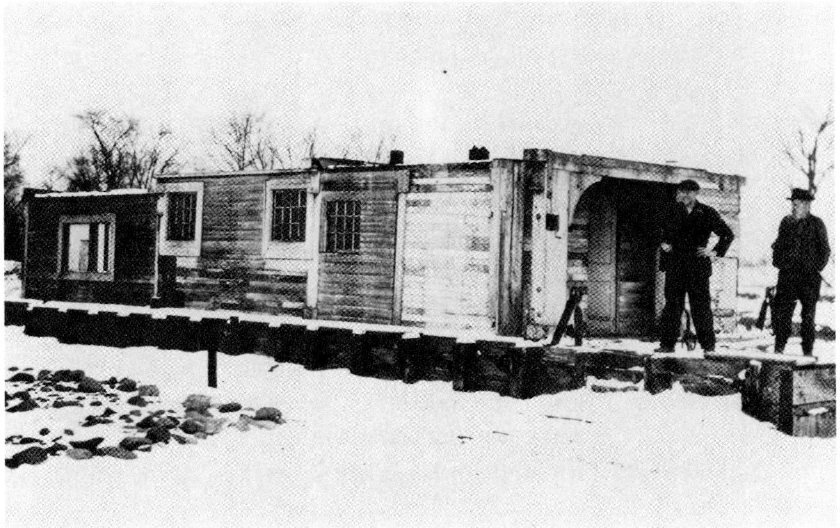

The End—With a Bang 13

March 1913 brought Ohio's canal system to an abrupt end. After a winter of record snowfall, spring rains had been abnormally heavy. All the reservoirs were gorged with water, and the excess spilled over into the canals, turning them into torrents in many places—destroying aqueducts, washing out banks, and flooding adjacent towns and countryside.

Dayton, Zanesville, Columbus, and Marietta were particularly hard hit. Situated as it is on the continental divide separating the Mississippi and St. Lawrence River systems, Akron would seem to be a most unlikely location for flood damage. But such was not the case that year. With its fifteen locks in a single mile, Akron's story exemplifies the disaster that visited many communities in the spring of 1913.

The heavy rains began early Sunday morning, March 23. They continued without letup all day, through the night, and into Monday. The canal was not the only waterway out of control. Akron's diminutive Little Cuyahoga River underwent an amazing transformation. As the hours passed it overflowed its banks in East Akron, submerging the lower floors of Goodyear, and quenching the fires in the company's power house. On Monday afternoon a man was drowned trying to cross Case Avenue. Akron's fire department began moving families from their homes along its banks—some of them forcibly. The next morning, a huge crowd gathered along the river in the valley to watch the raging torrent eat into the banks and wash away the foundations of homes. Houses began floating down the river. At 9:00 A.M. the Howard Street bridge was washed away.

As the level of the Portage Lakes reservoirs continued to rise, many fashionable summer cottages were inundated. At midnight on Monday, when the storm was at its height, the dyke on Akron's East Reservoir gave way at Paddy Ryan's Inn. Some insisted it was dynamited.

As the dyke gave way a massive quantity of impounded water cascaded into Long Lake. At Manchester Road (at Young's Tavern) the flood divided, part going toward Barberton and the rest rushing into South Akron, down the Ohio Canal toward the city's downtown. With the torrent roaring over the tops of the gates to a depth of eight feet at some of the locks, it was impossible to open them against the head of water. Scores of warehouses and store basements were flooded in the flats west of Main Street. The west end of the Alexander Block (a complex of stores bordering the canal) collapsed. The boilers of the electric power plant in the Cascade valley were drowned, and downtown lights went out. Several locks were dynamited in an effort to save the city.

Just before noon on Tuesday, John Henry Vance, a B. F. Goodrich engineer who was former chairman of the Waterways Commission in Akron, received an urgent telephone call from the lockmaster at Lock One, a block north of the Goodrich complex, which bordered the canal. Although the upper (south) gates of the lock had been opened on Monday with three blasts of dynamite, a new emergency threatened as the water level continued to mount behind the fifteen-foot dam created by the lower gates—which were holding back the entire nine-mile stretch of swollen Summit Level waters. The following detail is based on Vance's own account, during an interview with the author in 1951:

When he saw what was happening, Vance recruited an army of workers from the Goodrich plant, whose own boilers were now in danger of being inundated. The men scattered to collect flour sacks, feed bags and burlap sacks; and they hauled in

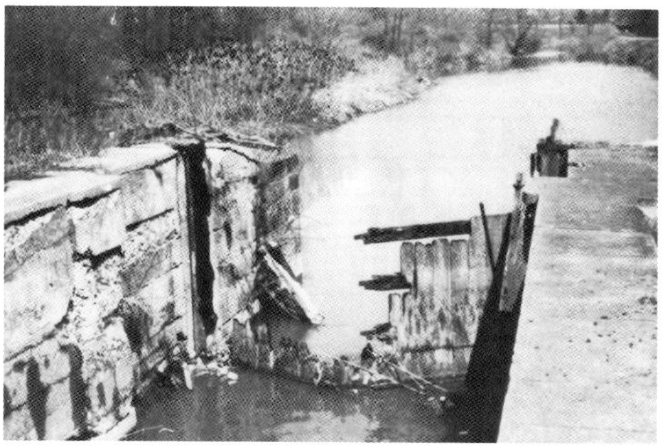

wagon loads of dirt to fill them. All afternoon and well into the night, without pausing for meals, they built their levee in a continuous downpour of rain.

By 9:00 P.M. the makeshift levee had begun to crumble, and the wall of water threatened to engulf Goodrich and adjacent businesses at Main and Exchange streets. Vance telegraphed the state offices at Columbus, and was given executive authority in the emergency. He called his friend Tom McShaffrey for assistance.

McShaffrey, an Akron contractor, knew of a construction site nearby where some dynamite was stored. Together, by the light of a lantern, they located the dynamite chest in pouring rain and broke open the lock. They removed six sticks, and took along a length of pipe lying nearby.

Returning to Lock One, the pair tied the dynamite to the end of the pipe, together with a block of concrete. Vance inserted an electric blasting cap into one of the sticks and handed its coil of wire to McShaffrey, who took refuge behind the sandbags.

Vance plunged the business end of the pipe into the depths at the base of the lower gates and rolled over the levee—throwing up his arm, signalling McShaffrey to fire the charge. A geyser of water shot into the night air, illuminated by the flash at its source, and millions of tons of unleashed water—the entire Summit Level volume—crashed through the remaining gates below, ending the canal in Akron for all time.

Dawn on Wednesday brought with it a cold wave that ended further flood danger. A total of 9.55 inches of rain had fallen in Akron. Across the state, 752 lives had been lost, and property damage mounted into the hundreds of millions. Although portions of the canal system would be restored for industrial water supply, its commercial traffic came to an end on March 26, 1913.

In a few places, remnants of Ohio's canal system have been preserved for industrial water supply. In others, tourists can ride for short distances on replica canal boats towed by draft animals trodding towpaths that have survived a century and a half. A daring nineteenth-century engineering achievement—a humble work of immigrant laborers—this was the ditch that brought the world to the wilderness.

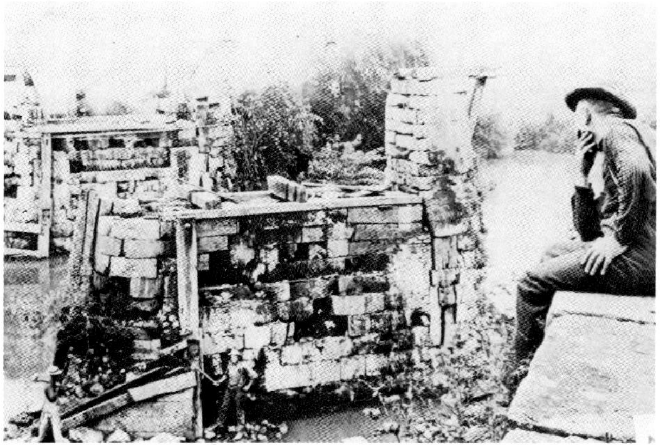

Although spring repairs of Roscoe's Walhounding Aqueduct were an expected annual chore, this construction foreman appears overwhelmed by the magnitude of the job, as his workmen examine the foundation of one of the aqueduct's piers; photo date uncertain. (RVF)

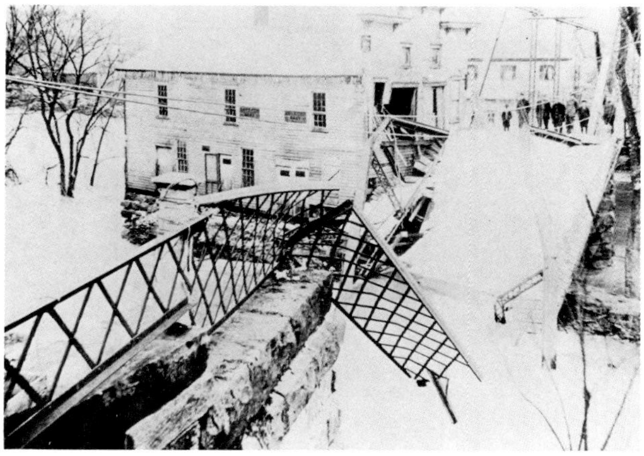

A washed-out bridge in Peninsula after the 1913 flood. (Pockrandt Collection, SCHS)

AKRON BEACON JOURNAL.

FORTY-FOURTH YEAR. NO. 95 X AKRON, OHIO, TUESDAY EVENING, MARCH 25, 1913. PRICE ONE CENT.

FIVE LIVES LOST; 500 HOMELESS; MILLIONS OF DAMAGE; AS FLOOD SWEEPS OVER CUYAHOGA VALLEY

FIRE DEPARTMENT RESCUES FAMILIES FROM MANY HOMES

Torrential Rain, Starting Sunday and Continuing All Day Monday, Sent River Over Its Banks Early Last Evening, Big Damage Being Done in East and South Akron.

Greatest Individual Loss Will be at the Goodyear Plant, Where It Has Been Found Necessary to Lay Off the Entire Rubber Working Force.

All of Akron gave thanks this morning shortly after 10 o'clock when the rain ceased, the Sun made a struggling effort to break through the clouds and the most strenuous 12 hours in Akron's history came to a close. With the stopping of the rain came a slack in the rising flood in the Cuyahoga Valley and the climax of the disaster had been

An Apology

It will be hardly necessary on a day like this, when the industry of the city is paralyzed by the flood horror, for the Beacon Journal to apologize for going to its readers in abbreviated form and a story of the flood that is but a makeshift, when the possibilities of the occasion are considered.

But without electricity Linotype machines cannot run and printing presses are but lifeless machinery. By the aid of motorcycles rigged up as a power plant we were able to run one type setting machine and this small paper has been printed at the plant of the Werner company.

We know our readers will overlook the manifest weakness of this issue and accept it as the best that was possible under the circumstances.

Should electric power be supplied later in the day, a more complete paper, with the usual advertising. will be printed.

RUMOR HAS 1,500 DEAD IN DAYTON

Cincinnati, O., March 25.---An unconfirmed report here says that 1,500 people lost their lives in the flood at Dayton today when a levee along the banks of the Miami river went out, and the waters inundated the town.

Cincinnati, March 25.---It is reported that the Miami river at Dayton has broken and flooded the city.

At 9 o'clock this morning a Western Union operator working with an operator in this city abruptly cut off a dispatch he was sending, and said:

"Good-bye, the levee is broken."

Dayton is on low ground, and the river levees rise 25 feet above the level of the town. Right in the heart of the city the Stillwater and Mad rivers merge into the Miami and during high water

PROBE DRAWS TO A CLOSE

Today Will be Spent Visiting the Plants of the Various Rubber Companies, After Which the Senators Will Depart for Columbus to Draft Their Report.

At last the probe into conditions in the Akron rubber plants is drawing to a close.

One more day and the Green committee will depart from Akron and prepare its report. Today will be spent in visiting the various rubber plants. This will require the larger part of the day as it is the plan to visit all of them and take up in each place the things which have been complained of by the strikers.

After the inspection tour is finished one more witness will be heard and then the probe will be at an end. The witness remaining to be heard is H. O. Parsons, secretary of the Manufacturers' association. He will be placed on the stand to tell of the existence of the alleged blacklist system in vogue in the Akron plants.

The large part of the afternoon session, Monday, was taken up with the direct examination of E. C. Shaw, general manager of the B. F. Goodrich company and the cross examination. Mr. Shaw told his story in a straight forward manner without frills and answered all questions put to him by the attorneys.

At one time during his cross examination he drew dramatic and rising to his feet he pointed his finger at Mr. Cook and said, "As an American citizen I am proud of the men and women who stayed in my factory in the face of all this, those who went out did so through intimidation without any grievance or cause whatever." This came in reply to a question by Mr. Cook as to the cause of the strike. Mr. Shaw had just finished telling about the marching around his plant and the mass picketing there.

Mr. Shaw refused absolutely to deal with the I. W. W. and said that unionism in his plant was not practical. He

With Akron's power plant inundated, the *Beacon Journal* was without electricity and barely got to press on March 25, 1913. But with the aid of motorcycles whose rear tires were belted to a dynamo, resourceful printers got one linotype machine operating. Undersize page make-ups of set type were carried to Werner Printing Company, where this abbreviated edition, without advertising, was produced. (Akron/Summit County Public Library)

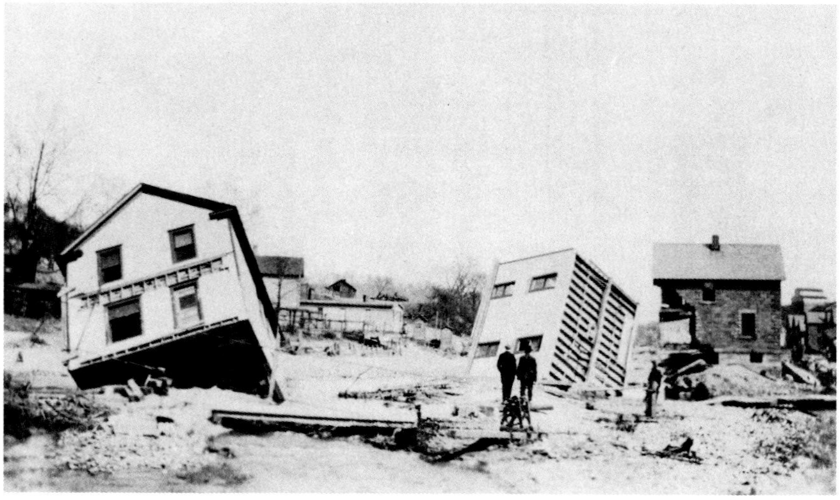

Aftermath of Akron's flood in March, 1913; the scene is near Case Avenue. (SCHS)

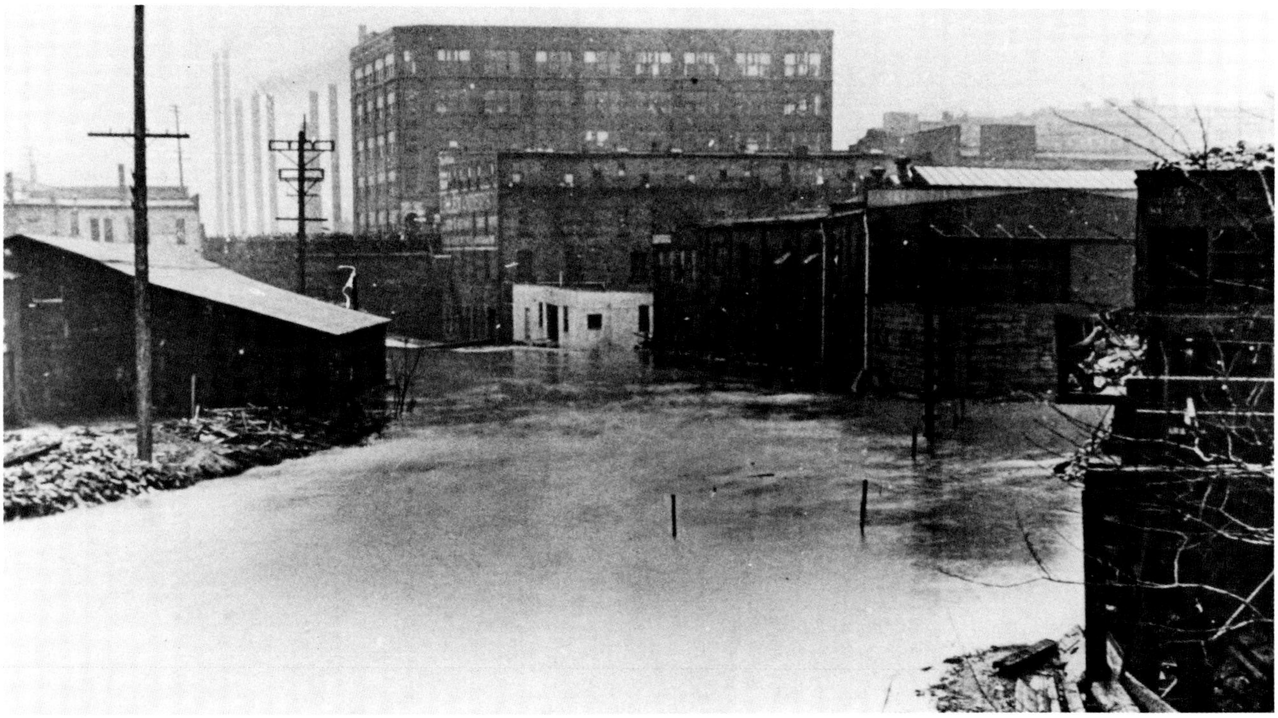

The swollen Ohio & Erie Canal in downtown Akron during the height of the flood, from Cherry Street Bridge looking northwest toward West Market; March 1913. (CSO)

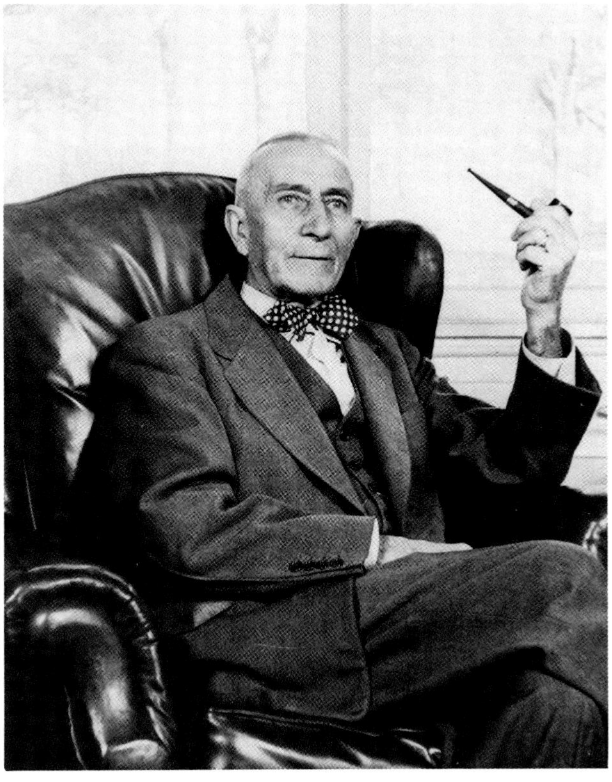

Above, John Henry Vance in 1951; then 78, he recounted how he and Tom McShaffrey dynamited Lock One in Akron in 1913. (1951 photo by Jack Gieck.)

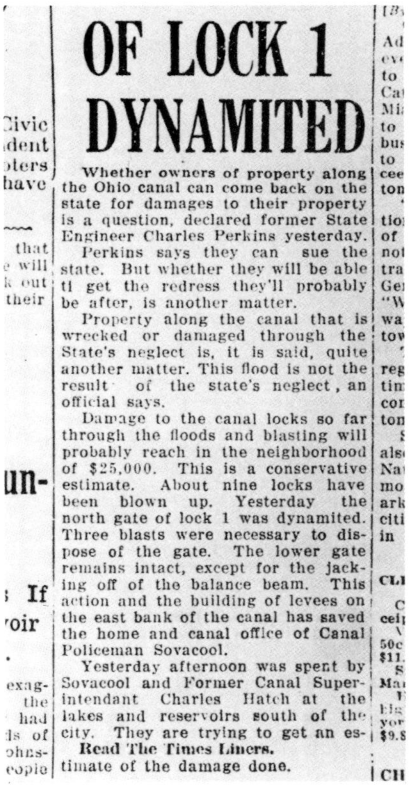

OF LOCK 1 DYNAMITED

Whether owners of property along the Ohio canal can come back on the state for damages to their property is a question, declared former State Engineer Charles Perkins yesterday.

Perkins says they can sue the state. But whether they will be able tl get the redress they'll probably be after, is another matter.

Property along the canal that is wrecked or damaged through the State's neglect is, it is said, quite another matter. This flood is not the result of the state's neglect, an official says.

Damage to the canal locks so far through the floods and blasting will probably reach in the neighborhood of $25,000. This is a conservative estimate. About nine locks have been blown up. Yesterday the north gate of lock 1 was dynamited. Three blasts were necessary to dispose of the gate. The lower gate remains intact, except for the jacking off of the balance beam. This action and the building of levees on the east bank of the canal has saved the home and canal office of Canal Policeman Sovacool.

Yesterday afternoon was spent by Sovacool and Former Canal Superintendant Charles Hatch at the lakes and reservoirs south of the city. They are trying to get an es-

Read The Times Liners.

timate of the damage done.

Right, On March 25, 1913, the *Akron Times* reported the dynamiting of Lock One's south gates the day before. That night, Vance blew up the north gates, releasing the entire Summit Level of backed-up water. (CSO)

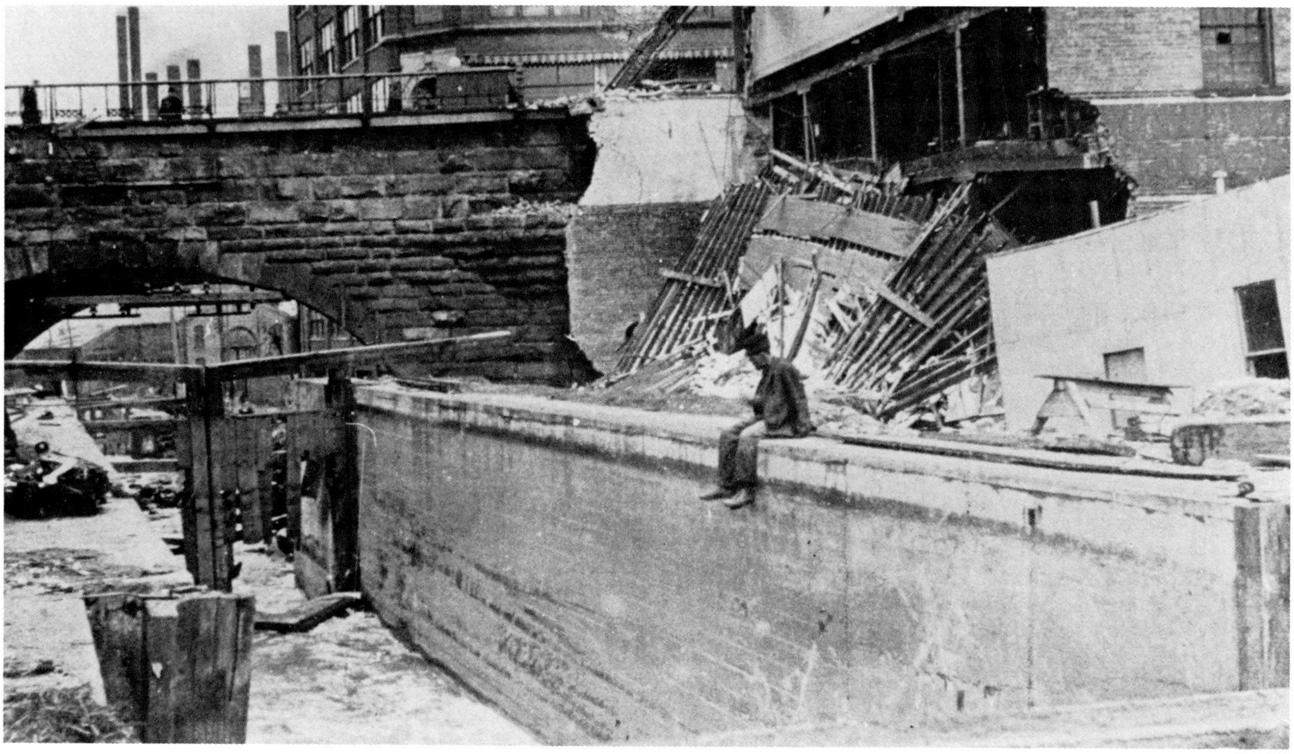

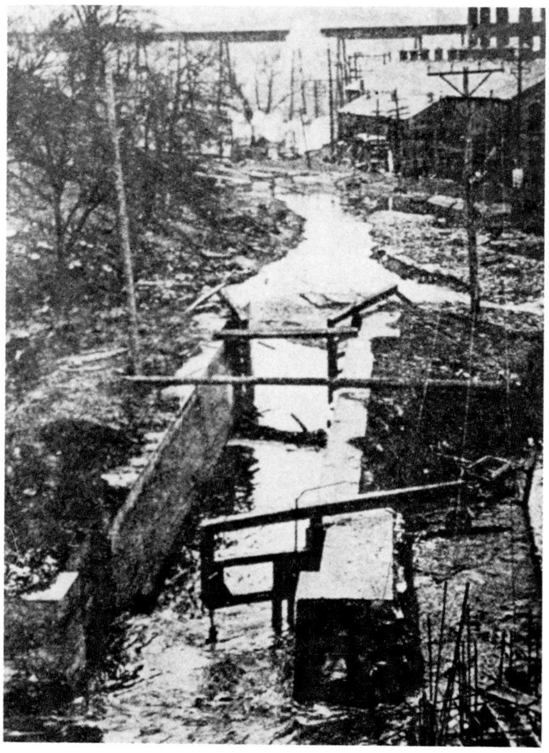

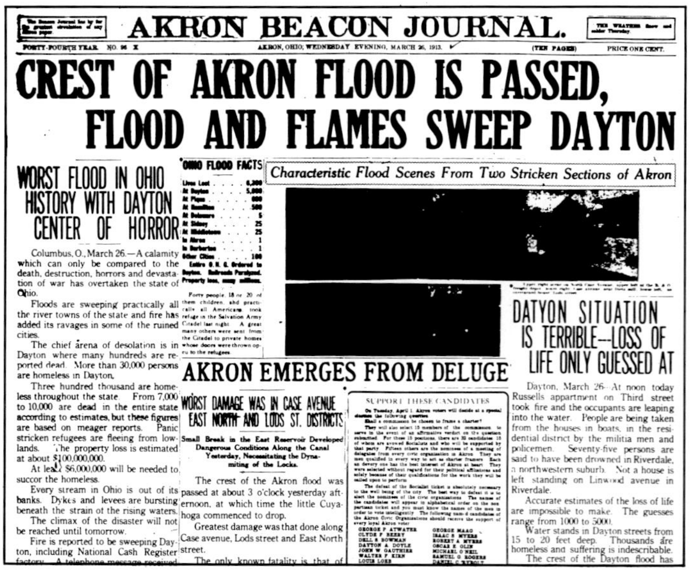

AKRON BEACON JOURNAL.

FORTY-FOURTH YEAR. NO. 96 X — AKRON, OHIO, WEDNESDAY EVENING, MARCH 26, 1913. — (TEN PAGES) — PRICE ONE CENT.

CREST OF AKRON FLOOD IS PASSED, FLOOD AND FLAMES SWEEP DAYTON

WORST FLOOD IN OHIO HISTORY WITH DAYTON CENTER OF HORROR

Columbus, O., March 26.—A calamity which can only be compared to the death, destruction, horrors and devastation of war has overtaken the state of Ohio.

Floods are sweeping practically all the river towns of the state and fire has added its ravages in some of the ruined cities.

The chief arena of desolation is in Dayton where many hundreds are reported dead. More than 30,000 persons are homeless in Dayton.

Three hundred thousand are homeless throughout the state. From 7,000 to 10,000 are dead in the entire state according to estimates, but these figures are based on meager reports. Panic stricken refugees are fleeing from lowlands. The property loss is estimated at about $100,000,000.

At least $6,000,000 will be needed to succor the homeless.

Every stream in Ohio is out of its banks. Dykes and levees are bursting beneath the strain of the rising waters.

The climax of the disaster will not be reached until tomorrow.

Fire is reported to be sweeping Dayton, including National Cash Register factory. A telephone message received

OHIO FLOOD FACTS

Entire O. N. G. Ordered to Dayton. Railroads Paralyzed. Property loss, many millions.

Forty people, 18 or 20 of them children, and practically all Americans, took refuge in the Salvation Army Citadel last night. A great many others were sent from the Citadel to private homes whose doors were thrown open to the refugees.

Characteristic Flood Scenes From Two Stricken Sections of Akron

AKRON EMERGES FROM DELUGE

WORST DAMAGE WAS IN CASE AVENUE EAST NORTH AND LODS ST. DISTRICTS

Small Break in the East Reservoir Developed Dangerous Conditions Along the Canal Yesterday, Necessitating the Dynamiting of the Locks.

The crest of the Akron flood was passed at about 3 o'clock yesterday afternoon, at which time the little Cuyahoga commenced to drop.

Greatest damage was that done along Case avenue, Lods street and East North street.

The only known fatality is that of

SUPPORT THESE CANDIDATES

DATYON SITUATION IS TERRIBLE—LOSS OF LIFE ONLY GUESSED AT

Dayton, March 26—At noon today Russells appartment on Third street took fire and the occupants are leaping into the water. People are being taken from the houses in boats, in the residential district by the militia men and policemen. Seventy-five persons are said to have been drowned in Riverdale, a northwestern suburb. Not a house is left standing on Linwood avenue in Riverdale.

Accurate estimates of the loss of life are impossible to make. The guesses range from 1000 to 5000.

Water stands in Dayton streets from 15 to 20 feet deep. Thousands are homeless and suffering is indescribable.

The crest of the Dayton flood has

Top, The West Market Bridge over the canal, looking north at chain of ruined locks; March 1913. *Above,* The ruins of locks north of West Market and Akron's power house, whose boilers were quenched by the flood, plunging Akron into darkness and silencing its industry; March 1913. (Pockrandt Collection, SCHS)

Appendix

Physical Data
and
Tables of Distances
PUBLIC-FUNDED CANALS OF OHIO

by Frank W. Trevorrow
and Barnett L. Golding

Originally compiled by Frank W. Trevorrow as an Appendix in *Ohio's Canals*, published in 1973 at Oberlin, Ohio.

Revised and augmented by Barnett L. Golding
Columbus, Ohio
July 1986

NOTE OF EXPLANATION

Data for these tabulations were gleaned from the various Annual Reports of the Board of Canal Commissioners and of the Board of Public Works, and from Department of Public Works Canal Plat Books, and occasionally from other sources. The intent was to list all major main line structures of the public-funded navigable canals in Ohio—wasteways and small culverts excepted. Details of un-navigable feeders are not included. Local names of locks and other features are far from complete and some possibly inaccurate. Most villages are listed, but probably not all, and certainly not all of their names over time. Only a smattering of mills and other industrial water users are included. Prominent basins or other canal-related features (e.g., Pinery Narrows) are mentioned at times but there was no attempt to balance or give equal treatment throughout the tabulations.

1859 mileages were used simply because this was the most complete tabulation extant. Mileage figures do not always agree with other records or modern mapping and there appears to be at least one "bust" of about a mile just south of the Licking Summit. These record distances were reported here with no attempt to rectify them. Some accurately scaled distances were added here and there.

Alternative canal era names of a given structure or place are separated with a virgule (/). Explanatory notes and post-canal era places and names are shown in parentheses.

Generally, no distinction was made between original features and those that came on line at a later time. For features which had an early demise, dates are given when known. For features which changed form over time (e.g., culvert to aqueduct), generally only the last form is listed, as only so much can be done on a single-line tabulation. This work is not intended to be definitive but merely a general inventory of the public-funded system. The authors welcome readers' corrections, additions, or other comments.

Barnett L. Golding

MIAMI & ERIE CANAL
Physical Data

Section	Miles Int.	Miles From Toledo	Number of Lift Locks	Number of Guard Locks	Rise/Fall in Ft.
Toledo (Manhattan) to Toledo (Swan Creek)	5	5	2	0	15 rise
Toledo (Swan Creek) to Defiance (Slack Water 4.5 miles)	56	61	11	0	85 rise
Defiance to Junction	9	70	7	0	57 rise
Junction to Bremen	58	128	32	0	217 rise
Loramie Summit	23	151	0	0	—
Lockington to Dayton	33	184	20	0	203 fall
Dayton to Cincinnati	66	250	34	0	309 fall
Totals	250	250	106	0	886 rise/fall
Navigable Branches and Feeders					
Swan Creek Side-Cut	1		2	0	15 rise
Maumee Side-Cut	1.5		6	0	63 fall
Gilead Side-Cut	0.2		0	0	—
Wabash & Erie Canal	18		6	0	28 rise
St. Marys Feeder	3		1	0	—
Sidney Feeder	14		0	0	—
Hamilton Side-Cut	0.6		0	0	—

Canal Prism:	*Toledo to Junction*	*Junction to Dayton**	*Dayton to Cincinnati*
Width at Water Line	60 feet	50 feet	40 feet
Width at Bottom	46 feet	36 feet	26 feet
Min. Depth of Water	6 feet	5 feet	4 feet

*Also Wabash & Erie Canal (Junction to State Line) and St. Marys and Sidney Feeders. Hamilton Side-Cut was 80 feet wide.

Lock Dimensions:	*Standard*	*Locks 44–53 Cincinnati*	*Bulkhead Lock St. Marys Feeder*
Length in Chamber	90 feet	90 feet	100 feet
Width in Chamber	15 feet	18 feet	20 feet
Lift	Varying	11 feet	

Distances are those given in the 21st (1859) Annual Report of the Board of Public Works. Several sources state the ascent from Lake Erie to Loramie Summit to be 395 feet. Based on 573 feet elevation of Lake Erie water level above sea level, this would make the Loramie Summit level 968 feet. The correct level is 947 feet, and the corresponding rise is 374 feet.

MIAMI & ERIE CANAL
Table of Distances
Toledo to Cincinnati

Miles from Lake Erie at Toledo	*Lock No.*	*Name of Lock or Structure*	*Place Name or Location*
		JUNCTION — MAUMEE RIVER	
0	1	Manhattan, River (until 1864)	Manhattan*
	2	Manhattan (until 1864)	Manhattan
4			Toledo*
5		SWAN CREEK SIDE-CUT	Side-cut*
		#52	Toledo, Newton St.
		#51	Toledo, Logan St.
	50	Armada Mills	Toledo, Mill St.
	49	Armada Mills	Toledo, Mill St.
	48	Armada Mills	Toledo, Mill St.
	47	Sash Factory	Toledo, Railroad Ave.
	46		
12	45	Port Miami (1887) (1909)	Port Miami*
13			Maumee City*
14		MAUMEE SIDE-CUT	Side-cut*
		6 Lift Locks	
		1.5 miles to Maumee River	
		2.5 miles to Perrysburg*	
18			Waterville*
25			Otsego*
	44	Providence Guard & Lift	Providence Mill/Ludwig Mill
30		PROVIDENCE DAM	Providence*
		SLACK WATER	Maumee River
		GILEAD SIDE-CUT	Grand Rapids
	43	Bucklin's/Bulkhead (1887)	
	42	Rice's /Rees' (1887)	
		BAD CREEK CULVERT	
34	41	Texas	Texas*
		DRY CREEK CULVERT	
		TURKEYFOOT CREEK CULVERT	
37			Damascus*
		OBERHAUS CREEK CULVERT	
44			Napoleon*
		GARRETT CREEK CULVERT	
		BENIEN CREEK CULVERT	
52			Florida*
57	40	Independence Guard & Lift	Independence*
		INDEPENDENCE DAM	
		SLACK WATER	Maumee River
61	39	River	Defiance*
	38	Defiance	Defiance
	37	Defiance	Defiance
	36	Defiance Mills/Wilhelm's (1909)	Defiance
	35	Palamo/Erie Mill (1887)	Defiance
	34	Paper Mill (1887)	Defiance
		THREEMILE CREEK CULVERT	
	33	Schooley's (1887) (1909)	
		FIVEMILE CREEK CULVERT	
		SIXMILE CREEK CULVERT	

Continued on next page

MIAMI & ERIE CANAL

Table of Distances — *Continued*

Miles from Lake Erie at Toledo	*Lock No.*	*Name of Lock or Structure*	*Place Name or Location*
70		JUNCTION — WABASH & ERIE CANAL	Junction*
		LITTLE FLATROCK CREEK CULVERT	
		FLATROCK CREEK AQUEDUCT	
	32	Vial's (1909)	
75		CHARLOE CULVERT	Charloe*
		BLUE CREEK AQUEDUCT	
78			St. Andrews*
79			Tiffin & Fort Wayne RR.,* Melrose
		LITTLE AUGLAIZE RIVER AQUEDUCT	
81			Murat*
82	31	Ridenhour's/Hipps' Flouring Mill	Ridenhour's Mill,* Timberville
			Mandale
86			Hamer*
87	30	Stone*	
	29		
91	28	Ottoville	Ottoville
	27		
	26		
	25		
		WEST FLAT FORK AQUEDUCT	
		JENNINGS CREEK AQUEDUCT	
	24	Stone	
97	23	Delphos (1880)	Delphos*
	22		
	21		
	20		
101	19		
	18		
104	17		
		SPENCERVILLE CULVERT	Sixmile Creek
106	16	Acadia	Acadia*
	15	Spencerville	Spencerville
		DEEP CUT	
108			Deep Cut Bridge*
110			Kossuth*
		PRAIRIE CREEK CULVERT	
112.5		SIX MILE AQUEDUCT*	Sixmile Creek
115	14	Saw Mill	
117			Large Basin* (Forty Acre Pond)
	13	St. Marys (1886)	St. Marys
120	12	St. Marys	St. Marys*

Continued on next page

		ST. MARYS RIVER AQUEDUCT	
121		JUNCTION — ST. MARYS FEEDER	Mercer County Reservoir*
		3 Miles to Bulkhead Lock	(Grand Lake St. Marys)
		11 Miles to Celina	
	11		
	10		
	9		
	8	Stone	
	7		
		DRY FEEDER to MERCER CO. RES.	Montezuma Tpk.
124	6		
	5		
	4	Saw Mill	
	3		
126.5	2	Flouring Mill	Lock Two
128	1	Bremen	Bremen* (New Bremen)
			N. End Loramie Summit
131			Minster*
132			Vocke's Distillery*
133		LORAMIE RESERVOIR FEEDER*	
		LORAMIE CREEK AQUEDUCT	
134			Berlin*
136			Stare Landing*
138			Newport*
		PAINTER CREEK CULVERT	
143			Patrick Station*
143.5			B. & I. RR Crossing*
		TURTLE CREEK CULVERT	
147			Woodbourne* (Newbern)
		MILL CREEK CULVERT	
		FOX CREEK CULVERT	
151		JUNCTION — SIDNEY FEEDER*	
			S. End Loramie Summit
151	1	Big	Lockington*/Lockport
		TURNING BASIN & DRY DOCK	
	2		Lockington
	3		Lockington
		PIQUA HYDRAULIC BASIN	
	4		Lockington
	5		Lockington
		LORAMIE CREEK AQUEDUCT	
	6	Crooked	
152	7	Landman's Mill	Loramie Mills*
	8	State Dam	
		PIQUA DAM & LORAMIE FEEDER	Miami River
		GUARD GATE	
155	9	Piqua	Piqua*
		PIQUA HYDRAULIC TAIL RACE	
		TROY HYDRAULIC HEAD RACE	
158	10	Farrington & Slosson	Farrington's Distillery*
160	11	Allen's Mill	Allen's Mill*
		TROY HYDRAULIC TAIL RACE	
163	12	Troy	Troy*

Continued on next page

MIAMI & ERIE CANAL

Table of Distances — *Continued*

Miles from Lake Erie at Toledo	*Lock No.*	*Name of Lock or Structure*	*Place Name or Location*
165	13	John's Mills	Culbertson's Distillery*
166	14	Feeder*	
		TROY DAM & FEEDER	Miami River
167			Cowlesville*
169	15	Tippecanoe	Tippecanoe* (Tipp City)
172	16	Picayune*	
174			National Road*
175		MIAMI AQUEDUCT*	Miami River
176	17	Smith's Distillery	Marysville*
	18	Distillery	
178	19	Flouring Mill	Lockville*
		MAD RIVER AQUEDUCT	
	20	Mad River	Dayton
		MAD RIVER DAM & FEEDER	
184	21	Dayton - Old Line	Dayton*
	21	Dayton - Cooper's New Line	Dayton
190	22	Snider's Mill	Snyder's Mills*
		HOLES CREEK CULVERT	
191	23	Dryden's Mill	Dryden's Mills*
192			Alexandersville*
193	24	Upper Carrollton (1909)	Carrollton,* (West Carrollton)
	25	Lower Carrollton (1909)	West Carrollton
		MIAMISBURG DAM & HYDRAULIC FEEDER	Miami River
		MIAMISBURG HYDRAULIC EXITS	
		MIAMISBURG AQUEDUCT	Sycamore Creek
195			Miamisburgh*/Miamisburg
	26	Miamisburg	
		SUNFISH AQUEDUCT	Crane's (Crains) Run
	27	Sunfish/Smiley's	
201			Franklin*
		CLEAR CREEK AQUEDUCT	
	28	Franklin/Clutch's Paper Mill	
	29	Upper Greenland (1909)	
205			Miami Dam*
		FEEDER to WARREN COUNTY CANAL	
	30	Lower Greenland	
	31	Dine's (1909)	
		MIDDLETOWN DAM & FEEDER	Miami River
		MIDDLETOWN HYDRAULIC	
208		JUNCTION — WARREN COUNTY CANAL	Middletown*
	32	Middletown/Doty's/Thomas Sons' Paper Mill	
210	33	Amanda's Mills	Amanda*
		AMANDA AQUEDUCT	Dicks Creek
	34	Excello Mills	Excello

Continued on next page

213			Leonardsville*/LeSourdsville
		LESOURDSVILLE AQUEDUCT	Gregory Creek
	35	Swindler's/Tangeman's	Rockdale
	36	Upper Hamilton/Snider's Sons	Hamilton
	37	Lower Hamilton/Carl's Flouring Mill	Hamilton
221		HAMILTON BASIN SIDE-CUT	Hamilton Sidecut*
		CRAWFORD RUN AQUEDUCT	
		PORT UNION AQUEDUCT	
229			Port Union*
230	38	Rialto/Friend & Fox Paper Mill	Swift's Mills*/Rialto
	39	Crescent Paper Mill	Crescentville
		CRESCENTVILLE AQUEDUCT	
237	40	Upper Lockland (1909)	Lockland*
	41	Collector's (1909)	Lockland
	42	Flour Mill (1909)	Lockland
		WEST FORK AQUEDUCT	West Fork Mill Creek
	43	Lower Lockland/Halderman's	Lockland
			Howell's Basin/Hartwell's Basin
		MILL CREEK AQUEDUCT	
240			Carthage*
241			Ruffner's Basin*
245			Cumminsville*
		CHEAPSIDE BASIN SIDE-CUT	
	44	(until 1863)	Cincinnati
	45	(until 1863)	Cincinnati
	46	(until 1863)	Cincinnati
	47	(until 1863)	Cincinnati
	48	(until 1863)	Cincinnati
	49	(until 1863)	Cincinnati
	50	(until 1863)	Cincinnati
	51	Congress St. (until 1863)	Cincinnati
	52	Front St. (until 1863)	Cincinnati
250	53	Ohio River Outlet (until 1863)	Cincinnati*
		JUNCTION — OHIO RIVER	

Place names marked * and all distances are from 1859 Board of Public Works Report.

Lock numbers are in accordance with the numbering system adopted in 1883.

Lock names are from 1883 Board of Public Works Report, except where other date is indicated.

WABASH AND ERIE CANAL
Table of Distances
Junction to Indiana State Line

Miles from Junction	*Lock No.*	*Name of Lock or Structure*	*Place Name or Location*
0*		JUNCTION — MIAMI & ERIE CANAL	Junction*
1	8	One Mile	
	9		
	10		
11*		RESERVOIR*	Six Mile Creek
	11		
14*			Antwerp*
	12		
18*	13		State Line*

Lock numbers are those given in the 1845 Board of Public Works Report.

Distances and place names marked * are from the 1859 Board of Public Works Report.

SIDNEY FEEDER CANAL
Table of Distances
Port Jefferson to Lockington

Miles from Port Jeff.	*Lock No.*	*Name of Lock or Structure*	*Place Name or Location*
0*		MIAMI DAM* & GUARD GATE	Miami River
1*			Port Jefferson*
4*		PLUM CREEK AQUEDUCT	Plum Creek*
5		WELLS CREEK CULVERT	
6*			Sidney*
9*			Barkalow's Saw Mill*
11		RUMLEY CREEK CULVERT	Mill Branch
12*			Wise's Pond*
14*		JUNCTION — MIAMI & ERIE CANAL	Lockington*

Distances and place names marked * are from 1859 Board of Public Works Report.

OHIO & ERIE CANAL
Physical Data

Section	Int.	Miles From Cleveland	Number of Lift Locks	Number of Guard Locks	Rise/ Fall in Ft.
Cleveland to Akron	38	38	44	0	395 rise
Portage Summit	9	47	0	0	—
Barberton to Webbsport	102	149	31	2	238 fall
Webbsport to Licking Summit (Slack Water 2 miles)	32	181	19	1	160 rise
Licking Summit	14	195	2	0	6 rise 6 fall
Licking Summit to Lockbourne	26	221	30	1	202 fall
Lockbourne to Portsmouth	88	309	25	3	211 fall
Totals	309	309	151	7	1,218 rise/ fall
Navigable Branches and Feeders					
Trenton Feeder	3		0	1	—
Dresden Side-cut	3		3	0	29 fall
Granville Feeder	6		1	1	10 fall
Columbus Feeder	12		2	2	14 fall

	Standard	Feeders
Canal Prism: Width at Water Line:	40 feet	32 feet
Width at Bottom:	26 feet	18 feet
Min. Depth of Water:	4 feet	4 feet

Lock Dimensions:	Standard	Sloop Locks 43 & 44 Cleveland
Length in Chamber:	90 feet	100 feet
Width in Chamber:	15 feet	25 feet
Lift:	Varying	6 feet

Weigh Locks: One at Cleveland

Distances are those given in the 1859 Board of Public Works Report.

OHIO & ERIE CANAL
Table of Distances
Cleveland to Portsmouth

Miles from Lake Erie at Cleveland	Lock No.	Name of Lock or Structure	Place Name or Location
		JUNCTION — CUYAHOGA RIVER	
0*	44	River (until 1872)	Cleveland,* Merwin St.
	43	Sloop (until 1872)	Cleveland, Merwin St.
3	42	3 Mile (before 1872)	Cleveland, N. of Dille St.

Continued on next page

Table of Distances — *Continued*

Miles from Lake Erie at Cleveland	*Lock No.*	*Name of Lock or Structure*	*Place Name or Location*
3	42	River (after 1872) Weigh Lock	Cleveland, Ft. of Dille St.
5*	41	Rathbun's*/5 Mile	Cleveland, Harvard Ave.
8	40	8 Mile	Cleveland, Independence Rd. & E. 71 St.
9*+		MILL CREEK AQUEDUCT	Mill Creek*
11	39	11 Mile	S. of Rockside Road
12	38	12 Mile	S. of Hillside Road
13*+		TINKERS CREEK AQUEDUCT	Tinkers Creek*
	37	14 Mile/Alexander	Alexander's Mill/Wilson's Mill
			Pinery Narrows
17*+		PINERY (DAM &) FEEDER*	Cuyahoga River
	36	Pinery/17 Mile/Brecksville	N. of S. R. 82
			Chaffee
	35	Kettlewell/Whiskey	S. of S. R. 82
	34	Red/Jaite	Vaughn Road
		BRANDYWINE CREEK CULVERT	
	33	Lower Boston/Wallace	
21*	32	Boston	Boston*
			Stumpy Basin
	31	Lonesome/Lost	
23.5+		PENINSULA DAM & FEEDER	Cuyahoga River
	30	Feeder	
	29	Peninsula	
24+		PENINSULA AQUEDUCT	Cuyahoga River
24*			Peninsula*
	28	Deep	
27	27	Johnny Cake	Everett
		FURNACE RUN AQUEDUCT	
	26	Pancake	
	25	Mudcatcher	
	24	Niles	
		YELLOW CREEK CULVERT	
30*			Niles*/Botzum
	23	Lower Portage	N. of N'hampton-Portage Twp. Line
	22	Upper Portage	S. of N'hampton-Portage Twp. Line
32*			Old Portage* (Cranmer)
35+		LITTLE CUYAHOGA DAM & FEEDER	
	21		N. of Memorial Pky.
	20		N. of Memorial Pky.
	19		S. of Memorial Pky.
	18		S. of Memorial Pky.
	17		S. of Memorial Pky.
	16		Akron, Ft. of Maple St.

Continued on next page

36.5+		LITTLE CUYAHOGA DAM & FEEDER	
	15		Akron, N. of North St.
	14	Cascade Mills	Akron, S. of North St.
	13		Akron, Ft. of Cascade St.
	12	Aetna Mills	Akron, Ft. of Furnace St.
	11	Aetna Mills	Akron, N. of Beech St.
	10		Akron, S. of Beech St.
	9	City Mills	Akron, N. of Market St.
37*			North Akron*
	8	Stone Arch	Akron, S. of Market St.
	7	Woolen Mill	Akron, S. of Cherry St.
	6	Allen Mill	Akron, N. of Ash St.
	5	Stone Mill	Akron, S. of Ash St.
	4	Tannery	Akron, At Bowery St.
	3	Boat Yard	Akron, N. of Center St.
	2	Dry Dock	Akron, N. of Buchtel St.
38*			South Akron*
		JUNCTION — PENNA. & OHIO CANAL	Lower Basin
	1		Akron, S. of Exchange St.
38+			N. End Portage Summit
			Upper Basin
39			Summit Lake
42		PORTAGE LAKES FEEDER	
44*			New Portage*
			Barberton
47+			S. End Portage Summit
47*	1	Wolf Creek*	Barberton, N. of Snyder Ave.
47+		SLACK-WATER CROSSING	Wolf Creek
		WOLF CREEK DAM	
	2	Upper Clinton	
	3	Lower Clinton	
52+		SLACK-WATER CROSSING	Tuscarawas River
		CLINTON DAM & GUARD LOCK	
52*			Clinton*
		NIMISILA CREEK CULVERT	
55+		MESELEE (NIMISILA) CREEK FEEDER	
56*			Fulton*/Canal Fulton
	4	Mill	
61*			Wellman's Mill* (Crystal Springs)
62+		MUD BROOK FEEDER	Mudbrook Creek
65*			Massillon*
		SIPPO CREEK FEEDER (until 1848)	
	5	Rolling Mill	Massillon, S. of Walnut St.
	5A	Massillon/Cemetery/Old No. 5	Massillon
			Rochester
71*	6	Navarre/Bethlehem	Navarre*
			Bethlehem
78+		TUSCARAWAS AQUEDUCT	Tuscarawas River
80*		JUNCTION — SANDY & BEAVER CANAL	Bolivar*
	7		
	8		
	9		
	10		

Continued on next page

OHIO & ERIE CANAL

Table of Distances — *Continued*

Miles from Lake Erie at Cleveland	*Lock No.*	*Name of Lock or Structure*	*Place Name or Location*
83+		ZOAR DAM & FEEDER	Tuscarawas River
83*		ZOAR SIDE-CUT	Zoar*
86*			Jenning's Bridge*
	11	Quarry	
	12	Jones	
93*			Dover*
93+		SLACK-WATER CROSSING DAM & GUARD LOCK	Sugar Creek
97*	13	Blake's Mills	Lockport/Blake's Mills* (New Philadelphia)
		OLDTOWN CREEK CULVERT	
99*	14	New Castle	New Castle*
		MUD RUN CULVERT	
103*			Trenton* (Tuscarawas)
	15	Upper Trenton	
	16	Lower Trenton	
104+		JUNCTION — TRENTON FEEDER	Tuscarawas River
		3 Miles to Eastport*	
		9 Miles to Uhricks Mill	
		FRYS CREEK CULVERT	
108*			Gnadenhutten*
	17		Seventeen P.O.
	18	Port Washington/Bremer	
112*			Port Washington*
	19	Hill's	
	20	Garfield/Daugherty/Tucker's	
118*			Newcomerstown*
		BUCKHORN CREEK CULVERT	
	21	Newcomerstown	
	22	Sedam/Suydam's	
		BLUE RIDGE RUN CULVERT	
	23	Mackey	
122*			Evansburg*/Orange
		EVANS CREEK AQUEDUCT	
	24	Reed's/Blue Hole/Felver	
		WHITE EYES CREEK CULVERTS	
	25	Wild Turkey	
132*			Lewisville*/Canal Lewisville
			Upper Basin/Mudport Basin
			Middle Basin
135+		WALHONDING AQUEDUCT	Walhonding River
	26		
	27		
			Lower Basin/Roscoe Basin
		WALHONDING FEEDER	Walhonding River
		JUNCTION — WALHONDING CANAL	

Continued on next page

135*			Roscoe*
			Conesville
	28	Adam's Mills	Adam's Mills
145*	29	Adam's Mills	Adam's Mills*
	30	Adam's Mills	Adam's Mills
			Munroe Basin
149*+		JUNCTION — DRESDEN SIDE-CUT	Webbsport*
	19	Lower Webb	
		PADDY'S FORK CULVERT	Little Wakatomika Creek
	18	Upper Webb	
152*			Hartford's*
155*	17	Frazeysburg	Frazeysburg*
156+		WAKATOMIKA CREEK AQUEDUCT	
156.25	16	Vickers	
161*			Nashport*
166*+		LICKING DAM* & GUARD LOCK	Licking River
		SLACK WATER	Licking Narrows (Black Hand Gorge)
	15	Outlet	
	14	Rocky Fork	
		ROCKY FORK CULVERT	
170*			Licking Town*
	13	Bowling Green	
		BOWLING GREEN RUN CULVERT	
	12	Lower Stadden	
	11	Upper Stadden	
		SHAWNEE RUN CULVERT	
	10	White Mill	Newark, Mill St.
176+		NORTH FORK AQUEDUCT	
176+		NORTH FORK FEEDER	
176*	9	Newark Town	Newark,* First St.
177+		RACCOON CREEK AQUEDUCT	
177+		RACCOON CREEK FEEDER	
	8	Lower Lockport	Newark, 11th St.
	7	Second Lockport	Newark, N. of Wilson St.
	6	Third Lockport	Newark, S. of B.&O. RR
	5	Upper Lockport	Newark
	4	Fairground	Newark, Indian Fort
		CRAB CREEK CULVERT	Ramp Creek
	3	Lower Taylor/Four Mile	(Heath)
	2	Second Taylor	(Heath)
	1	Upper Taylor	(Heath)
181+			N. End Licking Summit
181*+		JUNCTION — GRANVILLE FEEDER*	Raccoon Creek
		5 Miles to Guard Lock	
		6 Miles to Granville	
182*			Hand's Basin*
		BEAVER RUN CULVERT	
185*+		HEBRON CULVERT	Hebron,* National Road
187+		LAKE RUN AQUEDUCT	
188+		SOUTH FORK AQUEDUCT	
	0	North/Minthorn	West Embankment
189*		LICKING RESERVOIR	Licking Summit* (Buckeye Lake)

Continued on next page

OHIO & ERIE CANAL

Table of Distances — *Continued*

Miles from Lake Erie at Cleve- land	*Lock No.*	*Name of Lock or Structure*	*Place Name or Location*
190+		SOUTH FORK FEEDER	
191*			Millersport*
		DEEP CUT	
	0	South/Deep Cut/Pugh/Summit	S. of Bickel Church Rd.
195			S. End Licking Summit
	1	King Watson	E. of S. R. 37
	2	David Miller/White's Mill	W. of S. R. 37
	3	Norris Mill	
		PAWPAW CREEK CULVERT	
	4	Short Level	
	5	Dry Dock	
196*	6	Mulnix Mill/Smeck Mill	Baltimore*
	7	Well's Mills	
197			Basil
	8	Bibler	
198+		DAM & FEEDER	Walnut Creek
201+		WALNUT CREEK AQUEDUCT	
			Highbanks
202*			Havensport*
204*+		JUNCTION — HOCKING CANAL	Carroll*
	9		
	10		
		KISTLER RUN CULVERT	
206*	11	Upper Lockville/Grist Mill	Lockville,* Chestnut St.
	12	Tennis/Tenant	Lockville, Lock St.
	13	Row/Rowe	Lockville, S. of Walnut St.
	14	Smallwood	Lockville, N. of Walnut St.
	15	Short Level/Fickle Mill	Lockville, N. of Maple St.
	16	Rover, Short Level	Lockville
	17	Swimmer	
	18	Creek	
207+		SLACK-WATER CROSSING	Walnut Creek
		WALNUT CREEK DAM & GUARD LOCK	
209*			Waterloo*
210*			Winchester*/Canal Winchester
	19	Chaney's Mill	
	20	Woolen Factory	
	21	George's	
212+		BLACK LICK FEEDER	Rager's Run/Georges Creek
212		GEORGES CREEK CULVERT	
	22		
214*			Rareysport/Groveport*
216*			Sharp's Landing*
	23		
	24		
	25		

Continued on next page

	26		
	27		
	28		Lockbourne, under C&0 RR
	29		Lockbourne, W. of Lockbourne Rd.
221*	30		Lockbourne*
221+		JUNCTION — COLUMBUS FEEDER	Scioto River, Big Belly Creek
224*			Holmes' Landing*
228*			Millport and Bloomfield*
	31		Campbell's Mill
228+		SLACK-WATER CROSSING	Walnut Creek
		WALNUT CREEK DAM & GUARD LOCK	
		HARGUS CREEK CULVERT	
236*			Circleville*
236+		CIRCLEVILLE AQUEDUCT	Scioto River
	32	Aqueduct	
			Spunkeytown
	33		
	34	Eighteen Mile	
238+		CIRCLEVILLE DAM & FEEDER	Scioto River
241*			Westfall*
		LICK RUN CULVERT	
245*			Yellowbud*
		YELLOWBUD CREEK AQUEDUCT	
249*		DEER CREEK AQUEDUCT	Deer Creek*
251*			Andersonville*
		DRY RUN CULVERT	
			Four Mile Basin
256*	35	Upper Marfield	Clinton Mills*/Marfield Mills
	36	Lower Marfield	
258*	37	Fourth St.	Chillicothe,* N. of Fourth St.
	38	Fifth St.	Chillicothe, N. of Fifth St.
		HYDRAULIC CANAL SIDECUT	
261+		PAINT CREEK AQUEDUCT	
	39	Upper Lunbeck's	
	40	Lower Lunbeck's	
		INDIAN CREEK CULVERT	
264*	41	Tomlinson's Three Locks	Tomlinson's*
	42	Tomlinson's Three Locks	
	43	Tomlinson's Three Locks	
265+		TOMLINSON'S DAM & FEEDER	Scioto River
266*			Stony Creek*
266*		STONY CREEK AQUEDUCT	Stony Creek*
269*		Higby's Guard Lock	Head of Big Bottom*
		SANDY BOTTOM CREEK	
		MOORE RUN AQUEDUCT	
		WILSON RUN CULVERT	
275*			Sharonville*/Omega
279*		CROOKED CREEK AQUEDUCT	Waverly*
	44	Waverly	Waverly
		PEE PEE CREEK AQUEDUCT	
	45	Upper Pee Pee	
	46	Lower Pee Pee	

Continued on next page

OHIO & ERIE CANAL

Table of Distances — *Continued*

Miles from Lake Erie at Cleveland	*Lock No.*	*Name of Lock or Structure*	*Place Name or Location*
283*			Trimble's Bridge*
			Jasper Basin
286*			Jasper*
		NO NAME CREEK CULVERT	
		GUARD LOCK	
287+		SUNFISH CREEK AQUEDUCT	
291*	47	Howard's*	Brook's Mill
292			Coopersville
		CAMP CREEK AQUEDUCT	
294*			Cutler's Station*
		BEAR CREEK AQUEDUCT	
		BIG RUN CULVERT	
	48	Herod's	
301*		BRUSH CREEK AQUEDUCT	Brush Creek*
	49	Rush's/Brush Creek	Rushtown
		POND CREEK CULVERT	
		DRY RUN CULVERT	
	50	Union Mills/Moss' Mills	Bertha (West Portsmouth)
	51	Union Mills/Moss' Mills	Bertha
	52	Union Mills/Moss' Mills	Bertha
	53	Elbow	Portsmouth
	54	Red Bridge	Portsmouth
309*	55	Outlet	Portsmouth*
		JUNCTION — OHIO RIVER	

Distances and place names marked * are from 1859 Board of Public Works Report.

Distances marked + are from 1832 Board of Canal Commissioners Report.

DRESDEN SIDE-CUT
(Muskingum Side-cut)
Table of Distances
Webbsport to Dresden

Miles from Cleve-land	*Lock No.*	*Name of Lock or Structure*	*Place Name or Location*
149*		JUNCTION — OHIO & ERIE CANAL	Webbsport*
		WAKATOMIKA CREEK AQUEDUCT	
151*			Dresden*
	31	Side-cut	
	32	Side-cut	
	33	Side-cut	
152		JUNCTION — MUSKINGUM RIVER	

Distances and place names marked * are from 1859 Board of Public Works Report.

COLUMBUS FEEDER CANAL
(Columbus Lateral)
Table of Distances
Columbus to Lockbourne

Miles from Colum-bus	*Lock No.*	*Name of Lock or Structure*	*Place Name or Location*
		SLACK WATER	Scioto River
0		DAM & GUARD LOCK	Columbus, Friend (Main) St.
			Lake Park
4.5	1	4 Mile	
9			Shadeville
	2		
11.5		SLACK-WATER CROSSING	Big Belly Creek/Big Walnut Creek
		DAM & GUARD LOCK	
11.75		JUNCTION — OHIO & ERIE CANAL	Lockbourne

WALHONDING CANAL
Physical Data

Section	Miles Int.	Miles From Roscoe	Number of Lift Locks	Number of Guard Locks	Rise/ Fall in Ft.
Roscoe to Walhonding Dam	6	6	3	1	24 rise
Walhonding Dam to Rochester	19	25	8	1	66 rise
Totals	25	25	11	2	90 rise

Canal Prism: Width at Water Line: 40 feet
Width at Bottom: 26 feet
Min. Depth of Water: 4 feet

Lock Dimensions: Length in Chamber: 90 feet
Width in Chamber: 15 feet
Lift: Varying

Distances are those given in the 1859 Board of Public Works Report.

WALHONDING CANAL
Table of Distances

Miles from Roscoe	Lock No.	Name of Lock or Structure	Place Name or Location
0*		JUNCTION — OHIO & ERIE CANAL	Roscoe,* Roscoe Basin
	1-2-3	Triple	
			T.W.V. & O. RR
2*		CROOKED RUN CULVERT	Crooked Run Bridge*
4			Hosfelt's Waste Way
	4	Walhonding Guard	
6*		SIX MILE DAM	Walhonding Dam*
		SLACK-WATER CROSSING	Walhonding River
	5		
	6	Crawford's	
8*	7	Warsaw	Warsaw*
9*		SIMMONS RUN CULVERT	Simmon's Bridge*
	8		
10*			Bedford Road (crossing)*
11*		DARLING RUN CULVERT	Darling's Bridge*
12*	9	Gamble's	Gamble's Lock*
13*			Butler's (lower crossing)* (Mohawk Dam)
		DUTCH RUN CULVERT	
14*			Turner's Saw Mill*
		HONEY RUN CULVERT	
15*	10	Fry's	Butler's (upper crossing)*
18*	11		Walhonding (town)*
	12		
19*			Mohican (road crossing)*
19*			Gamble's Saw Mill*
21*		CUMMINGS RUN CULVERT	Cummin's Bridge*
			T.W.V. & O. RR.

Continued on next page

	13	Mohican Guard	
23*		MOHICAN FEEDER DAM	Mohican Dam*
		SLACK WATER	Mohican River
24*			Cavallo*
25*			Rochester*

Distances, structure names, and place names marked * are from the 1859 Board of Public Works Report. All other names are from "Twenty Five Miles to Nowhere" by Terry K. Woods, 1978.

All of the Walhonding Canal above Six Mile Dam was officially abandoned in 1896. The dam and lower six miles of canal were retained as the Walhonding Feeder to the Ohio & Erie Canal.

HOCKING CANAL
Physical Data

Section	*Miles Int.*	*Miles From Carroll*	*Number of Lift Locks*	*Number of Guard Locks*	*Rise/ Fall in Ft.*
Carroll to Lancaster	9	9	0	0	
Lancaster to Sugar Grove	7	16	10	0	
Sugar Grove to Logan (Slack Water 6 mi.)	12	28	4	2	
Logan to Nelsonville	13	41	6	1	
Nelsonville to Chauncey	8	49	5	0	
Chauncey to Athens (Slack Water 1 mi.)	7	56	1	2	
Totals	56	56	26	5	202 fall

Canal Prism: Width at Water Line: 40 feet
Width at Bottom: 26 feet
Min. Depth of Water: 4 feet

Lock Dimensions: Length in Chamber: 90 feet
Width in Chamber: 15 feet
Lift: Varying

Distances are those given in the 1859 Board of Public Works Report.

HOCKING CANAL
Table of Distances

Miles from Carroll	*Lock No.*	*Name of Lock or Structure*	*Place Name or Location*
0*		JUNCTION — OHIO & ERIE CANAL	Carroll*
3		KISTLER RUN FEEDER	
		CLAYPOOL RUN FEEDER	
5		GARY RUN CULVERT	

Continued on next page

HOCKING CANAL

Table of Distances — *Continued*

Miles from Carroll	*Lock No.*	*Name of Lock or Structure*	*Place Name or Location*
7		BUSH RUN CULVERT	
			Reservoir
9*	1		Lancaster,* S. of Chestnut St.
	2		Lancaster, S. of Walnut St.
	3		Lancaster, N. of RR Crossing
	4		Lancaster, E. of Columbus St.
		DAM & FEEDER	Hocking River
10		EWING BASIN SIDE-CUT	Lancaster, E. of Maple St.
		BALDWIN RUN CULVERT	
11	5		
	6		
		DAM & FEEDER	Hocking River
13		PLEASANT RUN CULVERT	
13	7		
14*	8	Ream's	Ream's Mills*/Horns Mill
16	9		Sugar Grove, N. of Bridge St.
16*	10		Sugar Grove, Rush Creek*
		SLACK WATER	Rush Creek, Hocking River
17*		BUCKEYE DAM & GUARD LOCK	McCann's Mills*
19	11		Millville (Rockbridge)
20			
21	12	Sheep Pen	
22*	13	Green's	Green's Mills*
		SLACK WATER	Hocking River
23*		DAM & CANAL	
	14	Bonner's/Bowner's	Bonner's Lock*
		SLACK WATER	Hocking River
24			Stiversonville (Enterprise)
26			Rock-cut Towpath
27*		DAM & GUARD LOCK	Hocking Falls,* Falls Mill
28*			Logan,* Mulberry St.
29		OLD TOWN CREEK CULVERT	
30	15	Boardman's	
	16	Dew's	
32		THREEMILE CREEK CULVERT	
33*			Wright's*
34		FIVEMILE RUN CULVERT	
34*	17	Wright's	Pattonsville*
35			Haydensville
36*			Seven Mile Run*
	18	Wolf's	
			Wolf's Basin
38	19		
39	20		
		FEEDER DAM	Hocking River
41*		GUARD LOCK	Nelsonville*
42	21		
44*		MONDAY CREEK AQUEDUCT	Monday Creek*
	22		

Continued on next page

46	23		
48	24		
49*	25		Chauncey*
		SLACK-WATER CROSSING	Hocking River
		1 Mile to Salina	
		DAM & GUARD LOCK	
51*			Wolf's* (The Plains)
54	26		
		SLACK WATER	Hocking River
55		DAM & GUARD LOCK	
56*		TERMINUS	Basin, Athens*

Distances and place names marked * are from 1859 Board of Public Works Report.

Portion of canal below Chauncey abandoned in 1873. Abandoned below Nelsonville in 1876. Remainder of Canal from Carroll to Nelsonville abandoned in 1894.

MUSKINGUM IMPROVEMENT
Physical Data

Type of Improvement:	Canalized river
Project Length:	91* miles
Channel Width:	60 feet
Project Depth:	4.5 feet
Number of Dams:	11
Number of Locks:	12
Total Lift:	125.5 feet

Lock Dimensions:	*Lock 11* *1834-1910*	*Standard* *1841*	*Lock 1* *1841*	*Lock 1* *1879*
Length in Chamber:	120 feet	160 feet	175 feet	400 feet
Width in Chamber:	22 feet	35 feet	35 feet	56 feet
Lift:	8 feet	Varying	12 feet	

*From 1859 Board of Public Works Report. Later records call for 93 miles.

MUSKINGUM RIVER IMPROVEMENT
Table of Distances
Marietta to Dresden

Miles from Ohio River at Marietta	*Lock & Dam No.*	*Name of Lock or Structure*	*Place Name or Location*
		JUNCTION — OHIO RIVER	
0*	1	HARMAR/MARIETTA	Marietta,* Harmar*
5*	2	DEVOL'S/DEVOLA	Devoll's*/Devola
12*	3	LOWELL (Canal)	Lowell*
23*	4	BEVERLY (Canal)	Beverly,* Waterford*
33*	5	LUKE CHUTE	Luke's Chute*
38*	6	WINDSOR/STOCKPORT	Windsor*/Stockport
48*	7	McCONNELSVILLE (Canal)	McConnellsville,* Malta*
55*	8	EAGLEPORT/ROKEBY	Rokeby,* Eagleport*
61			Gaysport
65*	9	TAYLORSVILLE/PHILO (Canal until 1896)	Taylorsville*/Philo, Duncan's Falls*/Duncan Falls
75*	10	ZANESVILLE (Canal)	Zanesville*
83	11	ELLIS (after 1908)	Ellis
85*	11	SYMMES CREEK (until 1908)	Symme's Creek*
91*		JUNCTION — DRESDEN SIDE-CUT	Dresden

Distances and place names marked * are from 1859 Board of Public Works Report.

Locks are located abreast of the dam or, where (Canal) is indicated, at the lower end of a short canal leading around and below the dam. Lock No. 10 is a double lock.

The State Lock, No. 1, on the west bank of the river, was replaced beginning in 1879 by a large "Ice Harbor Lock" under the jurisdiction of the U.S. Engineers. This lock, on the east bank, and the dam were removed in 1968.

Symmes Lock was replaced in 1908 by Ellis Lock at about the same location.

In 1886, the Federal Government assumed control of the Muskingum Improvement. In 1958, this canalized river system reverted to the State of Ohio and currently exists as part of the Muskingum River Parkway.

Index

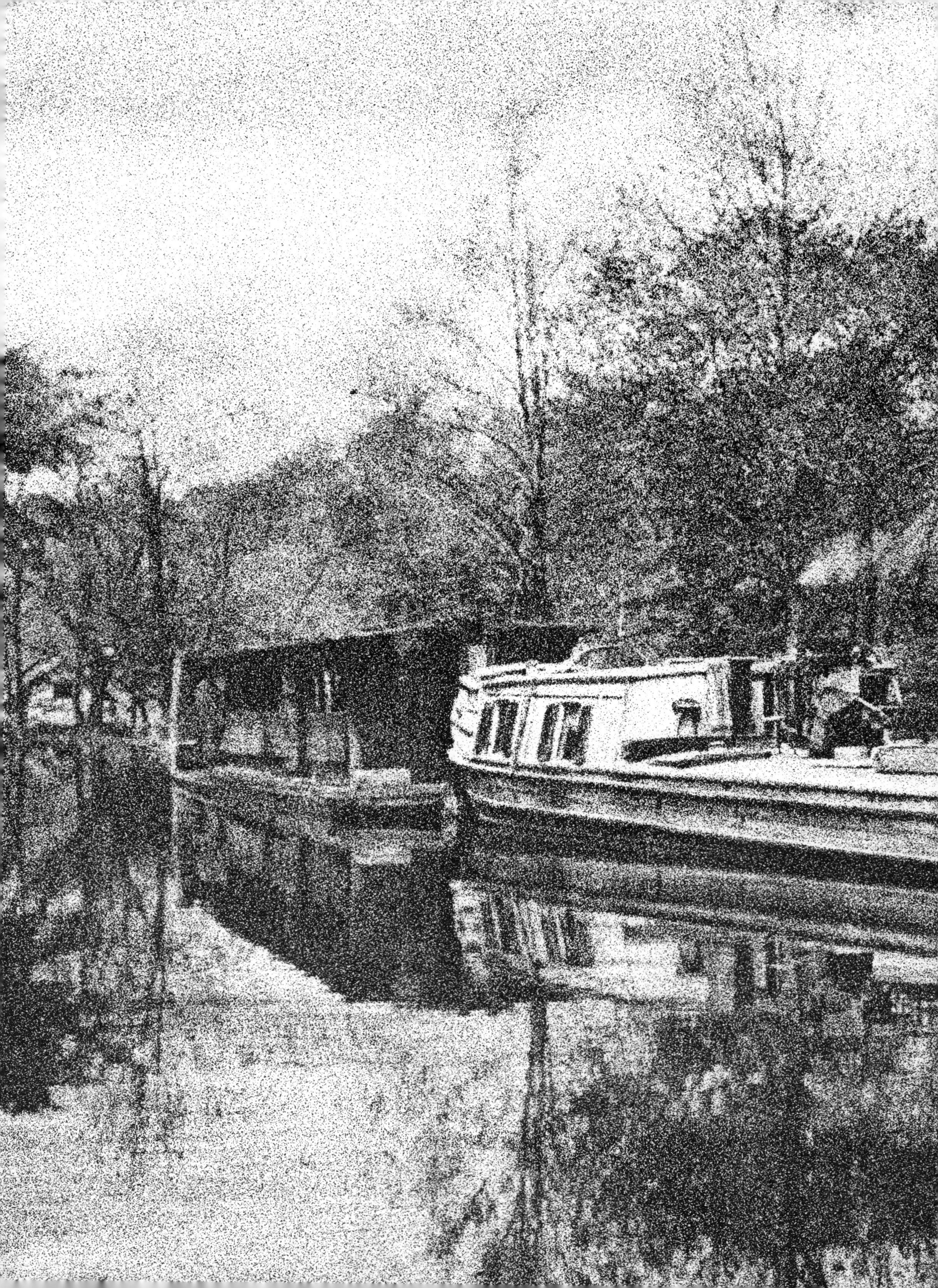